Advances in
ATOMIC AND MOLECULAR PHYSICS

VOLUME 19

CONTRIBUTORS TO THIS VOLUME

K. BLUM

B. H. BRANSDEN

N. A. CHEREPKOV

MANFRED FAUBEL

WAYNE M. ITANO

R. K. JANEV

F. JENČ

H. KLEINPOPPEN

J. T. PARK

D. G. THOMPSON

R. S. VAN DYCK, JR.

D. J. WINELAND

ADVANCES IN ATOMIC AND MOLECULAR PHYSICS

Edited by

Sir David Bates

DEPARTMENT OF APPLIED MATHEMATICS AND THEORETICAL PHYSICS
THE QUEEN'S UNIVERSITY OF BELFAST
BELFAST, NORTHERN IRELAND

Benjamin Bederson

DEPARTMENT OF PHYSICS
NEW YORK UNIVERSITY
NEW YORK, NEW YORK

VOLUME 19

1983

ACADEMIC PRESS

A Subsidiary of Harcourt Brace Jovanovich, Publishers

New York   London
Paris   San Diego   San Francisco   São Paulo   Sydney   Tokyo   Toronto

COPYRIGHT © 1983, BY ACADEMIC PRESS, INC.
ALL RIGHTS RESERVED.
NO PART OF THIS PUBLICATION MAY BE REPRODUCED OR
TRANSMITTED IN ANY FORM OR BY ANY MEANS, ELECTRONIC
OR MECHANICAL, INCLUDING PHOTOCOPY, RECORDING, OR ANY
INFORMATION STORAGE AND RETRIEVAL SYSTEM, WITHOUT
PERMISSION IN WRITING FROM THE PUBLISHER.

ACADEMIC PRESS, INC.
111 Fifth Avenue, New York, New York 10003

United Kingdom Edition published by
ACADEMIC PRESS, INC. (LONDON) LTD.
24/28 Oval Road, London NW1 7DX

LIBRARY OF CONGRESS CATALOG CARD NUMBER: 65–18423

ISBN 0–12–003819–6

PRINTED IN THE UNITED STATES OF AMERICA

83 84 85 86   9 8 7 6 5 4 3 2 1

Contents

CONTRIBUTORS ix

Electron Capture in Collisions of Hydrogen Atoms with Fully Stripped Ions

B. H. Bransden and R. K. Janev

I.	Introduction	2
II.	Basic Properties of the One-Electron Two-Coulomb Centers System	3
III.	The Calculation of Cross Sections by Expansion Methods Using a Molecular Orbital Basis	10
IV.	Atomic and Pseudostate Expansions	35
V.	Charge Exchange at High Velocities	48
VI.	Classical Descriptions	50
VII.	Conclusions	61
	References	62

Interactions of Simple Ion–Atom Systems

J. T. Park

I.	Introduction	67
II.	Experimental Methods	68
III.	Theoretical Methods	78
IV.	Excitation	82
V.	Electron Capture	100
VI.	Ionization	117
VII.	Elastic Scattering	123
VIII.	Conclusions	127
	References	129

High-Resolution Spectroscopy of Stored Ions

D. J. Wineland, Wayne M. Itano, and R. S. Van Dyck, Jr.

I.	Introduction	136
II.	Ion Storage Techniques	137
III.	Lepton Spectroscopy	149
IV.	Mass Spectroscopy	159
V.	Atomic and Molecular Ion Spectroscopy	166
VI.	Negative Ion Spectroscopy	176
VII.	Radiative Lifetime Measurements	180
	References	181

Spin-Dependent Phenomena in Inelastic Electron–Atom Collisions

K. Blum and H. Kleinpoppen

I.	Introduction	188
II.	Exchange Effects in Inelastic Collisions between Electrons and Light Atoms	189
III.	Excitation of Intermediate and Heavy Atoms: Spin Polarization and Asymmetry Studies	192
IV.	Influence of Spin-Dependent Interactions on Coherence Parameters	204
V.	Excitation of Heavy Atoms by Polarized Electrons: Stokes Parameter Analysis	225
VI.	Electron–Photon Coincidence Experiments with Polarized Electrons	236
VII.	Ionization	241
	Appendix A	259
	Appendix B	260
	References	261

The Reduced Potential Curve Method for Diatomic Molecules and Its Applications

F. Jenč

I.	Introduction	266
II.	Methods for Construction of Potential Curves	267
III.	Reduced (Internuclear) Potential of Diatomic Molecules	271
IV.	Applications of RPC Method	294
V.	Comments on Some Misunderstandings about the RPC Method	302
VI.	Conclusions	305
	References	306

The Vibrational Excitation of Molecules by Electron Impact

D. G. Thompson

I.	Introduction	309
II.	Theoretical Considerations	311
III.	Applications and Comparison with Experiment	323
	References	340

Vibrational and Rotational Excitation in Molecular Collisions

Manfred Faubel

I.	Introduction	345
II.	Theoretical Methods	347
III.	Experimental Techniques	354
IV.	Studies of Rotational Scattering Cross Sections	362
V.	Studies of Vibrational Excitation	380
VI.	Summary of Detailed Scattering Experiments and Concluding Remarks	385
	References	389

Spin Polarization of Atomic and Molecular Photoelectrons

N. A. Cherepkov

I.	Introduction	395
II.	Theory of Spin Polarization Phenomena in Atoms	397
III.	Comparison with Experiment and Applications	416
IV.	Spin Polarization of Molecular Photoelectrons	434
V.	Conclusion	442
	References	443

INDEX	449
CONTENTS OF PREVIOUS VOLUMES	461

Contributors

Numbers in parentheses indicate the pages on which the authors' contributions begin.

K. BLUM, Institut für Theoretische Physik I, Universität Münster, 4400 Münster, Federal Republic of Germany (187)

B. H. BRANSDEN, Department of Physics, University of Durham, Durham DH1 3LE, Great Britain (1)

N. A. CHEREPKOV, A. F. Ioffe Physico-Technical Institute, Academy of Sciences of the USSR, 194021 Leningrad, USSR (395)

MANFRED FAUBEL, Max-Planck-Institut für Strömungsforschung, 3400 Göttingen, Federal Republic of Germany (345)

WAYNE M. ITANO, Time and Frequency Division, National Bureau of Standards, Boulder, Colorado 80303 (135)

R. K. JANEV, Institute of Physics, Belgrade, Yugoslavia (1)

F. JENČ, Department of Theoretical Physics, University of Marburg, 335 Marburg, Federal Republic of Germany (265)

H. KLEINPOPPEN, Atomic Physics Laboratory, University of Stirling, Stirling FK9 4LA, Scotland (187)

J. T. PARK, Department of Physics, University of Missouri, Rolla, Missouri 65401 (67)

D. G. THOMPSON, Department of Applied Mathematics and Theoretical Physics, The Queen's University of Belfast, Belfast BT7 1NN, Northern Ireland (309)

R. S. VAN DYCK, JR., Department of Physics, University of Washington, Seattle, Washington 98195 (135)

D. J. WINELAND, Time and Frequency Division, National Bureau of Standards, Boulder, Colorado 80303 (135)

ELECTRON CAPTURE IN COLLISIONS OF HYDROGEN ATOMS WITH FULLY STRIPPED IONS

B. H. BRANSDEN

Department of Physics
University of Durham
Durham, Great Britain

R. K. JANEV

Institute of Physics,
Belgrade, Yugoslavia

I. Introduction	2
II. Basic Properties of the One-Electron Two-Coulomb Centers System	3
A. The Eigenvalue Problem for the (Z_1eZ_2) System	3
B. Expressions for the Adiabatic Energy Splitting	7
III. The Calculation of Cross Sections by Expansion Methods Using a Molecular Orbital Basis	10
A. The Expansion Method	10
B. The Molecular Orbital Basis	14
C. Numerical Calculations with a Molecular Orbital Basis	22
D. Approximate Treatments of the Diabatic PSS Equations	24
IV. Atomic and Pseudostate Expansions	35
A. The Two-Center Atomic Orbital Basis	35
B. Numerical Calculations	37
C. Approximate Treatments of AO–CC Equations	39
V. Charge Exchange at High Velocities	48
VI. Classical Descriptions	50
A. Classical Trajectory Monte Carlo Method	51
B. Analytical Models	56
VII. Conclusions	61
References	62

I. Introduction

Electron capture in collisions of highly charged ions with atoms has in recent years attracted a great deal of attention due to its extremely important role in many laboratory and astrophysical plasmas. Particularly strong motivations for the study of this process have come from research in controlled thermonuclear fusion, in the creation of short-wavelength lasers, and in heavy-ion sources. The progress achieved in the experimental and theoretical studies of this process has been reflected in several review articles (Olson, 1980; Salzborn and Müller, 1980; Bransden, 1980; Gilbody, 1981; Janev and Presnyakov, 1981; Greenland, 1982).

Charge exchange in collisions between hydrogen atoms and fully stripped ions, i.e., the reaction

$$H(1s) + A^{Z+} \rightarrow H^+ + A^{(Z-1)+} \tag{1}$$

where Z is the nuclear charge, takes a special place among all the electron capture processes involving multiply charged ions. This process forms the basis of spectroscopic and corpuscular fusion plasma diagnostics and plays an important role in the supplementary neutral beam heating of tokamak plasmas (Harrison, 1978, 1980, 1983; Hogan, 1978, 1983; Post, 1983). From a theoretical point of view, the hydrogen atom–fully stripped ion system provides a case in which different theoretical models for the electron-capture collision dynamics can be tested in a pure way, since the electronic structure problem for this system can be considered as solved exactly (see Section II). On the other hand, this one-electron collision system, under certain circumstances, can also serve as an appropriate model for describing the more complex, many-electron colliding systems (the one-electron approximation).

The purpose of the present article is to give a survey, and a critical analysis, of the theoretical methods currently used for description of charge exchange in hydrogen atom–fully stripped ion collisions. Most of these methods are, of course, also applicable to the study of charge exchange in many-electron systems, but in those cases the accuracy of the methods is obscured by the approximations made in treating the electronic structure problem of the collision system. Our main objective is to discuss the adequacy and the intrinsic power of theoretical methods for describing the collision dynamics of electron transfer processes. For this reason, particular attention is paid to the problem of state-selective electron capture, which provides the most sensitive test for the theoretical models. Emphasis is also placed on those aspects of theoretical descriptions which need further clarification or refinement. The results of the theoretical cross-section calculations for electron capture in hydrogen atom–fully stripped ion collisions will be presented only for illustrative purposes; a detailed critical account of the data has been given elsewhere (Janev *et al.,* 1983). The experimental cross-section data for this collision system are rather limited (see, e.g., Gilbody, 1981) but, where available, useful comparisons with the theoretical predictions can be made.

In charge-exchange collisions of multiply charged ions with atoms, a natural expansion parameter is the reduced velocity $\tilde{v} = vZ^{-1/4}$, where v is the relative collision velocity and Z is the ionic charge.[1] For $\tilde{v} \ll 1$, the molecular aspects of the collision system are strongly pronounced, and the dynamical problem can be formulated using an expansion of the total electron wave function over a molecular basis set. For $\tilde{v} \geq 1$, an atomic basis provides a more appropriate representation of the electronic states, however, in a wide-velocity region $0.1 \leq \tilde{v} \leq 1$ the distinction between atomic and molecular expansion methods is not great. At high velocities for $\tilde{v} \gg 1$ methods based on a perturbational approach can be used in principle. For not excessively large values of Z, $\tilde{v}$ and v are close, and in that case we shall not differentiate between these two quantities.

The plan of our article is the following. In Section II we shall give some basic information about the electronic properties of the colliding system (p, e, Z), which will be used in the subsequent sections. In Sections III and IV we shall discuss the theoretical methods based on molecular- or atomic-state expansions, respectively. In Section V we shall briefly outline the high-energy approximations for the charge-exchange process and in Section VI we shall discuss in more detail the classical methods for description of the process. Finally, in Section VII we shall summarize our conclusions.

II. Basic Properties of the One-Electron Two-Coulomb Centers System

A. THE EIGENVALUE PROBLEM FOR THE $(Z_1 e Z_2)$ SYSTEM

In order to facilitate the formulation of the collision problem defined by Eq. (1), especially at low energies, we shall first recall some of the properties of the one-electron two-Coulomb centers system, $(Z_1 e Z_2)$. (In this section, the charges of the proton and fully stripped ion are denoted by Z_1 and Z_2, respectively.) The quantum mechanics of this system is well understood and an extensive account of the subject can be found in the monograph by Komarov *et al.* (1976).

---

[1] Atomic units ($e = \hbar = m_e = 1$) are used throughout unless otherwise indicated.

The Schrödinger equation for the electronic motion at a fixed internuclear distance R is

$$\left[\frac{1}{2}\nabla_r^2 + \frac{Z_1}{r_1} + \frac{Z_2}{r_2} + E(R)\right]\Phi(\mathbf{r}, \mathbf{R}) = 0 \tag{2}$$

where $\mathbf{r}_1$, $\mathbf{r}_2$, and $\mathbf{r}$ are the position vectors of the electron with respect to the centers "1," "2," and the midpoint of the internuclear line, respectively, and $\Phi(\mathbf{r}, \mathbf{R})$ and $E(R)$ are the electronic wave function and the energy, respectively. In the limit $R \to \infty$, the wave function $\Phi(\mathbf{r}, \mathbf{R})$ goes over into the corresponding atomic wave function ϕ_i ($i = 1, 2$), depending on to which of the two nuclei the electron is bound.

It is well known (see, e.g., Morse and Feshbach, 1953) that the variables in Eq. (2) can be separated in prolate spheroidal coordinates

$$\begin{aligned}
\xi &= (r_1 + r_2)/R, & 1 &\leq \xi < \infty \\
\eta &= (r_1 - r_2)/R, & -1 &\leq \eta \leq 1 \\
\phi &= \arctan(y/x) & 0 &\leq \phi < 2\pi
\end{aligned} \tag{3}$$

Representing $\Phi(\mathbf{r}, \mathbf{R})$ in the form

$$\Phi(\mathbf{r}, \mathbf{R}) = \frac{X(\xi)Y(\eta)}{(\xi^2 - 1)^{1/2}(1 - \eta^2)^{1/2}} \frac{\exp(\pm im\phi)}{(2\pi)^{1/2}} \tag{4}$$

one obtains the following equations for $X(\xi)$ and $Y(\eta)$:

$$X''(\xi) + \left[-p^2 + \frac{a\xi - \lambda}{\xi^2 - 1} + \frac{1 - m^2}{(\xi^2 - 1)^2}\right]X(\xi) = 0 \tag{5a}$$

$$Y''(\eta) + \left[-p^2 + \frac{b\eta + \lambda}{1 - \eta^2} + \frac{1 - m^2}{(1 - \eta^2)^2}\right]Y(\eta) = 0 \tag{5b}$$

where

$$p = \tfrac{1}{2}(-2E)^{1/2}R, \qquad a = (Z_1 + Z_2)R, \qquad b = (Z_2 - Z_1)R \tag{6}$$

and λ is the separation constant. For $E < 0$ (discrete part of the spectrum), $X(\xi)$ and $Y(\eta)$ satisfy the following boundary conditions:

$$X(1) = 0, \qquad X(\xi) \xrightarrow[\xi \to \infty]{} 0 \tag{7a}$$
$$Y(\pm 1) = 0 \tag{7b}$$

In order for Eqs. (5a) and (5b) to be equivalent to the initial Eq. (2), the eigenvalues $\lambda^{(\xi)}(p, a)$ and $\lambda^{(\eta)}(p, b)$, as obtained, respectively, from the boundary-value problems (5a), (7a) and (5b), (7b), must be equal:

$$\lambda^{(\xi)}(p, a) = \lambda^{(\eta)}(p, b) \tag{8}$$

If the eigenstates of the Sturm–Liouville problem (5)–(7) are characterized (for $E < 0$) by the set of quantum numbers $j = \{kqm\}$, where k and q coincide, respectively, with the number of nodes of $X(\xi)$ and $Y(\eta)$, and $m = 0, 1, 2 \ldots$, then for fixed values of a and b, Eq. (8) has a unique solution:

$$p = p_{kqm}(a, b) \tag{9}$$

Using the relation (6) between p and E, the energy spectrum of the $(Z_1 e Z_2)$ system can be found from Eq. (9):

$$E_j(R) = E_{kqm}(Z_1, Z_2, R) \tag{10}$$

For $R \to \infty$, the prolate spherodial coordinates ξ, η, ϕ, go over into the parabolic coordinates μ, ν, ϕ, and it is convenient to characterize the electronic states in this limit by the set of parabolic quantum numbers $[n'n_1'n_2'm]$, when the electron is around center Z_1, and by $[nn_1n_2m]$, when it is around center Z_2. (The quantum number m remains the same since the projection of the angular momentum on the internuclear axis is conserved.) For $R = 0$, the two-center Coulomb system reduces to a hydrogen-like one (with charge $Z_1 + Z_2$) and, accordingly, in this limit electronic states are characterized by the usual spherical quantum numbers (Nlm). The connection between the electronic energies of the system at finite values of R with those in the "separated atom limit" ($R \to \infty$) and those in the "united atom limit" ($R \to 0$) (which is displayed in "correlation diagrams") can be established in a unique manner (Gershtein and Krivchenkov, 1961; Power, 1973). The corresponding quantum numbers are related by

$$N = k + q + m + 1, \quad N = 1, 2, 3, \ldots \tag{11a}$$

$$l = q + m, \quad l = 0, 1, 2, \ldots, N - 1 \tag{11b}$$

$$k = n_1 = n_1' \tag{11c}$$

$$q = \begin{cases} n_2, & n_2 < n(Z_2 - Z_1)/Z_2 \\ n_2 + 1 + \text{Ent}[n_2 - n(Z_2 - Z_1)/Z_2], & n \geq n(Z_2 - Z_1)/Z_2 \end{cases} \tag{11d}$$

for the (eZ_2) levels, and

$$q = \begin{cases} 2n_2' + n'(Z_2 - Z_1)/Z_1, & n'Z_2/Z_1 = \text{integer} \\ 2n_2' + 1 + \text{Ent}[n'(Z_2 - Z_1)/Z_1], & n'Z_2/Z_1 \neq \text{integer} \end{cases} \tag{11e}$$

for the (eZ_1) levels, where $\text{Ent}(x)$ is the entire part of x.

Several numerical algorithms have been developed for solving Eqs. (5)–(10) (Madsen and Peek, 1971; Ponomarev and Puzynina, 1967, 1968,

1970). However, in the regions of small and large values of R, analytical treatments of the two-center Coulomb problem are also possible by using perturbational and asymptotic methods, respectively (see, for example, Komarov et al., 1976). In particular, numerous analytical results are known for the properties of the $(Z_1 e Z_2)$ system in the asymptotic region ($R \gg 1$). Some of these, which are of direct relevance to the electron transfer problem, will be given below.

The electronic energies of the system $(Z_1 e Z_2)$ in the limits $R \to 0$ and $R \to \infty$ are given by (Power, 1973)

$R \to 0$:

$$E_{Nlm}(Z_1, Z_2, R) = -\frac{(Z_1 + Z_2)^2}{2N^2}$$

$$- \frac{2Z_1 Z_2 [l(l+1) - 3m^2](Z_1 + Z_2)^2}{N^3 l(l+1)(2l+3)(2l+1)(2l-1)} R^2 + O(R^3) \quad (12)$$

$R \to \infty$:

$$E_{nn_1 n_2 m}(Z_1, Z_2, R) = -\frac{Z_2^2}{2n^2} - \frac{Z_1}{R} + \frac{3nZ_1 \Delta}{2Z_2 R^2} + O(R^{-3}) \quad (13a)$$

$$\Delta = n_1 - n_2$$

$$E_{n' n_1' n_2' m}(Z_1, Z_2, R) = E_{nn_1 n_2 m}(Z_1 \leftrightarrow Z_2, n \to n', \Delta \to \Delta') \quad (13b)$$

Analogous expansions have been derived for the quantity $p = (R/2)(-2E)^{1/2}$. For $R \to \infty$ one has

$$p_{nn_1 n_2 m}(R) = \frac{Z_2 R}{2n} + \frac{nZ_1}{2Z_2} - \frac{n^2 Z_1}{4Z_2^3 R}(3 \Delta Z_2 + nZ_1) + O(R^{-2}) \quad (14a)$$

$$p_{n' n_1' n_2' m}(R) = P_{nn_1 n_2 m}(Z_1 \leftrightarrow Z_2, n \to n', \Delta \to \Delta') \quad (14b)$$

One of the specific features of the $(Z_1 e Z_2)$ system is that, besides the geometrical group of symmetry, it possesses an additional (dynamical) symmetry (Alliluev and Matveenko, 1967). This additional symmetry (as well as the separation of the variables in prolate spheroidal coordinates) results from the commutation of the electronic Hamiltonian with the operator λ of the separation constant [see also Eq. (8)]. As a consequence of this higher intrinsic symmetry of the $(Z_1 e Z_2)$ system, the well-known von Neumann–Wigner theorem on the noncrossing of the adiabatic energies of diatomic molecules with variation of the parameter R is now much more restrictive. For two energies $E_{kqm}(R)$ and $E_{k'q'm'}(R)$, a pseudocrossing can exist (for $R < \infty$) only when the following conditions are fulfilled:

$$k = k', \quad q = q' + 1, \quad m = m' \tag{15}$$

For the H(1s) + Z system ($Z_1 = 1$, $Z_2 = Z$), this noncrossing rule specifies that the initial-state potential energy curve $E_{n'n_1'n_2'm}(Z_1, Z_2, R) = E_{1000}(1, Z, R)$ has an "avoided crossing" only with the final-state energy curve $E_{nn_1n_2m}(Z, 1, R)$, corresponding to the Stark state $[n, 0, n-1, 0]$. Assuming that the pseudocrossing takes place at large internuclear distances where the asymptotic expansions (13a) and (13b) are valid, the distance R_c at which the curves $E_{1000}(1, Z, R)$ and $E_{n,0,n-1,0}(Z, 1, R)$ pseudocross is given by (in the lowest approximation)

$$R_c = 2n^2(Z-1)/(Z^2 - n^2), \quad n < Z \tag{16}$$

The existence of a real solution (even approximate) of the equation $E_{kqm}(R) = E_{k,q+1,m}(R)$ contradicts the noncrossing rule and signifies that the corresponding energies in the region of that solution are not adequately determined. In the pseudocrossing region the functions Φ_{kqm} and $\Phi_{k,q+1,m}$ have the same order of magnitude in the vicinity of either of the centers Z_1 and Z_2; accordingly, the electronic motion in this region is strongly delocalized. This characteristic of the electronic motion leads to exponentially small corrections of the energy, which are not included in the asymptotic expansions (13). The splitting $\Delta_{kqm} = E_{kqm}(R) - E_{k,q+1,m}(R)$ of the adiabatic energies in the region of the pseudocrossing is, however, determined just by these exponential corrections. In the next section we shall discuss the adiabatic energy splitting in more detail.

B. Expressions for the Adiabatic Energy Splitting

The appearance of exponentially small terms in the adiabatic energies of the ($Z_1 e Z_2$) system results from the existence of such terms in the eigenvalues $\lambda_{qm}^{(n)}(p, b)$ of the boundary-value problem (5b), (7b). Physically, these terms reflect the under-barrier electron transitions in the two-well potential of Eq. (5b). Allowing for these corrections, the following representation of the adiabatic energies in the pseudocrossing region is obtained:

$$E_{\{n\}} = E_{\{n\}}^{(0)} + \delta E_{\{n\}}, \quad E_{\{n'\}} = E_{\{n'\}}^{(0)} - \delta E_{\{n'\}} \tag{17}$$

where $\{n\}$ designates the set of parabolic quantum numbers $[nn_1n_2m]$, and $E_{\{n\}}^{(0)}$, $E_{\{n'\}}^{(0)}$ in the asymptotic region are given by Eqs. (13). At the point $R_c = R_{n_2n_2'}^{km}$ ($k = n_1 = n_1'$) defined by the equation $E_{\{n\}}^{(0)}(R_c) = E_{\{n'\}}^{(0)}(R_c)$, the difference of the adiabatic energies is

$$\Delta E(R_{n_2n_2'}^{km}) = \delta E_{\{n\}}(R_{n_2n_2'}^{km}) + \delta E_{\{n'\}}(R_{n_2n_2'}^{km}) \tag{18}$$

The exponentially small corrections $\delta E_{(n)}$, $\delta E_{(n')}$ can be calculated analytically in the asymptotic region ($R \gg 1$) often by the "comparison equation" method (described by Komarov et al., 1976; Power, 1973).

For the case $Z_1 \sim Z_2$, the result for $\Delta E(R_{n_2 n_2'}^{km})$ with the asymptotic method is (Komarov and Slavyanov, 1968; Greenland, 1978)

$$\Delta E(R_{n_2 n_2'}^{km}) = 4E_c(R_{n_2 n_2'}^{km}) \frac{(4p)^{n_2 + n_2' + m + 2} e^{-2p}}{[n_2!(n_2 + m)! n_2'!(n_2' + m)!]^{1/2}} [1 + O(1/p)] \quad (19)$$

where

$$E_c(R_{n_2 n_2'}^{km}) = -\frac{1}{2}\left[\frac{Z_2 - Z_1}{n_2 - n_2'}\right]^2, \quad p(R_{n_2 n_2'}^{km}) = \left[-\frac{1}{2} E_c(R_{n_2 n_2'}^{km})\right]^{1/2} R_{n_2 n_2'}^{km} \quad (20)$$

For $Z_2 \gg Z_1$, another large parameter appears in the problem and the analysis is more complex. The result for this case and for $n_2 \gg 1$ is (Komarov and Solov'ev, 1979)

$$\Delta E(R_{n_2 n_2'}^{km}) = \left\{\frac{Z_1^2}{n'^3}\left[1 + \frac{Z_1^2(n - n')^3}{n'^3(Z_2 - Z_1)^2}\right]^{-1/2} \right.$$
$$\left. + \frac{Z_2^2}{n^3}\left[1 - \frac{Z_2^2(n - n')^3}{n^3(Z_2 - Z_1)^2}\right]^{-1/2}\right\} \delta(R_{n_2 n_2'}^{km}) \quad (21)$$

where

$$\delta(R_{n_2 n_2'}^{km}) = \frac{(2p)^{n_2' + (1+m)/2}}{[2\pi n_2'!(n_2' + m)!]^{1/2}} \exp\left\{-[\sqrt{2} - \ln(\sqrt{2} + 1)]p\right.$$
$$\left. + 2B \ln(\sqrt{2} + 1) + \frac{5}{2}\left[n_2' + \frac{1+m}{2}\right](\ln 2)\right\}\left[1 + O(1/p)\right] \quad (22)$$

with

$$B = \sqrt{2}\left[\frac{7}{8}(n_1' - n_2') + 1\right]$$

An asymptotically exact expression for $\delta(R_{nn'}^{km})$ has recently been derived for the highly excited states of the ($Z_1 e Z_2$) system (Janev et al., 1982b). For the homonuclear case ($Z_1 = Z_2$), all these expressions determine the splitting between the energies of those gerade and ungerade states which are degenerate at $R \to \infty$.

For the system (p, e, Z), systematic numerical calculations of the potential energy curves have been carried out for $4 \leq Z \leq 54$ by Olson and Salop (1976). From these numerical results they derived an analytical fit giving the adiabatic splittings at $R = R_c$ [see Eq. (16)] of the energies corresponding to the initial [1, 0, 0, 0] and final [n, 0, n − 1, 0] states. Within a 17% accuracy these splittings can be represented by the expression

$$\Delta E(R_c) = 18.26 Z^{-1/2} \exp(-1.327 R_c Z^{-1/2}) \tag{23}$$

All the above expressions for $\Delta E(R_c)$ are meaningful only for R_c satisfying the condition

$$R_c > R_0 = 2n'[(n' - \Delta')(2n'Z - n' - \Delta')^{1/2} - \Delta'] \tag{24}$$

where R_0 is the internuclear distance at which the energy level in Eq. (5b) becomes equal to the top of the potential barrier. For the H(1s) + Z system $R_0 = 2(2Z - 1)^{1/2}$. The number of pseudocrossings in the region $R > R_0$ is approximately given by (Ponomarev, 1969)

$$r_0 = \text{Ent}\{n'(x - 1)[x + 1 - (1 + 2x)^{1/2}]\}, \qquad x = (Z_2/Z_1)^{1/2} \tag{25}$$

According to Eq. (25) the first crossing appears for $Z_2/Z_1 = 5$. With increasing x^2, the number of crossings increases rapidly. In the region $R \gg R_0$ and for $|Z_2 - Z_1| \gg 1$, the motion of the electron under the barrier can be described quasi-classically.

In this case one can apply the quasi-classical method to calculate the energy splitting $\Delta E(R)$ (Chibisov, 1979, 1981; Duman and Men'shikov, 1979). The quantity $\Delta E(R)$ is then directly related to the quantum penetrability of the potential barrier, thus revealing its physical nature in an obvious way (electron transition frequency, or exchange interaction). The calculation of the barrier penetrability integral can be carried out analytically by expanding the quasi-classical momentum in terms of the small quantity $R_0/R \sim Z^{1/2}/R$. The result for the H(1s) + Z system is (Duman and Men'shikov, 1979) [we denote here $\Delta E(R)$ by $\Delta(R)$].

$$\Delta_{nlm}(R) = \delta_{m0} G(l) \Delta_{n0}(R) \tag{26}$$

$$G(l) = (2l + 1)^{1/2} \exp[-l(l + 1)/2Z] \tag{27}$$

$$\Delta_{n0}(R) = 2(2/\pi)^{1/2} (R/n^{3/2}) \exp[-R F(Z, R)] \tag{28}$$

where

$$F(Z, R) = 1 - \frac{2(Z-1)}{R(1 + 2Z/R)^{1/2}} \ln\left[\left(\frac{R}{2Z}\right)^{1/2} + \left(\frac{R}{2Z} + 1\right)^{1/2}\right] \tag{29}$$

and (n, l, m) are the spherical quantum numbers of the electron in its state around the ion Z.

In two limiting cases, the expression (28) for $\Delta_{n0}(R)$ can be simplified:

$$\Delta_{n0}(R) = 2\left(\frac{2}{\pi}\right)^{1/2} \frac{R}{n^{3/2}} \exp\left(-\frac{R^2}{3Z}\right), \qquad 4Z^{1/2} < R < 2Z \tag{30a}$$

$$\Delta_{n0}(R) = \frac{2}{Z\Gamma(Z)} \left(\frac{2R}{e}\right)^Z e^{-R}, \quad R > 2Z \tag{30b}$$

The result (30a) can be related with the decay probability $\Gamma(R)$ of the electronic state in the constant electric field of the ion Z/R^2 by

$$\Gamma(R) = (\pi/2)\Delta_{n0}^2/\Delta E, \quad \Delta E = Z^2/n^3 \tag{31}$$

The result (30b) does not depend on n, meaning that the electron transition has a resonant character. Quasi-classical expressions for $\Delta_{nlm}(R)$, analogous to those given by Eqs. (26)–(29), can be derived for the system (Z_1, e, Z_2) both for the $1s \to nlm$ transition (Duman and Men'shikov, 1979) and for the $n_0 l_0 m_0 \to nlm$ transition (Chibisov, 1981).

III. The Calculation of Cross Sections by Expansion Methods Using a Molecular Orbital Basis

In this section, we shall discuss the expansion method for obtaining appropriate solutions of the Schrödinger equation for the dynamical system composed of one electron and two nuclei, going on to discuss the particular case in which the expansion is in terms of the molecular orbitals as described in Section II.

A. The Expansion Method

In the center-of-mass system, the Schrödinger equation for the system of one electron and two nuclei of charges Z_1 and Z_2 is

$$\left(T - \frac{Z_1}{r_1} - \frac{Z_2}{r_2} + \frac{Z_1 Z_2}{R} - \epsilon\right)\psi(\mathbf{r}, \mathbf{R}) = 0 \tag{32}$$

where T is the kinetic energy operator. If $\mathbf{x}$ is the coordinate joining the electron to the center of mass of the nuclei 1 and 2, T can be expressed as (approximating the reduced mass of the electron by $m_e = 1$)

$$T = -\frac{1}{2}\nabla_x^2 - \frac{1}{2\mu}\nabla_R^2 \tag{33}$$

where $\mu = M_1 M_2/(M_1 + M_2)$ is the reduced mass of the two nuclei. Different expressions for T are (ignoring m_e in comparison with M_1 or M_2)

$$T = -\frac{1}{2}\nabla_{r_1}^2 - \frac{1}{2\mu}\nabla_\sigma^2 = -\frac{1}{2}\nabla_{r_2}^2 - \frac{1}{2\mu}\nabla_\rho^2 \tag{34}$$

$$= -\frac{1}{2}\nabla_r^2 - \left(\frac{1}{M_1} - \frac{1}{M_2}\right)\nabla_r \cdot \nabla_R - \frac{1}{2\mu}\nabla_R^2 \tag{35}$$

where σ is the vector joining nucleus 2 to the center of mass of the atom composed of the electron and nucleus 1, and ρ is the vector joining nucleus 1 to the center of mass of the atom composed of the electron and nucleus 2. The cross term in Eq. (35) arises because $\mathbf{r}$ and $\mathbf{R}$ do not form a set of center-of-mass coordinates, except when $M_1 = M_2$.

The expansion method starts by choosing some set of linearly independent functions $\chi_j(\mathbf{r}, \mathbf{R})$ and approximating the wave function by the finite expansion

$$\psi(\mathbf{r}, \mathbf{R}) = \sum_{j=0}^{N} \chi_j(\mathbf{r}, \mathbf{R}) F_j(\mathbf{R}) \tag{36}$$

The function χ_j will be assumed to be normalized to unity for all values of $\mathbf{R}$. The choice of the functions χ_j is dictated by physical considerations. For example, at low velocities for which the motion of the nuclei is slow compared with the electronic motion ($\tilde{v} < 1$), the χ_j can be associated with molecular orbitals Φ_j introduced in Section II. We shall return to the choice of functions below, but first we notice that the functions can be divided into two sets, those labeled with $0 \leq j \leq L$ which, when R becomes large, approach (up to a phase) atomic orbitals $\phi_{1,j}$ representing the target atom, and those labeled with $L < j \leq N$ which approach atomic orbitals $\phi_{2,j}$ representing the rearranged situation in which the electron is bound to nucleus 2:

$$\chi_j \xrightarrow[R\to\infty]{} \begin{cases} \phi_{1,j}(\mathbf{r}_1)e^{i\alpha_j(\mathbf{r},\mathbf{R})} & \text{for } 0 \leq j \leq L \\ \phi_{2,j}(\mathbf{r}_2)e^{i\gamma_j(\mathbf{r},\mathbf{R})} & \text{for } L < j < N \end{cases} \tag{37}$$

For large R, the functions $F_j(\mathbf{R})$, representing the relative motion of the nuclei in the different channel j, satisfy the boundary condition

$$F_j(\mathbf{R}) \sim \delta_{j0} \exp i\mathbf{k}_1 \cdot \mathbf{R} + f_j(\Theta, \Phi)R^{-1} \exp ik_j R \tag{38}$$

where $j = 0$ is the incident channel and k_j is the heavy particle momentum in channel j. The scattering amplitudes $f_j(\Theta, \Phi)$ determine the cross sections for elastic scattering, excitation, and charge exchange. Although it is convenient to work in terms of the variable $\mathbf{R}$, rather than σ and ρ, the penalty to be paid is that the individual terms in Eq. (36) do not satisfy the Schrödinger equation in the limit $R \to \infty$ unless the phase functions α_j, γ_j in Eq. (37) are nonzero, and satisfy (again ignoring m_e in comparison with M_1 and M_2)

$$\alpha_j(\mathbf{r}, \mathbf{R}) \xrightarrow[R\to\infty]{} \frac{1}{M_1}\mathbf{k}_j \cdot \mathbf{r}_1; \quad \gamma_j(\mathbf{r}, \mathbf{R}) \xrightarrow[R\to\infty]{} -\frac{1}{M_2}\mathbf{k}_j \cdot \mathbf{r}_2 \tag{39}$$

Coupled partial differential equations can be obtained for the functions $F_j(\mathbf{R})$, by requiring

$$\sum_{j=0}^{N} \int dr\, \chi_i^*(\mathbf{r}, \mathbf{R})\left(T - \frac{Z_1}{r_1} - \frac{Z_2}{r_2} + \frac{Z_1 Z_2}{R} - \epsilon\right) \chi_j(\mathbf{r}, \mathbf{R})\, F_j(\mathbf{R}) = 0$$

$$i = 0, 1, 2, \ldots, N \quad (40)$$

At low energies, for example, at energies less than 100 eV/amu, it is convenient to make a partial wave decomposition of the function $F_j(\mathbf{R})$ to obtain coupled second-order ordinary differential equations for radial functions $F_{j,k}(R)$. At higher energies the number of partial waves becomes very large and an impact parameter method is usually used. At energies above a few hundred electron volts per atomic mass unit, the heavy particle scattering is confined to a narrow cone about the forward direction and the heavy particle motion can be described by a classical straight-line trajectory (see, for example, Bransden, 1983), $\mathbf{R} = \mathbf{R}(t)$, where

$$\mathbf{R}(t) = \mathbf{b} + \mathbf{v}t; \quad \mathbf{b} \cdot \mathbf{v} = 0 \quad (41)$$

where t is the time, $\mathbf{v}$ is the (constant) relative velocity of the heavy particles, and $\mathbf{b}$ is the impact parameter vector. We shall take the Z axis as being parallel to $\mathbf{v}$ and the plane of the heavy particle motion to be the XY plane. The electronic motion is determined by the time-dependent Hamiltonian

$$H(t) = -\frac{1}{2}\nabla_r^2 - \frac{Z_1}{r_1} - \frac{Z_2}{r_2} + \frac{Z_1 Z_2}{R(t)} \quad (42)$$

where $r_1 = |\mathbf{r} + \mathbf{R}(t)/2|$; $r_2 = |\mathbf{r} - \mathbf{R}(t)/2|$. The electronic wave function satisfies the Schrödinger equation:

$$(H(t) - i\, \partial/\partial t)\psi(\mathbf{r}, t) = 0 \quad (43)$$

where the time derivative is to be taken with $\mathbf{r}$, the position vector of the electron with respect to the midpoint of the internuclear line, held constant.[2] The wave function ψ can be expanded in terms of the function χ_j as

$$\psi(\mathbf{r}, t) = \sum_{j=0}^{N} a_j(b, t)\, \chi_j(\mathbf{r}, \mathbf{R}) \quad (44)$$

and the expansion functions must again satisfy Eq. (37) for large R (large $|t|$). In view of Eq. (37), the functions χ_j are mutually orthogonal at large $|t|$, so that the coefficients $a_j(b, t)$ can be interpreted as probability amplitudes in

---

[2] For a simple derivation of Eq. (43) from Eq. (32), see Bransden (1983).

the large $|t|$ limit. If the target atom is initially in the level $j = 1$, we have

$$\lim_{t \to -\infty} a_j(b, t) = \delta_{j0} \qquad (45)$$

and the cross sections for excitation, or charge exchange, are

$$\sigma_j = 2\pi \int_0^\infty b \, db \lim_{t \to +\infty} |a_j(b, t)|^2 \qquad (46)$$

It is easily verified that, if the individual terms of the expansion (44) are to satisfy Eq. (43) in the limit $|t| \to \infty$, the functions α_j and γ_j must obey the following conditions as $|t| \to \infty$.

$$\alpha_j \to (-\tfrac{1}{2}\mathbf{v} \cdot \mathbf{r} + \tfrac{1}{8}v^2 t + \eta_j)$$
$$\gamma_j \to (\tfrac{1}{2}\mathbf{v} \cdot \mathbf{r} + \tfrac{1}{8}v^2 t + \eta_j) \qquad (47)$$

where η_j is the binding energy of the atom to which the system separates. The translational factors $\exp i\alpha_j$, $\exp i\gamma_j$ represent the translational motion of the electron, relative to the chosen origin. When attached to centers 1 or 2 the electron has a momentum $-v/2$ or $+v/2$, respectively, and kinetic energy $v^2/8$. The theory is translationally invariant in the sense that the same probability amplitudes are obtained with any other choice of origin (situated on the internuclear line), provided the momenta $\pm v/2$ are replaced by the appropriate momenta of the electron when attached to 1 or 2, relative to the new origin.

By projecting Eq. (43) with the functions χ_j, coupled equations are found for the amplitudes $a_j(b, t)$. Using a matrix notation,

$$i\mathbf{S} \cdot \dot{\mathbf{a}} = \mathbf{V} \cdot \mathbf{a} \qquad (48)$$

where $\mathbf{S}$ is the overlap matrix with elements

$$S_{ij} = \int \chi_i^*(\mathbf{r}, \mathbf{R}) \chi_j(\mathbf{r}, \mathbf{R}) \, d\mathbf{r} \qquad (49)$$

and $\mathbf{V}$ is a potential matrix with elements

$$V_{ij} = \int \chi_i(\mathbf{r}, \mathbf{R})(H - i \, \partial/\partial t) \chi_j(\mathbf{r}, \mathbf{R}) \, d\mathbf{r} \qquad (50)$$

The first-order coupled differential equations can be integrated subject to the boundary condition (45) by any standard technique. The cross sections are subsequently computed from Eqs. (46).

As we shall see, with a sufficiently well-chosen basis set, accurate cross sections for charge exchange can be computed by the expansion method for a range of energies. The practical upper limit on the energy range arises because the charge-exchange cross section becomes very small at high energies compared with the ionization cross section, and is therefore difficult to compute accurately even if the basis set is capable of describing the important ionization channels. Below ~ 500 eV/amu, the assumption of a linear heavy particle trajectory becomes less accurate, particularly for small values of the impact parameter. In this case, the impact parameter method can be retained, but the linear trajectory (41) has to be replaced by a more general trajectory:

$$\mathbf{R} = \mathbf{R}(\mathbf{b}, \mathbf{v}, t) \tag{51}$$

obtained by assuming that the heavy particle motion is determined by some average potential. It is often sufficient to employ a screened Coulomb potential, which behaves like $+Z_1 Z_2/R$ at small R. A greater degree of refinement can be obtained if different potentials are used in each channel j. New developments along these lines have been discussed recently by Green (1981a,b) following earlier work by Riley (1973).

It should be noted that although the internuclear potential is important in determining the appropriate nonlinear trajectory at lower velocities, the term $Z_1 Z_2/R(t)$ appearing in Eq. (42) can be removed from Eq. (43) by the phase transformation

$$\psi'(\mathbf{r}, t) = \psi(\mathbf{r}, t) \exp\left[-i \int^t dt' Z_1 Z_2/R(t)\right] \tag{52}$$

This implies that total cross sections calculated in the straight-line impact parameter method are completely independent of the internuclear potential. This is not the case for differential cross sections (Greenland, 1982; Bransden, 1983) since these depend on the phase as well as the magnitude of the amplitudes.

B. The Molecular Orbital Basis

When the relative motion of the nuclei is slow compared with the electronic motion, it is natural to suppose that the wave function retains much of its molecular character during the collision, and an expansion can be made in terms of the adiabatic solutions Φ_j of Eq. (2). Using plane wave translations factors introduced by Bates and McCarroll (1958), defined by

$$\begin{aligned} \alpha_j &= \tfrac{1}{2}\mathbf{v} \cdot \mathbf{r} - \tfrac{1}{8}v^2 - \eta_j \\ \gamma_j &= -\tfrac{1}{2}\mathbf{v} \cdot \mathbf{r} - \tfrac{1}{8}v^2 - \eta_j \end{aligned} \tag{53}$$

the expansion functions χ_j become

$$\chi_j = \begin{cases} \Phi_j \exp i\alpha_j & \text{for } 0 \leq j \leq L \\ \Phi_j \exp i\gamma_j & \text{for } L < j \leq N \end{cases} \quad (54)$$

The adiabatic functions are orthonormal for all R, however, because of the translational factors, the off-diagonal elements of the overlap matrix **S** are nonzero when i and j belong to differential arrangements. For the same reason, whereas H is diagonal with respect to the adiabatic functions Φ_j, it is not so with respect to the functions χ_j. Thus, the "direct" matrix elements for $0 \leq i \leq L$, $0 \leq j \leq L$ or $L < i \leq N$, $L < j \leq N$ become

$$S_{ij} = \delta_{ij}; \quad V_{ii} = E_i(R) + \frac{Z_2 Z_2}{R} + \eta_i$$

$$V_{ij} = -ie^{i(\eta_i - \eta_j)t} \int dr\, \Phi_i^* (\partial/\partial t + \mathbf{v} \cdot \nabla_r) \Phi_j; \quad i \neq j \quad (55)$$

and the "exchange" matrix elements connecting states which separate to atomic states on different centers $0 \leq i \leq L$, $L < j \leq N$ or $0 \leq j \leq L$, $L < i \leq N$ become

$$S_{ij} = e^{i(\eta_i - \eta_j)t} \int dr\, e^{\pm i \mathbf{v} \cdot \mathbf{r}} \Phi_i^* \Phi_j$$

$$V_{ij} = e^{i(\eta_i - \eta_j)t} \int dr\, e^{\pm i \mathbf{v} \cdot \mathbf{r}} \Phi_i^* \left[E_j(R) + \frac{Z_1 Z_2}{R} - \eta_j \right. \quad (56)$$

$$\left. - i\frac{\partial}{\partial t} - i \mathbf{v} \cdot \nabla_r \right] \Phi_j$$

The operation $\partial/\partial t$ is to be carried out with **r** fixed in the laboratory coordinate system; however, molecular wave functions are generally determined in a body-fixed system in which the internuclear line is taken as the axis of quantization. The body-fixed axes rotate with respect to the space-fixed axis and if **r'** is the electron position vector in the rotating system, it is easily shown (Bransden, 1983) that

$$\frac{\partial}{\partial t} = \left.\frac{\partial}{\partial t}\right|_{\mathbf{r}'} + \dot{\Theta}\left(z'\frac{\partial}{\partial x'} - x'\frac{\partial}{\partial z'}\right) \quad (57)$$

where $\partial/\partial t|_{\mathbf{r}'}$ is the time derivative taken with **r'** fixed and $\Theta(t)$ is the angle between **R** and **v**. Since $\dot{\Theta} = -bv/R^2$, we have

$$\frac{\partial}{\partial t} \Phi_j[(\mathbf{r}', R(t)] = \dot{R}\frac{\partial}{\partial R}\Phi_j + \frac{bv}{R^2} L_{y'} \Phi_j \quad (58)$$

where $L_{y'}$ is the y' component of the angular momentum operator. The coupling matrix elements V_{ij} containing $\partial/\partial R$ are known as "radial coupling" terms and those containing $L_{y'}$ as "rotational coupling" terms. In terms of the spheroidal coordinates introduced in Section II the operators $\partial/\partial R$ and $L_{y'}$ are (Piacentini and Salin, 1974)

$$\frac{\partial}{\partial R} = -\frac{2/R^2}{E_j - E_{j'}} \frac{(Z_2 + Z_1)\xi + (Z_2 - Z_1)\eta}{(\xi^2 - \eta^2)}$$

$$+ \frac{1}{R} \frac{\xi(1 - \xi^2) \partial/\partial\xi + \eta(\eta^2 - 1) \partial/\partial\eta}{(\xi^2 - \eta^2)} \quad (59)$$

$$iL_{y'} = \frac{[(\xi^2 - 1)(1 - \eta^2)]^{1/2}}{\xi^2 - \eta^2} \cos\phi \left(\eta \frac{\partial}{\partial\xi} - \xi \frac{\partial}{\partial\eta}\right)$$

$$- \xi\eta \sin\phi[(\xi^2 - 1)(1 - \eta^2)]^{1/2} \frac{\partial}{\partial\phi}$$

General properties of the matrix elements of these operators with respect to the orbitals Φ_j are given by Janev and Presnyakov (1981).

1. The Perturbed Stationary State Approximation

In the original form of the MO expansion method, called the perturbed stationary state (PSS) approximation (Massey and Smith, 1933), the translational factors were omitted, which has the effect of eliminating the terms in $\mathbf{v} \cdot \nabla_r$ in the matrix elements V_{ij} and of replacing the factors $\exp(\pm i\mathbf{v} \cdot \mathbf{r})$ by unity. This approximation is most likely to be accurate for small velocities ($\tilde{v} \ll 1$) and has the advantage that the approximate matrix elements V_{ij} become much easier to evaluate. No distinction need be made between direct and exchange matrix elements and for all $i, j \leq N$, we have

$$\begin{aligned} S_{ij} &= \delta_{ij}; & V_{ii} &= E_i(R) - \eta_i + Z_1 Z_2/R \\ V_{ij} &= V_{ij}^R - V_{ij}^L; & i &\neq j \end{aligned} \quad (60)$$

where

$$V_{ij}^R = -i\dot{R} e^{i(\eta_i - \eta_j)t} \int d\mathbf{r}\, \Phi_i^* \frac{d}{dR} \Phi_j$$

$$V_{ij}^L = -i \frac{bv^2}{R^2} e^{i(\eta_i - \eta_j)t} \int d\mathbf{r}\, \Phi_i^* L_{y'} \Phi_j \quad (61)$$

Making the phase transformation

$$A_i(b, t) = a_i(b, t) \exp\left\{-i \int_{-\infty}^{t} dt'\, V_{ii}(t')\right\} \quad (62)$$

the equations for the amplitudes $A_i(b, t)$ take the simple form

$$i\dot{A}_i(b, t) = \sum_{i \neq j} V_{ij}(t) \exp\left\{i \int_{-\infty}^{t} dt'[V_{ii}(t') - V_{jj}(t')]\right\} A_j(t') \tag{63}$$

The price to be paid if the PSS approximation is used is twofold. The model now depends on the choice of origin of the coordinate system, and spurious long-range couplings are encountered, because the individual terms in the expansion no longer satisfy the Schrödinger equation in the limit $|t| \to \infty$. The probability amplitudes for large $|t|$ are not all well defined, but these ambiguities need not be very important in practice if v is small enough.

The quantum version of the PSS method is obtained by replacing the expansion function χ_i in Eq. (40) by the molecular orbitals Φ_i, and setting the factors α_i and γ_j equal to zero. The coupled equations for the functions $F_j(\mathbf{R})$ become

$$[\nabla_R^2 + 2\mu(\epsilon - E_j)]F_i(\mathbf{R}) = \sum_j (2\mathbf{A}_{ij} \cdot \nabla_R + B_{ij})F_j(\mathbf{R}) \tag{64}$$

where the coupling matrices $\mathbf{A}_{ij}$ and B_{ij} are defined as

$$\begin{aligned} \mathbf{A}_{ij}(R) &= \int \Phi_i^* \nabla_R \Phi_j \, d\mathbf{x} \\ B_{ij}(R) &= \int \Phi_i^* \nabla_R^2 \Phi_j \, d\mathbf{x} \end{aligned} \tag{65}$$

and the differential operations are to be taken with x constant. After a partial wave decomposition and neglecting the matrix elements of $\partial^2/\partial R^2$, the coupling matrix elements reduce to combinations of the radial and rotational couplings appearing in Eqs. (61) (see, for example, Delos, 1981).

In the PSS approximation, the radial coupling only connects states having the same symmetry, whereas the rotational coupling only connects states having different symmetries. The selection rule obeyed by the rotational coupling is

$$m_j - m_i = \pm 1 \tag{66}$$

where m_j is the component of the angular orbital momentum along the internuclear line in the level j. Qualitatively, the radial coupling is most important near an avoided crossing (pseudocrossing) of two molecular potential curves of the same symmetry, whereas rotational coupling is often large at small values of R, because groups of levels with different values of m become degenerate in the united atom limit.

2. The Choice of the Basis Orbitals

In choosing a set of molecular orbitals as a basis, it is necessary to include all those orbitals which separate to the important exit channels. For example, at low energies, the important channels in the collision between He^{2+} and $H(1s)$ are (i) elastic scattering, (ii) charge exchange to the 2s, $2p_0$, and $2p_{\pm 1}$ levels of He^+ and a *minimum* basis must include four terms separating to these atomic levels. As the energy increases, charge exchange to the $n = 3$ levels of He^+ becomes important and the basis must be increased accordingly. At still higher energies, further charge-exchange channels, excitation channels, and ionization channels become important and an adequate basis may become rather large. The basis size also increases with the charge of the incident ion; for Li^{3+} both $n = 2$ and $n = 3$ levels of Li^{2+} are important in the final state at low energies and a 10-state minimum basis is required. For C^{+6} the $n = 4$ and $n = 5$ levels of C^{5+} are the most important and a 26-state basis is needed. If only total cross sections are required, the basis may be restricted. In the case of C^{6+}, below 1 keV/amu capture to the $n = 4$ levels is more important than to the $n = 5$ levels and an 11-state basis is satisfactory. It is very important to recognize that the minimum basis set is sensitive to the incident channel of the system. For example, we have discussed the $(HeH)^{2+}$ system with the initial channel $He^{2+} + H(1s)$. The same system has been studied starting from the initial channel $H^+ + He^+(1s)$. In this case, the most easily populated charge-exchange channel is $H(1s) + He^{2+}$, but because of the large change in electronic energy this is a weak process, and excitation and ionization are relatively more important than for the previous case. This implies that a larger basis set is required and accurate cross sections are much more difficult to obtain. In the absence of external information, how can the important channels be predicted? From Eq. (63) it is seen that if the energy difference $(V_{ii} - V_{jj})$ is large, the exponential factor oscillates rapidly and the effective coupling between the states i and j is small. On the other hand, if $(V_{ii} - V_{jj})$ is small, the effective coupling may be strong. In particular, if there is a pseudocrossing between the potential energy curves at some value of R, $R = R_c$, for the pair of levels i and j, the effective coupling is strong in a localized region about R_c. An example is the case of $(CH)^{6+}$ for which the correlation diagram is shown in Fig. 1. The 1s ground state of hydrogen correlates with the $6h\sigma$ level in the united atom limit. However, there is an avoided crossing at about $R = 16$ a.u. (not shown on the figure) between the $6h\sigma$ and the $5g\sigma$ levels and the system effectively transfers entirely to the $5g\sigma$ level at this point. The major source of charge transfer is the avoided crossing between $5g\sigma$ and $4f\sigma$ levels near $R = 8$ a.u., and coupling between these levels and the $4p\pi$, $4d\pi$, and $4f\pi$ levels is also strong. All these levels correlate with C^{5+} in the separated atom limit, and

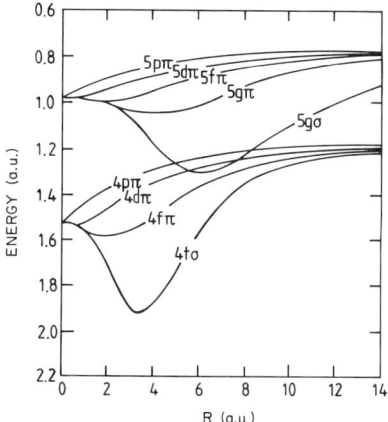

FIG. 1. A correlation diagram for (CH)⁶⁺.

below 1 keV/amu a satisfactory basis comprises the 5gσ, 4fσ, 4sσ, 4pσ, 4pπ, 4dπ, 4dσ, 4fπ, 4fδ, and 4fφ orbitals.

These general remarks about the choice of basis set are not confined to the PSS model, but apply equally to improved calculations which take account of the momentum transfer of the electron by using translational factors of one kind or another.

3. Diabatic Basis States

In the vicinity of a pseudocrossing, the radial coupling matrix element often changes very rapidly and is numerically difficult to calculate. For this reason an orthogonal transformation is often made on the basis to produce a new basis in which the radial coupling vanishes. The new basis is termed a diabatic basis (Smith, 1969; Baer, 1975; Heil and Dalgarno, 1979; Garrett and Truhlar, 1981). If the new basis functions are denoted by Φ_j^D then

$$\Phi_j^D = \sum_i \Phi_i T_{ij} \qquad (67)$$

where $T_{ij}(R)$ is an element of the transformation matrix T, which is subject to the condition $T_{ij} \to \delta_{ij}$ as $R \to \infty$, so that the basis is unchanged in the separated atom limit. It is then easy to show that the matrix elements of the operator $\partial/\partial R$ are zero taken with respect to the basis Φ_j^D provided that

$$\sum_k \left(\int d\mathbf{r}\, \Phi_j^* \frac{\partial}{\partial R} \Phi_k \right) T_{ki}(R) + \frac{\partial T_{ji}(R)}{\partial R} = 0 \qquad (68)$$

In the diabatic basis H is no longer diagonal, so that interaction matrix elements V_{ij} in the PSS approximation are replaced by $\overline{V}_{ij}$ given by

$$\overline{V}_{ij} = H_{ij} - \eta_i \delta_{ij} - i e^{i(\eta_i - \eta_j)t} \int d\mathbf{r} \, \Phi_i^{D*}(L_{y'})\Phi_j^{D} \tag{69}$$

where H_{ij} is the matrix element of H with respect to the diabatic states. It should be noted that levels i and j which have an avoided crossing in the adiabatic basis have a real crossing, $H_{ii}(R_c) = H_{jj}(R_c)$, in the diabatic basis.

Apart from easing the numerical solution of the coupled equations, the diabatic basis is important because, particularly for the one-electron system, approximate diabatic orbitals can be constructed from perturbed atomic wave functions (Janev, 1976). An interesting summary of the use of approximate diabatic orbitals in many-electron systems has been given by McCarroll (1982).

4. Alternative Translational Factors

As we have seen, except possibly at very low velocities, it is necessary to take account of the momentum transfer of the captured electron by the introduction of the translational factors. Although the plane wave forms (53) are satisfactory in many respects, these factors effectively associate the electron with one or another of the nuclei, even in the united atom limit, which seems unphysical except in the high-velocity region. To obtain a wave function with greater flexibility, and to remove this supposed defect, more general translational factors $\alpha_j(\mathbf{r}, \mathbf{R})$, $\gamma_j(\mathbf{r}, \mathbf{R})$ can be introduced which satisfy the asymptotic conditions (47) and which vanish in the united atom limit

$$\alpha_j(\mathbf{r}, \mathbf{R}) \to 0; \quad \gamma_j(\mathbf{r}, \mathbf{R}) \to 0 \quad \text{as} \quad R \to 0 \tag{70}$$

In principle, an optimal choice of α_j and γ_j can be made by the variational method (Riley and Green, 1971). The appropriate Euler–Lagrange equations for α_j and γ_j have to be satisfied simultaneously with the coupled equations for the amplitudes, but this is a very complicated numerical problem. A more restricted problem is obtained if α_j and γ_j are written as

$$\alpha_j = -\tfrac{1}{2} f_j(R) \mathbf{v} \cdot \mathbf{R} + \tfrac{1}{8} v^2 t - \eta_j$$
$$\gamma_j = \tfrac{1}{2} f_j(R) \mathbf{v} \cdot \mathbf{R} + \tfrac{1}{8} v^2 t - \eta_j \tag{71}$$

where f_j are functions of R only which satisfy

$$f_j(R) \to 0, \quad R \to 0; \quad f_j(R) \to 1, \quad R \to \infty \tag{72}$$

Starting from somewhat different viewpoints, Crothers and Hughes (1978; see also Crothers and Todd, 1981a,b: Crothers, 1981b) and Green

(1981a,b) have obtained variational estimates of $f_j(R)$, which allow these functions to be predetermined before solving the coupled equations for the amplitudes. It should be noted that the coupled equations are modified in two ways: (1) by the replacement of the factors $\exp(\pm i\mathbf{v} \cdot \mathbf{r})$ in the matrix elements (55) and (56) by $\exp[\pm i(\alpha_i - \gamma_j)]$ and (2) by the appearance of extra terms proportional to $\dot{f_j}(R)S_{ij}$, resulting from the $\partial/\partial t$ operation. In detailed studies of the $(CH)^{6+}$ system Green and his co-workers (1982a,b) have shown that it is sufficient to introduce just two functions $f_j(R)$, one for the direct and one for the exchange channels, and that the calculated cross sections are not very different from those using plane wave factors (with the same basis set).

An alternative approach, introduced by Schneiderman and Russek (1969), is to use a single translational factor, a function of both $\mathbf{r}$ and $\mathbf{R}$, such that

$$\alpha_j = \gamma_j = \tfrac{1}{2} f(\mathbf{r}, \mathbf{R})\mathbf{v} \cdot \mathbf{r} + \tfrac{1}{8} v^2 t - \eta_j, \qquad 1 < j < N \tag{73}$$

with the properties

$$\begin{aligned} f &\to -1 & &\text{if } r_1 \ll r_2 \text{ and } R \to \infty \\ f &\to 1 & &\text{if } r_2 \ll r_1 \text{ and } R \to \infty \\ f &\to 0 & &\text{if } R \to 0 \end{aligned} \tag{74}$$

The original function suggested by Schneiderman and Russek is

$$f(\mathbf{r}, \mathbf{R}) = \cos\theta \left(\frac{R^2}{a^2 + R^2} \right) \tag{75}$$

where θ is the angle between $\mathbf{r}$ and $\mathbf{R}$, and a is a cutoff parameter. The advantage of this approach is that the interaction matrix elements no longer contain exponential factors depending on $\mathbf{r}$, with the consequence that (*i*) if the basis set Φ_j is orthonormal, then the overlap integral S_{ij} is equal to δ_{ij} and (*ii*) the matrix elements can be reduced to the standard integrals of molecular structure calculations. Of course, since f depends on both $\mathbf{r}$ and t, additional interaction terms are encountered, proportional to $f(\mathbf{r}, \mathbf{R})$, $\nabla_r f(\mathbf{r}, \mathbf{R})$, and $\nabla_r^2 f(\mathbf{r}, \mathbf{R})$. For small v some simplification can be achieved by omitting terms proportional to v^2, in which case the coupling matrix elements reduce to

$$V_{ij} = e^{i(\eta_i - \eta_j)t} \int d\mathbf{r}\, \Phi_i^* \left(-i\frac{\partial}{\partial t} + i(\epsilon_i - \epsilon_j) f(\mathbf{v}, \mathbf{r}) \right) \Phi_j\, d\mathbf{r}, \qquad i \neq j \tag{76}$$

$$V_{ii} = E_i(R) - \eta_i + Z_1 Z_2/R$$

and the coupled equations take the form of Eq. (63).

Clearly the adoption of the form (75) is arbitrary, and several other functions have been put forward based on physical intuition [some of these have been discussed by Delos (1981)]. A particular form suitable for fully stripped ions of charge Z incident on hydrogen atoms has been suggested by Vaaben and Taulbjerg (1981). It is

$$f(\mathbf{r}, \mathbf{R}) = \frac{Zr_1^3 - r_2^3}{Zr_1^3 + r_2^3} + \frac{1-Z}{1+Z} \tag{77}$$

This is obtained by studying the electrostatic properties of the system and ensures that the electron shares the translational motion of the centers 1, 2 and the center of charge in the limiting cases $r_1 \ll r_2$, $r_2 \ll r_1$ and the united atom limit, respectively.

Evidently, since these functions do not provide a factor like $\exp(\pm i\mathbf{v} \cdot \mathbf{r})$ in the integrands of the interaction matrix elements, the correct high-energy limit of the coupled-channel model cannot be obtained. Despite this, the single translational (switching) factor method may be useful at low velocities, but the method would be more useful if less arbitrary methods for the choice of f could be devised.

C. Numerical Calculations with a Molecular Orbital Basis

Some examples of calculations using a molecular orbital basis of adequate size will now be given. At very low energies, for which a partial wave treatment is appropriate, there have not been extensive investigations into the fully stripped ion–hydrogen atom system. On the other hand, several calculations have been made for certain cases of partially stripped ions in collisions with atomic hydrogen which are of astrophysical interest. A recent study of this kind by Heil et al. (1981) has been made for N^{2+} + H(1s) and C^{3+} + H(1s), for a range of energies extending down to 2.7×10^{-4} eV (for N^{2+}) and up to 8.1 eV. Translational factors were omitted [which according to Gargaud et al. (1981) should not be serious below 100 eV] and to avoid spurious couplings in the final state, the origin of coordinates was taken to be on the heavy ion. Starting from numerically determined adiabatic states, a transformation was made to a diabatic basis to avoid difficulties in solving the coupled equations. At these energies, the number of levels which are strongly coupled is small—two for the $(NH)^{2+}$ and three for the $(CH)^{3+}$ system. In the case of the $(NH)^{2+}$ system, the diabatic curves for the $2^3\Pi$ and $3^3\Pi$ cross twice. Whereas the coupling matrix in the adiabatic representation varies rapidly in the two crossing regimes (near $R = 3$ and $R = 6$), in the diabatic representation the coupling is both smooth and weak, which shows

ELECTRON CAPTURE

that the diabatic states already provide a good zero-order description of the system. The diabatic couplings for $(CH)^{3+}$ are also well behaved.

Turning to fully stripped ions, the most-studied case is for $Z = 2$—the $(HeH)^{2+}$ system. Detailed studies of the convergence of the PSS expansion and also of the MO equation with Bates–McCarroll plane wave translational factors have been made by Winter and Lane (1978), Hatton et al. (1979), and Winter and Hatton (1980). They found that at ~ 1 keV/amu and below, the PSS model was satisfactory and agreed closely with the Bates–McCarroll model, using the same set of molecular orbitals. As the energy increases, the results of the two models become significantly different. By ~ 25 keV/amu, about 10 terms were required in the basis set, if plane wave translational factors are used, but the PSS model converged much more slowly, at least twice that number of terms being required. For $Z > 2$, PSS calculations have been reported by Harel and Salin (1977) (for $Z = 4, 5, 8$) and by Salop and Olson (1977, 1979) (for $Z = 6, 8$), but above ~ 1 keV/amu only results of limited accuracy can be expected from this model. More recently, a number of calculations including translational factors of different kinds have been performed and a list of those which have basis sets of adequate size is shown in Table I. Comparison with experiments (where these exist) suggests that the MO expansion method is accurate, at least for total capture cross sections for velocities of at least up to $\tilde{v} = 1$, good agreement is obtained for $Z = 2$ and $Z = 6$, but only fair agreement for $Z = 3$ (illustrated in Fig. 2), perhaps because only a limited basis of 6 terms was used (Kimura and Thorson, 1981b). Partial cross sections to particular final states would provide a more sensitive test but apart from the reaction $H^{2+} + H(1s) \rightarrow H^+(2s) + H^+$ these have not been measured. On general

TABLE I

SOME RECENT MO COUPLED-CHANNEL CALCULATIONS FOR $A^{Z+} + H(1s) \rightarrow A^{(Z-1)+} + H^{+\,a}$

A^{Z+}	Number of terms in basis	Translational factor	Energy range (keV/amu)	Reference
He^{2+}	Up to 10	Plane wave	0.25–17.5	Hatton et al. (1979)
				Winter and Hatton (1980)
	Up to 12	Switching factor	1–20	Kimura and Thorson (1981a)
	5	Optimized	0.5–6.5	Crothers and Todd (1981a,b)
Li^{3+}	6	Switching factor	2–10	Kimura and Thorson (1981b)
C^{6+}	10 or 33	Optimized	0.013–27[b]	Green et al. (1982a,b)
O^{8+}	33	Optimized	0.02–34[b]	Shipsey et al. (1983)

[a] A^{Z+} is a fully stripped ion.
[b] The linear trajectory was accurate above $v = 0.1$ a.u.; the nonlinear trajectory was used at lower energies.

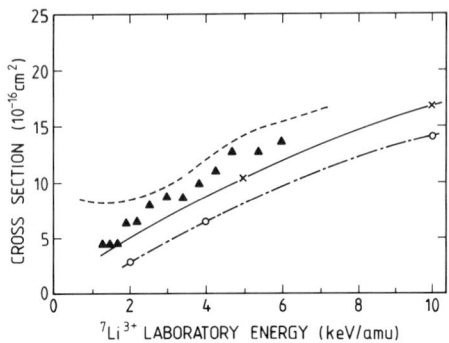

FIG. 2. Cross sections for electron capture by Li^{3+} from $H(1s)$ at low velocities. (▲) Experimental data of Seim et al. (1980). Theoretical cross sections: (——) atomic expansion (Bransden and Noble, 1982); (—·—) molecular orbital expansion (Kimura and Thorson, 1981b); (----) UDWA (Ryufuku and Watanabe, 1979a,b).

grounds one would expect the larger partial cross sections to be accurate, but smaller cross sections might not be so reliable.

D. Approximate Treatments of the Diabatic PSS Equations

In a diabatic representation of electronic states, the coupled equations in the PSS approximation for the expansion coefficients have the following forms (assuming no rotational coupling and denoting the direct and exchange amplitudes by a_i and b_j) respectively:

$$i \frac{da_i}{dt} = \sum_i V^*_{ij}(R) \exp\left[-i \int^t \omega_{ij}(\tau) \, d\tau\right] b_j \quad (78a)$$

$$i \frac{db_j}{dt} = \sum_i V_{ji}(R) \exp\left[i \int^t \omega_{ij}(\tau) \, d\tau\right] a_i \quad (78b)$$

$$V_{ij} = \langle i|H|j\rangle, \qquad \omega_{ij} = V_{ii} - V_{jj} - (\eta_i - \eta_j) \quad (78c)$$

where $\mathbf{R} = \mathbf{R}(t)$ is the classical trajectory and η_i (η_j) is the binding energy of the electron in the state i (j) when $R \to \infty$. In obtaining Eqs. (78), it is assumed that the basis set is orthonormal (or if not, then the corresponding overlap integrals are negligibly small).

For the system $H + Z$, when Z is sufficiently high (say $Z \gg 10$), one can neglect all nondiagonal elements in the interaction matrix V_{ij}, except those which couple the initial state $|i\rangle = |0\rangle$ with the final states $|j\rangle$. Then, the system (78) reduces to

$$i \frac{da_0}{dt} = \sum_j V_{0j}^*(R) \exp\left[-i \int^t \omega_{0j}(\tau)\, d\tau\right] b_j \tag{79a}$$

$$i \frac{db_j}{dt} = V_{j0}(R) \exp\left[i \int^t \omega_{0j}(\tau)\, d\tau\right] a_0 \tag{79b}$$

The initial conditions are

$$|a_0(-\infty)| = 1, \quad b_j(-\infty) = 0 \tag{80}$$

We also assume that the unitarity is preserved:

$$|a_0(t)|^2 + \sum_j |b_j(t)|^2 = 1 \tag{81}$$

and that V_{0j} is real ($V_{0j}^* = V_{j0}$).

Under certain assumptions about the R (or t) dependences of the functions $V_{0j}(R)$ and $\omega_{0j}(R)$, the system (79) can be solved exactly. For example, when the diabatic potentials $V_{jj}(R) + \eta_j$ are parallel and crossed by the potential $V_{00}(R) + \eta_0$ at the same angle, and the nondiagonal matrix elements V_{0j} are constant at the crossing points R_{0j}, one can find a solution which is a multilevel generalization of the Landau–Zener model (Osherov, 1966a,b; Demkov and Osherov, 1968). For the H(1s) + Z collision system, however, Eqs. (79) can be solved under much more general and realistic assumptions (see below). In this section we shall consider several models for the H + Z charge-exchange problem, based (directly or indirectly) on approximations of these equations.

1. Separable-Interaction Model

Let us first note that nondiagonal (coupling) matrix element $V_{0j}(R)$ in the region of a diabatic potential curve crossing is connected with the splitting $\Delta(R)$ of the corresponding adiabatic energies (pseudocrossing) by the relation (Landau and Lifshitz, 1963)

$$\Delta(R) = 2(V_{0j} - SV_{00}) \approx 2V_{0j}(R) \tag{82}$$

where $S (\ll 1)$ is the overlap integral. Let us assume that the functions $V_{0j}(R)$ and $\omega_{0j}(R)$ in Eq. (81) have the following properties (taking the initial state "0" to be a 1s state and the final state j to have quantum numbers n, l, m)

(i) $\omega_{100,nlm}(R) = \omega_n(R)$ \hfill (83a)

(ii) $V_{100,nlm}(R) = \delta_{m0} f(n, l)\, V_n(R)$ \hfill (83b)

Referring to the relation (81) and the expression (26) for $\Delta_{nlm}(R)$, we see that the above assumptions are satisfied for the H(1s) + Z system, if $Z \gg 1$. Introducing new quantities $\tilde{V}_n$ and $\tilde{b}_n$ by the relations

$$b_{nl} = \left\{ f(n, l) \left[\sum_{l'} f^2(n, l) \right]^{-1/2} \right\} \tilde{b}_n \tag{84a}$$

$$V_n = \left[\sum_l f^2(n, l) \right]^{-1/2} \tilde{V}_n \tag{84b}$$

the system (79) can be transformed in the form

$$i \frac{da_0}{dt} = \sum_n \tilde{V}_n(R) \exp\left[-i \int^t \omega_n(\tau) \, d\tau\right] \tilde{b}_n \tag{85a}$$

$$i \frac{d\tilde{b}_n}{dt} = \tilde{V}_n(R) \exp\left[i \int^t \omega_n(\tau) \, d\tau\right] a_0 \tag{85b}$$

For capture into a group of sufficiently high levels n, and having in mind the expression (27) for $f(n, l)$, one immediately obtains from Eq. (84a) the following result for the partial cross section σ_{nl} (Presynakov et al., 1981):

$$\sigma_{nl} = W_{nl} \sigma_n \tag{86}$$

$$W_{nl} = \frac{f^2(n, l)}{\sum_l f^2(n, l)} \simeq \frac{2l + 1}{Z} \exp\left(-\frac{l(l + 1)}{Z}\right) \tag{87}$$

Thus, within the present model, the relative populations $W_{nl} = \sigma_{nl}/\sigma_n$ of the l-sublevels do not depend on the collision dynamics [i.e., the solution of Eqs. (85)]. The result (87) has also been obtained by a perturbation method, using the form (26) of $\Delta_{nlm}(R)$ (Duman et al., 1979).

For electron capture into a group of highly excited final levels n, the system of coupled equations (85) can be solved exactly (Presnyakov et al., 1981), and the probability $P_n = |b_n(+\infty)|^2$ is given by

$$P_n = n^{-3} \left| \int_{-\infty}^{\infty} dt \, V(t) \exp\left[-\Gamma(t) + i \int^t \omega_n(\tau) \, d\tau\right] \right|^2 \tag{88}$$

where

$$\Gamma(t) = \frac{\pi}{Z^2} \int_{-\infty}^{t} V^2(t') \, dt' \tag{89}$$

and for the H(1s) + Z system

$$V(R) = \left(\frac{2Z}{\pi}\right)^{1/2} R \exp\left(-\frac{R^2}{3Z}\right), \quad \omega_n(R) = \frac{1}{2} - \frac{Z^2}{2n^2} + \frac{Z - 1}{R} \tag{90}$$

The partial cross section σ_n is now

$$\sigma_n = 2\pi \int_0^\infty P_n b\, db \qquad (91)$$

b being the impact parameter. It is interesting to note that this analytical model for the charge-transfer problem predicts strong Z-oscillations of the total cross section in the region $Z < 30$. Such Z-oscillators are found also in some other descriptions of the charge-exchange process and will be discussed in more detail in the subsequent sections.

Further investigation of the separable interaction model (Presnyakov *et al.*, 1982) has shown that it contains both the multichannel Landau–Zener and the decay models for the electron capture in the H + Z system as special cases. However, these models need to be considered in a broader context.

2. *The Multichannel Landau–Zener Model*

Straightforward generalizations of the two-state Landau–Zener model to a multi-curve-crossing case have been derived many times in the past (see, e.g., Bates and Lewis, 1955; Gershtein, 1962). The underlying assumption is that the effective transition zones around each of the crossings do not overlap. For the H + Z system, the distance between the two consecutive crossing points R_n and R_{n+1} [$\Delta R_{n,n+1} \simeq 2(Z-1)(2n+1)/(Z^2 - n^2)$] usually increases with increasing n and Z, while the transition region $\delta R_n \simeq \Delta(R_n)/[2|\Delta F(R_n)|] \simeq R_n^2 \Delta(R_n)/2Z$ decreases. [Dynamical criteria (see, e.g., Abramov *et al.*, 1978a) lead to the same conclusions.] Therefore, the local dynamics in each δR_n region can be described by the Landau–Zener model, and for the transition probability P_n one has

$$P_n = \exp\left(-\frac{\pi \Delta^2(R)}{2\dot{R}\,\Delta F(R)}\right)_{R=R_n} \qquad (92)$$

where $\Delta(R)$ is the adiabatic energy splitting, $\Delta F(R) = (Z-1)/R^2$ is the difference of the slopes of the corresponding diabatic potentials, and $\dot{R} = v(1 - b^2/R^2)^{1/2}$ is the radial velocity, using the straight-line trajectory approximation for the nuclear motion. The transitions, described by Eq. (92), take place only between the states [1000] and [$n, 0, n-1, 0$] of the system, having "noncrossing" adiabatic energies. In other words, the initial state [1000] does not interact with the other $n^2 - 1$ Stark states of the [nn_1n_2m] manifold by the radial coupling mechanism. However, when the system is already in the ionic [$n, 0, n-1, 0$] state, transitions may occur between this and the other Stark states due to the rotation of the internuclear axis.

Since the rotational coupling is most pronounced at smaller values of R (the angular velocity of nuclear motion is $\dot{\Theta} = -v/R^2$ for a linear trajectory) and since the curve crossing takes place at large R, it is sufficient to consider the effect of the Stark mixing only in the region $R < R_n$. Furthermore, if the collision velocity is not extremely small, the radial coupling (in the region δR_n) and the rotational coupling (in the region $R < R_n$ between two passes of the crossing point R_n) may be considered independently. Under such conditions, the probability for the rotational decay of the $[n, 0, n-1, 0]$ parabolic states to all other $n^2 - 1$ Stark states (taken to be degenerate) in the region $R < R_n$ is given by (Abramov et al., 1978a)

$$q_n = 1 - (1 - \sin^2 \beta \sin^2 \alpha)^{2(n-1)} \tag{93}$$

$$\beta = \tan^{-1}\left(\frac{2Zbv}{3n}\right), \quad \alpha = \frac{\Delta\chi}{2}\left[1 + \left(\frac{3n}{2Zbv}\right)^2\right]^{1/2} \tag{94}$$

where $\Delta\chi$ is the angle of rotation of the internuclear axis,

$$\Delta\chi/2 = \cos^{-1}(b/R_n) \tag{95}$$

The probability q_n of rotational transitions depends on the value of parameter $\delta = 2ZR_n V/3n$. When δ is small, the rotational mixing is also small ($\beta \to 0$, $q_n \to 0$), and vice versa. It can be easily verified that the total electron capture probability for a single curve-crossing case is enhanced because of the rotational transition. For a multi-curve-crossing situation, the enhancement may be quite significant.

Within the above model, the multichannel charge-exchange problem has been first considered by Abramov et al. (1978a). For N reaction channels, with crossing points $R_1 > R_2 > \cdots > R_k \cdots > R_N$, the probability that after the collision a particular nth ionic level will be populated is given by (Janev et al., 1983)

$$\begin{aligned}
P_n = \; & p_1 p_2 \cdots p_n (1 - p_n)\{1 + (p_{n+1} p_{n+2} \cdots p_N)^2 \\
& + (p_{n+1} p_{n+2} \cdots p_{N-1})^2 \\
& \times (1 - p_N)^2 (1 - q_N) \\
& + (p_{n+1} p_{n+2} \cdots p_{N-2})^2 (1 - p_{N-1})^2 (1 - q_{N-1}) \\
& + \cdots + p_{n+1}^2 (1 - p_{n+2})^2 (1 - q_{n+2}) \\
& + (1 - p_{n+1})^2 (1 - q_{n+1})\} \\
& + p_1 p_2 \cdots p_{n-1} (1 - p_n)^2 q_n
\end{aligned} \tag{96}$$

By setting $q_k = 0$ ($k = 1, \ldots, N$) in Eq. (96), one obtains the result for the pure multi-curve-crossing Landau–Zener problem (Salop and Olson,

1976). Due to the sharp dependence of p_k on $\Delta(R_k)$, the number of curve-crossing regions contributing to P_n is limited. The distant crossings are passed by the system diabatically, whereas the ones with small R_k [and therefore large $\Delta(R_k)$] are avoided adiabatically.

The particular cross section σ_n for population of a particular product state n is given by

$$\sigma_n = 2\pi \int_0^{R_n} P_n b \, db \qquad (97)$$

and the total reaction cross section is

$$\sigma = \sum_{n=1}^{N} \sigma_n \qquad (98)$$

Extensive partial (σ_n) and total (σ) cross section calculations for the H + Z system ($Z = 5-54$) have been performed (Janev et al., 1983) using the multichannel Landau–Zener model (with rotational coupling included). The expression (23) for the adiabatic energy splitting has been used in these calculations. An example is given in Fig. 3, where the total cross section for the H + O^{8+} → H$^+$ + O^{7+} electron-capture reaction is presented. It is

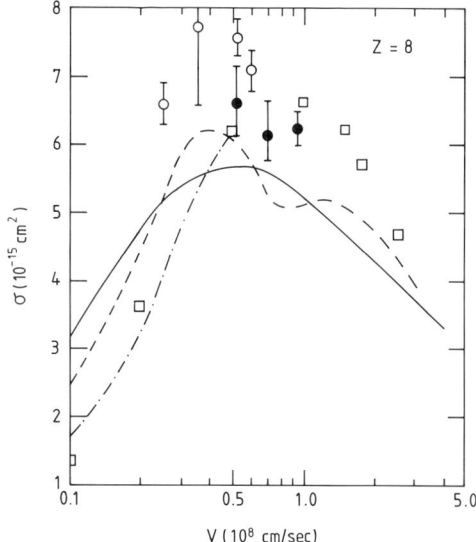

FIG. 3. Total cross sections for the H + O^{8+} → H$^+$ + O^{7+} reaction: (———) M–LZ model (Janev et al., 1983); (----) UDWA result (Ryufuku and Watanabe, 1978); (— · — ·) 8-state PSS calculations (Salop and Olson, 1979); (□) 33-state MO–CC calculation (Shipsey et al., 1983). The experimental points are data from Crandall et al. (1980) for H + Ar^{8+} (●) and H + Xe^{8+} (○) collisions.

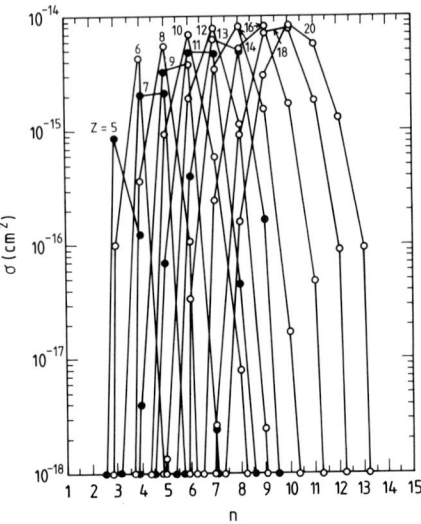

FIG. 4. n-Distributions of captured electrons in the H + Z collision system at $E = 0.49$ keV/amu. (From Janev et al., 1983.)

compared with the theoretical results obtained in an 8-state PSS calculation (Salop and Olson, 1979), in a 33-state MO–CC calculation (Shipsey et al., 1983), and in the unitarized distorted wave approximation (UDWA) (Ryufuku and Watanabe, 1978). The experimental data are for Xe^{8+} and Ar^{8+} on hydrogen (Crandall et al., 1980). The partial cross sections were also calculated. For ionic charges $Z = 5-20$ and an energy of 0.5 keV/amu, the n-distributions are shown in Fig. 4. It is seen from this figure that, in this range of Z, only a few final levels are significantly populated: one for even Z (except for $Z = 20$) and two for odd Z (except for $Z = 5$). The principal quantum number of the dominantly populated level is given by $n_m \simeq Z^{0.768}$ and changes by one unit when Z changes by two units. This gives rise to a specific oscillatory structure of the maxima of n-distributions, which remains also in the Z dependence of the total cross section (see Fig. 5). For high values of Z (above certain $Z = Z_c$), the oscillations are smeared out, since in this case the ionic levels are dense and a group of levels around n_m is significantly populated. It has been shown (Janev et al., 1983) that, even in the adiabatic region, n_m depends on the collision velocity (albeit slowly). The v dependence of n_m gives rise to a change in the extrema of Z-oscillations (see Fig. 5). The increase of v also leads to a significant population of several final ionic levels (transition zones δR_n are passed faster), which results in smearing out of the oscillations, starting with some smaller value Z_c. Analytic relations between Z_c, n_m, and v have not been derived as yet. The

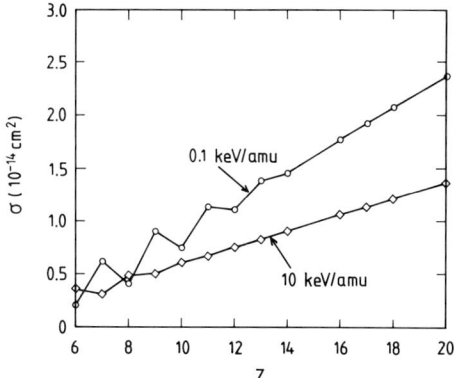

FIG. 5. Z-oscillations of the total electron capture cross section from the M–LZ calculations: $H + Z \rightarrow H^+ + (Z-1)$. (From Janev et al., 1983.)

calculations of a large number of total electron-capture cross sections in the multichannel Landau–Zener model have shown that for $Z > Z_c(v)$, the Z dependence of the cross section is linear whereas its v dependence is logarithmic. For $Z \geq 16$ and $0.04 \leq v \leq 1$, the total cross sections can be represented (within 5% accuracy) by the following analytical fit (Janev et al., 1983):

$$\sigma_{m-LZ} = 2.25 \times 10^{-16} Z \ln(15/v) \quad (cm^2) \tag{99}$$

The electron capture into specific final (n, l)-substates cannot be adequately treated within the multichannel Landau–Zener model. However, in two limiting cases, one can derive some conclusions about the relative populations of different l-substates within the same n. At the first passage of the transition region δR_n, the radial coupling populates only the $[n, 0, n-1, 0]$ Stark state. If the rotation of internuclear axis is small (the rate of the rotation being $\omega_n \sim v/R_n$), the state $[n, 0, n-1, 0]$ is little depopulated by the rotational mixing of the states, and after the second passage of the transition zone, the system will remain predominantly in this Stark state. (We neglect the mixing effects in the region $R > R_n$ due to the long-range dipole interaction.) By expanding this state in terms of the angular momentum states, the relative population W_{nl} of the l-substates is obtained in the form (Abramov et al., 1978b):

$$W_{nl} = (2l + 1)[(n-1)!]^2/(n+l)!(n-1-l)! \tag{100}$$

In the opposite, strong mixing regime, the final l-substates are expected to be populated according to their statistical weights,

$$W_{nl}^{st} = n^{-2}(2l + 1) \tag{101}$$

The mixing rate of the Stark states depends on $\omega_n \sim v/R_n$, i.e., on v and n (via R_n). Generally speaking, for energies below ~ 1 keV/amu, the l-distribution of captured electrons is given by Eq. (100) and has a maximum at some $l = l_m$ ($l_m = 1$ for $n \leq 8$, $l = 2$ for $8 \leq n < n_c$, etc.). For energies above 10 keV/amu and not too high values of n, the l-distribution is close to the statistical one [Eq. (101)], with $l_m = n - 1$ as a dominantly populated substate. With increasing n, the mixing parameter $\omega_n \sim v/R_n \sim v/n^{2+\alpha}$ ($\alpha > 0$) again decreases, which leads to a shift of the l_m toward lower values of l ($l_m < n - 1$). More detailed discussions of the final-state distributions of captured electrons (including the m-distributions) in the multichannel Landau–Zener model, as well as in other theoretical treatments of the charge-exchange process, have been given by Janev (1982a,b).

3. Decay Models

If the charge of multiply charged ions is very large, such that the density of final states available for electron capture is also high, the process can be considered as an underbarrier electron transition (tunneling) into a quasi-continuum of states. Alternatively, the same picture can be expressed in terms of the decay of the initial electronic state, which, due to its interaction with a quasi-continuum of states, becomes quasi-stationary in character. Within such a concept, the electron-capture process is analogous to the problem of decay of an electronic state in an external electric field. This analogy has been used by Chibisov (1976), who treated the ionic field as being homogeneous, with a strength Z/R^2. The decay probability per unit time of the hydrogen atom ground state in such a field is given by (Landau and Lifshitz, 1963)

$$W^{(F)}(R) = \frac{4R^2}{Z} \exp\left(-\frac{2}{3}\frac{R^2}{Z}\right) \tag{102}$$

The decay of a bound electronic state in the inhomogeneous Coulomb field of a multicharged ion has been treated by Grozdanov and Janev (1978a) and the result for the transition probability per unit time is (for the H + Z system)

$$W(R) = \frac{4R^2}{Z} \exp\left(-\frac{R^2}{Z} f(\gamma) - 2g(\gamma)\right), \quad \gamma = \frac{R}{2Z} \tag{103}$$

$$f(\gamma) = \frac{1}{\gamma}\left[1 - \frac{1}{\gamma} g(\gamma)\right]$$

$$g(\gamma) = \gamma \ln[\gamma^{1/2} + (\gamma + 1)^{1/2}][\gamma(\gamma + 1)]^{-1/2} \tag{104}$$

For $(2Z)^{1/2} \ll R \ll 2Z$, $W(R)$ reduces to $W^{(F)}(R)$, whereas for $R \gg 2Z$, $W(R) = W^{(O)}(R) = 2R \exp(-2R)$. $W^{(O)}(R)$ does not depend on Z and describes a resonant electron transition. An expression for $W(R)$, analogous to that given by Eq. (103), has been derived (in a different way) by Duman and Smirnov (1978).

The cross section in the adiabatic approximation and for straight-line trajectories is given by

$$\sigma = 2\pi \int_0^\infty b \, db\{1 - \exp[-\Gamma(b, v)]\} \quad (105)$$

where

$$\Gamma(b, v) = 2 \int_b^\infty \frac{W(R)R \, dR}{v(R^2 - b^2)^{1/2}} \quad (106)$$

Using the fact that in the asymptotic region of R (which gives the main contribution to the charge-exchange cross section), $W(R)$ depends strongly on R, one can represent the cross section σ for the H + Z charge-exchange system in the form (Grozdanov and Janev, 1978a)

$$\sigma = \pi Z \chi_0 \quad (107)$$

where χ_0 is determined by the equation

$$\chi_0 \exp[-\chi_0 f(\gamma_0)] = \frac{v}{4} \left[\frac{f(\gamma_0)}{\pi Z} \right]^{1/2} \exp[2g(\gamma_0) - C]$$

$$\gamma_0 = \frac{1}{2} \left(\frac{\chi_0}{Z} \right)^{1/2} \quad (108)$$

and $C = 0.577 \ldots$ is the Euler's constant. As seen from Eq. (108), the quantity $\chi_0 = \chi_0(v, Z)$ depends logarithmically on v and is almost independent of Z. The total cross section calculations, performed by Duman and Smirnov (1978) for a large number of fully stripped ions ($4 \leq Z \leq 30$) in the energy range below 100 keV/amu, can be represented (within a 10% accuracy and for $Z \geq 10$) by the analytical expression

$$\sigma_{DM} = 1.5 \times 10^{-16} Z \ln(314/v) \quad (cm^2) \quad (109)$$

The basic idea of the decay model, namely the existence of a large number of ionic states available for the reaction, may also be formulated in terms of a dense multi-curve-crossing system. In that case one can introduce an internuclear distance R_a, such that for $R > R_a$ all crossings are passed diabatically (with zero transition probability) and for $R \leq R_a$ the capture probability is equal to unity. In this absorbing sphere model, the reaction cross section is given by

$$\sigma_a = \pi R_a^2 \qquad (110)$$

The radius R_a of the absorbing sphere may be determined using various physical arguments. Salop and Olson (1976) have determined R_a from the condition that at this point the Massey parameter for some equivalent single-crossing problem attains the value corresponding to the maximum of the Landau–Zener cross section. This condition leads to the relation

$$\pi R_a^2 \, \Delta^2(R_a)/2v(Z-1) = 0.15 \qquad (111)$$

If in Eq. (111), one uses the expression (23) for $\Delta(R)$, then the cross section σ_a is obtained in the form

$$\sigma_a = \pi Z \ln^2\left[\frac{a}{Zv} \ln^2\left(\frac{a}{Zv}\cdots\right)\right] \qquad (112)$$

where a is a constant and the ellipses in the parentheses indicate that an iterative procedure has been used in obtaining R_a from the transcendental equation (111). With increasing Z, the logarithmic terms in Eq. (112) become increasingly important, which leads to a significant departure of the Z dependence of σ_a from linear. One should, however, note that the Z dependence of σ_a depends on the form of $\Delta(R, Z)$ used in Eq. (111), or more precisely, on the way in which the parameter Z appears in $\Delta(R, Z)$. Using the parameters of the splitting $\Delta(R)$ from Eq. (23), as well as those appearing in Eq. (111), one can put Eq. (112) in the form

$$\sigma_a = 1.1 \times 10^{-17} \, Z \ln^2\left[\frac{690}{Zv} \ln^2\left(\frac{690}{Zv}\right)\right] \quad (\text{cm}^2) \qquad (113)$$

The question of the final-state distributions of captured electrons cannot be answered within the decay models. This results from the very concept of the models. However, within the electron tunneling formalism, one can still determine the level which will be preferentially populated in the process (Chibisov, 1976; Grozdanov and Janev, 1978b). This is possible because of two circumstances: the resonant character of the process (in the particular model) and the strong variation of the decay probability with R. Neglecting the weak (logarithmic) dependence of the effective radius R_0 for the charge-exchange process [then $R_0 \simeq 2(2Z)^{1/2}$] and using the resonant energy condition at this distance, one gets

$$n_m = Z\left[1 + \frac{Z-1}{(2Z)^{1/2}}\right]^{-1/2} \xrightarrow[Z \gg 1]{} 2^{1/4} Z^{3/4} \qquad (114)$$

Up to $Z \simeq 50$, the values n_m practically coincide with those obtained from the expression $n_m = Z^{0.768}$, which is an analytical fit of the results for n_m

obtained in the unitarized distorted wave approximation (Ryufuku and Watanabe, 1979b).

IV. Atomic and Pseudostate Expansions

A. The Two-Center Atomic Orbital Basis

The expansion functions χ_j from which the coupled equations (48) are obtained are by no means confined to molecular orbitals. In fact, as velocity increases, it is expected that the electron will be increasingly associated with one or another of the heavy particle centers and expansion functions centered on the nuclei will be more appropriate. In its simplest form, the two-center expansion method takes as the basis states the particular atomic orbitals centered about nuclei 1 and 2, which are important asymptotically as $|t| \to \infty$. Thus, in place of Eqs. (54), we have

$$\chi_j(\mathbf{r}, \mathbf{R}) = \begin{cases} \phi_{1,j}(\mathbf{r}_1)[\exp i\alpha_j(\mathbf{r}, \mathbf{R})] & \text{for } 0 \leq j \leq L \\ \phi_{2,j}(\mathbf{r}_2)[\exp i\gamma_j(\mathbf{r}, \mathbf{R})] & \text{for } L < J \leq N \end{cases} \quad (115)$$

where $\mathbf{r}_1$ and $\mathbf{r}_2$ are to be expressed in terms of the independent variables $\mathbf{r}$ and t. The translational factors $\exp(i\alpha_j)$ and $\exp(i\gamma_j)$ are usually taken as plane waves, but there is no reason why more general, optimized functions should not be used.

The connection between the molecular orbital method and the atomic expansion method is very close. If the atomic orbitals in Eqs. (115) are constructed from a given set of Slater orbitals, then, in the absence of translational factors, if a set of approximate molecular orbitals are constructed from the same set of Slater orbitals, identical cross sections will be obtained from the atomic and molecular basis sets (Bates and Lynn, 1959). For this reason it is not surprising that the results of calculations using atomic and molecular basis sets containing the same asymptotic states agree over a wide velocity range (at least $\tilde{v} = 0.1–1.0$), even if the molecular orbitals are not constructed from the atomic orbitals.

The two-center basis can be improved by the addition of further functions, centered on nuclei 1 and 2, which are not solutions of the unperturbed atomic Schrödinger equations, but are of a more general nature. To be able to determine the required cross sections easily, it is convenient to choose these "pseudostate" functions to be orthogonal to the atomic orbitals which are necessary to describe the important entrance, excitation, and charge-exchange channels. There are two rather different motives for the addition of such functions. The first arises because at low velocities it is important to represent the united atom limit, of the pseudomolecule formed during the collision, accurately. An expansion in two-center atomic orbitals is known to be slowly convergent at small values of R in the static problem. This prompted Cheshire et al. (1970) in studying the $H^+ + H(1s)$ system to add pseudostates chosen to overlap the low-lying united atom orbitals of $He^+(nl)$ as much as possible. This idea has been taken up again by Fritsch and Lin (1982a), who have studied the $(HH)^+$, $(HeH)^{2+}$, and $(NeO)^+$ systems. An alternative approach, which is perhaps less convenient from a numerical point of view, is to place orbitals about the center of charge, in addition to those about centers 1 and 2 (Anderson et al., 1974). The importance of these improvements is expected to decrease as Z increases because the difference between orbitals determined by the charge $(Z + 1)$ and those determined by the charge Z diminish. The second main reason for the addition of pseudostates lies in the increasing importance of ionization as v increases. Since continuum hydrogenic orbitals are difficult to bring into a coupled channel scheme, the continuum can be approximated by a suitable set of discrete functions which are chosen to overlap the continuum functions as much as possible over the relevant range of energies. How this can be done consistently has been studied by Reading et al. (1979). Both main objectives can be achieved by taking as the expansion functions members of some complete set of discrete functions and attempting to increase the size of the basis set to convergence. Among such functions are Gaussian (Dose and Semini, 1974) and Sturmians (Gallaher and Wilets, 1968; Shakeshaft, 1976; Winter, 1982).

In order to allow for all the important discrete and continuum channels, the size of the basis set can become unmanageably large as v increases, and since a single center expansion is formally complete and numerically easy to use, it might be questioned whether results could be obtained by representing the wave function by a very large single center basis, obtaining charge-exchange cross sections from an integral expression. This approach does not work well in practice since, although over a finite region of space the wave function can be represented accurately by a large, but finite, single centered basis, the representation of the rearranged system in the asymptotic region $|t| \to \infty$ requires an infinite number of terms. A way to overcome this difficulty and to retain the speed of a single centered expansion has been suggested by Reading et al. (1981), who represented the wave function in the form

$$\psi(\mathbf{r}, t) = \sum_{j=0}^{L} a_j(b, t) \phi_{1,j}(\mathbf{r}_1)(\exp i\alpha_j)$$

$$+ \sum_{j=L+1}^{N} a_j(b, \infty) Q_j(t) \phi_{2,j}(\mathbf{r}_2)(\exp i\gamma_j) \qquad (116)$$

In this "one and a half centered" expansion (OHCE), the first sum is about the nucleus labeled 1 and is in terms of a pseudostate set of functions $\phi_{1,j}$ large enough to represent the wave function accurately in the interaction region. The second sum is required to represent the rearranged system for large $|t|$ and contains wave functions centered about 2. The functions $Q_j(t)$ are arbitrary, subject to the conditions $Q_j(t) \to 0$ as $t \to -\infty$, $Q_j(t) \to 1$ as $t \to \infty$. The constants $a_j(b, \infty)$, $j > L$, are the required charge-exchange amplitudes. By projecting the Schrödinger equation with the functions $\phi_{1,j}$ or $\phi_{2,j}$, coupled differential equations are found for the amplitudes $a_j(b, t)$ ($1 \leq j \leq L$) together with algebraic equations for the constant amplitudes $a_j(b, \infty)$ ($j > L$). Because only single centered integrals are required to construct the potential matrix connecting the amplitudes for $1 \leq j \leq L$, a very large basis set, drawn from a complete set of discrete functions, can be used and the expansion can be pushed to convergence.

B. Numerical Calculations

In Table II, some recent calculations based on two-center expansions are listed. In each case, plane-wave translational factors have been employed, although as noted above there is no reason why improved optimized factors should not be used in conjunction with these basis sets. In general, at the lowest energies ($E < 1$ keV/amu) the integration of the coupled equations is carried out along nonlinear trajectories. At the higher energies ($\tilde{v} > 1$), capture into a wide range of final state is significant and it is not possible to include all the corresponding orbitals in the basis set. To overcome this in some calculations, as noted in Table II, the cross section is corrected by making an allowance for capture into states not included in the basis set. This correction is often based on the $1/n^3$ rule of Oppenheimer but, as pointed out by Crothers (1981a), this procedure may lead to an overestimate of the total cross section.

For the $He^{2+} + H(1s)$ reaction, the cross section obtained by different two-center expansion calculations for $v < 1$ agree closely among themselves, with those of the molecular orbital method, and with experiment. At higher velocities, the computed cross sections appear to decrease a little less rapidly than the experimental data. The situation is similar for the $Li^{3+} + H(1s)$ system, and this is illustrated in Figs. 2 and 6. It might be thought that the rather large cross sections obtained at higher energies were due to the increasing importance of ionization. The OHCE model, which uses a large basis overlapping the continuum and which should represent ionization well, does succeed for $He^{2+} + H(1s)$ in reducing the high-energy charge-exchange cross section to values closer to the data, but in contrast, for the $Li^{3+} + H(1s)$ reaction no corresponding improvement is obtained.

TABLE II

SOME RECENT COUPLED-CHANNEL CALCULATIONS FOR $A^{Z+} + H(1s) \rightarrow A^{(Z-1)+} + H^+$ USING TWO-CENTER ATOMIC ORBITAL BASIS SETS[a,b]

A^{Z+}	Size and type of basis	Comment	Energy range (keV/amu)	Reference
He^{2+}	All orbitals with $n \leq 2$ on each center, no pseudofunctions.	Allowance for capture into higher states by $1/n^3$ rule	2.5–250	Bransden and Noble (1981)
	Same, with all orbitals $n \leq 3$			Bransden et al. (1983)
	Sturmian expansion with 24 terms		5–50	Winter (1982)
	Pseudostate basis centered on He^{2+}	OHCE method	25–130	Reading et al. (1982)
Li^{3+}	All orbitals with $n \leq 3$ on each center, no pseudofunctions	Allowance for capture into excited states by $1/n^3$ rule	1.4–200	Bransden and Noble (1982)
	Up to 34 orbitals including united atom pseudostates	United atom effects not very significant for this system	0.2–20	Fritsch and Lin (1982b)
C^{6+}	Pseudostate basis centered on Li^{3+}	OHCE method	50–200	Ford et al. (1982)
	10 orbitals on C^{6+}, 1 orbital on H^+	Studies the effects of nonlinear trajectories	0.1–1.0	Fritsch (1982)

[a] A^{Z+} is a fully stripped ion.
[b] Exploratory two-state calculations for $Z = 2, 3, 4, 5$ have been published by Bransden et al. (1980), and for the same values of Z calculations have been made in the range 2–50 keV/amu using a converged pseudostate basis of Hylleraas type by Lüdde and Dreizler (1982).

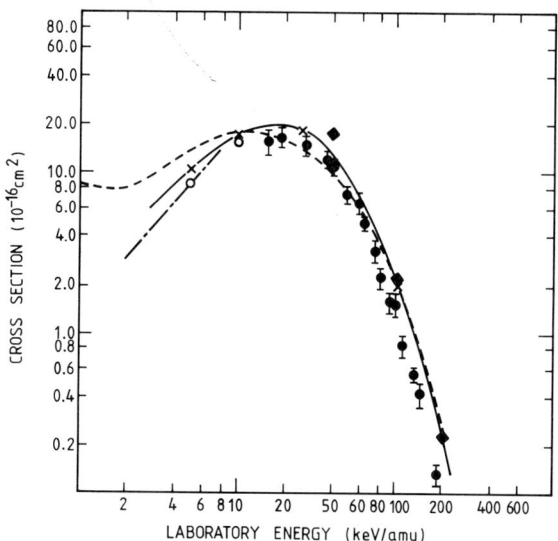

FIG. 6. Cross sections for electron capture by Li^{3+} from $H(1s)$ at intermediate velocities. (I) Experimental data (Shah et al., 1978). Theoretical cross sections: (———) atomic expansion (Bransden and Noble, 1982); (—·—) molecular orbital expansion (Kimura and Thorsen, 1981b); (----) UDWA (Ryufuku and Watanabe, 1979a,b); (♦) OHCE model (Ford et al., 1982).

The calculation using 11 AOs for the reaction $C^{6+} + H(1s)$ by Fritsch (1982) extends to low energies (100 eV/amu) and is significant because it agrees closely with the elaborate MO calculation of Green et al. (1982a,b), showing that the theory is not sensitive to any specific molecular binding effects, unless perhaps at even lower energies.

C. Approximate Treatments of AO–CC Equations

The system of diabatic equations (78) or (79) may be used for description of the charge-exchange problem in the intermediate energy region, provided the wave functions $|i\rangle$ and $|j\rangle$ contain translational phase factors and are appropriately orthogonalized, in which case the unitarity of the S-matrix is preserved. When using a one-center expansion with only one term describing the initial electronic state [i.e., the coupled system (79)], then the continuum states of the electron must also be included in the basis. Only for the one-center expansions is the system of coupled equations Hermitian (Polvéktov and Presnyakov, 1968).

Approximate treatments of the system of coupled equations (78), with orthogonalized atomic orbitals and plane-wave translational factors included, have been performed in the context of the charge-exchange atom–highly charged ion problem by Presnyakov and Ulantsev (1975) and by Ryufuku and Watanabe (1978, 1979a,b). In both cases the charge-exchange probability has been expressed in terms of the two-state transition probability (or the corresponding transition matrix elements) for the coupling of the initial state with each of the final states. Whereas Presynakov and Ulantsev (1975) calculated the two-state probability in the Vainshtein–Presynakov–Sobel'man (VPS) approximation (Vainshtein et al., 1964), Ryufuku and Watanabe (1978) used the distorted wave Born approximation (Bates, 1958). We shall refer to these methods as the multichannel–Vainshtein–Presnyakov–Sobel'man (M–VPS) approximation and the unitarized distorted wave approximation (UDWA), respectively. Below, we shall give a brief account of these two approximations and of the main results.

1. Multichannel VPS Method

Let us introduce the notation ($j \to n$)

$$V_n(t) = V_{0n} \exp\left[-i \int^t \omega_{0n}(\tau)\, d\tau\right] \tag{117}$$

and rewrite Eqs. (79) in the form ($V_{0n}^* = V_{n0}$)

$$i\, da_0/dt = \sum_n V_n^*(t)\, b_n(t) \tag{118a}$$

$$i\, db_n/dt = V_n(t)\, a_0(t) \tag{118b}$$

$$|a_0(-\infty)| = 1, \quad b_n(-\infty) = 0, \quad |a_0(t)|^2 + \sum_n |b_n(t)|^2 = 1 \tag{119}$$

Introducing, further, a new function $K_n(t)$ by the relation

$$K_n(t) = b_n(t)/a_0(t) \tag{120}$$

the system of coupled equations is reduced to a system of nonlinear equations:

$$dK_n/dt + i(V_n - V_n^* K_n^2) = iQ_n K_n \tag{121}$$

$$K_n(-\infty) = 0, \quad Q_n = \sum_{m \neq n} V_m^* K_m \tag{122}$$

The amplitudes a_0 and b_n are related to K_n by

$$|b_n(t)|^2 = \frac{|K_n(t)|^2}{1 + \sum_m |K_m(t)|^2}, \quad |a_0(t)|^2 = \frac{1}{1 + \sum_m |K_m(t)|^2} \tag{123}$$

These relations show that, independent of the approximation to which K_n is calculated, the unitarity in the problem is strictly preserved. For example, the first-order solution of Eq. (121) is

$$K_n(t) = -i \int_{-\infty}^{t} V_n(t') \, dt'$$

$$= -i \int_{-\infty}^{t} V_{0n}(t') \exp\left[-\int^{t'} \omega_{0n}(\tau) \, d\tau\right] dt' \quad (124)$$

which, together with Eq. (123), gives a normalized transition probability in the distorted wave approximation (i.e., a version of the UDWA). Such a unitarization procedure corresponds to the K-matrix approach in general scattering theory. The general equations (121) cannot be solved exactly. However, the terms on the right-hand sides in these equations are expected to be small, since these terms describe higher-order transitions of the system. If this is so, a perturbation treatment of these terms is appropriate. If the right-hand sides of Eqs. (121) are set equal to zero, then a system of uncoupled Riccati-type equations is obtained, which can be solved exactly. The solution as $t \to \infty$ is

$$|K_n(+\infty)|^2 = P_{0n}/(1 - P_{0n}) \quad (125)$$

where P_{0n} is the exact probability for the two-state problem. Inserting Eq. (125) into Eqs. (123), one obtains the probability W_{0n} for population of the nth state in the charge transfer process:

$$W_{0n} = T^{-1} P_{0n} \bar{P}_n, \qquad \bar{P}_n = \prod_{m \neq n} (1 - P_{0m}) \quad (126)$$

$$T = \prod_n (1 - P_{0n}) + \sum_n P_{0n} \bar{P}_n \quad (127)$$

Thus, apart from the normalization factor T^{-1}, probability W_{0n} is expressed as a product of the two-state probability P_{0n} for the $0 \to n$ transition and the probability $\bar{P}_n$ that during the collision, transition to other states does not occur. Equations (126) are obtained without any assumptions about the form of the coupling matrix elements V_{0n} or the potentials V_{nn}. In that sense it is a generalization of the Demkov–Osherov result on the transitions in a system of parallel diabatic potential curves, crossed at the same angle by another potential curve (Demkov and Osherov, 1968). The validity of Eq. (126) is restricted by the condition $\frac{1}{4}|\ln \bar{P}_n| < 1$ and $\frac{1}{2}P_{0n}^{1/2}|\ln \bar{P}_n| \ll 1$.

In the VPS approximation, the probability P_{0n} is given (Vainshtein et al., 1964)

$$P_{0n} = \left| \int_{-\infty}^{+\infty} dt \, V_{0n}(t) \exp\left[i \int^{t} dt' (\omega_{0n}^2 + 4V_{0n}^2)^{1/2}\right] \right|^2 \quad (128)$$

For the H + Z system ω_{0n} is given by

$$\omega_{0n} = V_{00}(R) - V_{nn}(R) - \Delta\epsilon_{0n} \approx (Z-1)/R(t) - \Delta\epsilon_{0n} \qquad (129a)$$

where

$$\Delta\epsilon_{0n} = Z^2/2n^2 - \tfrac{1}{2} \qquad (129b)$$

The main contribution to P_{0n} comes from the regions where the function $\Omega(t) = (\omega_{0n}^2 + 4V_{0n}^2)^{1/2}$ is zero. If V_{0n} exponentially decreases with R, then $\Omega(R)$ has the following zeros in the complex R-plane (Presnyakov and Ulantsev, 1975):

$$R^{(1)} \simeq 1.2\frac{n^2}{Z} + i\left(\frac{\pi}{2\gamma} + 4\pi k\right), \qquad k = 0, 1, 2, \ldots \qquad (130)$$

$$R^{(2)} \simeq R_c + i\frac{2R_c}{\Delta\epsilon_{0n}} V_{0n}(R_c) \qquad (131)$$

$$R_c \simeq \frac{Z-1}{\Delta\epsilon_{0n}}, \qquad \gamma = \frac{1}{2}\left[\left(\frac{Z^2}{n^2} + \frac{v^2}{4}\right)^{1/2} + \left(1 + \frac{v^2}{4}\right)^{1/2}\right] \qquad (132)$$

The singular point $R^{(2)}$ corresponds to a pseudocrossing; its real part R_c is positive only for $n < Z$. For all n between Z and $Z/2$, one has

$$\text{Re}\{R^{(2)}\} > \text{Re}\{R^{(1)}\}, \qquad \text{Im}\{R^{(2)}\} \ll \text{Im}\{R^{(1)}\} \qquad (133)$$

and the contributions from these two points to the transition probability P_{0n} can be separated:

$$P_{0n} = P_{0n}^{(1)} + P_{0n}^{(2)} \qquad (134)$$

The transitions around the point $R^{(2)}$ are of Landau–Zener type. They can be described by the usual Landau–Zener model with the parameters $\omega_{0n} = [(Z-1)/R_c^2](R - R_c)$ and $V_{0n} = V_{0n}(R_c)$. The transitions around the point $R^{(1)}$ have underbarrier character and can be described by the VPS formula. Due to the relations (133), these two mechanisms operate in different velocity regions and may cause structure (two maxima) in the partial cross section σ_n. Neglecting some interference effects, the representation (134) of P_{0n} leads to an analogous representation for W_{0n}, and to a two-term form of the partial cross section:

$$\sigma_n = \sigma_{0n}^{(1)} + \sigma_{0n}^{(2)} \qquad (135)$$

For $n \geq Z$, the singular point $R^{(2)}$ does not exist, and σ_n is given by the VPS cross section $\sigma_{0n}^{(1)}$ only. As is well known, $\sigma_{0n}^{(1)}$ gives the correct first-order result of the perturbation theory when $v \gg 1$ (the Brinkman–Kramers approximation), has a maximum at $v \simeq 1$, and exponentially decreases with decreasing collision velocity. For electon capture into the resonant level $n = Z$ and for $v \leqslant 1$, the cross section has the form

$$\sigma_{n\approx Z} = \sigma_{0n}^{(1)} \simeq \pi Z^2 \exp\left[-\frac{\pi}{v}\left(1 - \frac{1}{Z}\right)\right] \quad (136)$$

In the same velocity region $\sigma_{0n}^{(2)}$ gets its maximum for smaller values of n ($n_m \simeq Z^{0.768}$) and it prevails over $\sigma_{0n}^{(1)}$. Thus, a structure (two maxima) is expected in the n-distribution of captured electrons for $v \leqslant 1$.

A characteristic result of the M–VPS approximation for charge exchange in the H + Z system is the quadratic dependence of the cross section on Z for $v \leqslant 1$ (in contrast to the linear one, predicted by the M–LZ and electron tunneling models). For the $\sigma_{0n}^{(1)}$ component of the cross section, the Z^2 dependence in the considered range follows from the VPS model itself. However, for the $\sigma_{0n}^{(2)}$ cross section, the quadratic Z dependence is an artifact of inappropriate R and Z dependences of the coupling matrix element $V_{0n}(R, Z)$. For $V_{0n}(R, Z)$, Presnyakov and Ulantsev (1975) assumed $V_{0n} \sim F(R, Z) \exp(-aR/Z)$, where $F(R, Z)$ is some polynomial and a is constant. By scaling transformations, it can be easily shown that the ratio R/Z in the exponent of V_{0n} leads to the Z^2 dependence of $\sigma_{0n}^{(2)}$. However, in the region of R [namely $(2Z)^{1/2} < R < 2Z$] giving the main contributions to $\sigma_{0n}^{(2)}$, $V_{0n}(R, Z) \simeq (\frac{1}{2}) \Delta(R, Z) \sim \exp(-aR^2/Z)$ [see Eq. (30a)], and the ratio R^2/Z in the exponent gives rise to a linear Z dependence of the cross section. The same result is obtained also if $V_{0n}(R, Z) \sim \exp(-aRZ^{-1/2})$. One of the main merits of the M–VPS theory for charge exchange is that it is able to include both the Landau–Zener and the tunneling transitions. It is therefore applicable in a very broad energy range.

2. Unitarized Distorted Wave Approximation (UDWA)

Ryufuku and Watanabe (1978, 1979a) have shown that unitarization of the approximate solutions for the transition amplitudes in the H + Z collision system can be easily achieved by using the interaction representation for the S-matrix. If $\{\xi_n\}$ is a complete basis, consisting of atomic orbitals centered on the proton ($\{\xi_n^{(a)}\}$) and on the ion ($\{\xi_n^{(b)}\}$) (and hence, $\{\xi_n\} = \{\xi_n^{(a)}\} \cup \{\xi_n^{(b)}\}$), the state vector $|c(t)\rangle$ of the expansion amplitudes $\{c_n(t)\}$ satisfies the equation

$$i\frac{d}{dt}|c(t)\rangle = \mathbf{H}|c(t)\rangle \quad (137)$$

with

$$\mathbf{H} = \mathbf{S}^{-1}\mathbf{h} \quad (138)$$

$$S_{mn} = \langle \xi_n | \xi_m \rangle, \quad h_{mn} = \left\langle \xi_n \left| H - i \frac{\partial}{\partial t} \right| \xi_m \right\rangle \tag{139}$$

where H is the electronic Hamiltonian of the system. The S-matrix, defined by

$$|c(\infty)\rangle = \mathcal{S} |c(-\infty)\rangle \tag{140}$$

can be written as

$$\mathcal{S} = \exp\left(-i \int_{-\infty}^{+\infty} H^{(0)} \, dt \right) \mathcal{S}^{\text{int}} \tag{141}$$

where $H^{(0)}$ is the diagonal part of the matrix H,

$$\mathcal{S}^{\text{int}} = T \exp\left[-i \int_{-\infty}^{\infty} H^{\text{int}}(t) \, dt \right] \tag{142}$$

$$H^{\text{int}}(t) = \exp\left(i \int_{-\infty}^{t} H^0 \, dt'\right)(H - H^0) \exp\left(-i \int_{-\infty}^{t} H^0 \, dt'\right) \tag{143}$$

and T is the chronological operator. If $|0\rangle$ is the initial electronic state, the probability that at $t = \infty$ the electron will be in the state $|n\rangle$ is given by

$$P_{0n} = |\langle n | \mathcal{S} | 0 \rangle|^2 = |\langle n | \mathcal{S}^{\text{int}} | 0 \rangle|^2 \tag{144}$$

In calculating $\langle | \mathcal{S}^{\text{int}} | 0 \rangle$, one can usually expand the exponent containing H^{int} and take a time-ordered product. The first major approximation made by Ryufuku and Watanabe in dealing with the expansion of $\langle n | \mathcal{S}^{\text{int}} | 0 \rangle$ was to ignore time ordering. This approximation is expected to be reasonable only for $v \geqslant 1$. A second major approximation was the neglect of all matrix elements except those involving the initial state $|0\rangle$, which reduces the solution to a distorted wave approximation. Retaining, thus, only the first nontrivial term in the expansion of $\langle n | \mathcal{S}^{\text{int}} | 0 \rangle$, one obtains (Ryufuku and Watanabe, 1978)

$$\langle 0 | \mathcal{S}^{\text{int}} | 0 \rangle = \cos p^{1/2} \tag{145}$$

$$\langle n | \mathcal{S}^{\text{int}} | 0 \rangle = i t_{0n} p^{1/2} \sin p^{1/2} \tag{146}$$

where

$$p = \sum_n |t_{0n}|^2, \quad t_{0n} = \int_{-\infty}^{+\infty} \langle n | H^{\text{int}} | 0 \rangle \, dt \tag{147}$$

It is clear from the above expressions that $\sum \langle n | \mathcal{S}^{\text{int}} | 0 \rangle^2 = 1$ independent of the fact that the higher-order terms in the expansion of $\langle n | \mathcal{S}^{\text{int}} | 0 \rangle$ have been omitted. However, the omitted higher-order terms, describing transitions via intermediate states, may lead to significant errors in P_{0n} at higher collision velocities.

The matrix element t_{0n} in Eqs. (147) may describe an excitation transition ($|n\rangle = \xi_n^{(a)}$), electron capture into a specific state ($|n\rangle = \xi_n^{(b)}$), or ionization ($|n\rangle = \xi_v^{(b)}$, $v^2/2$ is the energy of the ejected electron, and its wavefunction is centered on the ion to account for the "capture" to the continuum"). Thus, the sum $p = \Sigma_n |t_{0n}|^2$ can be decomposed into three parts:

$$p = p_{\text{exc}} + p_{\text{ion}} + p_{\text{CT}} \tag{148}$$

related to the excitation, ionization, and charge transfer processes. The two-state matrix elements t_{0n} can be calculated for each of these processes by using atomic electron wave functions, with plane-wave translational factors included. For the electron transfer process, for instance, t_{0n}^{CT} is given by ($S_{0n}^2 \ll 1$)

$$t_{0n}^{\text{CT}} = \int_{-\infty}^{+\infty} dt \left(V_{0n}^{a,b} - S_{0n} V_{00}^{a,a} \right) \exp\left\{ i \int_{-\infty}^{t} dt' \, \omega_{0n}^{a,b} \right\} \tag{149}$$

where

$$\omega_{0n}^{a,b} = V_{nn}^{b,b} - V_{00}^{a,a} - (\eta_n^b - \eta_0^a) \tag{150}$$

$$V_{kj}^{\alpha,\beta} = \langle \xi_k^{(\alpha)} | H - H_0^{(\beta)} | \xi_j^{(\beta)} \rangle$$

$$H_0^{(\beta)} | \xi_j^{(\beta)} \rangle = \eta_j^{(\beta)} | \xi_j^{(\beta)} \rangle \tag{151}$$

The probability for electron capture into a final state n is now [Eqs. (146)–(148)]

$$P_{0n}(b) = \frac{|t_{0n}^{\text{CT}}|^2}{p} \sin^2 p^{1/2} \tag{152}$$

and for capture to all final states, one has

$$P(b) = \frac{\Sigma_n |t_{0n}^{\text{CT}}|^2}{p} \sin^2 p^{1/2} = \frac{p_{\text{CT}}}{p} \sin^2 p^{1/2} \tag{153}$$

where b is the impact parameter, and n denotes the set of final-state quantum numbers (n, l, m).

In view of the particular approximations made, the UDWA is expected to be valid in the interval $1 \leqslant v \leqslant 3\text{–}4$. A large number of cross section calculations (including $Z = 1\text{–}6, 8, 10, 14, 20$) have been performed, however, far beyond these limits (Ryufuku and Watanabe, 1978, 1979a,b; Ryufuku, 1982). The UDWA total cross section results can be adequately represented in a scaled form (Ryufuku, 1982):

$$\sigma(E) = Z^\alpha \tilde{\sigma}(\tilde{E}), \quad \tilde{E} = E/Z^\beta, \quad \alpha = 1.07, \quad \beta = 0.35 \tag{154}$$

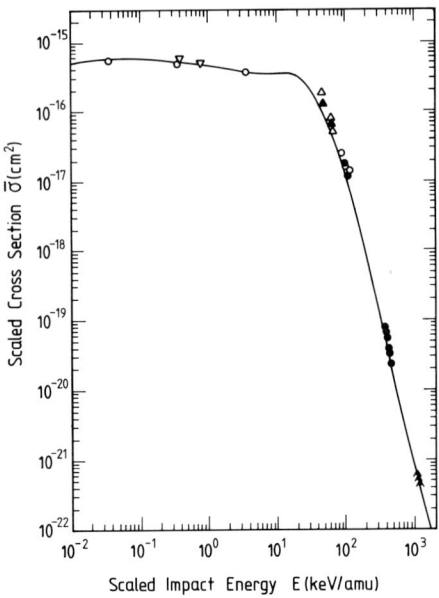

FIG. 7. Dependence of the reduced capture cross section $\tilde{\sigma}$ on the reduced energy $\tilde{E}$ [see Eq. (154)] for the H + Z collision system in UDWA (Ryufuku, 1982). The experimental data shown are for Ca^{20+} (○), Fe^{12+} to Fe^{22+} (○, △), Xe^{12+} (▲), and for Fe^{12+} to Fe^{22+} on H_2 (●, ▲).

in which for $\tilde{E} \geqslant 25$ keV/amu, the reduced cross sections $\tilde{\sigma}(\tilde{E})$ differ very little from each other. For $\tilde{E} < 25$ keV/amu, the difference between them for different values of Z ($\geqslant 3$) is within a factor of about two. In this region $\tilde{\sigma}(\tilde{E})$ varies very slowly with the reduced energy $\tilde{E}$, down to $\tilde{E} \simeq 10^{-2}$ keV/amu (see Fig. 7). The scaling of Eqs. (154) has been derived from the charge-exchange cross section calculations based on Eqs. (148)–(153), taking account of excitation and ionization channels in the total probability p.

In Fig. 8 the partial cross sections σ_n, for capture into a specific ionic principal shell in the H + Si^{14+} and H + Ne^{10+} collision systems, are shown. It is seen from these figures that the n-distributions in the energy range below 100 keV/amu have maxima at the same principal quantum number n_m given by $n_m \simeq Z^{0.768}$ (Ryufuku and Watanabe, 1979b); and that with increasing collision energy, n_m decreases rapidly. The partial cross sections σ_{nl}, for capture into a specific nl-substate for the reaction H + $Ne^{10+} \rightarrow$ $H^+ + Ne^{9+}(nl)$ are shown in Fig. 9 for $E = 25$ and 500 keV/amu. The l-distributions at $E = 25$ keV/amu have a maximum for $n < n_m$ (=6 for $Z = 10$) at $l = l_m < n - 1$, whereas for $n > n_m$, the substrate $l = n - 1$ is dominantly populated. At the energy of 500 keV/amu, the l-distributions

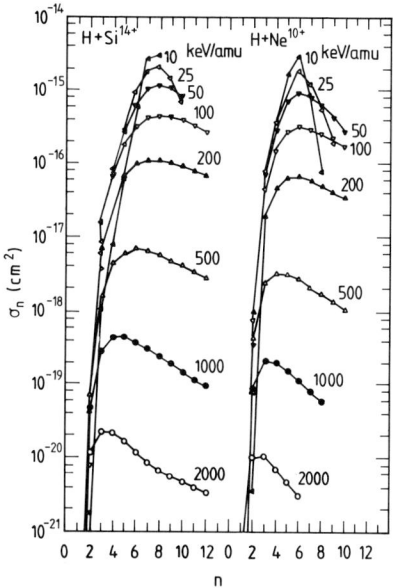

FIG. 8. UDWA n-distributions of captured electrons in the H + Si^{14+} and H + Ne^{10+} collision sytems for different energies. (From Ryufuku and Watanabe, 1979b.)

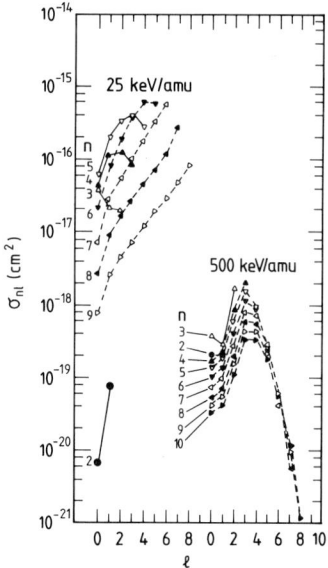

FIG. 9. UDWA l-distributions of the captured electrons in the H + Ne^{10+} → H + Ne^{9+}(nl) reaction for two different energies: 25 and 500 keV/amu. (From Ryufuku and Watanabe, 1979b.)

have a maximum at $l_m < n - 1$. The reasons for such behavior of σ_{nl} are discussed elsewhere (Janev, 1982a).

V. Charge Exchange at High Velocities

At velocities in excess of 2–4 a.u. ($E = 100-200$ keV/amu), the coupled-channel approximations become unwieldy, because many pseudostates are needed to describe the dominant ionization channels accurately and simpler methods are required. It is well established that at asymptotically high velocities the charge-exchange cross section is given by the second Born approximation. Indeed, there is no velocity region in which first-order Born methods can be expected to be useful (Shakeshaft and Spruch, 1979). In the energy region of interest (up to a few MeV/amu) the Brinkman–Kramers series, obtained by omitting the internuclear potential and expanding in terms of the free particle Green's function, has been shown to be slowly convergent and not to be of practical use as a calculation method. For instance, explicit calculations for the p + H(1s) → H(1s) + p reaction (Simony and McGuire, 1981) show that the second-order cross section is larger than the first-order cross section, which already exceeds the experimental data. Fortunately, other types of perturbation theory, based on distorted waves, are possible and one of these, the continuum distorted wave approximation of Cheshire (1964) appears to provide accurate cross sections for $v \geqslant 2.5$. This method has been reviewed in detail by Belkić *et al.* (1979) and we will only make a few brief remarks about it here. The idea is to take as a zero-order wave function for the initial state (within the straight-line impact parameter method), the function

$$\psi_i = \phi_{1,i}(\mathbf{r}_1) L_i(\mathbf{r}_2, t) \exp i\alpha_i \tag{155}$$

where $\phi_{1,i}$ is the initial wave function of the target and α_i is a plane-wave translational factor. The function L_i is a Coulomb function which represents the interaction between the projectile and the electron in the initial state:

$$L_i(\mathbf{r}_2, t) = N \exp iv \ln(vR - v^2 t) F(i\bar{v}, i, iv_2 r_2 - i\mathbf{v} \cdot \mathbf{r}_2) \tag{156}$$

where N is a normalization factor, $v = Z_1 Z_2/v$ and $\bar{v} = Z_2/v$. A similar distorted wave ψ_f is employed for the final channel of interest, and the scattering amplitude is expressed as a matrix element with respect to ψ_i and ψ_f of a corresponding perturbing potential operator. This method has not, as yet, been extensively applied to the $Z + H(1s)$ system for large Z, but for small Z it is very successful. For example, the experimental data for the $Li^{3+} + H$ reaction are very well represented by the computed cross sections

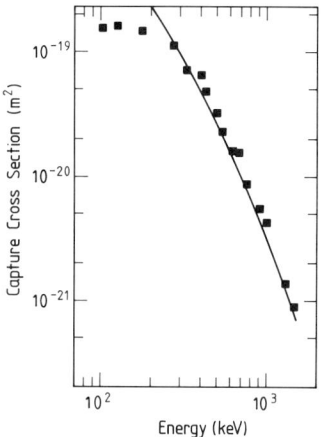

FIG. 10. Cross section for electron capture by Li^{3+} from H(1s) at high energies: (———) CDW approximation (Belkić et al., 1979); (■) experimental data (Shah et al., 1978).

(see Fig. 10) (Belkić et al., 1979; Crothers and Todd, 1980). As Crothers (1982) has shown, the CDW model is capable of further development, by using the functions ψ_i and ψ_f as the basis of a two-state coupled-channel calculation.

A different approach, suggested by Mittleman and Quong (1968) and developed by Dewangan (1975, 1977), is based on the eikonal, or Glauber, approximation. The exact amplitude for charge exchange to a final level can be represented as (in the notation of Section III,B)

$$a_f(b, \infty) = -i \int_{-\infty}^{\infty} dt \int \phi_{2,f}^*(\mathbf{r}_2) e^{-i\gamma_f} (Z_1/r_1) \psi(\mathbf{r}, t) \, d\mathbf{r} \qquad (157)$$

In the eikonal approximation (see, for instance, Bransden, 1983), the exact wave function for the system, $\psi(\mathbf{r}, t)$, is approximated by

$$\psi(\mathbf{r}, t) = \phi_{1,i}(\mathbf{r}_1) e^{i\alpha_i} \exp\left\{ i \int_t^{\infty} dt' \, (Z_2/r_2) \right\} \qquad (158)$$

The phase factor represents scattering between the projectile of charge Z_2 and the bound electron in the initial state, treating the bound electron as a fixed charge, which seems physically reasonable when the velocity of the projectile is large. If the phase factor is replaced by unity, Eq. (157) reduces to the unsatisfactory first-order Brinkman–Kramers cross section, but if the phase factor is retained, by expanding in powers of Z_2, it is seen that some allowance is made for every term in the perturbation series. Dewangan (1977), Chan and Eichler (1979a,b,c), and Ho et al. (1981) have shown that the eikonal cross section can be evaluated analytically. It takes the form of a product of the first-order Brinkman–Kramers cross section σ_{BK} and a factor which varies slowly with the quantum numbers of the final state, and the ratios of partial cross sections in the eikonal approximation are given approximately by σ_{BK}, although the absolute magnitude is very different. The Brinkman–Kramers cross section σ_{BK} has been given in closed form by Sil (1954; see Basu et al., 1978). As can be seen from Eqs. (157) and (158), the second-order term in the expansion in powers of Z_2 of $a_f(b, \infty)$ is purely real, whereas the corresponding term in the Brinkman–Kramers series is complex. For this reason the eikonal approximation must fail in the high-energy limit. Since Eq. (157) is only of first order in the perturbation (Z_1/r_1), the eikonal approximation must also fail at low energies. Empirically, the computed total cross sections for $Z_2 \leq 3$ ($Z_1 = 1$) appear to be accurate in an energy range from ~ 50 to ~ 500 keV/amu (Eichler, 1981), but it seems that for $Z_2 = 5$ and 6, the eikonal cross sections may be too large. Further work is required to establish for what ranges of v and Z_2 the eikonal model can be expected to be accurate.

VI. Classical Descriptions

There have been many attempts in the past to describe the charge-exchange process by the methods of classical mechanics (see, e.g., Mapleton, 1972). Their recent application to the charge transfer in hydrogen atom–highly charged ion collisions has been remarkably successful (Olson et al., 1977, 1978; Phaneuf et al., 1977).

It seems that the origin of the success of the classical description of the collision processes in the H + Z system lies in the specific nature of the Coulomb interaction. Both in the hydrogen atom and in the two-center Coulomb problem there are additional (dynamical) symmetries which are not observed in other systems. In the case of the hydrogen atom, dynamical symmetry is connected with the conservation of the Runge–Lentz vector, where in the (Z_1eZ_2) system, it is connected with the conservation of the separation constant. As a result, the momentum distribution of a bound electron in the Coulomb field is the same both in quantum (after summation over the degeneracies) and classical mechanics (Fock, 1935). Similarly, the equation of motion of the electron in the field of the two fixed Coulomb charges allows separation of the variables (in the prolate spheroidal coordinates) both in the quantum mechanical and classical cases. Since at high collision energies, the charge-transfer process is largely determined by the electron momentum distributions in its initial and final state (Mapleton,

1972), and in the adiabatic region one can always reach an internuclear distance R_0 such that for $R < R_0$ the electronic motion becomes classically allowed, the above-mentioned properties of the considered collision system might be a plausible basis to justify the application of classical concepts in describing the H + Z charge-exchange problem.

In the applications, two approaches have been followed: numerical solution of classical equations of motion by using the Monte Carlo method to determine the initial conditions, and construction of approximate analytical models.

A. Classical Trajectory Monte Carlo Method

The motion of the three-particle system (p, e, Z) in the classical mechanics is described by Hamilton's equation (Landau and Lifshitz, 1973):

$$\partial H/\partial p_i = \dot{q}_i, \qquad \partial H/\partial q_i = -\dot{p}_i \tag{159}$$

where q_i and p_i are the coordinates and momenta of the particles and H is the Hamilton's function. Let (q_1, q_2, q_3) be the Cartesian coordinates of the electron with respect to the proton, (q_4, q_5, q_6) the coordinates of the projectile nucleus (Z) with respect to the center of mass of the hydrogen atom (p, e), and (q_7, q_8, q_9) the coordinates the center of mass of the entire system. The $p_i (i = 1, \ldots, 9)$ are the corresponding conjugate momenta. By subtracting the center of mass motion, the system (159) is reduced to 12 coupled equations for $q_i(i = 1, \ldots, 6)$ and $p_i(i = 1, \ldots, 6)$ (Olson and Salop, 1977):

$$\dot{q}_i = \left(1 + \frac{1}{m_p}\right) p_i, \qquad i = 1, 2, 3 \tag{160a}$$

$$\dot{q}_i = \left(\frac{1}{m_z} + \frac{1}{1 + m_p}\right) p_i, \qquad i = 4, 5, 6 \tag{160b}$$

$$\dot{p}_i = \frac{1}{1 + m_p} \left(\frac{1}{1 + m_p} q_i + q_{i+3}\right) \frac{Z}{R_1^3} + q_i \frac{1}{R_2^3}$$

$$+ \frac{m_p}{1 + m_p} \left(\frac{m_p}{1 + m_p} q_i - q_{i+3}\right) \frac{Z}{R_3^3}, \qquad i = 1, 2, 3 \tag{160c}$$

$$\dot{p}_i = \left(\frac{1}{1+m_p} q_{i-3} + q_i\right) \frac{Z}{R_1^3}$$

$$- \left(\frac{1}{1+m_p} q_{i-3} - q_i\right) \frac{Z}{R_3^3}, \quad i = 4, 5, 6 \quad (160d)$$

where R_1 is the internuclear (p–Z) separation, R_2 is the e–p separation, and R_3 is the e–Z distance. The $R_i (i = 1, 2, 3)$ are given by

$$R_1^2 = \sum_{i=1}^{3} \left(\frac{1}{1+m_p} q_i + q_{i+3}\right)^2 \quad (161a)$$

$$R_2^2 = \sum_{i=1}^{3} q_i^2 \quad (161b)$$

$$R_3^2 = \sum_{i=1}^{3} \left(\frac{m_p}{1+m_p} q_i - q_{i+3}\right)^2 \quad (161c)$$

In order to solve Eqs. (160), the initial conditions on q_i^0 and p_i^0 must be specified. If the axis z of the Cartesian coordinate system is chosen in the direction of the initial relative velocity vector v, then

$$p_4^0 = p_5^0 = 0, \quad p_6^0 = \left(\frac{1}{1+m_p} + \frac{1}{m_z}\right) v \quad (162)$$

Further, if the coordinate system is oriented so that the projectile and the center of mass of the hydrogen atom lie in the $y - z$ plane, then

$$q_4^0 = 0, \quad q_5^0 = b, \quad q_6^0 = -(R_*^2 - b^2)^{1/2} \quad (163)$$

where b is the impact parameter and R_* is the initial distance of Z from the center of mass of the hydrogen atom. (In the calculations R_* is usually taken to be about 10–20 times Z.)

The initial conditions for the electronic coordinates are taken to be

$$q_1^0 = r_e \sin\theta \cos\phi$$
$$q_2^0 = r_e \sin\theta \sin\phi \quad (164)$$
$$q_3^0 = r_e \cos\theta$$

where (r_e, θ, ϕ) are the spherical coordinates of the electron. Introducing the angle ξ between the plane of the orbit and the plane containing both the z axis and the major axis of the orbit, the initial conditions for the electron momenta can be written as

$$p_1^0 = -p_e(\cos\theta \cos\phi \cos\xi + \sin\phi \sin\xi)$$
$$p_2^0 = -p_e(\cos\theta \sin\phi \cos\xi - \cos\phi \sin\xi) \quad (165)$$
$$p_3^0 = p_e \sin\theta \cos\xi$$

where p_e is the magnitude of the electron momentum in the hydrogen atom. The values r_e and p_e can be expressed in terms of the eccentricity ϵ of the orbit and the eccentric angle u by

$$r_e = (Z/2|E_H|)(1 - \epsilon \cos u) \quad (166a)$$
$$p_e = (2|E_H|)^2(1 - \epsilon^2 \cos^2 u)^{1/2}/(1 - \epsilon \cos u) \quad (166b)$$

where E_H is the electron binding energy in the hydrogen atom. The orbit itself is defined by Kepler's equation:

$$\theta_n = u - \epsilon \sin u \quad (167)$$

where θ_n is the angle swept by the radius vector r_e along the orbit.

Since the energy of the electron in its initial state is fixed, only 5 of the variables, specifying the electronic state in the phase space, are free. Abrines and Percival (1966a,b) have shown that the multiplicity of the Kepler orbits with fixed major axis (or, equivalently, fixed electron binding energy) define a microcanonical ensemble, equivalent to the hydrogen atom ground state. In other words, by picking orbits from this ensemble, one can reproduce the quantum state of H(1s). Moreover, the quantities $\cos\theta$, ϕ, ξ, θ_n, and ϵ^2 for the members of this ensemble are uniformly distributed in the following intervals: $-1 \leq \cos\theta \leq 1, 0 \leq \phi \leq 2\pi, 0 \leq \xi \leq 2\pi, 0 \leq \epsilon^2 \leq 1, 0 \leq \theta_n \leq 2\pi$. In the calculations, numbers from these intervals are selected by the Monte Carlo method. Additionally, the initial condition q_6^0 is also randomly determined by the Monte Carlo method, by selecting the impact parameter b in the interval $0 \leq b \leq b_{\max}$. The maximum impact parameter $b_{\max}$ is chosen in such a way that the region $b > b_{\max}$ does not give a contribution to the cross section greater than the statistical error of the calculations. [Note that in order to obtain r_e and p_e, needed for Eqs. (163) and (164), first one must solve Kepler's equation (167) for chosen ϕ_n and E.]

With the specified initial conditions, Hamilton's equations are solved numerically. In order to get good statistics, the number of trajectories must be between 1000 and 2000 (for the total cross section), or even larger (for the partial cross sections). The integration is carried out for distances between the proton and the ion on the order of $10Z$ or larger. After calculating a sufficient number of trajectories, the total charge-exchange cross section is obtained as

$$\sigma = \frac{N_{CT}}{N} \pi b_{max}^2 \left[1 \pm \left(\frac{N - N_{CT}}{N \, N_{CT}} \right)^{1/2} \right] \quad (168)$$

where N is the total number of trajectories, and N_{CT} is the number of trajectories for which after the collision the electron is found in the vicinity of the highly charged ion. The second term in the brackets in Eq. (168) represents the standard deviation of the results.

The trajectories leading to charge transfer can be further analyzed. If the electron after the collision is found in a Kepler's orbit around the ion with energy E and angular momentum J, then with these quantities one can associate continuous classical numbers n_c and l_c by the relations

$$E = -Z^2/2n_c^2, \quad J = [l_c(l_c + 1)]^{1/2} \quad (169)$$

The classical numbers n_c and l_c can be "quantized" according to relations

$$n - \tfrac{1}{2} \leq n_c \leq n + \tfrac{1}{2}, \quad l \leq l_c \leq l + 1 \quad (170)$$

in which the quantities n and l can be interpreted as principal and angular quantum numbers, respectively. (In order to obtain the statistical distribution of l within the same n, it is necessary to renormalize the l_c values obtained from the calculations, by multiplying them by n/n_c, to reflect the fact that l_c corresponds to a classical n_c.) Classical trajectories leading to n_c and l_c values, defined by Eqs. (170), normalized appropriately, give the partial σ_n and σ_{nl} cross sections for the electron capture. Let us note that the "correspondence relations" (169) and the "quantization" conditions (170) can be justified only if both n_c and l_c are large enough.

Systematic classical trajectory–Monte Carlo (CTMC) calculations have been performed for the H + Z ($Z = 1 - 36$) system, both for the total (Olson and Salop, 1977) and partial (Salop, 1979; Olson, 1981) cross sections. The energy range in which the CTMC method is expected to give reliable results is from ~ 30 to $250-300$ keV/amu. In Figs. 11–13 are given some of the results of σ, σ_n, and σ_{nl}, obtained by the CTMC method. In this energy region and for $Z \geq 3$, the CTMC total cross sections scale (roughly) linearly with Z. The partial cross sections σ_n, for a given energy, exhibit maxima at $n_m \simeq Z^{3/2}$, in accordance with UDWA theory. The n-distributions in the CTMC method have been calculated only for two energies, 50 and 100 keV/amu (Olson, 1981), and no conclusion can be drawn about the velocity dependence of n_m. The partial cross sections σ_{nl} have properties similar to those obtained in the UDWA, although appearing at somewhat lower energies. Namely, for $n < n_m$, the CTMC l-distributions have maxima at $l = n - 1$, whereas for $n > n_m$ they have maxima at $l = l_m \simeq n_m$. From a classical

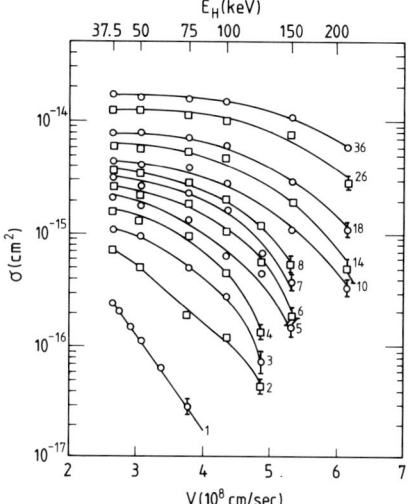

FIG. 11. CTMC total electron capture cross sections for the H + Z collision system. (From Olson and Salop, 1977.)

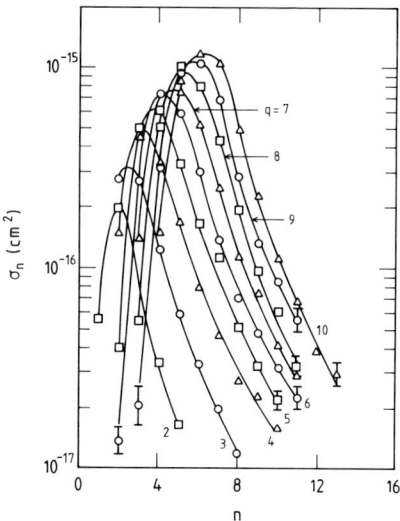

FIG. 12. CTMC n-distributions of captured electrons in H + Z collisions at 50 keV/amu. (From Olson, 1981.)

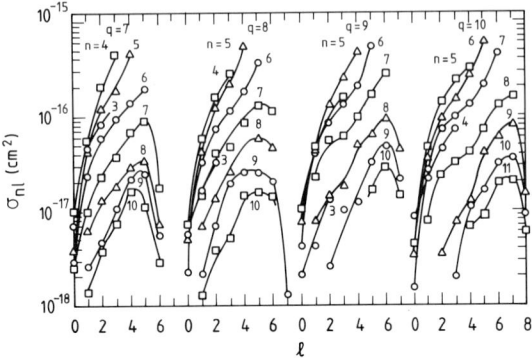

FIG. 13. CTMC l-distributions of captured electrons in H + Z collisions at 50 keV/amu. (From Olson, 1981.)

point of view these features can easily be understood. The levels with $n < n_m$ are populated in close collisions, providing capture into orbits with a small eccentricity (i.e. large angular momentum). The levels with $n > n_m$ are populated in collisions with large impact parameters, and the electrons are easily captured into orbits with a high eccentricity (small angular momentum).

B. ANALYTICAL MODELS

Two analytical models for the H + Z charge-exchange problem have been developed within the concepts of classical mechanics. One of them describes the overbarrier transitions in the (p, e, Z) system using some statistical arguments and is applicable in the velocity region $v \leq 1$ (Komarov and Solov'ev, 1980; Grozdanov, 1980). The other model, developed by Bohr and Lindhard (1954), is based on some elementary relations for the forces and energies involved in the colliding system. This model is applicable in the range of intermediate collision velocities ($1 \leq v \leq 3-4$). Below, we shall briefly describe these models and present their most characteristic results.

1. Overbarrier Capture Model

The motion of the electron in the overbarrier region of the two-Coulomb center potential in the classical picture is described by (see, e.g., Messiah, 1972)

$$\partial S/\partial t + H = 0 \qquad (171)$$

ELECTRON CAPTURE

$$\partial f/\partial t + \nabla(f\,\nabla S) = 0 \tag{172}$$

where $S(\mathbf{r}, t)$ is the classical "principle function" of Hamilton, $H(\mathbf{p}, \mathbf{r}, t)$ is the Hamiltonian function, $\mathbf{p} = \nabla S$ is the electron momentum, and $f(\mathbf{r}, t)$ is the spatial probability density. Equations (171) and (172) are, respectively, the Hamilton–Jacobi and the continuity equations. For the two-Coulomb center system $(Z_1 e Z_2)$, the Hamilton function is

$$H = \mathbf{p}^2/2 - Z_1/r_1 - Z_2/r_2 \tag{173}$$

[$Z_1 = 1$, $Z_2 = 2$ in the (p, e, Z) case]. It is convenient to introduce the prolate spheroidal coordinates ξ, η, ϕ [see Eq. (3)] and their conjugate momenta p_ξ, p_η, p_ϕ ($p_\xi = \partial S/\partial \xi$, $p_\eta = \partial S/\partial \eta$, $p_\phi = \partial S/\partial \phi$), so that H can be written as (Landau and Lifshitz, 1973)

$$H = \frac{1}{2}\left(\frac{p_\xi^2}{h_\xi^2} + \frac{p_\eta^2}{h_\eta^2} + \frac{p_\phi^2}{h_\phi^2}\right) - \frac{2Z}{R(\xi+\eta)} - \frac{2Z}{R(\xi-\eta)} \tag{174}$$

where h_ξ, h_η, h_ϕ are the Lamé coefficients:

$$h_\xi = \frac{R}{2}\left(\frac{\xi^2-\eta^2}{\xi^2-1}\right)^{1/2}, \qquad h_\eta = \frac{R}{2}\left(\frac{\xi^2-\eta^2}{1-\eta^2}\right)^{1/2}$$

$$h_\phi = \frac{R}{2}[(\xi^2-1)(1-\eta^2)]^{1/2} \tag{175}$$

Since H does not depend explicitly on t (in the adiabatic region R is treated as a parameter), the variables in the Hamilton–Jacobi equation (171) can be separated in the (ξ, η, ϕ) coordinates and the solution for S is (Landau and Lifshitz, 1973)

$$S = -Et + \mu\phi \pm \int p_\xi\,d\xi \pm \int p_\eta\,d\eta \tag{176}$$

$$p_\xi^2 = \frac{R^2 E}{4} + \frac{a\xi - \lambda}{\xi^2 - 1} - \frac{\mu^2}{(\xi^2-1)^2}, \qquad a = R(Z_1 + Z_2) \tag{177a}$$

$$p_\eta^2 = \frac{R^2 E}{4} + \frac{b\eta - \lambda}{1 - \eta^2} - \frac{\mu^2}{(1-\eta^2)}, \qquad b = R(Z_2 - Z_1) \tag{177b}$$

and λ is the separation constant.

Knowing S, and separating the variables in Eq. (172) in the (ξ, η, ϕ) coordinates, one can find the (stationary) spatial probability density in the form (Grozdanov, 1980)

$$f(\mathbf{r}) = C/(1-\eta^2)p_\eta(\xi^2-1)p_\xi \tag{178}$$

where C is a normalization constant.

Let us now consider in more detail the one-dimensional motions of the electron in ξ and η directions for the H(1s) + Z system ($\mu = 0$, for this case). Keeping in mind that the term $RE/4$ in Eqs. (177) plays the role of the energy for one-dimensional motions, the corresponding potentials for the particle in the ξ and η directions are [$a = R(Z + 1)$, $b = R(Z - 1)$]

$$V(\xi) = -\frac{R(Z + 1)\xi - \lambda}{2(\xi^2 - 1)}, \quad V(\eta) = -\frac{R(Z - 1)\eta + \lambda}{2(1 - \eta^2)} \quad (179)$$

In the asymptotic region ($R \gg 1$), the energy E and the separation constant $\lambda \simeq 2p_\eta$ are, respectively, given by the expansions (13) and (14). Taking the leading terms in this expansion (i.e., $E \simeq -\frac{1}{2} - Z/R$, $\gamma \simeq RZ$), from $p_\xi = 0$ and $p_\eta = 0$ one obtains the turning points for the motion in the ξ and η directions:

$$\xi_{1,2} = \frac{2(Z + 1) \pm (R^2 + 8Z + 4)^{1/2}}{R + 2Z} \quad (180a)$$

$$\eta_{1,2} = -\frac{2(Z - 1) \pm (R^2 - 8Z + 4)^{1/2}}{R + 2Z} \quad (180b)$$

In the ξ direction, the motion is classically allowed (for all R) in the region $1 \leq \xi \leq \xi_1$ ($\xi_2 < 1$ always), whereas in the η direction, it is classically allowed in the interval $-1 \leq \eta \leq 1$ only for $R < R_0$, where R_0 is defined by the coalescence condition of the points η_1 and η_2 ($\eta_1 = \eta_2$), and given by

$$R_0 = 2(2Z - 1)^{1/2} \quad (181)$$

For $R = R_0$ the energy $ER/4$ becomes equal to the maximum of the potential $V(\eta)$. Note that this value is different from the value $\tilde{R}_0 = 2(2Z^{1/2} + 1)$ obtained by equating the electron binding energy with the maximum of the three-dimensional potential of the two-Coulomb center system taken along the internuclear axis. The latter way of determining R_0 (see, e.g., Ryufuku et al., 1980) is incorrect. The situation is analogous to that when one, considering radial motion in a spherically symmetric field, neglects the centrifugal potential.

The classical probability for electron capture in the adiabatic region is given by (we assume rectilinear trajectories)

$$P_{cl} = 1 - \exp\left[-\frac{2}{v} \int_b^{R_0} \frac{W_{cl}(R) R \, dR}{(R^2 - b^2)^{1/2}}\right] \quad (182)$$

where $W_{cl}(R)$ is the transition probability of the electron from the region of the proton into the region of multicharged ion, and b is the impact parameter. W_{cl} can be calculated as the flux of particles through a surface Σ placed at the top of the potential $V(\eta)$ and dividing the proton and ion field regions (Grozdanov, 1980):

$$W_{cl} = \frac{1}{2} \int_{\Sigma} f(\mathbf{r}) \frac{p_\eta}{h_\eta} d\Sigma$$

$$= \frac{2}{R^2} \int_1^{\xi_1} \frac{d\xi}{(\xi^2 - 1)p_\xi} \left[\int_1^{\xi_1} \int_{-1}^{\eta_m} \frac{(\xi^2 - \eta^2) d\xi \, d\eta}{(\xi^2 - 1)p_\xi (1 - \eta^2)p_\eta} \right]^{-1} \quad (183)$$

where $\eta_m = -2(Z - 1)/(R + 2Z)$ is the coordinate of the maximum of $V(\eta)$. The cross section for the overbarrier electron capture is thus

$$\sigma_{cl} = 2\pi \int_0^{R_0} P_{cl}(b, v) b \, db \quad (184)$$

If one takes that for $R < R_0$ the capture probability P_{cl} is $P_{cl}^{(0)} = 1$, then (Komarov and Solov'ev, 1980)

$$\sigma_{cl}^{(0)} = \pi R_0^2 = 4\pi(2Z - 1) \quad (185)$$

When account is taken of the finite value of W_{cl}, σ_{cl} is found to depend on velocity, although relatively weakly (see below).

It is interesting to compare the results of the above analytical model of the overbarrier transitions with those obtained by the CTMC method. Figure 14 gives a comparison of the capture probabilities $P_{cl}(b)$ for the case of $Z = 18$ and at a collision energy of 50 keV/amu. The agreement is excellent. (Note,

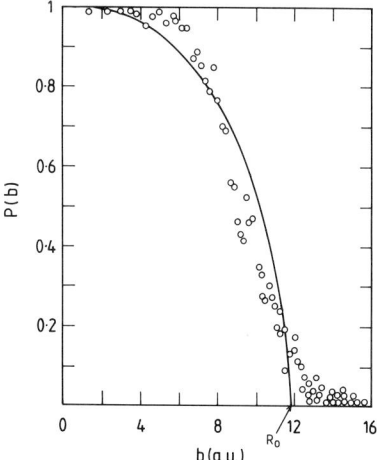

FIG. 14. Dependence of the electron capture probability on the impact parameter for the H + Ar^{18+} collisions at $E = 50$ keV/amu: (———) result of the analytical classical model (Grozdanov, 1980); (O) CTMC calculations (Olson and Salop, 1977).

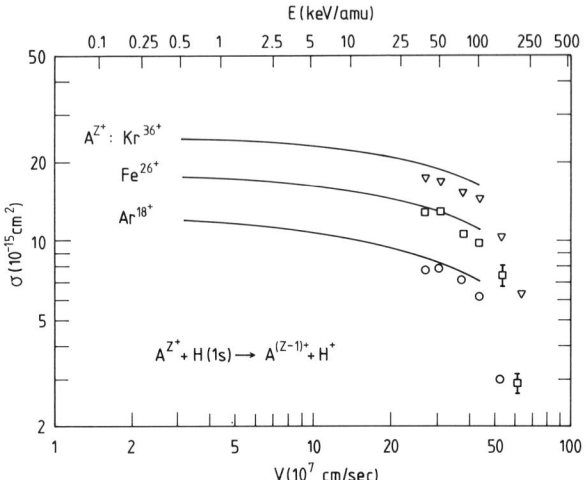

FIG. 15. Total electron capture cross sections for H + Z ($Z = 18, 26, 36$) collisions: (———) classical analytical model (Grozdanov, 1980); CTMC results (Olson and Salop, 1977) for Kr^{36+} ($\triangledown$), Fe^{26+} ($\square$), and Ar^{18+} ($\bigcirc$).

however, that the CTMC method predicts nonzero classical capture probability in the quantum tunneling region, $b > R_0$.) A comparison of the total cross sections provided by the CTMC and the analytical classical methods is given in Fig. 15.

2. The Bohr–Lindhard Classical Model

In considering the electron capture from the hydrogen atom by a bare nucleus, Bohr and Lindhard (1954) introduced two important internuclear distances. The first, R_1, is defined by the equality of the force which binds the electron in the atom with the attractive force of the projectile, i.e.,

$$u^2/a = Z/R_1^2 \tag{186}$$

where a and u are the orbital radius and velocity of the electron in the atom, respectively. R_1 is called the "release distance," since for $R < R_1$, the Coulomb attractive force of the ion becomes stronger than the binding force, and the electron is released and eventually captured. From Eq. (186) one has [for the H(1s) + Z system]

$$R_1 = Z^{1/2} \tag{187}$$

The other critical distance, R_2, is defined by the equality of the potential energy of the electron in the ionic field with the kinetic energy in the form frame, i.e.,

$$\frac{1}{2} v^2 = \frac{Z}{R_2} \qquad (188)$$

where v is the collision velocity. Capture is possible when the internuclear distance is smaller than R_2 given by

$$R_2 = 2Z/v^2 \qquad (189)$$

When $R_1 < R_2$, the condition for capture is automatically satisfied and (assuming a probability of unity) the cross section is

$$\sigma_1 = \pi R_1^2 = \pi Z \qquad (190)$$

In the case when $R_1 > R_2$, however, whether a released electron will be captured depends on the probability for release u/a (considered as a gradual process) and the time R_2/v during which the capture can take place. The cross section is now given by ($u = v_0 = 1$, $a = a_0 = 1$)

$$\sigma_2 = \pi R_2^2 (R_2/v) = 8\pi Z^3/v^7 \qquad (191)$$

Detailed investigations of this model (Knudsen *et al.*, 1981) have shown that the cross section σ_1 is applicable in the velocity region $v < 2^{1/2} Z^{1/4}$, whereas the expression σ_2 holds for $v \geqslant 2^{1/2} Z^{1/4}$. The linear dependence of the capture cross section σ_1 on ionic charge Z in the region $v < 2^{1/2} Z^{1/4}$ is in accordance with the analytical classical model for the overbarrier transitions and with the CTMC model. In the energy region 10–200 keV/amu, Goffe *et al.* (1979) have experimentally confirmed both the Z^3 scaling as well as the velocity v^{-7} dependence of the cross section σ_2 (more precisely, their cross sections were successfully fitted by the expression $\sigma = 9\pi Z^3/v^{7.3}$).

Although derived from very elementary classical arguments, the Bohr–Lindhard model has the very important feature of predicting the change of the scaling law for the capture cross section. The change of the scaling leads to a saturation effect in the Z dependence of the reduced cross section σ/Z (Janev and Hvelplund, 1981).

VII. Conclusions

For energies below ~ 100 keV/amu, there seems no doubt that for moderate values of the charge Z on the incident ion, the expansion method for the calculation of charge-exchange cross sections is both practical and reliable, whether an atomic or molecular basis is used, and provided that proper allowance for translational invariance is made. Below 1 keV/amu, it is reasonable to drop the "translational factors" and the simpler perturbed stationary state method can be used. To date, the expansion method has only been applied for $Z \leq 8$, but it will certainly be possible to extend this range in the future, perhaps up to $Z = 20$ or more. At the higher velocities, the "one and a half" center expansion method has great potentaility and above $v = 3$ or 4 a.u. the continuum distorted wave approach is also both practical and accurate. Turning to more approximate methods, perhaps the most useful at intermediate velocities is the classical trajectory–Monte Carlo method, although the limits of its applicability are not altogether clear. The unitarized distorted wave approximation may be less useful, because it is probably accurate over a rather limited range of velocities, and as it requires the "difficult" exchange matrix elements to be computed, not much extra labor is required to perform a full atomic basis coupled-channel evaluation.

For large Z and low velocities, approximations based on the multicrossing Landau–Zener method are important, but limited in accuracy because of the neglect of translational invariance. One can envisage much-improved versions of multicrossing theory in which the translational factors are included, and in each crossing region a two-state calculation is made, avoiding any kind of Landau–Zener approximation. Although a general picture of the behavior of the charge-exchange cross sections as a function of Z and v has emerged, more work is required to establish accurate cross sections and, in particular, partial cross sections for capture into individual levels, for the larger values of Z.

REFERENCES

Abramov, V. A., Baryshnikov, F. F., and Lisitsa, V. S. (1978a). *Sov. Phys.—JETP* (*Engl. Transl.*) **47**, 469.
Abramov, V. A., Baryshnikov, F. F., and Lisitsa, V. S. (1978b). *JETP Lett.* (*Engl. Transl.*) **27**, 464.
Abrines, R., and Percival, I. C. (1966a). *Proc. Phys. Soc., London* **88**, 861.
Abrines, R., and Percival, I. C. (1966b). *Proc. Phys. Soc., London* **88**, 873.
Alliluev, S. P., and Matveenko, A. V. (1967). *Sov. Phys.—JETP* (*Engl. Transl.*) **24**, 1260.
Anderson, D. G. M., Antal, M. J., and McElroy, M. B. (1974). *J. Phys. B* **8**, 1513.
Baer, N. (1975). *Chem. Phys. Lett.* **35**, 112.
Basu, D., Mukherjee, S. C., and Sural, D. P. (1978). *Phys. Rep.* **42**, 145.
Bates, D. R. (1958). *Proc. R. Soc. London, Ser. A* **247**, 294.
Bates, D. R., and Lewis, J. T. (1955). *Proc. Phys. Soc., London, Sect. A* **68**, 173.
Bates, D. R., and Lynn, N. (1959). *Proc. R. Soc. London, Ser. A* **253** 141.
Bates, D. R., and McCarroll, R. (1958). *Proc. R. Soc. London, Ser. A* **247**, 158.

Belkić, D. Ž., Gayet, R., and Salin, A. (1979). *Phys. Rep.* **56**, 279.
Bohr, N., and Lindhard, J. (1954). *Mat.-Fys. Medd.—K. Dan. Vidensk. Selsk.*, **28** (7), 1.
Bransden, B. H. (1980). *In* "Atomic and Molecular Processes in Thermonuclear Fusion Research" (M. R. C. McDowell and A. Ferendeci, eds.) p. 185. Plenum, New York.
Bransden, B. H. (1983). "Atomic Collision Theory," 2nd ed. Addison-Wesley, Benjamin, Reading, Massachusetts.
Bransden, B. H., and Noble, C. J. (1981). *J. Phys. B* **14**, 1849.
Brandsen, B. H., and Noble, C. J. (1982). *J. Phys. B* **15**, 451.
Bransden, B. H., Newby, C. W., and Noble, C. J. (1980). *J. Phys. B* **13**, 4245.
Bransden, B. H., Chandler, J., and Noble, C. J. (1983). *J. Phys. B* (submitted for publication).
Chan, F. T., and Eichler, J. (1977a). *Phys. Rev. Lett.* **42**, 145.
Chan, F. T., and Eichler, J. (1977b). *Phys. Rev. A* **20**, 104.
Chan, F. T., and Eichler, J. (1977c). *Phys. Rev. A* **20**, 184.
Cheshire, I. M. (1964). *Proc. Phys. Soc., London* **84**, 89.
Cheshire, I. M., Gallaher, D. F., and Taylor, A. J. (1970). *J. Phys. B* (*Engl. Transl.*) **3**, 813.
Chibisov, M. I. (1976). *JETP Lett.* (*Engl. Transl.*) **24**, 46.
Chibisov, M. I. (1979). *Sov. Phys.—JETP* (*Engl. Transl.*) **49**, 962.
Chibisov, M. I. (1981). *Sov. Phys.— Tech. Phys.* (*Engl. Transl.*) **26**, 284.
Crandall, D. H., Phaneuf, R. A., and Meyer, F. W. (1980). *Phys. Rev. A.* **22**, 379.
Crothers, D. S. F. (1981a). *J. Phys. B* **14**, 1035.
Crothers, D. S. F. (1981b). *Adv. At. Mol. Phys.* **17**, 55.
Crothers, D. S. F. (1982). *J. Phys. B* **15**, 2061.
Crothers, D. S. F., and Hughes, J. G. (1978). *Proc. R. Soc. London, Ser. A* **359** 345.
Crothers, D. S. F., and Todd, N. R. (1980). *J. Phys. B* **13**, 2277.
Crothers, D. S. F., and Todd, N. R. (1981a). *J. Phys. B* **14**, 2233.
Crothers, D. S. F., and Todd, N. R. (1981b). *J. Phys. B* **14** 2251.
Delos, J. B. (1981). *Rev. Mod. Phys.* **53**, 287.
Demkov, Yu. N., and Osherov, V. I. (1968). *Sov. Phys.—JETP* (*Engl. Transl.*) **26**, 916.
Dewangan, D. P. (1975). *J. Phys. B* **8**, L119.
Dewangan, D. P. (1977). *J. Phys. B* **10**, 1083.
Dose, V., and Semini, C. (1974). *Helv. Phys. Acta* **47**, 609.
Duman, E. L., and Men'shikov, L. I. (1979). *Sov. Phys.—JETP* (*Engl. Transl.*) **50**, 433.
Duman, E. L., and Smirnov, B. M. (1978). *Sov. J. Plasma Phys.* (*Engl. Transl.*) **4**, 650.
Duman, E. L., Men'shikov, L. I., and Smirnov, B. M. (1979). *Sov. Phys.—JETP* (*Engl. Transl.*) **49**, 260.
Eichler, J. (1981). *Phys. Rev. A* **22**, 498.
Fock, V. (1935). *Z. Phys.* **98**, 145.
Ford, A. L., Reading, J. F. and Becker, R. L. (1982). *J. Phys. B* **15**, 3257.
Fritsch, W. (1982). *J. Phys. B* **15**, L389.
Fritsch, W., and Lin, C. D. (1982a). *J. Phys. B* **15**, 1255.
Fritsch, W., and Lin, C. D. (1982b). *J. Phys. B* **15**, L281.
Gallaher, D. F., and Wilets, L. (1968). *Phys. Rev.* **169**, 139.
Gargaud, M., Hanssen, J., McCarroll, R., and Valiron, P. (1981). *J. Phys. B* **14**, 2259.
Garrett, B. L., and Truhlar, D. G. (1981). *Theor. Chem.: Period. Chem. Biol.* **6A**, 215.
Gershtein, S. S. (1962). *Sov. Phys.—JETP* (*Engl. Transl.*) **16**, 501.
Gershtein, S. S., and Krivchenkov, V. D. (1961). *Sov. Phys.—JETP* (*Engl. Transl.*) **13**, 1044.
Gilbody, H. B. (1981). *Phys. Scr.* **23**, 143.
Goffe, T. V., Shah, M. B., and Gilbody, H. B. (1979). *J. Phys. B* **12**, 3763.
Green, T. A. (1981a). *Phys. Rev.* **23**, 519.
Green, T. A. (1981b). *Phys. Rev.* **23**, 532.

Green, T. A., Shipsey, E. J., and Browne, J. C. (1982a). *Phys. Rev. A* **25**, 546.
Green, T. A., Shipsey, E. J., and Browne, J. C. (1982b). *Phys. Rev. A.* **25**, 1346.
Greenland, P. T. (1978). *J. Phys. B* **11**, 3563.
Greenland, P. T. (1982). *Phys. Rep.* **81**, 131.
Grozdanov, T. P. (1980). *J. Phys. B* **13**, 3835.
Grozdanov, T. P., and Janev, R. K. (1978a). *Phys. Rev. A* **17**, 880.
Grozdanov, T. P., and Janev, R. K. (1978b). *Phys. Lett. A* **66A**, 191.
Harel, C., and Salin, A. (1977). *J. Phys.* **10**, 3571.
Harrison, M. F. A. (1978). *Phys. Rep.* **37**, 59.
Harrison, M. F. A. (1980). In "Atomic and Molecular Processes in Thermonuclear Fusion Research" (M. R. C. McDowell and A. Ferendeci, eds.), p. 15. Plenum, New York.
Harrison, M. F. A. (1983). In "Atomic and Molecular Physic of Controlled Thermonuclear Fusion" (C. J. Joachin and D. E. Post, eds.). Plenum, New York (in press).
Hatton, G. J., Lane, N. F., and Winter, T. G. (1979). *J. Phys. B* **12**, L571.
Heil, T. G., and Dalgarno, A. (1979). *J. Phys. B* **12**, L557.
Heil, T. G., Butler, S. E., and Dalgarno, A. (1981). *Phys. Rev. A* **23**, 1100.
Ho, T. S., Umberger, D., Day, R. L., Lieber, M., and Chan, F. T. (1981). *Phys. Rev. A* **24**, 705.
Hogan, J. T. (1978). *Phys. Rep.* **37**, 83.
Hogan, J. T. (1980). In "Atomic and Molecular Processes in Controlled Thermonuclear Fusion" (C. T. Joachain and D. E. Post, eds.). Plenum, New York (In press).
Hogan, J. T. (1983). In "Atomic and Molecular Physics of Controlled Thermonuclear Fusion" (C. T. Joachin and D. E. Post, eds.). Plenum, New York (in press).
Janev, R. K. (1976). *Adv. At. Mol. Phys.* **12**, 1.
Janev, R. K. (1982a). *Comments At. Mol. Phys.* (submitted for publication).
Janev, R. K. (1982b). *Phys. Scr.* (submitted for publication).
Janev, R. K., and Hvelplund, P. (1981). *Comments At. Mol. Phys.* **11**, 75.
Janev, R. K., and Presnyakov, L. P. (1981). *Phys. Rep.* **70**, 1.
Janev, R. K., Bransden, B. H., and Gallagher, J. W. (1982a). *J. Phys. Chem. Ref. Data* (submitted for publication).
Janev, R. K., Joachain, C. J., and Nedeljkovic, N. N. (1982b). *Phys. Rev. A* **26**, 116.
Janev, R. K., Belić, D. S., and Bransden, B. H. (1983). *Phys. Rev. A* (submitted for publication).
Kimura, M., and Thorson, W. (1981a). *Phys. Rev. A* **24**, 3019.
Kimura, M., and Thorson, W. (1981b). *ICPEAC, 12th, 1981* Abstracts, p. 638.
Knudsen, H., Haugen, H. K., and Hvelplund, P. (1981). *Phys. Rev. A* **23**, 597.
Komarov, I. V., and Slavyanov, S. Yu. (1968). *J. Phys. B* **1**, 1066.
Komarov, I. V., and Solov'ev, E. A. (1979). *Teor. Mat. Fiz.* **40**, 130.
Komarov, I. V., and Solov'ev, E. A. (1980). *Vopr. Teor. At. Stolknovenii* **2**, 234.
Komarov, I. V., Ponomarev, L. I., and Slavyanov, S. Yu. (1976). "Spheroidal and Coulomb Spheroidal Functions." Nauka, Moscow (in Russian).
Landau, L. D., and Lifshitz, E. M. (1963). "Kvantovaya Mekhanika" ("Quantum Mechanics"). Fizmatgiz, Moscow (transl. by J. B. Sykes and J. S. Bell. Pergamon, Oxford).
Landau, L. D., and Lifshitz, E. M. (1973). "Mekhanika," Nauka, Moscow (transl by J. B. Sykes and J. S. Bell. Pergamon, Oxford).
Lüdde, H. J., and Dreizler, R. M. (1982). *J. Phys. B* **15**, 2713.
McCarroll, R. (1982). In "Atomic and Molecular Collision Theory" F. A. Gianturco, ed.), pp. 165–231. Plenum, New York.
Madsen, M. M., and Peek, J. M. (1971). *At. Data* **2**, 171.
Mapleton, R. A. (1972). "Theory of Charge Exchange." Wiley, New York.
Massey, H. S. W., and Smith, R. A. (1933). *Proc. R. Soc. London, Ser. A* **142**, 142.
Messiah, A. (1972). "Quantum Mechanics," p. 222. North-Holland, Publ., Amsterdam.

Mittleman, M. H., and Quong, J. (1968). *Phys. Rev.* **167**, 76.
Morse, P., and Feshbach, H. (1953). "Methods of Theoretical Physics." McGraw-Hill, New York.
Olson, R. E. (1980). *In* "Electronic and Atomic Collisions" (N. Oda and K. Takayanagi, eds.), p. 391. North-Holland Publ., Amsterdam.
Olson, R. E. (1981). *Phys. Rev. A* **24**, 1726.
Olson, R. E., and Salop, A. (1976). *Phys. Rev. A* **14**, 579.
Olson, R. E., and Salop, A. (1977). *Phys. Rev. A* **16**, 531.
Olson, R. E., Salop, A., Phaneuf, R. A., and Meyer, F. W. (1977). *Phys. Rev. A* **16**, 1867.
Olson, R. E., Berkner, K. H., Graham, W. G., Pyle, R. V., Schlachter, A. S., and Stearns, J. W. (1978). *Phys. Rev. Lett.* **41**, 163.
Osherov, V. I. (1966a). *Sov. Phys.—JETP (Engl. Transl.)* **22**, 804.
Osherov, V. I. (1966b). *Sov. Phys.—Dokl. (Engl. Transl.)* **11**, 528.
Phaneuf, R. A., Meyer, F. W., McKnight, R. H., Olson, R. E., and Salop, A. (1977). *J. Phys. B* **10**, L425.
Piacentini, R. D., and Salin, A. (1974). *J. Phys. B* **7**, 1666.
Poluéktov, I. A., and Presnyakov, L. P. (1968). *Sov. Phys.—JETP (Engl. Transl.)* **27**, 95.
Ponomarev, L. I. (1969). *Sov. Phys.—JETP (Engl. Transl.)* **28**, 971.
Ponomarev, L. I., and Puzynina, T. P. (1967). *Sov. Phys.—JETP (Engl. Transl.)* **25**, 846.
Ponomarev, L. I., and Puzynina, T. P. (1968). *USSR Comput. Math. Math. Phys. (Engl. Transl.)* **8** (No. 6), 94.
Ponomarev, L. I., and Puzynina, T. P. (1970). *J. Inst. Nucl. Res., Dubna. USSR, Prepr.* **JINR-P4-5040**.
Post, D. E. (1983). *In* "Atomic and Molecular Physics of Controlled Thermonuclear Fusion" (C. J. Joachain and D. E. Post, eds.). Plenum, New York (in press).
Power, J. D. (1973). *Philos. Trans. R. Soc. London* **274**, 663.
Presnyakov, L. P., and Ulantsev, A. D. (1975). *Sov. J. Quantum. Electron. (Engl. Transl.)* **4**, 1320.
Presynakov, L. P., Uskov, D. B., and Janev, R. K. (1981). *Phys. Lett. A* **84A**, 243.
Presynakov, L. P., Uskov, D. B., and Janev, R. K. (1982). *Zh. Eksp. Teor. Fiz.* **83**, 933.
Reading, J. F., Ford, A. L., Swafford, G. L., and Fitchard, A. (1979). *Phys. Rev. A* **20**, 130.
Reading, J. F., Ford, A. L., and Becker, R. L. (1981). *J. Phys. B* **14**, 1995.
Reading, J. F., Ford, A. L., and Becker, R. L. (1982). *J. Phys. B* **15**, 625.
Riley, M. E. (1973). *Phys. Rev. A* **8**, 742.
Riley, M. E., and Green, T. A. (1971). *Phys. Rev. A* **4**, 619.
Ryufuku, H. (1982). *Phys. Rev. A* **25**, 720.
Ryufuku, H., and Watanabe, T. (1978). *Phys. Rev. A* **18**, 2005.
Ryufuku, H., and Watanabe, T. (1979a). *Phys. Rev. A* **19**, 1538.
Ryufuku, H., and Watanabe, T. (1979b). *Phys. Rev. A* **20**, 1828.
Ryufuku, J., Sasaki, K., and Watanabe, T. (1980). *Phys. Rev. A* **21**, 745.
Salop, A. (1979). *J. Phys. B* **12**, 919.
Salop, A., and Olson, R. E. (1976). *Phys. Rev. A* **13**, 1312.
Salop, A., and Olson, R. E. (1977). *Phys. Rev. A* **16**, 1811.
Salop, A., and Olson, R. E. (1979). *Phys. Rev. A* **19**, 1921.
Salzborn, E., and Muller, A. (1980). *In* "Electronic and Atomic Collisions" (N. Oda and K. Takayanagi, eds.), p. 707. North-Holland publ., Amsterdam.
Schneiderman, S. B., and Russek, A. (1969). *Phys. Rev. A* **181**, 311.
Seim, W., Muller, A., and Salzborn, E. (1980). *Phys. Lett. A* **80A**, 20.
Shah, M. B., Goffe, T. V., and Gilbody, H. B. (1978). *J. Phys. B* **11**, L233.
Shakeshaft, R. (1976). *Phys. Rev. A* **14**, 1626.

Shakeshaft, R., and Spruch, L. (1979). *Rev. Mod. Phys.* **51,** 369.
Shipsey, E. J., Green, T. A., and Browne, J. C. (1983). *Phys. Rev. A* **27,** 821.
Sil, N. C. (1954). *Indian J. Phys.* **28,** 232.
Simony, P. R., and McGuire, J. M. (1981). *J. Phys. B* **14,** L737.
Smith, F. T. (1969). *Phys. Rev.* **179,** 111.
Vaaben, J., and Taulbjerg, K. (1981). *J. Phys. B* **14,** 1815.
Vainshtein, L., Presnyakov, L., and Sobel'man, I. (1964). *Sov. Phys.—JETP (Engl. Transl.)* **18,** 1383.
Winter, T. G. (1982). *Phys. Rev. A* **25,** 697.
Winter, T. G., and Hatton, G. J. (1980). *Phys. Rev. A* **21,** 793.
Winter, T. G., and Lane, N. F. (1978). *Phys. Rev. A* **17,** 66.

INTERACTIONS OF SIMPLE ION–ATOM SYSTEMS

J. T. PARK

Department of Physics
University of Missouri
Rolla, Missouri

I. Introduction	67
II. Experimental Methods	68
A. Experimental Methods Involving the Detection of Secondary Particles	69
B. Experimental Methods Involving Detection of the Primary Particle	74
III. Theoretical Methods	78
A. Classical Techniques	78
B. Quantum Mechanical Techniques	79
IV. Excitation	82
A. Excitation of Atomic Hydrogen by Protons	82
B. Excitation of Helium Atoms by Protons	92
C. Excitation of Atomic Hydrogen by Helium Ions	96
V. Electron Capture	100
A. Electron Capture by Protons from Atomic Hydrogen	100
B. Electron Capture by Protons from Helium	113
VI. Ionization	117
A. Total Cross Sections for Ionization of Atomic Hydrogen	117
B. Energy-Loss Differential Cross Sections	121
VII. Elastic Scattering	123
VIII. Conclusions	127
References	129

I. Introduction

The systems which are the subject of this article are the simplest and most basic ions and atoms; however, the interactions are not simple. In the energy range under study (15–200 keV) the interactions change character. Theoretical approximations which are valid at one end of this energy range, are invalid at the other. In the case of proton–atomic hydrogen collisions, the crossover between the different theoretical approaches frequently occurs near the energy where the projectile has one atomic unit of velocity. This is where the proton projectile has the same velocity as the orbital velocity of the electron in the first Bohr orbit of the target hydrogen atom. However, there are numerous examples of theoretical approximations that provide reasonable results in energy ranges where the approximation is not strictly valid. There are other examples of poor agreement between theory and experiment for formally sound theoretical efforts.

The experiments on these interactions are also very difficult. To the experimenter attempting to work with atomic hydrogen, the system is not simple. The number of experiments on the truly basic interactions is very modest. Over much of the energy range, experimental measurements are only available from one laboratory. All of the experiments on atomic hydrogen are normalized in some way to theory or to an experiment involving a simple gaseous target.

The following discussion is focused on the most recent results and the most detailed measurements. In particular the recent differential cross section measurements are emphasized. A completely satisfactory theory should not only be able to provide correct total cross sections but should be able to meet the stricter demands required to yield the correct differential cross sections. Attention in the discussion is limited to the particularly difficult intermediate energy range of 15–200 keV, which brackets the energy wherein the projectile velocity matches the orbital electron velocity in the first Bohr orbit. For convenience of expression, the term "simple" signifies protons, atomic hydrogen, and helium.

The interactions of simple ion–atom systems provides a unique opportunity for experimental tests of the foundations of quantum scattering theory. In the collisions of simple ion–atom systems, comparison of the experimental and theoretical results tests the quantum mechanical model of the process in situations where the fundamental interaction forces are known. The simplest three-body collision system, which consists of a proton incident on a hydrogen atom, holds the key for understanding three-body collision physics. The proton–atomic hydrogen collision system has been the subject of numerous theoretical studies, many of which will be discussed in this article. However, the three body-problem has not been solved. In the intermediate energy range, a great deal remains to be done both experimentally and theoretically before even the simplest three-body scattering problem involving the interactions between two protons and a single electron is fully understood.

II. Experimental Methods

The experimental techniques involved in the interactions of these simple systems has been reviewed thoroughly in earlier publications, and only a brief summary will be presented here (see Park, 1978; Thomas, 1972). The techniques are no longer new. The most recent techniques have not proven to be particularly useful for studying collisions with atomic hydrogen because of difficulties related to the dissociation of the molecular hydrogen to produce the atomic hydrogen target.

The experimental methods employed in the study of ion–atom collisions can be divided into two classes. The classes are distinguished by the particle detected after the collision process. In one case, the primary particle is detected. In the other case, a secondary particle is detected. The secondary particle may be a photon or an electron ejected from the target or the projectile, or the nucleus of the target.

A. Experimental Methods Involving the Detection of Secondary Particles

The experimental method involving the detection of photons emitted from excited states of the target or projectile has the advantage of high resolution and sensitivity. The major disadvantage is that the measurement is of the emission process, not the excitation process itself. Because direct excitation is only one of the many possible reactions that could leave the target in an excited state following a collision, it may be difficult to identify with certainty the process responsible for the excitation. Even in a straightforward excitation measurement, the detected photon comes from a two-step process: excitation and emission, and often the detected emission is only one of the possible transitions from the excited state.

The technique involved in optical detection of photons emitted from collision products is thoroughly reviewed by Thomas (1972) and Park (1978). Only a brief discussion of the techniques is repeated here.

To obtain an excitation cross section, a series of measurements are required in addition to the primary measurement of the number of photons emitted from a transition from the excited state. The radiative depopulation of the excited state due to alternate transitions must be determined. The effects of the cascade transitions from higher states into the state being studied must be measured or calculated. Possible competitive excitation processes, which could also populate the state, must be evaluated. Secondary collision processes with other target atoms, which could depopulate the state before it radiates, must be studied. The effects of polarization and the possible nonisotropic nature of the radiation must be determined. The quantum efficiency of the detector must be ascertained. The target density and the incident ion flux must be accurately measured and the effects of the finite lifetime of the radiating state must be calculated. Because an error in

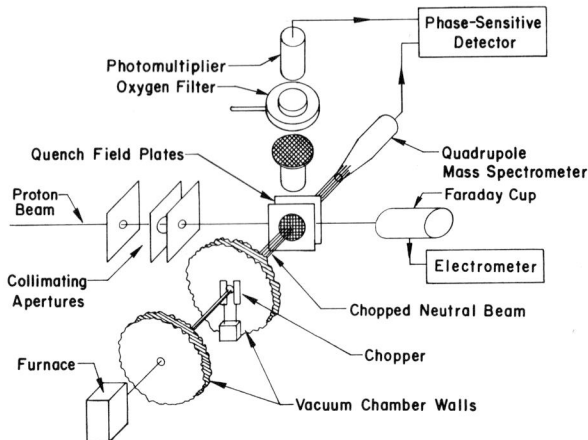

Fig. 1. Schematic drawing of an optical crossed-beam experimental apparatus.

any of these measurements can produce erroneous excitation cross section results, the literature contains many examples of large differences between the experimental measurements of the same cross section by different investigators.

The classic, proton–atomic hydrogen, crossed-beam experiment of Gilbody's group (Morgan et al., 1973) is an excellent example of the class of experiments involving the detection of the secondary particle. This experiment provides cross sections for electron capture by protons from atomic hydrogen into the 2s and 2p states of atomic hydrogen, $H^+ + H(1s) \rightarrow H(2s$ or $2p) + H^+$, and excitation of the atomic hydrogen target to the 2s or 2p state by protons, $H^+ + H(1s) \rightarrow H^+ + H(2s$ or $2p)$. The experimental apparatus is shown in Fig. 1.

In a crossed-beam experiment of this type, a mass-analyzed proton beam intersects a thermal energy beam of dissociated atomic hydrogen. The target beam of highly dissociated hydrogen is produced in a tungsten tube furnace heated to 2700 K. The beam is effused from a small aperture and is collimated by circular apertures mounted in the vacuum chamber walls. The apertures serve both to define the atomic hydrogen beam and to limit molecular gas flow between vacuum chambers. This differential pumping arrangement makes it possible to maintain the vacuum chamber at about 2×10^{-7} Torr. The neutral atomic hydrogen beam is modulated by an electromechanical chopper located in the differential pumping region. A quadrupole mass spectrometer is employed to determine the fraction of atomic hydrogen in the neutral beam. The collimated proton beam current is detected by a Faraday cup, which is carefully designed to ensure complete collection of the proton beam. The intersecting proton and atomic hydrogen beams are carefully aligned so that the 7-mm diameter of the neutral beam fully encompasses the smaller 3-mm-diameter proton beam.

The Lyman-α radiation from the region of intersection between the two beams is detected by an open-ended multiplier, which is sensitive in the ultraviolet. An oxygen filter with lithium fluoride windows provides an extremely narrow-band filter with transmission at 1215.7 Å. This filter is located in front of the multiplier. Together, the oxygen filter and the multiplier make a detector that is sensitive to the Lyman-α radiation resulting from the atomic hydrogen 2p–1s transition. Apertures and shields restrict the view of the Lyman-α detector to the beam interaction region.

Even with differential pumping, high-speed vacuum pumps, and strict adherence to good vacuum technique, the residual background gas density in crossed-beam experiments is of the same magnitude as the density of the target atomic hydrogen beam. Phase-sensitive detection techniques are used to separate the Lyman-α photons from the desired collision process from the photons resulting from collisions with the background gas. The Lyman-α photons from the modulated beam are separated from the unmodulated signal from the background gas using the specific frequency and phase of the modulated Lyman-α signal.

The detected Lyman-α photons are emitted from spontaneously decaying H(2p) atoms in the beam interaction region. To allow for the detection of metastable H(2s) atoms, a quench field is applied across the interaction region. An electric field applied to a H(2s) atom causes Stark mixing of the 2s and 2p states. The resulting emission of the Lyman-α photons comes from the spontaneous emission of the excited hydrogen atom, when it is in the 2p state.

The field of 600 V/cm used by Morgan *et al.* (1973) in an apparatus of the type described above, completely quenched the H(2s) atoms. Stray fields were reduced so that they did not observe any accidental quenching in the absence of an applied quenching field.

These investigations made six different measurements of the Lyman-α signal. With the detector set at 90° to the beam intersection region, measurements were made with the quench field on and off. With the detector set at 54.7°, the Lyman-α radiation emitted by the fast hydrogen atoms formed by electron capture was Doppler-shifted sufficiently to assure strong attenuation of the radiation in the oxygen filter. Measurements were, therefore, made with the detector set at 54.7° with the quench field on and off and, in each case, with the oxygen in and out of the filter. The Lyman-α photons measured with the quench field off were emitted from H(2p) atoms. With the quench field off the H(2s) atoms passed out of the field of view of the detector without radiating. Measurements at 90° with the quench field on recorded the Lyman-α radiation from both slow and fast excited hydrogen atoms in the 2s and 2p states. These six measurements plus another measurement of the transmission fraction of the oxygen filter for the Lyman-α radiation provided data for two different methods of calculation for each of the cross sections for excitation and electron capture to the 2s and 2p states of atomic hydrogen. The different calculation methods yielded results within 5%.

The absolute efficiency of the detector was not known. Morgan et al. (1973) normalized their cross sections to the 2p capture cross sections for $H^+ + H_2$ collisions measured by Birely and McNeal (1972), whose measurements were in turn normalized to absolute cross sections for Lyman-α production in $H^+ + Ne$ collision determined by Andreev et al. (1966).

The measurements were corrected to compensate for the fact that because of the finite lifetime of the excited states, some of the excited hydrogen atoms will pass out of the field of view of the detector before radiating.

The angle 54.7° was chosen because measurement of electric dipole radiation viewed at the "magic" angle of 54.7° with respect to the quantization axis is independent of the polarization fraction of the emitted radiation. The measurements at 90° and the measurements of the quenched H(2s) radiation, whose axis of quantization was the direction of the quenching field, were both dependent on polarization. Morgan et al. (1973) estimate that effects from polarization would change their results by less than 5%.

All experiments that require a measurement of emitted radiation must determine if the excitation process under study is the original source of the emitted radiation. To determine excitation cross sections, it is necessary to account for the effects of cascading from states $n \geq 3$. Morgan et al. (1973) estimated that cascading contributes less than 1% to the electron capture cross sections. However, the contribution of cascading to the excitation cross section could be as much as 10%.

Crossed-beam techniques for measuring ionization cross sections differ from those for measuring excitation in that the secondary ions and electrons resulting from the ionization are detected rather than the photons. The recent $H^+ + H$ ionization cross section measurements of Shah and Gilbody (1981) cover an impressive impact energy range for crossed-beam measurements, which have usually been restricted to low impact energies.

Figure 2 shows a schematic drawing of the apparatus used by Shah and Gilbody (1981), and the similarity to the crossed-beam experiment of Morgan et al. (1973) is obvious. Particle detectors used to detect both the ionized target atom and the electron replace the Lyman-α detector. A redesigned tungsten tube furnace is utilized to produce a very dense, slow atomic hydrogen beam. This arrangement removed the need to modulate

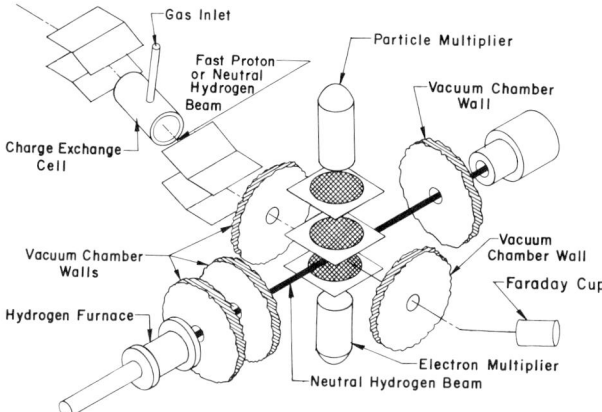

FIG. 2. Schematic drawing of a crossed-beam apparatus for the measurement of ionization cross sections.

the hydrogen beam. However, the basic crossed-beam experimental arrangement is unchanged.

In the Shah and Gilbody (1981) experiment, the particle multipliers used to detect the slow ions and electrons, which are extracted from the collision region by an electric field, replaced the parallel plate collectors used in the conventional "condenser method." The potentials of high transparency grids were adjusted to prevent secondary electrons from reaching the particle multipliers. The process studied, $H^+ + H \rightarrow H^+ + H^+ + e$, was unambiguously identified by recording the time between the detection of the ejected electron by one particle multiplier and detection of the corresponding proton produced by the ionization. Because the H_2^+ molecular ions had a very different time of flight than the H^+ ions, the products of the process $H^+ + H_2 \rightarrow H^+ + H_2^+ + e$ could be identified and studied separately.

The normal measurement technique involves the use of the signal from the particle multiplier, which detects the positive ion, to activate a time-to-amplitude converter (TAC) and a suitably delayed signal from the particle multiplier, which detects electrons, to stop the TAC. The resulting amplitude corresponding to the difference in time of flight for the ion and electron resulting from the ionization is recorded in a multichannel analyzer. An ionization event for a proton–atomic hydrogen collision is identified by the time of flight corresponding to the production of a $H^+ + e$, proton–electron pair. Because $H^+ + e$ pairs with the correct time of flight can also be produced by the dissociative ionization process $H^+ + H_2 \rightarrow H^+ + H + H^+ + e$, this signal is measured with a pure H_2 neutral beam from a cold furnace. Another source of H^+ + e pairs arises from dissociative ionization of the background gas, primarily from dissociative ionization of water. The production rate for this signal source is measured in the absence of gas flow through the furnace. The cross section for ionization of both atomic hydrogen and molecular hydrogen is calculated from the measured coincidence rates with corrections for the possible effects of dissociative ionization.

Because these ionization cross section measurements extend to high impact energies, they are normalized to theoretical results from the first Born approximation values for proton impact energies greater than 1 MeV (Bates and Griffing, 1953).

B. Experimental Methods Involving Detection of the Primary Particle

The second major class of experimental methods, which involves the detection of the incident particle itself, has been employed for many years. Early measurements of this type involved the total scattering of ions. The basic idea of energy-loss spectrometry dates from the Frank–Hertz experiment. Only fairly recently has the technique been utilized in the intermediate energy range of 15–200 keV to produce high-resolution cross sections for specific states. One advantage of the measurement of the energy loss of the projectile is the unambiguous identification of the source of the excitation process. The technique also makes it possible to measure differential cross sections without resorting to coincidence techniques. Elastic scattering in particular can only be measured by using a technique which involves detecting the primary particle.

The ion energy-loss technique has been applied to a wide range of cross section measurements including elastic scattering, excitation, and ionization. It does not depend on one's ability to detect a particular secondary particle. Energy-loss spectrometry provides an energy-loss spectrum, which yields simultaneous information on all of the energy-loss processes that do not involve electron capture. The dominant process can be clearly identified.

The experimental method utilized is reviewed in Park (1978, 1979). The apparatus used by Park et al. (1978b) is shown in Fig. 3. The energy-loss spectrometer consists of a mass-selected source of monoenergetic ions, an ion accelerator, a collision region, a mass selector behind the collision region, an ion decelerator, and an energy analyzer.

Monoenergetic mass-selected ions are accelerated and steered through an entrace collimator into the collision chamber. In the experiments involving

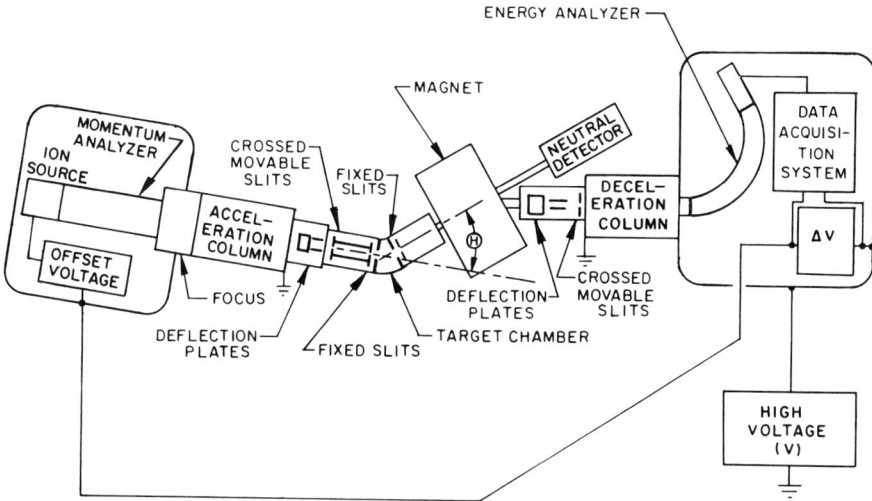

FIG. 3. Schematic drawing of an energy-loss spectrometer.

atomic hydrogen targets, the collision chamber consists of a furnace constructed of coaxial tungsten tubing. The collision takes place inside the target furnace (Park et al., 1978, 1980). Fast ions which are scattered by the collision into the angle set by the exit collimator, pass through the collimating slits into a magnetic mass analyzer, which removes any products of charge-changing collisions. If the charge/mass ratio of the ion has not been changed during the collision, the magnetic field directs them into the decelerator. The decelerated ions are electrostatically analyzed.

The apparatus is electrically connected so that the same voltage source is used for both acceleration and deceleration. This technique cancels out the effects of small variations in the high-voltage power supply. To accomplish this, the energy lost in the collision is offset by a small voltage, which is added to the accelerating voltage. This technique permits the energy-loss scale to be determined to an accuracy of ± 0.03 eV (York et al., 1972).

Spectra differential in energy loss are obtained by measuring the detected ion current as a function of the potential difference between the accelerator and decelerator terminals. The energy resolution of the energy-loss spectrum depends on the experiment, but is typically 0.7–1.5 eV.

Figure 4 shows an energy–loss spectrum for He^+ ions incident on atomic hydrogen. A number of ion–atom collision processes can be identified in this spectrum. The cross section for each identifiable process could be

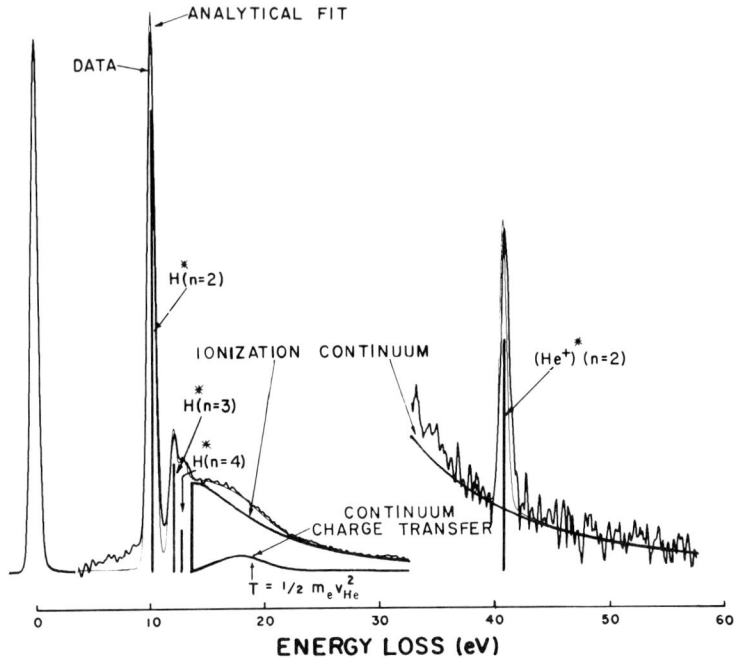

FIG. 4. Energy-loss spectrum of 40-keV helium ions incident on atomic hydrogen.

calculated from measurements on the spectrum itself and knowledge of the target line density and the scattering geometry of the apparatus.

The entire accelerator, ion source, and associated collimating slits can be rotated about the collision center. The relative position of the accelerator is known to $\pm 3.3 \times 10^{-6}$ rad (Park *et al.*, 1978b). The measured angular resolution of the apparatus is 120 μrad. Angular distributions of the scattered ion current corresponding to a particular scattering process can be measured by setting the energy loss to the calculated energy value, which incorporates the effects of both the internal and kinematic energy losses while rotating the accelerator about the scattering center. Alternatively, energy-loss spectra can be taken at a series of different scattering angles, and the differential cross sections can be calculated from the resulting data.

The data analysis techniques are fairly involved. They are discussed in detail by Park (1979) and Park *et al.* (1978b). The set of energy-loss spectra obtained at different scattering angles is a double convolution of the doubly differential cross section in energy loss and scattering angle with the apparatus resolution functions in energy and angle. To measure total cross sections, the set of energy-loss spectra taken at various angles is integrated over all angles, or the exit collimator is opened to accept the entire scattered ion beam. When the total cross section is being measured for a process that is resolved from other features in the energy-loss spectrum, the cross section can be obtained from exact solutions to the differential equations describing the data (Park, 1978). However, the data analysis is more complex when differential cross sections are desired. The doubly differential cross sections are extracted from the differential equations which describe the data. The solution to the equations is valid to second order (Park, 1979). Because the measurements are made under single collision conditions, the data analysis method should not introduce any errors into the derived cross sections; however, the angular differential cross sections must still be extracted from the resulting "apparent" cross section, which is a convolution of the angular differential cross section with the angular resolution function of the apparatus (Park *et al.*, 1978b). If the cross section varies relatively slowly, the apparent and true cross sections are nearly identical. When the differential cross section changes rapidly, the deconvolution process could potentially be the source of systematic error.

The most obvious experiments involving detection of the primary particle are electron capture experiments. There are many such experiments reported in the literature which quite simply involve the charge analysis of the fast primary particle after the collision. The technique has been applied with high angular resolution to yield differential electron capture cross sections (Martin *et al.*, 1981a,b; Bratton *et al.*, 1977). In these experiments a fast-charge-analyzed ion beam is collimated and directed onto a gas target. Fast particles which are scattered into the exit collimator are mass-analyzed and detected by a fast atom detector. The differential electron capture cross section can be calculated from the fast atom flux measured as a function of target line density and a knowledge of the angular resolution of the apparatus (Martin *et al.*, 1981a,b). Because the absolute detection efficiency of the atom detector is unknown, the differential cross sections are normalized to total cross sections for electron capture.

Each type of experiment has its own strengths and limitations. Some problems, such as the problem involved in determining absolute target line density, are common to any experiment involving a gaseous target. Atomic hydrogen targets provide a special set of problems in that the furnace used to dissociate the hydrogen usually has a brief lifetime, the absolute target density is unknown, and the purity of the atomic hydrogen target is always in doubt. However, the experiments are becoming increasingly more accurate. The basic cross sections are becoming well established. The agreement between the experiments using various techniques and normalization is generally very good.

III. Theoretical Methods

The theoretical techniques applied to ion–atom collisions are discussed in texts and reviews on scattering theory such as those presented by McDowell and Coleman (1970) and by Bransden (1970, 1979). A thorough discussion of the theoretical techniques is beyond the scope of this article; however, a brief discussion of the techniques commonly applied is given to aid in understanding the succeeding comparison of the theoretical and experimental results.

A. Classical Techniques

Any theoretical treatment of ion–atom collision processes must describe the transition from the initial condition of widely separated projectile ion and target atom through the collision to the final condition involving widely separated particles possessing new values of momentum, internal energy, and charge state. This problem can be approached classically. In the case of total scattering, cross sections can be obtained directly from the application of classical mechanics.

Recently several groups have applied modifications of the classical binary encounter approach (BEA) to electron capture in ion–atom collisions. The classical BEA theory of Thomas (1927) is valid only in the high-impact-energy region, and is usually only applicable for fast, light ions incident on heavier targets.

A great deal of success has been achieved through use of the classical trajectory Monte Carlo (CTMC) technique (Olson and Salop, 1977; Abrines and Percival 1966a,b). In the CTMC technique, Hamilton's equations of motion for a three-body system are solved numerically for numerous trajectories. A Monte Carlo method is used to randomly select the impact parameter of the incident ion relative to the target atom and the orientation and momentum of the bound electron about the atomic target. The forces between the three bodies are usually taken to be Coulombic.

The CTMC technique is essentially a theoretical experiment. The outcomes of thousands of collisions are evaluated, and cross sections are calculated from the relative probabilities of the various outcomes. Individual trajectories for the collision are evaluated by placing the incident ion a large distance from the target atom and integrating the equation of motion through the collision until the particles are widely separated again. The final states of the particles are evaluated, and the results of the collision are recorded. A large number of trajectories must be evaluated to reduce statistical errors to a reasonable level.

The CTMC technique has the advantage of providing the cross sections for several processes simultaneously. Unlike a real experiment, which depends on detector sensitivity, the CTMC can be used to evaluate simultaneously the elastic scattering, excitation, ionization, and electron capture. The value of this technique cannot help but increase as larger, faster, and less expensive computers become available.

B. Quantum Mechanical Techniques

Quantum mechanical approaches essentially involve the scattering solution of the Schrödinger equation. This solution involves the wave function that describes the initial system, the interaction potential between the projectile and the target, and the wave function that describes the final system after the collision has taken place. Mathematically, the problem involves the calculation of a transition matrix (T-matrix), which describes the collision process. In even the simplest case, this calculation includes an integral over all space. The incident ion is initially assumed to be a free particle with kinetic energy and momentum that can be adequately described by a plane wave; however, if the incident ion has structure, a more sophisticated wave function is required. A more formidable problem is encountered when one determines the wave function to represent the outgoing ion, because the wave function representing the final system contains information about the interaction. To solve the problem, some simplifying approximation is required.

The first Born approximation is the simplifying approximation most frequently applied. In the Born approximation, a plane wave, which has the same final momentum and kinetic energy as the scattered ion, is substituted for the more complex wave function, which truly represents the final system. The final kinetic energy is set equal to the initial kinetic energy minus the energy required for any inelastic process. One can expect the Born approximation to be valid only for collisions with ions of high incident energy, because the scattered ion is treated as a free particle when it is represented by a plane wave. The Born approximation is, therefore, only valid when the kinetic energy of the incident ion is very large relative to the interaction potential.

To be a true solution to the Schrödinger equation describing the collisional process, the wave function representing the scattered ion must incorporate the effects of the interaction potential. In an attempt to include the effects of the interaction potential, the scattered wave function may be expanded in a Born series, which represents successively higher-order corrections to the plane wave function. In the second Born approximation, the first correction term is retained as well as the plane wave. This second term is apparently essential in the case of high-energy electron capture, which appears to have the character of a two-step process. Although the inclusion of additional terms from the series expansion gives a better representation of the scattered particle, it also rapidly increases the computational difficulty. In addition, the Born series appears to be an asymptotic approximation. The increased difficulty in computation quickly outweighs the benefits of including successively higher terms.

A modification of the Born technique, the Coulomb-projected-Born (CPB) approximation (Geltman, 1971) employs a Coulomb-wave final state to represent the interaction between the scattered particle and the target atom. Like the first Born approximation, the CPB approximation is a high-energy approximation.

Another attempt to overcome some of the weaknesses in the Born approximation by partially correcting for the distortion of the wave function due to the interaction between the incident particle and the bound electrons is given the acronym VPSA (Vainstein, Presnyakov, Sobel'man approximation) after its originators (Vainstein et al., 1964). In the VPSA, an attempt is made to treat the interaction between the projectile and the active electron exactly. The interaction between the projectile and the target core is evaluated approximately but in a manner that satisfies the boundary conditions of the collision process.

Another approach to the problem of incorporating the effects of the interaction potential into the wave function describing the final state is the eikonal approximation. The eikonal approximation modifies or distorts the plane wave representing the projectile to take into account the influence of the interaction potential on the scattered particle both as it approaches and leaves the target. The eikonal approximation is expected to be valid at lower impact energies than the first Born approximation, because the incident and scattered wave functions are no longer plane waves. Several different formulations of the eikonal approximation that use different forms of the potential to distort the plane waves have been applied to ion-atom collisions. The most common, the Glauber approximation, is usually utilized when the incident ion has no electronic structure. In the Glauber approximation, the interaction potential itself is used as the distorting potential. The projectile ion is usually assumed to travel in a straight line, and the momentum transferred to the target is assumed to be perpendicular to the projectile's path. Because momentum is, therefore, not conserved, the application of the usual Glauber approximation is restricted to certain substates of the target. Other eikonal approximation techniques which usually employ an average potential to account for the distortion of the plane wave, have been developed to allow for conservation of momentum. The eikonal approximation results in general seem to give unexpectedly good agreement with experiment.

A powerful alternative to the first Born approximation is the close-coupling approximation. As mentioned above, there are a multitude of possible outcomes of an ion–atom collision. The probabilities of these alternate outcomes are not independent. The wave function representing the scattered particle should be a composite of the wave functions that represent the various reaction channels. The probability that the collision results in a particular reaction is related to the coefficient of the wave function related to that particular reaction.

The simplest application of the close-coupling approach utilizes only two reaction channels, the elastic scattering of the particle and the reaction under study. The resulting 2×2 matrix representing the potential interaction yields a coupled set of equations, which can be solved approximately. The restricted two-state system solved in this manner is called the distortion approximation.

The availability of large computers permits close-coupling studies involving several states beyond the one of particular interest. The set of coupled equations result in a potential interaction matrix in which each element couples two of the possible reaction channels. Even with large computers, the coupled equations are usually too complicated to solve without the use of additional approximations. The projectile is usually assumed to follow a straight line, which passes a fixed distance (impact parameter) from the target. The resulting equations are solved as a function of the impact parameter.

The close-coupling procedure can be used in conjunction with a complete set of square integrable functions called pseudostates. The pseudostates are chosen so that the dominant bound states are accurately represented. The complete set of basis functions can be truncated in such a way that the continuum is adequately represented. The resulting finite basis set makes it possible to solve the coupled differential equations.

The close-coupling procedure is not satisfactory in cases where it does not allow for continuum states. Inadequate representation of continuum states can have significant effects on the derived cross sections from the coupled states. If the basis set used overlaps the continuum, it is possible to calculate ionization cross sections. The ionization cross section is equated to the difference between the total cross section and the sum of cross sections for discrete excitation and elastic scattering. It should be noted that this approach should involve both target and projectile continuum states.

Coupled-state impact parameter calculations can provide differential cross sections if a Fraunhoffer integral technique is used to convert the phase amplitudes given as a function of impact parameter to functions of scattering angles. Through the use of this technique, Bransden and Noble (1979) and Shakeshaft (1978) have been able to report differential cross sections for excitation of the $n = 2$ level of atomic hydrogen by proton impact. They obtained the differential cross sections from the phase information contained in their calculated transition amplitudes by using the procedure discussed in detail by Wilets and Wallace (1968). Contrary to the classical transformation, there is not a one-to-one relationship between the scattering angle and the impact parameter. In this procedure the transition amplitudes are multiplied by an appropriate Bessel function and integrated over all of the impact parameters. Thus, all of the impact parameters contribute to a given scattering angle. (Note that a one-to-one correspondence between scattering angle and impact parameter is not observed in CTMC calculations because in the CTMC calculations the scattering angle depends both on the impact parameter and the initial position and angular momentum of the electron.)

There are a bewildering number of variations, combinations and extensions of these basic theoretical approaches. The details of some of the more recent theoretical efforts will be discussed as the theoretical results are compared with the experimental measurements in the sections which follow.

IV. Excitation

A. Excitation of Atomic Hydrogen by Protons

Both theoretically and experimentally, a great deal of effort has been applied to the excitation of simple targets by ions. The research on collisional excitation of simple systems prior to 1976 is summarized by Park (1978). A representation of recent efforts is shown in Figs. 5 and 6, which give the total cross sections for excitations of the $n = 2$ and $n = 3$ levels of atomic hydrogen by incident protons. Figure 5 shows data obtained with both crossed-beam and energy-loss techniques along with representative theoretical curves.

The crossed-beam experiments provide cross sections to specific substates. Morgan *et al.* (1973) measured the cross sections for both the H(2s) and H(2p) states. Stebbings *et al.* (1965; see also Young *et al.*, 1968) and Kondow *et al.* (1974) only measured the cross sections for the H(2p) excitation. The experimental results for $n = 2$ excitation shown for Steb-

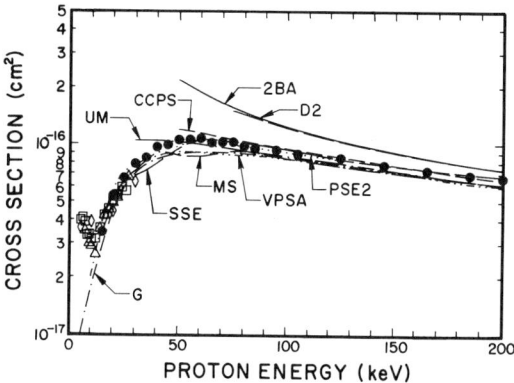

FIG. 5. The cross section for the excitation of atomic hydrogen to its $n = 2$ level [$H^+ +$ $H(1s) \rightarrow H^+ + H^*(n = 2)$] by 10- to 200-keV protons. Experimental data: ●, Park et al. (1976); □, Morgan et al. (1973); ◇, Stebbings et al. (1965); △, Kondow et al. (1974). Theoretical results: —, 2 BA, Bransden and Dewangan (1979); — —, D2, Bransden et al. (1979); —·—; UM, Fitchard et al. (1977); — - — -, SSE, Shakeshaft (1976); ····, G, Franco and Thomas (1971); - —·, MS, Shakeshaft (1978); -··-··, VPSA, Theodosiou (1980); ----, CCPS, Bransden and Noble (1979); ·····, PSE2, Morrison and Öpik (1979).

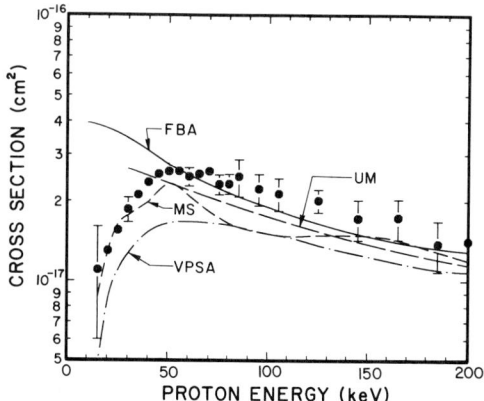

FIG. 6. The cross section for the excitation of atomic hydrogen to its $n = 3$ level [$H^+ +$ $H \rightarrow H^+ + H^*(n = 3)$] by 15- to 200-keV protons. Experimental data: ●, Park et al. (1976). Theoretical results: ———, FBA, Bates and Griffing (1953); - - -, UM, Fitchard et al.(1977); ----, MS, Shakeshaft (1978); -·-·, VPSA, Theodosiou (1980).

bings et al. (1965) and Kondow et al. (1974) include the H(2s) cross sections of Morgan et al. (1973).

The energy-loss data (Park et al., 1975, 1976) cover an energy range that has been a source of difficulty for theorists. In this energy range, neither approximations suitable for very low-velocity collisions nor those suitable for very high-velocity collisions are valid. Also in this range, the cross section reaches a maximum. Matching both the magnitude and shape of the $n = 2$ excitation cross section provides the theorist with a major challenge.

The measurements of Park et al. (1975, 1976) were normalized to the Born approximation results of Bates and Griffing (1953) for 200-keV proton impact excitation of atomic hydrogen to its $n = 2$ level. The Born approximation result for excitation of atomic hydrogen to its $n = 2$ level is reasonably accurate at a proton impact energy of 200 keV, although significant differences between the Born approximation and more recent and reliable theoretical calculations are observed. Considering the major differences in normalization method and experimental technique, the agreement between the energy-loss and crossed-beam experiments is very good.

There are numerous theoretical studies of the excitation of atomic hydrogen by protons. This collision process is the primary testing ground for any new theory or variation of a theoretical method. Because many of the earlier efforts are reviewed by Park (1978) and Thomas (1972), they are not discussed in detail here.

Figures 5 and 6 show only the most recent results. The excellent agreement between the recent theoretical treatments and the experimental results for excitation of atomic hydrogen to its $n = 2$ level represents a significant improvement over earlier theoretical efforts (see Park, 1978). The improvement is the result of a great deal of effort applied both to new theoretical approaches and to extensions of previously studied approximations. The agreement between experiment and theory for proton impact excitation of atomic hydrogen to its $n = 3$ level is not quite so good as for the $n = 2$ level; however, the improvement of recent efforts over earlier theoretical treatments is significant (see Park et al., 1976).

Measurements of the differential cross section for the excitation of atomic hydrogen to its $n = 2$ level by protons are possible if one uses energy-loss spectrometry (Park et al., 1978a, 1980). Differential cross section measurements provide a more meaningful test of theory than do total cross section measurements. The process of integrating the differential cross sections to obtain total cross sections can mask significant differences between theory and experiment. As can be seen in Figs. 7–9, the differential cross sections decrease dramatically with increasing scattering angle. If the theory and experiment agree at the particular angle where $(\sin\theta)d\sigma/d\Omega$ is at a maxi-

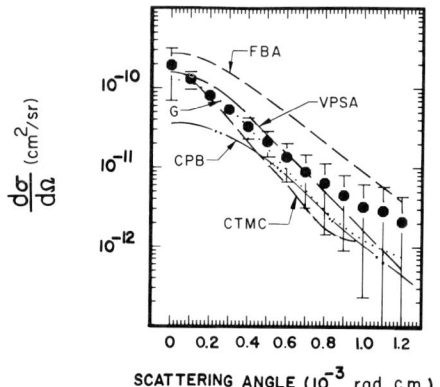

FIG. 7. Angular differential cross section for the excitation of atomic hydrogen to its $n = 2$ level $[H^+ + H(1s) \rightarrow H^+(\theta) + H^*(n = 2)]$ by 25-keV protons. Experimental data: ●, Park et al. (1980). Theoretical results: — — —, FBA, Franco and Thomas (1971), R. E. Thomas (private communication, 1978); - - -, VPSA, Theodosiou (1980); —··—, CPB, Datta and Mukherjee (1981); —·—·—, CTMC, R. E. Olson (private communication, 1982); ····, G, Franco and Thomas (1971), B. K. Thomas (private communication, 1978).

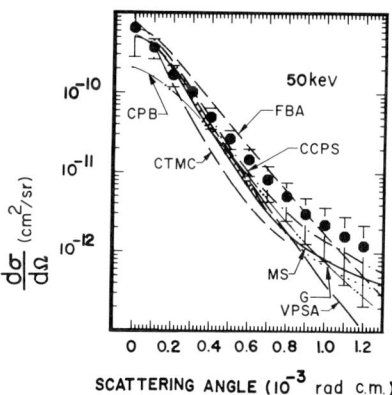

FIG. 8. Angular differential cross section for the excitation of atomic hydrogen to its $n = 2$ states $[H^+ + H(1s) \rightarrow H^+(\theta) + H^*(n = 2)]$ by 50-keV protons. Experimental data: ●, Park et al. (1980). Theoretical results: — — —, FBA, Franco and Thomas (1971), B. K. Thomas (private communication, 1978); - - -, VPSA, Theodosiou (1980); —··—, CPB, Datta and Mukherjee (1981); ——, MS, Shakeshaft (1978); —·—·—, CTMC, R. E. Olson (private communication, 1982); ····, G, Franco and Thomas (1971), B. K. Thomas (private communication, 1978); —·—·, CCPS, Bransden and Noble (1979).

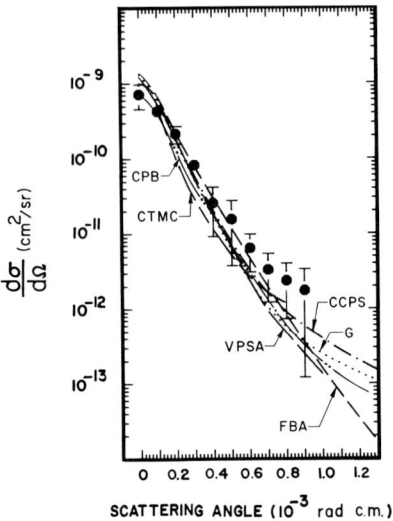

FIG. 9. Angular differential cross section for the excitation of atomic hydrogen to its $n = 2$ level [$H^+ + H(1s) \rightarrow H^+(\theta) + H^*(n = 2)$] by 100-keV protons. Experimental data: ●, Park et al. (1980). Theoretical results: ———, FBA, Franco and Thomas (1971), B. K. Thomas (private communication, 1978); - - -, VPSA, Theodosiou (1980); —··—, CPB, Datta and Mukherjee (1981); —·—·, CTMC, R. E. Olson (private communication, 1982); ----, G, Franco and Thomas (1971), B. K. Thomas (private communication, 1978). —·—, CCPS, Bransden and Noble (1979).

mum, significant differences in the shape of the theoretical and experimental differential cross sections may not be reflected in the total cross sections.

The display of differential cross sections using multicycle, logarithmic scales is dictated by the rapid decrease of the cross section with increasing scattering angle. The logarithmic scale, however, tends to obscure the differences in magnitude between the experiment and the various theoretical results. The differences may be an order of magnitude at some angles.

Born approximation calculations are available for the process $H^+ + H \rightarrow H^+ + H(n = 2)$. The total Born approximation cross sections for excitation of atomic hydrogen to its $n = 2$ and 3 levels by protons greatly overestimates the cross sections at the lower velocities. Because the data of Park et al. (1975, 1976) are normalized to the Born approximation results for the $n = 2$ excitation at 200 keV, exact agreement at that energy is not significant. As is expected for a basically high-energy approximation, the comparison between the experimental cross section and the Born approximation improves with increasing collision velocity.

The Born calculation of differential cross sections for proton impact excitation of atomic hydrogen to its $n = 2$ level is from B. K. Thomas (private communication 1978; Franco and Thomas, 1971). The Born approximation results are not expected to be valid at 25 keV. As anticipated, the 25-keV Born approximation results are clearly too high and display a different curvature than is observed in the experimental data. The agreement for proton energies greater than 50 keV is fairly good, although at impact energies higher than 70 keV the Born approximation results are more sharply peaked than the experimental results. At the larger scattering angles the Born approximation differential cross sections are well below the experimental results and decrease rapidly in magnitude as the scattering angle continues to increase.

The Born approximation is considered to be fairly accurate for proton impact energies greater than 200 keV for total cross sections for excitation of low-lying strongly coupled states. This does not carry over to the calculation of differential cross sections at large scattering angles where it is necessary to include the internuclear interaction by either solving the coupled equation or incorporating the effect of the internuclear interaction by using a distorted wave approximation (Bransden, 1979).

The CPB results of Datta and Mukherjee (1981) are shown in Figs. 7–9. The CPB results are generally lower than the experimental differential cross sections at the lower impact energies. At 100 keV the CPB results provide good agreement with experiment.

Impact parameter formulations (Crothers and Holt, 1966; Van den Bos and de Heer, 1967) of the first Born approximation for excitation of the $n = 2$ level of atomic hydrogen by proton impact are not shown. This approach does not provide improved agreement with experiment over the results of the first Born approximation. Several second Born calculations (Kingston et al., 1960; Moiseiwitsch and Perrin, 1965; Holt and Moiseiwitsch, 1968) for this cross section have been published. Shown in Fig. 5 is the recent second Born calculation of Bransden and Dewangan (1979). The second Born approximation results are typically 50% or more higher than the experimental cross sections. Also not shown are the distortion approximation results (Bates, 1959, 1961) which, although better than the first Born approximation results, provide poor agreement with experiment.

Eikonal calculations of proton impact excitation of atomic hydrogen provide surprisingly good agreement with the experiment. The results of the eikonal distorted wave calculation for $H^+ + H \rightarrow H^+ + H(n = 2)$ of Joachain and Vanderpoorten (1973) differs only slightly from the Glauber approximation results of Franco and Thomas (1971) which are shown in Fig. 5 (see also Bandra and Ghosh, 1971). The eikonal calculations provide quite good agreement over the proton energy range of 20–200 keV, especially when the relative simplicity of these calculations is considered. The results of the eikonal-type calculations tend to fall below the experimental results for excitation of atomic hydrogen by protons, particularly near the maximum in the total cross section curves. The eikonal calculations follow the experimental results at the lower energies and provide good results at impact energies as low as 15 keV.

The Glauber approximation results (Franco and Thomas, 1971; B. K. Thomas, private communication, 1978) are also in reasonably good agreement with the experimental differential cross sections of Park et al. (1978a, 1980) over the entire range of impact energies studied (see Figs. 7–9). The Glauber approximation results are slightly lower than experiment for both differential and total cross sections. The Glauber approximation does remarkably well considering its relative computational simplicity.

Coupled-state calculations of the total excitation cross section of atomic hydrogen to its $n = 2$ level by proton impact have been made by several groups. There are several four-state calculations (Flannery, 1969a; Wilets and Gallaher, 1966; Cheshire et al., 1970; Rapp and Dinwiddie, 1972) that are noticeably different from each other but provide roughly equivalent fits to the experimental data. However, it is noted that the inclusion of the electron capture channel appears to be necessary to obtain reasonable agreement with the experimental data. There are also several seven-state calculations (see Rapp and Dinwiddie, 1972), but the inclusion of the additional three states does not appear to produce any dramatic changes. Both the four- and seven-state calculations provide reasonable agreement between experiment and theory (see Park, 1978).

The recent seven-state calculation of Bransden and Noble (1979) used a target-centered basis set expansion, which included eigenfunctions to represent the discrete target states and pseudostates chosen to represent the continuum. The agreement with experiment is good only at energies above 50 keV because electron capture has been neglected. However, the agreement with the experimental total cross sections for excitation of atomic hydrogen to its $n = 2$ level is very good over the range of validity of this calculation (see curve CCPS in Fig. 5), and the calculated differential cross sections provide excellent agreement with the experimentally determined differential cross sections at 50 keV (see Fig. 8). At 100 keV, the theoretical results of Bransden and Noble (1979) shown in Fig. 9 are slightly more sharply peaked than the experimental results but are inside the error bars except at the largest angles recorded by Park et al. (1978a, 1980).

For impact energies less than 20 keV, the seven-state close-coupling calculation of Cheshire et al. (1970) gives good results for excitation of the 2s and 2p states of atomic hydrogen by protons at low impact energies (see Park, 1978). This calculation includes electron capture and uses pseudostates to represent coupling to higher states, but it tends to overestimate the cross section at the peak of the cross section curve for excitation of the $n = 2$ level of atomic hydrogen by protons. Shakeshaft (1976) attributed the discrepancy to the large value of the ionization cross section at its peak near 55 keV. The pseudostates are, therefore, larger because the flux attributable to ionization resides in the pseudostates. Because the pseudostates cannot adequately represent ionization of an electron at infinity, some of this flux may "leak" back into the bound states as the collision progresses, thereby yielding spurious contributions to the excitation cross sections (Shakeshaft, 1976). It should also be noted that the calculation of Cheshire *et al.* (1970) drastically overestimates the cross section for excitation of the $n = 3$ state of atomic hydrogen by protons (Park *et al.*, 1976).

Shakeshaft (1976) published an impact parameter coupled-state calculation using an expansion in Sturmian functions. He used 12 Sturmian functions centered around each proton. This is an extension of an approach applied by Gallaher and Wilets (1968). A Sturmian expansion has merit in that the functions form a complete basis set of square integrable functions. Estimates of transition probabilities calculated with a Sturmian expansion, therefore, become exact as the basis set is enlarged in contrast to an expansion in terms of bound atomic orbitals (Shakeshaft, 1976). The results observed by Shakeshaft (1976) for proton impact excitation of the $n = 2$ state of atomic hydrogen are shown in Fig. 5 (curve SSE). The results are in good agreement with the experiment except for a slight shoulder between 20 and 50 keV that does not appear in the data. The earlier coupled-state calculation of Gallaher and Wilets (1968), in which the Sturmian functions were used, it not in good agreement with experiment. Their curve demonstrates a minimum at about 35 keV that is not observed in the experimental data.

The results of the sophisticated 70-state calculation (35 scaled hydrogenic states centered on each proton) of Shakeshaft (1978) is shown (curve MS) in Figs. 5 and 6. The dip in Shakeshaft's theoretical curve for excitation of the $n = 2$ level of atomic hydrogen by protons at 55 keV is almost certainly artificial. Shakeshaft (1978) notes that this feature may be due to the inability of the scaled hydrogenic functions to account adequately for ionization. The ionization cross section peaks at 58 keV. To make certain that the eigenfunctions are square integrable, they are extended for only a finite distance; hence, the eigenfunctions employed by Shakeshaft (1978) do not adequately represent the possibility of the electron ionizing by escaping to infinity.

Shakeshaft's (1978) 70-state theoretical results for the excitation of the $n = 3$ level of atomic hydrogen by protons also display a structure that is not observed in the experimental results. The dip in the $n = 3$ excitation cross section is at a higher energy than in the $n = 2$ excitation results, but perhaps this is also a result of the failure of the calculation to account adequately for ionization.

Shakeshaft (1978) applied the Fraunhoffer integral technique to his multistate calculation to provide differential cross sections for excitation of atomic hydrogen to its $n = 2$ level by proton impact. His results are in very good agreement with experiment except that the result for the 15-keV proton impact energy is slightly less peaked than the experimental result. At 50 and 60 keV, the fit to the experimental data is extremely good with only significant differences at the largest scattering angles. The Shakeshaft (1978) results for 145 keV are more sharply peaked than the experimental results of Park *et al.* (1978a, 1980), but are still in general agreement.

The relatively recent pseudostate expansion of Fitchard *et al.* (1977) (curve UM) and of Morrison and Öpik (1979) (curve PSE2) are also shown in Figs. 5 and 6. These calculations involve large basis sets and are expected to be very accurate. For example the time-development U-matrix pseudostate calculations by Fitchard *et al.* (1977) employed a pseudostate set with up to 10s, 22p, and 21d states. Fitchard *et al.* (1977) found that the d states make a significant contribution. The higher-order method employed by Fitchard *et al.* (1977) uses a target-centered expansion and is not suitable below ~ 50 keV because electron capture is important at low impact energies. However, the agreement between both of these pseudostate calculations and the experimental data is very good for impact energies greater than 50 keV.

The VPSA results (Theodosiou, 1980) are also shown in Figs. 5–9. The agreement with experiment is extremely good for total cross sections for excitation of atomic hydrogen to the $n = 2$, 3, and 4 levels by protons over the entire range covered by the experiments of Park *et al.* (1976). Theodosiou's calculation expands on earlier efforts (see the review by Coleman, 1969) to include an effective charge and the contribution of the projectile–target core interaction to the transition matrix element. The effective charge is introduced to minimize the effect of neglecting terms in the Schrödinger equation of the collision system. Theodosiou's (1980) VPSA results appear to provide an agreement with experiment (Park *et al.*, 1975, 1976) which is as good as that provided by the highly regarded U-matrix pseudostate calculation of Fitchard *et al.* (1977).

The good agreement observed for the energy dependence of the total cross sections is also observed for the dependence of the cross section on scattering angle. Except at the largest angles, the VPSA results of Theodosiou (1980) for differential excitation cross sections are in excellent agreement with the experimental results of Park *et al.*(1978a, 1980).

Although there are no data for the excitation of atomic hydrogen to the 2s and 2p substates for proton impact energies above 26 keV, the low-energy data can be artificially extended by multiplying the experimental results of Park et al. (1975, 1976) by the ratios of the cross sections for excitation to the specific substates of the $n = 2$ level. Cross sections taken from the theoretical calculations of Shakeshaft (1978) were used to obtain the necessary ratios.

Data for the excitation of atomic hydrogen to the 2p state of Stebbings et al. (1965), Young et al. (1968), Kondow et al. (1974), and Morgan et al. (1973) along with the data of Park et al. (1975, 1976), adjusted as discussed above, are given in Fig. 10. The data of Morgan et al. (1973), of Chong and Fite (1977), and the adjusted data of Park et al. (1975, 1976) for the excitation of atomic hydrogen to its 2s state are shown in Fig. 11.

Figures 10 and 11 also display the theoretical results for the 2s and 2p states separately. The excellent data of Morgan et al. (1973) provide a key to the validity of the calculations at low impact energies; however, those calculations, which do not fully account for electron capture, cannot be taken seriously at energies less than 50 keV. These figures illustrate the large variation in the theoretical cross sections for excitation of the 2s state compared to the cross sections for the 2p state. The second Born approximation (2BA) calculation of Bransden and Dewangen (1979) and the second-order distorted wave calculation (D2) of Bransden et al. (1979) are higher than the experimental results for both 2s and 2p excitation. However, the VSPA results of Theodosiou (1980), the Glauber results (G) of Franco and Thomas (1971), the pseudostate expansion (SSE) results of Shakeshaft

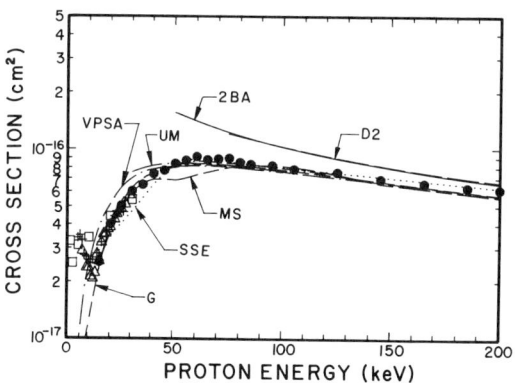

FIG. 10. The cross section for the excitation of atomic hydrogen to its 2p state [H⁺ + H(1s) → H⁺ + H*(2p)] by 15- to 200-keV protons. Experimental data: ●, Park et al. (1976) (see text); □, Stebbings et al. (1965); +, Morgan et al. (1973); △, Kondow et al. (1974); Theoretical results: ——, 2BA, Bransden and Dewangen (1979); ― ― ―, D2 Bransden et al. (1979); —·—·, UM, Fitchard et al. (1977); ----, SSE, Shakeshaft (1976); - - -, G, Franco and Thomas (1971); — ·· — ··, MS, Shakeshaft (1978); -·—·, VPSA, Theodosiou (1980).

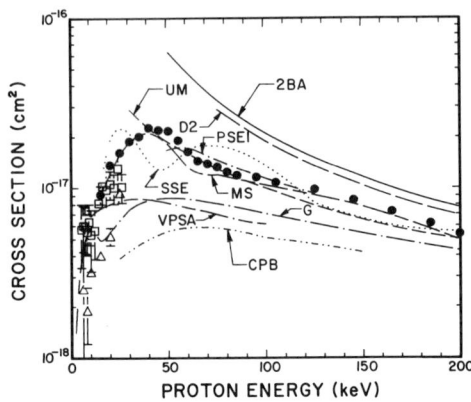

FIG. 11. The cross section for the excitation of atomic hydrogen to its 2s state [H⁺ + H(1s) → H⁺ + H*(2s)] by 15- to 200-keV protons. Experimental data: ●, Park et al. (1976) (see text); △, Chong and Fite (1977); □, Morgan et al. (1973). Theoretical results: ——, 2BA, Bransden and Dewangan (1979); — —, D2, Bransden et al. (1979); ····, PSE1, Morrison and Öpik (1979); — - — -, MS, Shakeshaft (1978); ----, SSE, Shakeshaft (1976); - — ·, VPSA, Theodosiou (1980); -···-··, CPB, Datta and Mukherjee (1981); —·—·, G, Franco and Thomas (1971).

(1976), and the CPB results of Datta and Mukherjee (1981) all display much better agreement with the experimental cross section for excitation of the 2p state of atomic hydrogen than for the excitation cross section of the 2s state.

The 2s excitation cross section is much more sensitive to the choice of approximation than is the 2p excitation (see Bransden, 1979). The Glauber approximation results for the excitation of the $n = 2$ level in the impact energy range 30–60 keV are probably lower than the experimental results primarily because of the low value of the cross section for excitation of the 2s state.

Only the most sophisticated and accurate calculations give good results for the 2s excitation. The calculations giving the best agreement are those of Fitchard et al. (1977), of Shakeshaft (1978), and of Morrison and Öpik (1979) which utilize large basis sets and are very involved calculations. The other approximations do not satisfactorily describe the cross section for excitation of atomic hydrogen to its 2s state.

B. Excitation of Helium Atoms by Protons

The addition of an extra electron to the ion–atom collision problem is a major complication for the theorist even though it adds little to the experimental difficulties. In fact, helium targets make for much simpler experiments than atomic hydrogen targets. The total cross sections for proton–helium excitation were reviewed by Park (1978) and by Thomas (1972). Their reviews include optical and energy-loss measurements and theoretical results for excitation of helium to many different states. Recently, there have been relatively few experiments or theoretical treatments related to total cross section measurements of excitation cross sections, but a significant number of the theoretical efforts have been directed toward the problem of calculating differential cross sections for proton excitation of helium.

Figures 12–14 display the angular differential cross sections for proton excitation of helium to the $n = 2$ level. The experimental data are taken from Park et al. (1978b) and recent unpublished results by the same group (T. J. Kvale, E. Redd, D. M. Blankenship, J. L. Peacher, and J. T. Park, private communication, 1982). The Born approximation results for $H^+ + H \rightarrow H^+(\theta) + He^*(n = 2)$ (Flannery and McCann, 1974b) are much higher than the experimental results for 25-keV proton impact. At higher impact energies, the Born approximation differential cross sections are in good agreement at small scattering angles, but are too small at large scattering angles. This is not unexpected because the large angle scattering is determined by

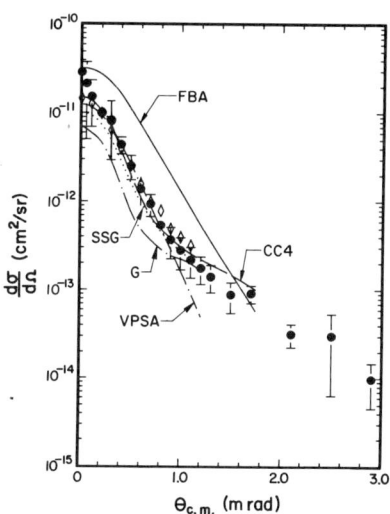

FIG. 12. Differential cross section for excitation of helium to its $n = 2$ level [$H^+ + He(n = 1) \rightarrow H^+(\theta) + He^*(n = 2)$] by 25-keV protons. Experimental data: ◇, Park et al. (1978b), ●, T. J. Kvale, E. Redd, D. M. Blankenship, J. L. Peacher, and J. T. Park, private communication (1982). Theoretical results: ——, FBA, Flannery and McCann (1974b); — —, CC4, Flannery and McCann (1974b); ····, SSG, Sur et al. (1981); —·—·, G, Sur et al. (1981); -—·, VPSA, Theodosiou (1981).

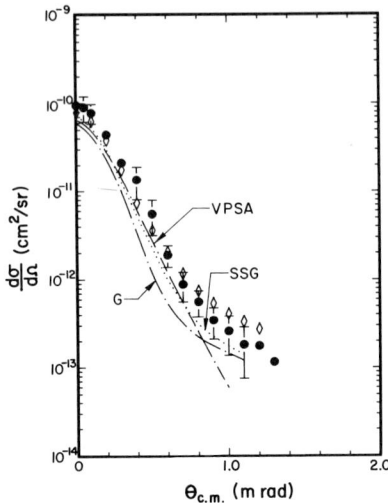

FIG. 13. Differential cross section for excitation of helium to its $n=2$ level [H$^+$ + He($n = 1$) → H$^+(\theta)$ + He*($n = 2$)] by 50-keV protons. Experimental data: ◊, Park et al. (1978b), ●, T. J. Kvale, E. Redd, D. M. Blankenship, J. L. Peacher, and J. T. Park, private communication (1982). Theoretical results: ····, SSG, Sur et al. (1981); —·—·, G, Sur et al. (1981); -—·, VPSA, Theodosiou (1981).

the Coulomb interaction between the nuclei, which does not contribute to the first-order Born calculation of inelastic scattering.

The CPB approximation is a high-energy approximation (Geltman, 1971) and cannot be expected to provide good agreement with the experimental data at low energies. Datta et al. (1980) have used the CPB approximation to calculate the differential cross section for excitation of helium to the $n = 2$ level by 100-keV protons. The results are significantly lower than experiment, but the CPB differential cross section displays a dependence on the scattering angle that has a curve shape very similar to that of the experimental data.

Sur et al. (1981) employed the Glauber approximation to calculate differential cross sections for proton–helium excitation. They determined differential cross sections in both the full Glauber (G) approximation and the single scattering Glauber (SSG) approximation. The addition of the double scattering term lowers the Glauber cross sections significantly at low impact energies. With increased impact energy, the G and SSG approximations approach each other in the forward scattering angles.

The 25- and 50-keV Glauber differential cross sections for excitation of helium to the $n = 2$ level show a curve shape that is similar to that of the experimental results of Park et al. (1978b). However, the Glauber results of

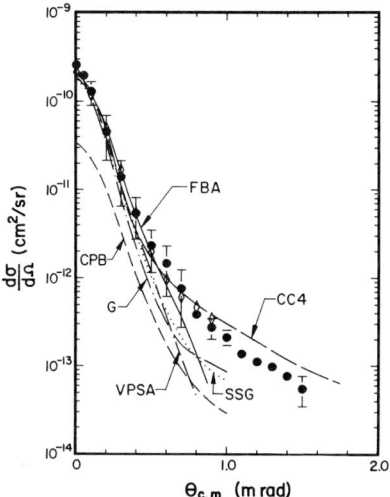

FIG. 14. Differential cross section for excitation of helium to its $n = 2$ level [H^+ + He($n = 1$) → $H^+(\theta)$ + He*($n = 2$)] by 100-keV protons. Experimental data: ◊, Park et al. (1978b), ●, T. J. Kvale, E. Redd, D. M. Blankenship, J. L. Peacher, and J. T. Park, private communication (1982). Theoretical results: ——, FBA, Flannery and McCann (1974b); — —, CC4, Flannery and McCann (1974b); ····, SSG, Sur et al. (1981); —·—·, G, Sur et al. (1981); ----, CPB, Datta et al. (1980); -—·, VPSA, Theodosiou (1981).

Sur et al. (1981) tend to underestimate the magnitude of the experimental differential cross sections results. In the forward direction there is improved absolute agreement between experiment and theory at higher energies. However, the Glauber differential cross section for excitation of the He(2 ^{1}S) state of helium that was calculated by Sur et al. (1981) displays a distinct minimum for 25- and 50-keV proton impact. This feature is masked by the much larger differential cross section for He(2 ^{1}P) excitation in the He($n = 2$) differential excitation cross sections displayed in Figs. 12 and 13.

The four-state eikonal results of Flannery and McCann (1974b) are in excellent agreement with the experimental results of Park et al. (1978b) with respect to both curve shape and absolute magnitude. It must be noted, however, that the agreement in magnitude is perhaps fortuitous because the integrated results of Park et al. (1978b) tend to overestimate the earlier and more accurate total cross sections measured by Park and Schowengerdt (1969).

The VPSA calculations of Theodosiou (1981) are also shown in Figs. 12–14. The VPSA calculation is considered a relatively simple calculation, although the calculation of Theodosiou (1981) includes several refinements to minimize the effects of neglecting terms in the Schrödinger equation of the collision system. The VPSA provides excellent agreement with experiment in the forward direction. The failure of the VPSA at large angles is thought to be a consequence of the peaking approximation employed (Theodosiou, 1980, 1981). Because the total cross section is dominated by the forward scattering angle results, the difference between the experimental and differential cross sections has little effect on the VPSA total cross section results, which are in excellent agreement with the total cross sections for proton excitation of helium to the $n = 2$ level measured by Park and Schowengerdt (1969).

The excitation of helium by protons has been a testing ground for the many theoretical approximations (see Park, 1978; Thomas, 1972). Because the wave functions of helium are known accurately, the calculations provide information on the ability of the various approximations to describe the collision dynamics. Many of these theoretical approximations remain to be tested against the strict standard set by the differential cross section measurements.

C. Excitation of Atomic Hydrogen by Helium Ions

The $He^+ + H$ collision system provides another means of studying two electron collision systems. In the case of $H^+ + He$ collisions, only electrons in the target are considered for the initial channel. The $He^+ + H$ collision system provides the possibility to study a system with an electron initially bound to each nuclei. As can be seen from Fig. 4, the $He^+ + H$ collision system opens the possibility of excitation of either the projectile or the target. Total cross sections for excitation of atomic hydrogen to the $n = 2$ level by singly charged helium ions have been calculated and measured by a number of groups (Young *et al.*, 1968; McKee *et al.*, 1977; Aldag *et al.*, 1981; Bell and Kingston, 1978; Franco, 1979a,b, 1982; Flannery, 1969a; Flannery and McCann, 1974a; J. M. Maidagan and R. D. Rivarola, private communication, 1982; Theodosiou, 1980). The experimental results of Young *et al.* (1968), McKee *et al.* (1977), and Aldag *et al.* (1981) are in good agreement considering the differences in normalization and experimental technique.

The Born approximation results for total excitation of atomic hydrogen to the $n = 2$ level by helium ions (Flannery, 1969a; Flannery and McCann, 1974a) are much higher than the experimental results over the entire energy range. The agreement in both curve shape and magnitude is poor for both the two- and four-state eikonal results (Flannery, 1969a; Flannery and McCann, 1974a). The first-order Glauber approximation of Franco (1979b) is also in poor agreement with experiment.

Recent theoretical efforts provide better agreement with the experimental total cross sections for excitation of atomic hydrogen by helium ions. The symmetrized first-order Glauber results of Franco (1979a) are in good agreement with experiment. The VPSA results of Theodosiou (1980) are in good agreement for impact energies greater than 50 keV.

A more detailed comparison between theory and experiment is provided by the angular differential cross section measurements for $He^+ + H$ (Aldag et al., 1981). The angular differential cross sections for $He^+ + H \rightarrow He^+(\theta) + H(n = 2)$ are shown in Figs. 15 and 16. The differential cross sections of Aldag et al. (1981) display a dramatic change in curve shape as a function of scattering angle as the impact energy is raised from 15 to 100 keV. While the differential cross section at 15 keV falls by a factor of five in 6 mrad, the differential cross section at 100 keV falls by almost six orders of magnitude in the same angular range. This large change in the dependence of the cross section on scattering angle presents a real challenge to the theorist. The recent work of Franco (1982), who used a full Glauber approximation calculation (curve G2 in Fig. 16) as well as the earlier

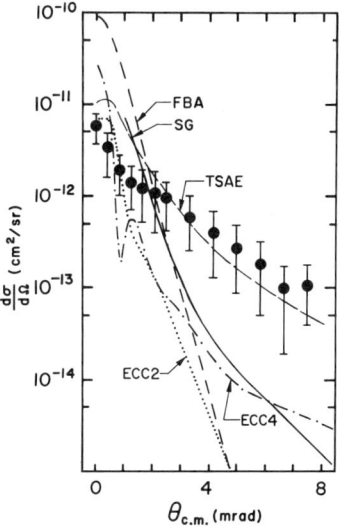

FIG. 15. Angular differential cross sections for the excitation of atomic hydrogen to its $n = 2$ level $[He^+ + H(1s) \rightarrow He^+(\theta) + H^*(n = 2)]$ by 25-keV helium ions. Experimental data: ●, Aldag et al. (1981). Theoretical results: ----, FBA, Flannery and McCann (1974a); ——, SG, Franco (1979a); — —, TSAE, J. M. Maidagan and R. D. Rivarola (private communication, 1982); ·····, ECC2, Flannery and McCann (1974a); -·-·, ECC4, Flannery and McCann (1974a).

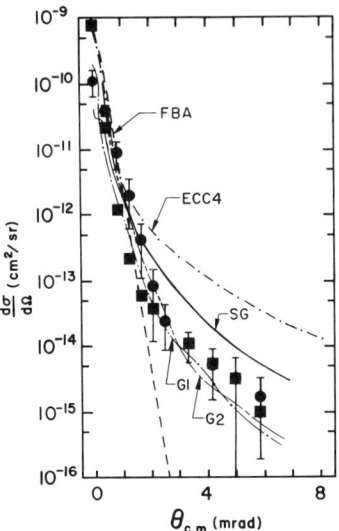

FIG. 16. Angular differential cross section for the excitation of atomic hydrogen to its $n = 2$ level [He$^+$ + H(1s) → He$^+(\theta)$ + H*($n = 2$)] by 100-keV helium ions. Experimental data: ●, ■, Aldag et al. (1981). Theoretical results: ----, FBA, Flannery and McCann (1974a); ——, SG, Franco (1979a); —·—·, G1, Franco (1982); —··—··, G2, Franco (1982). ····, ECC4, Flannery and McCann (1974a).

symmetrized Glauber approximation calculation (Franco, 1979a,b) (curve SG in Figs. 15 and 16), provides very satisfactory agreement with the 100-keV experimental results of Aldag et al. (1981). For the forward scattering angles, the full Glauber calculation of Franco (1982) provides very good agreement with experiment to energies as low at 50 keV. Franco (1982) demonstrates that the effect of neglecting the electron of the He$^+$ projectile (curve G1 in Fig. 16) becomes increasingly damaging as the impact energy decreases. He also suggests that to describe the large-angle angular differential cross sections for He$^+$ impact energies lower than 100 keV successfully, it may be necessary to consider higher-order multiple collisions. It must be noted that the data of Aldag et al. (1981) are normalized to the 200-keV Born approximation total cross section for H$^+$ + H → H$^+$ + H(n = 2), which is about 8% higher than the 200-keV Glauber approximation cross section for the same process. Renormalization to the Glauber approximation would make the agreement at the forward angles between the Glauber theory and experiment look even better.

The first Born approximation (FBA) differential cross sections for He$^+$ + H → He$^+(\theta)$ + H(n =2) (Flannery and McCann, 1974a) do not demonstrate the correct angular dependence at the larger scattering angles for any energy, and for lower impact energies the agreement is poor at all angles. The two- and four-state eikonal close-coupling calculations of Flannery and McCann (1974a) (ECC2 and ECC4 in Figs. 15 and 16) also fail to reproduce the experimental angular dependence of the cross section at large scattering angles.

The best agreement between experiment and theory at low impact energies was obtained by J. M. Maidagan and R. D. Rivarola (private communication, 1982), who employed a two-state atomic expansion (TSAE) using static potentials. They do not give any results at energies over 75 keV, but the agreement with the experimental results of Aldag et al. (1981) is very good over the ion impact energy range 25–75 keV. J. M. Maidagan and R. D. Rivarola (private communication, 1982) find that the He(1s, 2p) capture state has negligible influence on the excitation cross section, which is contrary to the experience of Franco (1982) using the Glauber approximation.

J. M. Maidagan and R. D. Rivarola (private communication, 1982) observed a maximum in the He$^+$ + H $\rightarrow$ He$^+(\theta)$ + H($n = 2$) differential cross section at nonzero angles. This is a result of a zero value of the calculated differential cross section for the $2p_{\pm 1}$ excited state at zero scattering angle. They anticipate that this effect would disappear if couplings between 2s, $2p_0$ and $2p_{\pm 1}$ states were included in the calculations, as would be done in a multistate atomic expansion calculation. The experimental differential cross sections of Aldag et al. (1981) do not display nonzero angular maxima.

The problem of excitation in ion–atom collisions is far from being solved. The good agreement between theory and experiment which is observed in total excitation cross sections to a given level is not as generally observed in the cross sections for excitation of substates, particularly optically forbidden substates. The theoretical differential cross sections also provide a less satisfactory comparison with experiment than do the total cross sections. Although the recent theoretical calculations provide improvements in reproducing the experimental cross sections, they seem to be valid only in an increasingly restricted energy range.

There is a need for state-selected differential excitation cross section measurements. Energy-loss techniques will not soon permit the separation of the atomic hydrogen 1s–2s and 1s–2p excitations. Such a differential excitation experiment might be possible using crossed-beam techniques. However, state-selected differential excitation cross sections for H$^+$ + He collisions would only require reasonable improvements in current experimental techniques and such experiments should be completed in the next few years.

V. Electron Capture

The technological implications of electron capture to controlled thermonuclear fusion research, high-power lasers, and energetic neutral beam devices (see Gilbody, 1979, 1981) as well as its importance in helping researchers understand basic ion–atom collision phenomena have led to a great deal of experimental and theoretical research. The theory is discussed in detail by Basu *et al.* (1978), Mapleton (1972), and McCarroll (1980). The theory of ion–atom charge exchange at high impact velocities has recently been reviewed by Belkic *et al.* (1979).

A. Electron Capture by Protons from Atomic Hydrogen

Electron capture experiments involving atomic hydrogen targets are very difficult. The lack of knowledge of the absolute hydrogen target density requires that the experimental results be normalized to some other cross section measurement or theory. The general agreement between the experiments is very good when the differences in technique and normalization are considered.

In their early electron capture measurements, Fite *et al.* (1958, 1960, 1962) employed modulated crossed-beam techniques. They used a fast proton beam to intersect a slow modulated atomic hydrogen beam formed by dissociated hydrogen emitted from a furnace. Their results were obtained from a study of the slow ion and electron currents generated in the beam-crossing region. Their measurements covered impact energies up to 40 keV. The cross section was normalized to the cross section for slow-ion production in $H^+ + H_2$ collisions (Stier and Barnett, 1956).

Gilbody and Ryding (1966) also used a modulated crossed-beam technique to study electron capture in proton–atomic hydrogen collisions in the energy range 40–130 keV. The cross section for electron capture becomes too small in this energy range to use the yield of slow ions as a means of accurate measurement, so Gilbody and Ryding (1966) based their results on postcollision charge state analysis of the fast beam. They also normalized their results using the Stier and Barnett (1956) electron capture cross section for $H^+ + H_2$ collisions.

McClure (1966) used a heated target chamber to dissociate the hydrogen. The collision between the ion beam and the atomic hydrogen took place inside the target chamber. The charge state of the ion beam exiting the target chamber was analyzed to determine the electron capture cross section. The target hydrogen atom density was determined from measurements of the angular scattering of the incident protons compared with theoretical values for Coulomb scattering (Lockwood and Everhart, 1962; Helbig and Everhart, 1964; Lockwood *et al.*, 1964). They used the double-capture dissociative process $H^+ + H_2 \rightarrow H^- + 2H^+$ to determine the relative density of molecular hydrogen in the target as a function of gas flow and target temperature.

Wittkower *et al.* (1966) extended the energy range of the electron capture measurements for protons from atomic hydrogen to 250 keV. They did not attempt to measure the electron capture cross sections directly but rather measured the ratio of the electron capture cross sections for atomic hydrogen targets to that for molecular-hydrogen targets over the energy range 50–250 keV. Their results were normalized to the ratio for the atomic-to-molecular electron capture cross sections obtained by McClure (1966) and by Gilbody and Ryding (1966) at 110 keV, an energy at which these two measurements are in fairly good agreement. To obtain electron capture cross sections, the resulting ratios of Wittkower *et al.* (1966) were multiplied by the Stier and Barnett (1956) cross sections for electron capture from molecular hydrogen.

All of these experiments are for electron capture by incident protons from atomic hydrogen and are for electron capture into all bound states $[H^+ + H \rightarrow H(\Sigma nl) + H^+]$. None of them had any method for determining the final state of the fast atomic-hydrogen atom following the capture process. The data, therefore, represents a statistical average of the cross sections for electron capture into all possible final bound states.

Figure 17 displays total cross sections for electron capture into all bound states $[H^+ + H \rightarrow H(\Sigma nl) + H^+]$ as a function of proton impact energy. The data from the experiments described above are in good agreement, but differences of a factor of two between the various experiments can be found at several energies. Such differences are not unreasonable considering the difficulty inherent in any experiment in which an atomic hydrogen target is used. It should be noted, however, that the various experimental results are not entirely independent because of the common normalization used in some of the experiments.

The availability of state-selected experimental results for electron capture to the 2s and 2p states permits a more rigorous test of the theory than does the total cross section for electron capture to all states. In particular, the cross section for capture of the electron into the 2s state of the projectile is quite sensitive to the approximation applied. Experimental results are available from the work of Ryding *et al.* (1966), of Bayfield (1969b), of Morgan *et al.* (1973, 1980), of Chong and Fite (1977), and of Hill *et al.* (1979a).

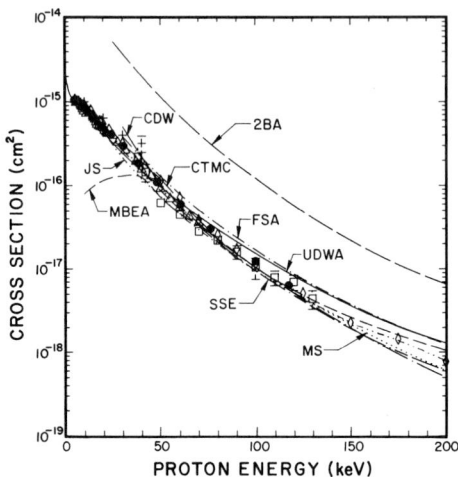

FIG. 17. Electron capture into all final bound states by protons from atomic hydrogen [$H^+ + H(1s) \rightarrow H(\Sigma nl) + H^+$]. Experimental data: ●, McClure (1966); +, Fite et al. (1960); □, Gilbody and Ryding (1966); △, Bayfield (1969b); ◇, Wittkower et al. (1966). Theoretical data: ——, 2BA, Wadehra et al. (1981); ----, FSA, Roy and Ghosh (1979); ——, UDWA, Ryufuku (1982); ·····, MS, Shakeshaft (1978); -·—·, SSE, Shakeshaft (1976); - - -, MBEA, Tan and Lee (1981a,b); -····-, JS, Jackson and Schiff (1953); —·—·-, CTMC, Olson and Salop (1977); —·—·, CDW, Belkic and Gayet (1977a,b).

The earliest measurements were those by Ryding et al. (1966). They measured the relative cross section for the 2s capture process $H^+ + H \rightarrow H(2s) + H^+$. The charge-exchanging collision took place inside a tungsten furnace target chamber which was Joule-heated to 2600 K to provide a dissociated hydrogen target. The cross section was determined from the intensity of the Lyman-α radiation emitted by the fast H(2s) atoms which emerged from the tungsten tube furnace target. The H(2s) atoms were quenched by an electric field to induce the Lyman-α radiation from the H(2p)–H(1s) transition. Because their measurements are relative, the data shown in Fig. 18 have been normalized to the later measurements of Morgan et al. (1980) at 60 keV.

Bayfield (1969a–c) also used a transmission target for making his measurements. He heated a tungsten target cell inside a vacuum furnace to temperatures of about 2400°C. Metastable H(2s) atoms were detected by the Lyman-α radiation which followed Stark-effect quenching of the fast metastable H(2s) atoms. The H(2s) electron capture cross section was normalized to the cross section for electron capture by protons from an argon gas target.

Morgan et al. (1973) employed a modulated crossed-beam technique. They used the Doppler shift of the Lyman-α radiation emitted to separate the radiation from the quenched metastable 2s atoms in the beam from radiation emitted by the target. Morgan et al. (1973) normalized their measurements to the 2p electron capture cross section in $H^+ + H_2$ collisions. The crossed-beam experimental design employed by Morgan et al. (1973) made it possible to measure the formation of the hydrogen 2s and 2p states which resulted from either excitation or electron capture.

Chong and Fite (1977) also employed a modulated crossed-beam technique. To separate the Lyman-α radiation emitted from the quenched metastable in the beam from those emitted from the target, they measured the change in the detected photon signal as a function of the quenched field voltage. Because the metastable H(2s) atoms produced through direct excitation of target hydrogen atoms traveled at thermal velocities, they quenched at lower quenching fields than did the fast metastable H(2s) produced by electron capture. The electron capture cross sections were normalized to the total cross section for H(2p) production as determined by Kondow et al. (1974).

Hill et al. (1979a,b) employed a tungsten tube furnace target. They determined the flux of the metastable H(2s) atoms in the beam emerging from the furnace from the intensity of the Lyman-α radiation emitted by the fast beam as it passed through an electric quenching field. The measurements were normalized using the cross sections for 2s capture in $H^+ + Ar$ collisions (Andreev et al., 1966).

Morgan et al. (1980) also utilized a tungsten tube furnance target containing thermally dissociated hydrogen. To detect the fast H(2s) atoms formed by the electron capture collisions, they used electric field quenching and Lyman-α photon counting techniques. They normalized their data to the average of the measurements of Bayfield (1969b), of Morgan et al. (1973), and of Chong and Fite (1977) at 24.5 keV.

The crossed-beam experiments emphasized the low-energy end of the energy range discussed in this article. As a result, only the data of Bayfield (1969b), Ryding et al. (1966), and Morgan et al. (1980) extend beyond 26 keV (see Fig. 18).

The main difficulty in all of these experiments is the calibration of the Lyman-α detector and the determination of the atomic hydrogen density in the target. All of the above experimenters resorted to some normalization procedure, and part of the explanation for the differences between the experimental results lies in the differences in the normalization techniques. The angular distribution of the fast H(2s) is broader than the angular distribution of the fast H atom beam, which is primarily in the ground state. The recent experiments of Hill et al. (1979a) and of Morgan et al. (1980) employ a wide angular acceptance and are therefore expected to be the most accurate.

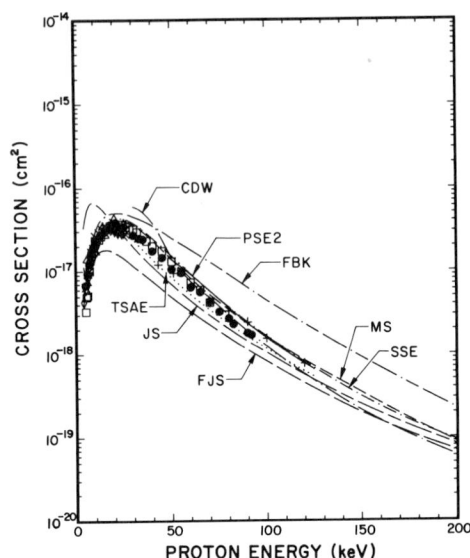

FIG. 18. Cross section for electron capture from atomic hydrogen to the 2s state [H$^+$ + H(1s) → H*(2s) + H$^+$] by protons. Experimental data: +, Ryding et al. (1966); □, Bayfield (1969b); △, Chong and Fite (1977); ●, Morgan et al. (1980); ×, Morgan et al. (1973); ◇, Hill et al. (1979a). Theoretical results: ——, PSE2, Morrison and Öpik (1979); ----, MS, Shakeshaft (1978); —·—·, CDW, Belkic and Gayet (1977a,b); ····, SSE, Shakeshaft (1976); ·····, TSAE, Band (1973b); —··—··-, JS, Jackson and Schiff (1953) taken from Sil et al. (1975); -—·, FBK, Sil et al. (1975); ———, FJS, Sil et al. (1975).

There are many theoretical studies of electron capture by protons from atomic hydrogen targets. The H$^+$ + H resonant electron capture problem has proven to be a much more difficult theoretical problem than the corresponding problem in excitation. The literature contains extended debates on the effects of including or omitting the internuclear potential, the necessity of the orthogonality of the wave functions, and the inability of first-order theories to obtain the correct asymptotic behavior at high energies.

In their early Born approximation results, Oppenheimer (1928) and Brinkman and Kramers (1930) (OBK) omitted the internuclear potential in their calculations. The resulting total electron capture cross sections were over a factor of four higher than the experimental results at some energies. In their calculations, Jackson and Schiff (1953) and Bates and Dalgarno (1953) retained the internuclear interaction in a first Born calculation. Because the Born approximation cannot be summed easily over all possible final states, Mapleton (1962) calculated electron capture for higher states using Born approximation techniques and obtained a total cross section for electron capture. [See also Hiskes (1965) for results on capture to higher states of atomic hydrogen in the OBK approximation.] The Jackson and Schiff (1953) Born approximation results for electron capture into all states are shown in Fig. 17. The results obtained by using variations of the first Born approximation are surprisingly good for proton impact energies greater than 25 keV. It is interesting that this model yields results that fit the experimental data so well. Whereas the Jackson–Schiff calculation is in reasonably good agreement with the experimental results for the total capture cross section, the differential cross sections obtained using the Jackson–Schiff (1953) approximation (Glembocki and Halpern, 1977) demonstrate a dramatic nonphysical minimum which is clearly not observed in the experimental results (Martin et al., 1981b).

Figure 18, which shows the experimental cross section as a function of impact energy for $H^+ + H(1s) \rightarrow H(2s) + H^+$, also shows the Jackson and Schiff (1953) cross sections as given by Sil et al. (1975). This calculation does not reproduce the energy dependence of the cross section at low impact energies. The excellent agreement between experiment and theory in the case of the $H^+ + H \rightarrow H(\Sigma nl) + H^+$ cross section is not observed for electron capture to the 2s state.

In the distorted wave Born approximation (DWBA) the interaction is modified by introducing distorting potentials. Bassel and Gerjuoy (1960) calculated the cross section for $H^+ + H \rightarrow H(1s) + H^+$ using the DWBA and, after correcting for electron capture into higher states, obtained fair agreement with experiment. The DWBA electron capture cross sections for $H^+ + H \rightarrow H(\Sigma nl) + H^+$ are generally higher than the experimental data. Grant and Shapiro (1965) and Geltman (1971) also employed variations of the distorted wave approximation. The DWBA results of both Grant and Shapiro (1965) and of Bassel and Gerjuoy (1960) tend to go to the OBK result at high impact energies.

Ryufuku (1982; see also Ryufuku and Watanabe, 1978) has used a unitarized distorted wave approximation (UDWA) to calculate a wide range of ion–atom cross sections including electron capture. The UDWA results shown in Fig. 17 are in good agreement with the low-energy experimental but tend to be higher than the experimental results at high energies.

Many theoretical studies of ion–atom collisions use only target-centered basis sets. Unless some alternate method of taking electron capture into account is introduced into the theory, these calculations may not only fail to provide electron capture cross sections but also may fail to describe adequately the process under study at impact energies for which the electron capture cross section is large. Bates (1958) proposed a two-center atomic expansion technique which takes into account the relative translational velocity of the projectile and the target. His theory was applied to the proton–atomic hydrogen electron capture process by McCarroll (1961). It has since been applied by others; for example, see the recent work of Lin (1978), Lin and Soong (1978), and Band (1973a–c), whose theories are variations of the two-center atomic expansion technique. The two-state expansions cannot be expected to be satisfactory where couplings with higher states and with the continuum are important.

When the electron capture process is not dominated by only a few states, multistate two-centered calculations are suggested. This is especially important at high energies at which coupling with continuum states is important (McCarroll, 1980). Coupled atomic-state calculations for proton–atomic hydrogen electron capture based on a two-center expansion have been carried out by a number of theorists. Wilets and Gallaher (1966), Rapp and Dinwiddle (1972), and Corigall and Wallace (1971) employed 1s/2s/2p close-coupling calculations which provide electron capture to the 2s and 2p states as well as to the ground 1s state. A pseudostate expansion, which was intended both to incorporate the effects of the continuum and to adequately represent the 3s state of hydrogen, was employed by Cheshire et al. (1970). Cheshire et al. (1970) explicitly included the 1s, 2s, and 2p hydrogenic states and introduced orthogonal pseudostates to represent the continuum. The cross section results of Cheshire demonstrate an energy dependence in the low energy range that is different from that of the experimental results.

Shakeshaft (1978) employed 35 scaled hydrogenic functions centered about each proton. He determined the scaling factors so that the resulting eigenvalues would almost coincide with the 1s, 2s, 2p, 3s, 3p, and 3d energies, and would overlap the rest of the discrete spectrum and the low-energy part of the continuous spectrum. The electron capture cross sections [$H^+ + H(1s) \rightarrow H(1s, 2s, 2p, 3s, 3p, 3d) + H^+$] is shown in Fig. 17 and the electron capture cross section to the H(2s) state is shown in Fig. 18. The results are in very good agreement with the experimental measurements.

In another approach to coupled multistate calculations for $H^+ + H$ electron capture, Gallaher and Wilets (1968) replaced the atomic wave functions with a basis set composed of traveling Sturmian functions centered around each proton. The Sturmian functions, which form a complete set without a continuum, are not equivalent to hydrogenic states, and some hydrogenic states are poorly represented by Sturmian functions. To solve this problem for $H^+ + H$ electron capture, Shakeshaft (1976) employed a large set of Sturmian basis functions centered about each proton, and constructed orthogonal linear combinations of the Sturmian basis functions to approximate hydrogenic states.

By using the above technique, Shakeshaft (1976) has calculated the cross sections for $H^+ + H$ electron capture for 12 Sturmian states centered around each proton. Shakeshaft's results for the cross section for electron capture [$H^+ + H \rightarrow H(1s, 2s, 2p_0, 2p_1) + H^+$] are shown in Fig. 17. The cross section for electron capture into the 2s state, $H^+ + H \rightarrow H(2s) + H^+$, is shown in Fig. 18. In both cases the agreement is very good.

Dewangan (1977a,b) has recently applied an eikonal approximation to electron capture in proton–atomic hydrogen collisions. He obtained cross sections for electron capture to the 1s, 2s, and 2p states of hydrogen. The theory of Dewangan is very similar to the fixed scattering approximation (FSA) of Ghosh (1977) and of Roy and Ghosh (1979) (see curve FSA in Fig. 17).

Morrison and Öpik (1978, 1979) applied a technique that involves a large basis set and the application of a first-order perturbation to the transition amplitudes. Their results for $H^+ + H \rightarrow H(2s) + H^+$ are only valid for impact velocities greater than 1 a.u. and they display only the restricted impact energy range from 49 to 125 keV. Their results for electron capture, which are shown in Fig. 18 (curve PSE2), are slightly higher than those of Shakeshaft (1978) over this energy range.

Electron capture appears to be a multiple-step process. It appears to be necessary to go to higher-order approximations to achieve satisfactory results at very high impact energies (see Shakeshaft and Spruch, 1979; Belkic et al., 1979). A calculation at least comparable to the second Born approximation is required to achieve the correct energy dependence of the electron capture cross sections at asymptotically high energies. Belkic et al. (1979) defined second-order calculations as those which simultaneously allow for some contribution from intermediate inelastic channels and which in the high-energy limit have a term with the same energy dependence as the second Born approximation (see Shakeshaft and Spruch, 1979). Although most of the investigators who have recently published articles on electron capture consider the effects of intermediate channels, most of their results have the same energy dependence at high energies as the first Born approximation. The second Born approximation, the continuum distorted wave (CDW) approximation, the impulse approximation, and the continuum intermediate state (CIS) approximation meet the requirements for second-order calculations suggested by Belkic et al. (1979).

The impulse approximation has been applied to electron capture of protons from hydrogen atoms by Cheshire (1963) and others (Pradhan, 1957; McDowell, 1961; Coleman and McDowell, 1965; Coleman and Trelease, 1971). The impulse approximation is appropriate when the collision time is short compared to the orbital period of the bound electron. The impulse approximation is thus a high-energy approximation and the results are markedly lower than the experimental results for proton impact energies less than 100 keV.

Recently Wadehra et al. (1981) calculated the cross section for $H^+ + H(1s) \rightarrow H(1s) + H^+$ in the second Born approximation. The results of this calculation are a factor of 10 or more higher than the experimental measurements (see curve 2BA in Fig. 17). Although the electron capture cross section of the second Born approximation has the correct asymptotic energy dependence, it does not provide reasonable agreement with the experimental data in the intermediate energy range.

A number of authors have applied the CDW approximation to the proton–atomic hydrogen resonant electron capture problem (Bassel and Gerjuoy, 1960; Cheshire, 1964; Bates, 1958; McCarroll, 1961; Shakeshaft, 1973, 1974; Belkic and Gayet, 1977a,b). The CDW technique is discussed in detail in Belkic et al. (1979). The results of Belkic and Gayet (1977a,b) are shown in Figs. 17 and 18. The CDW approximation can be shown to have the correct asymptotic behavior, however, the experimental data in the intermediate energy range are in better agreement with the multistate calculations which do not have the correct asymptotic energy dependence.

Belkic (1977, 1979) introduced the CIS approximation which treats one channel of the scattering problem in the same manner as in the CDW approximation, but introduces a different distorting potential in the other channel. The cross section for $H^+ + H$ electron capture using the CIS approximation is slightly higher than the CDW approximation results and has an almost identical energy dependence.

Sil et al. (1975) employed a Faddeev approach which appears to be of higher than second order because in the Faddeev method the integral equation for the three-body problem is solved numerically. They made this calculation with and without the internuclear interaction. Their results (shown in Fig. 18) bracket the experimental data. The results with the internuclear interaction included (FJS) are lower than the data and those in which the internuclear interaction was left out (FBK) are higher.

Tripathy and Rao (1978) employed a nonperturbative approach in which the wave function corresponding to the outgoing scattered wave in the incident channel is evaluated by solving the Schrödinger equation it satisfies. Their approach can introduce both the effects of the Coulomb distortion and the polarization of the target atom. The resulting approximation is similar to a distorted wave calculation (see Geltman, 1971). Their results, which are obtained by leaving out the polarization part of the potential, are in good agreement with the experimental cross section for $H^+ + H \rightarrow H(\Sigma nl) + H^+$ at intermediate energies, but appear to differ in energy dependence at energies greater than 80 keV.

A number of classical scattering models have also been applied to electron capture in proton–atomic hydrogen collisions. A modification of the binary encounter approach of Thomas (1927) employing the Gryzinski (1965) model has recently been studied by Roy and Rai (1979), by Kumar and Roy (1979), and by Tan and Lee (1981a,b). The results of Tan and Lee (1981a,b) are shown in Fig. 17 (curve MBEA). The agreement with experiment is quite good between 40 and 200 keV.

The classical trajectory Monte Carlo (CTMC) method produces results which are in very good agreement with the experimental data. The results of Olson and Salop (1977) are shown in Fig. 17. The total electron capture cross section obtained by Olson and Salop (1977) for proton–atomic hydrogen collisions are in agreement with those of Abrines and Percival (1966a,b) and of Banks et al. (1976). Molecular processes become important at collision velocities which are close to the orbital velocity of the electron in the hydrogen target. Hence, at velocities less than 1 a.u., the CTMC method is not expected to provide good agreement with experiment. At the highest collision energies the cross section for electron capture becomes very small, and as a result the statistical errors in the CTMC calculation increase.

Eichenauer et al. (1981) employed the use of cutoff Wigner functions for the initial distribution of the coordinates and momenta of the electron instead of the microcanonical phase distribution. This CTMC calculation also provides good agreement with the experimental results for $H^+ + H \rightarrow H(\Sigma nl) + H^+$.

Very recently, experimentally determined differential cross sections for proton–atomic hydrogen electron capture have become available (Martin et al., 1981b). The differential cross sections cover center-of-mass scattering angles of 0–3 mrad. The experiment employs the same basic apparatus and techniques developed by Park (1979) for inelastic scattering measurements on atomic hydrogen targets. A collimated beam of protons from a specially designed accelerator collide with atomic hydrogen in a tubular coaxial tungsten furnace. The beam of mixed ions and fast neutral hydrogen atoms exiting the target furnace is separated by an analyzing magnet. A fast neutral detector with appropriate angle-defining slits is mounted on the zero-degree port of the analyzing magnet. The angular resolution of the apparatus is 120 μrad in the laboratory system. Because the detection efficiency of the fast neutral particle detector is not accurately known, the results are normalized to the total cross section for electron capture into all bound states (Barnett et al., 1977).

The differential cross section for the process $H^+ + H(1s) \rightarrow H(\theta, \Sigma nl) + H^+$ is shown in Figs. 19–21. Impact parameter approximation calculations in a two-state two-center atomic expansion method (TSAE), a CDW approximation, and a multistate (MS) two-center coupled-state approximation are shown. Also shown are an optical eikonal approximation calculation (OE) and two CTMC calculations.

None of the theoretical calculations provide completely satisfactory

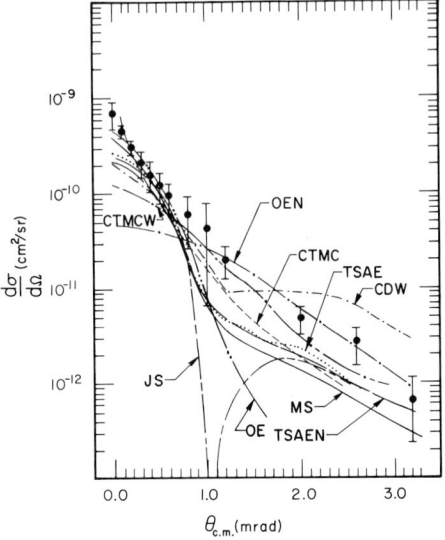

FIG. 19. Angular differential cross section for electron capture by 25-keV protons from atomic hydrogen [H⁺ + H → H(θ) + H⁺]. Experimental data: ●, Martin et al. (1981b). Theoretical results: — · — ·, JS, Jackson and Schiff (1953), Glembocki and Halpern, (1977); ·····, TSAE, Lin and Soong (1978), C. D. Lin (private communication, 1981); — —, TSAEN, J. M. Maidagan, J. M. Piacentini, and R. D. Rivarola, private communication (1982); ····, CDW, Belkic et al. (1979); ——, MS, Shakeshaft (1978); — ·· —, OE, Ho et al. (1982); — · —·, OEN, Ho et al. (1982); ----, CTMC, R. E. Olson (private communication, 1982); - —·, CTMCW, Eichenauer et al. (1981).

agreement with the experimental results of Martin et al. (1981b). The shape of the MS calculation of Shakeshaft (1978) is in reasonably good agreement with the shape of the experimentally derived differential cross section for electron capture. The experimental results of Martin et al. (1981b) are for capture into all states of the outgoing hydrogen atom, whereas the differential cross sections reported in the MS approximation are for electron capture to the ground state only. However, using the total cross section results obtained by Shakeshaft (1978) for all calculated states to provide a multiplicative factor does not bring the theoretical and experimental results into confluence.

The comparatively simple TSAE calculation (Lin and Soong, 1978; C. D. Lin, private communication, 1981) is in surprisingly good agreement with both the more sophisticated MS calculation and the experimental results of Martin et al. (1981b). A similar two-state atomic expansion which includes a variable nuclear charge (TSAEN) is shown at 25 keV (J. M. Maidagan, R. D.

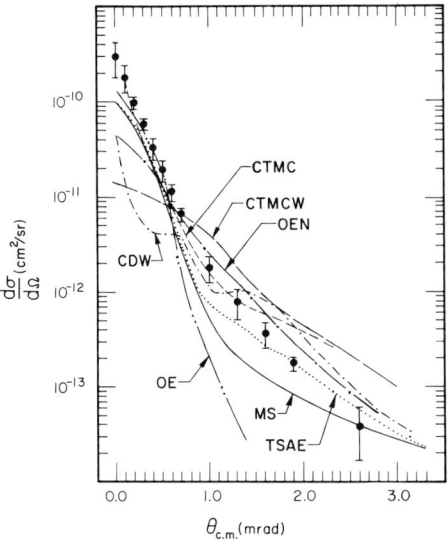

FIG. 20. Angular differential cross section for electron capture by 60-keV protons from atomic hydrogen [H$^+$ + H → H(θ) + H$^+$]. Experimental data: ●, Martin et al. (1981b). Theoretical results: ·····, TSAE, Lin and Soong (1978), C. D. Lin (private communication, 1981); ----, CDW, Belkic et al. (1979); ———, MS, Shakeshaft (1978); —··—, OE, Ho et al. (1982); —·—·, OEN, Ho et al. (1982); -----, CTMC, R. E. Olson (private communication, 1982); — —, CTMCW, Eichenauer et al. (1981).

Piacentini, and R. D. Rivarola, private communication, 1982). The difference between the TSAE and TSAEN calculations is at most 18%.

The CDW calculation (see Belkic et al., 1979) overestimates the experimental cross section at scattering angles greater than 1.0 mrad for all but the highest energy reported. At small scattering angles the CDW results underestimate the experimental results more than the results of either the TSAE or MS calculation. The most serious deficiency in the CDW calculation is the introduction of structure in the differential cross section curve. This structure, which is very evident (see Fig. 20) in the CDW differential cross section for electron capture by 60-keV protons, is not observed in either the experiment or the MS or TSAE calculations. The CDW approximation does give good total electron capture cross section results (see Figs. 17 and 18).

The optical eikonal approximation results (OE) of Ho et al. (1982) for electron capture are also shown in Figs. 19–21. The OE agreement with the experiment at ion velocities which are large relative to the velocity of the orbital electron is attributed to the fact that the OE theory includes the interaction of both the projectile nucleus and the target nucleus.

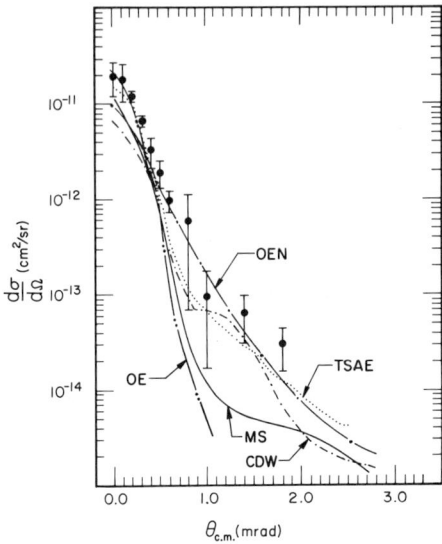

FIG. 21. Angular differential cross section for electron capture by 125-keV protons from atomic hydrogen [$H^+ + H \rightarrow H(\theta) + H^+$]. Experimental data: ●, Martin et al. (1981b). Theoretical results: ·····, TSAE, Lin and Soong (1978), C. D. Lin (private communication, 1981); ----, CDW, Belkic et al. (1979); ——, MS, Shakeshaft (1978); —··—, OE, Ho et al. (1982); —·—·, OEN, Ho et al. (1982).

The interaction with the projectile nucleus is only included to first order, but the interaction with the target nucleus is included to higher order.

Ho et al. (1982) calculated the differential electron capture cross section both with and without the internuclear interaction. The differential cross section for proton–atomic hydrogen electron capture into all states which includes the internuclear interaction is labeled OEN, whereas the calculated differential cross section which does not include the internuclear interaction is labeled OE. At the larger scattering angles where the internuclear interaction is expected to be very important, the agreement between the OEN theory including the internuclear interaction and the data of Martin et al. (1981b) is fairly good. At forward angles the results of the OEN theory which includes the internuclear interaction are significantly below the experimental results. The magnitude of the discrepancy in the forward angles decreases with increasing impact energy. The OE theory curve of Ho et al. (1982), which does not include the internuclear interaction, provides excellent agreement with the experimental data for forward-angle electron capture; however, as expected, it fails to approximate the differential electron capture cross section at larger angles. Both the OE and OEN differential cross sections give the same value for the total electron capture cross section.

At the small scattering angles which correspond roughly to large impact parameters, the projectile "sees" a screened target, and the calculations, which ignore the internuclear interaction, give excellent agreement with the experiment. At the small impact parameters which correspond to the larger angular scattering, the projectile "sees" a bare target nucleus, and the internuclear interaction becomes critical in the calculation. Although a more sophisticated approach will be required to provide good agreement over the entire angular range, the results of Ho *et al.* (1982) clearly demonstrate that a satisfactory theory must incorporate both the nuclear and electron interactions.

The OBK and Coulomb–Brinkman–Kramers (CBK) approximation calculations are not shown in Figs. 19–21. These calculations neglect the interaction between the captured electron and the target nucleus and are consistently too large. They do not, however, demonstrate the nonphysical resonance structure displayed by the Jackson–Schiff calculations (see Fig. 19, curve JS).

Chen and Hahn (1975) calculated differential cross sections for proton–atomic hydrogen collisions by using a distorted coupling Born approximation in a reduced matrix equation formulation. They also presented calculations in which they used the theory of Bassel and Gerjuoy (1960). The Bassel and Gerjuoy (1960) results are too low and too sharply peaked in the forward angles. The results of Chen and Hahn (1975), although less sharply peaked than the Bassel and Gerjuoy (1960) results, are much more sharply peaked than the data of Martin *et al.* (1981b).

Classical trajectory Monte Carlo calculations (CTMC) for differential electron capture have been completed by R. E. Olson (private communication, 1982) and by Eichenauer *et al.* (1981). Olson used the microcanonical distribution for the 1s electron in the Coulomb field of the proton, whereas Eichenauer *et al.* (1982) employed both microcanonical and Wigner functions. The CTMC calculations are for capture into all bound states and therefore provide an unambiguous comparison with the experimental results of Martin *et al.* (1981b). The agreement is generally good. Olson's CTMC results provide a better fit to the data in the forward angles. The differences between the CTMC calculations at the larger scattering angles are probably less than the statistical uncertainty in the calculations.

B. Electron Capture by Protons from Helium

Many of the theoretical treatments which have been explored for proton–atomic hydrogen electron capture, have also been tried for systems with more than one electron, in particular for proton–helium atom colli-

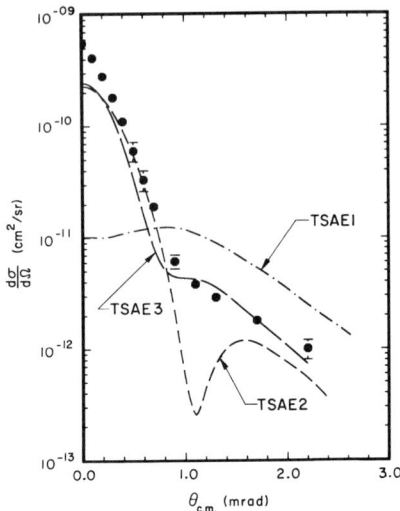

FIG. 22. Differential cross section for electron capture by 30-keV protons from helium atoms [$H^+ + He \rightarrow H(\theta) + He^+$]. Experimental data: ●, Martin et al. (1981a). Theoretical results from Martin et al. (1981a): -·-·-, TSAE1, TSAE calculation using a Coulomb potential; ----, TSAE2, TSAE calculation using a static potential; ———, TSAE3, TSAE calculation using the potential from Rivarola et al. (1980).

sions. Because H^+ + He electron capture is not resonant, the energy dependence of the total cross section is quite different than in the proton–atomic hydrogen case. The total cross sections for electron capture by protons from helium were reviewed by Basu et al. (1978) and by Belkic et al. (1979).

The recent differential cross section measurements of Martin et al. (1981a) are shown in Figs. 22 and 23. The experiment studied the process $H^+ + He \rightarrow H(\theta, \Sigma nl) + He^+(\Sigma nl)$ in the impact energy range 25–100 keV for center-of-mass scattering angles 0–2 mrad. The only other differential cross section measurement is the single measurement at 293 keV by Bratton et al. (1977).

Bratton et al. (1977) directed a highly collimated beam of protons into a gas cell containing helium gas. The fast excited hydrogen atoms resulting from electron capture were detected by a surface barrier detector which could be scanned in both directions perpendicular to the beam axis. Transmitted protons were electrostatically deflected and detected. The measurements were normalized by measuring the ratio of the differential cross sections for electron capture to a measurement of total scattering performed by removing the electrostatic deflection of the transmitted protons. The cross section for total scattering was then set equal to theoretical calculations for classical scattering from a screened Coulomb potential. The integrated

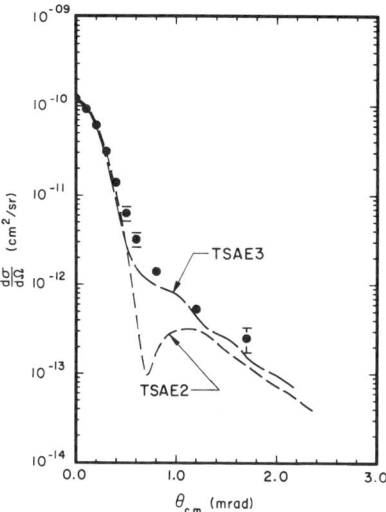

FIG. 23. Differential cross section for electron capture by 100-keV protons from helium atoms [$H^+ + He \rightarrow H(\theta) + He^+$]. Experimental results: ●, Martin et al. (1981a). Theoretical results from Martin et al. (1981a): ----, TSAE2, TSAE calculation using a static potential; — —, TSAE3, TSAE calculation using the potential from Rivarola et al. (1980).

value of the cross section for electron capture obtained in this manner is about 17% higher than the total cross section reported by Barnett and Reynolds (1958). The angular resolution was determined by scanning the detector across the unscattered beam.

The high-energy CDW approximation of Belkic and Salin (1978) provides very good agreement with the experimental results of Bratton et al. (1977) at 293 keV if the Coulomb internuclear repulsion is included. The Belkic and Salin (1978) results for $H^+ + He$ electron capture are only for the process $H^+ + He \rightarrow H(1s) + He^+(1s)$ whereas the experiment of Bratton et al. (1977) is for electron capture to any discrete state. Belkic et al. (1979) estimated that the contribution of capture into excited states is less than 20%.

The results of Rogers and McGuire (1977) using a static Brinkman–Kramers approximation are considerably higher than the data of Bratton et al. (1977). Belkic and Salin (1978) attribute this discrepancy to the inadequacy of first-order theories. However, the TSAE approximation of Lin and Soong (1978), which is also a first-order theory, is in very good agreement with the experimental data. The good agreement of the TSAE approximation of Lin and Soong (1978) is also observed at lower energies.

The differential cross sections of Martin et al. (1981a) for electron capture from helium by 25- to 100-keV incident protons were obtained by a technique very similar to that discussed earlier with respect to the measurement of differential cross sections for electron capture by protons from atomic hydrogen. Because the absolute efficiency of the neutral detectors was not known to sufficient accuracy, Martin et al. (1981a) normalized the differential cross sections for proton–helium electron capture by equating the integral of the differential cross sections to the average of experimental results for the cross section for electron capture from helium into all bound states of atomic hydrogen reported by Barnett et al. (1977).

Martin et al. (1981a) tested the TSAE results obtained using several different internuclear potentials. Using the TSAE approximation, they found that the internuclear potentials which gave satisfactory agreement with the high-energy H^+ + He electron capture cross section results were not appropriate at lower impact energies. In particular, the Coulomb potential applied by Belkic and Salin (1978) in their CBK approximation and the static potential employed by Rogers and McGuire (1977) in their static Brinkman–Kramers approximation were found to give results which did not agree with the experimental results at impact energies of 100 keV or less. The TSAE approximation results for H^+ + He → $H(1s)$ + $He^+(1s)$ (taken from Martin et al., 1981a) shown in Figs. 22 and 23, which use the internuclear potential reported by Rivarola et al. (1980), provide good agreement with the experimental results. Rivarola et al. (1980) derived the appropriate internuclear potential from the many-electron Hamiltonian within the independent electron approximation. The theoretical and experimental results of Martin et al. (1981a) indicate that the TSAE method can provide reasonably good differential cross section values at low impact energies if a realistic potential is used. This conclusion is consistent with the theoretical findings of Lin and Tunnell (1980). It should be noted that no significant differences between the TSAE results, for which the different potentials were used, were observed in their comparison with the experiment of Bratton et al. (1977) for H^+ + He differential electron capture at 293 keV. The TSAE results at lower energies, however, are clearly dependent on the correct choice of potential.

The TSAE theoretical results reproduce the dominant feature of the experimentally determined differential cross section, which is a distinct change in slope at 0.8 mrad. This feature is observed in both the H^+ + He differential cross section results of Martin et al. (1981a) and in the results of Bratton et al. (1977) at 293 keV.

The TSAE results are only for electron capture to the ground state, whereas the experimental results are for electron capture to all bound states. The inclusion of the higher bound states is not expected to greatly affect the agreement between the TSAE theory and experiment.

The large number of theoretical calculations of electron capture, particularly in the case of the proton–atomic hydrogen collision, prevented displaying all of them in the figures. Only a few of the most recent calculations were displayed. This fact, however, hides the degree of progress which has been made. As was true with excitation, the agreement between theory and experiment decreases as the process becomes more specific. The differences between the various theoretical efforts for electron capture to the 2s state are markedly larger than for total electron capture cross sections. The differential cross sections also displayed large differences.

More detailed experiments which can provide thorough tests of theory are possible. In particular, a measurement of proton-impact differential electron capture from atomic hydrogen targets to the 2s state should be possible using existing techniques.

The heated debate concerning the necessity of including the internuclear potential is answered by comparison with experimental differential electron capture cross sections. To obtain correct results for large angle scattering, the internuclear potential is required. However, because the largest contribution to the total cross section involves large impact parameter–small scattering angle collisions, the omission of the internuclear potential may not seriously affect the total electron capture cross section value.

It is also noted that a theory with the current asymptotic behavior does not necessarily provide good electron capture cross sections in the intermediate energy range. The theories which provided the best agreement with experimental electron capture cross sections also tended to have a restricted energy range of validity. Presumably as the theoretical techniques improve, the energy range of validity will also increase.

VI. Ionization

A. Total Cross Sections for Ionization of Atomic Hydrogen

There are only a few experimental studies of ionization of atomic hydrogen by protons. The experiments are very difficult. The usual difficulties resulting from the handling of atomic hydrogen are further complicated by the difficulties in collecting and detecting slow ions and electrons. The available data for proton–atomic hydrogen total ionization cross sections are shown in Fig. 24. The early measurements of Fite et al. (1960; see also Fite et al., 1958) employed a modulated crossed-beam technique. A fast proton beam collided at right angles with a modulated atomic hydrogen

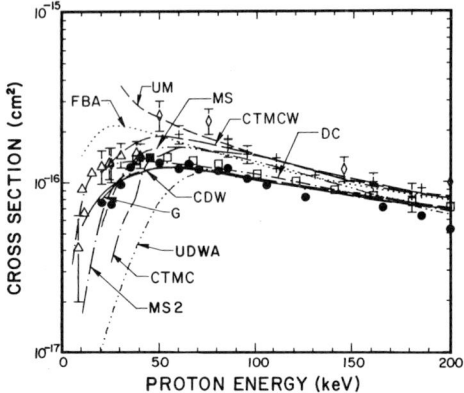

FIG. 24. Ionization cross section for protons incident on atomic hydrogen [$H^+ + H(1s) \to H^+ + H^+ + e$]. Experimental data: □, Shah and Gilbody (1981); ◊, Park et al. (1977); △, Fite et al. (1960); +, Gilbody and Ireland (1963); ●, J. T. Park (private communication, 1980). Theoretical results: ----, UM, Fitchard et al. (1977); — - — -, G, McGuire (1982); ·····, FBA, Bates and Griffing (1953); — ·· —, CTMC, Olson and Salop (1977); -·····, UDWA, Ryufuku (1982); ——, CDW, Belkic (1978); — ·· —, MS, Shakeshaft (1978); — · —, MS2, Garibotti and Miraglia (1982); -···-, DC, Janev and Presnyakov (1980). — —, CTMCW, Eichenauer et al. (1981).

beam from a tungsten tube source. The slow ions formed in the collision were detected by either a magnetic-sector mass spectrometer or a set of collection plates located above and below the interaction region. By using the phase and frequency of the detected signal, they were able to distinguish the slow ions and electrons resulting from collisions with the atomic hydrogen beam from the slow ions resulting from collisions with the background gas.

The mass spectrometer allowed Fite et al. (1960) to separate signals from the slow protons coming from the collisional ionization of atomic hydrogen from those coming from other ions. The mass spectrometer, however, could not distinguish between protons formed by ionization ($H^+ + H \to H^+ + H^+ + e$) and those produced by electron capture ($H^+ + H \to H + H^+$). The parallel-plate detector ("condenser method") enabled them to distinguish between ionization and electron capture because, in principle, the collection of positive particles yields the sum of cross sections for ionization and electron capture, whereas the collection of the negative electrons yields only the cross section for ionization.

Fite et al. (1960) normalized their cross section for ionization to the cross section for electron capture by protons from H_2 measured by Stier and Barnett (1956).

Measurements which used a very similar technique were also made by

Gilbody and Ireland (1963). These measurements were normalized to the cross section for ionization of H_2 by protons.

Park et al. (1977) measured differential energy-loss cross sections for ionization of atomic hydrogen by protons. They also reported the total ionization cross section obtained by integrating their differential cross section results. As a result, the total cross sections include the large errors inherent in a differential measurement as well as possible errors in the integration itself. Because the differential cross sections of Park et al. (1977) were not measured for energy losses greater than 50 eV, estimates of the differential cross sections for larger energy losses were used to complete the integrals.

Unpublished total ionization cross sections obtained at a larger number of impact energies and with higher accuracy were measured by J. T. Park (private communication, 1980). The same technique was applied but improved statistics were obtained. Longer energy-loss sweeps eliminated the need to make cross section extrapolations to obtain the total cross sections. The resulting measurements for which energy-loss techniques were used by J. T. Park (private communication, 1980) are in excellent agreement with the recent crossed-beam measurements of Shah and Gilbody (1981).

The most recent and accurate measurements by Shah and Gilbody (1981) was described in Section II. The experiment was carefully done and significantly more accurate and precise than earlier measurements. Shah and Gilbody (1981) normalized their results to the Born approximation (Bates and Griffing, 1953) at proton impact energies greater than 1 MeV.

The experimental measurements are in general accord, but the results of Park et al. (1977) and of Gilbody and Ireland (1963) appear to be too high between 50 and 100 keV. The 25-keV measurements of Park et al. (1977) are in good agreement with the measurements of Fite et al. (1960), and the high-energy measurement is in reasonable agreement with the accurate data of Shah and Gilbody (1981).

Figure 24 also shows various theoretical results. The high-energy data are normalized to the first Born approximation (Bates and Griffing, 1953), therefore, agreement at high energy is not significant. However, the first Born cross section is significantly higher than the data of Shah and Gilbody (1981) at 50 keV and is above all of the other data at 25 keV.

The Glauber approximation results of Golden and McGuire (1975) and of McGuire (1982) are in good agreement with experiment for impact energies greater than 100 keV. At lower impact energies the Glauber approximation falls below the experimental results.

The Born distorted wave calculation of Salin (1969) includes the effect of both the projectile and target protons. The long-range interaction between the ejected electron and the scattered proton enhances the cross section when the velocities of the scattered proton and the ionized electron are equal. The results of Salin (1969) are higher than the best experimental results except at very high energies. Overestimation of the cross section at low energies is inherent in a first Born-type approximation calculation and differences in theory and experiment are not necessarily due to the long-range interaction.

Classical trajectory Monte Carlo techniques have been applied to proton–atomic hydrogen collisions by a number of theorists (Abrines and Percival, 1966a,b; Banks et al., 1976; Eichenauer et al., 1981; Olson and Salop, 1977). The various CTMC theoretical results are in general agreement, although significant differences exist between the various derived cross sections. The CTMC results of Olson and Salop (1977) are shown in Fig. 24. The agreement is good for proton impact energies greater than 50 keV, but the CTMC results fall below the low-energy experimental results of Fite et al. (1960). The recent CTMC results of Eichenauer et al. (1981), which employ a cutoff Wigner distribution instead of a microcanonical ensemble to represent the ground state of the hydrogen atoms, are also shown in Fig. 24 (CTMCW). The results of Eichenauer et al. (1981) are in good agreement with experiment over the entire energy range.

The CDW approximation of Belkic (1978) is primarily valid at high impact energies. However, the CDW results are in good agreement with the experimental results of Shah and Gilbody (1981) over the entire energy range.

The proton–atomic hydrogen ionization cross sections obtained by Ryufuku (1982) in the UDWA are significantly below the experimental results in the low-energy region, but are in good agreement at impact energies greater than 75 keV.

Coupled-channel calculations do not provide ionization cross sections directly. If the basis set used is complete enough, then the ionization cross section can be identified with the difference between the total loss from the entrance channel and the cross sections for elastic scattering, excitation, and electron capture. This technique has been applied using basis sets comprised of Sturmian functions, pseudostate expansions, and scaled hydrogenic functions. With very large basis sets this technique can produce accurate cross sections.

Fitchard et al. (1977) applied this technique to obtain ionization cross sections with their pseudostate expansion calculation. This calculation is expected to be very accurate. However, the use of a single target-centered basis set restricts the validity of the calculations to energies where electron capture is small. The results of Fitchard et al. (1977) are generally higher than the experimental results of Shah and Gilbody (1981) in the low-impact-energy region (see curve UM in Fig. 24).

Shakeshaft (1976, 1978) specifically includes the effects of electron capture into continuum states (CTTC) in his calculations (see Sellin, 1982, for a review of CTTC). His two-centered calculation divides rather naturally into direct contributions to the ionization from eigenfunctions centered around the target nucleus and the probability for electron capture to the continuum obtained by summing over eigenfunctions centered around the projectile nucleus. Shakeshaft (1978) found that the cross section for CTTC is larger than the cross section for direct ionization for impact energies less than 60 keV. The MS results shown in Fig. 24 include both direct and CTTC ionization. The MS calculation for proton–atomic hydrogen ionization of Shakeshaft (1978), who used 35 scaled hydrogenic states centered on each proton, is generally in good agreement with the experimental results.

Garibotti and Miraglia (1982) have applied a first-order multiscattering theory to ionization of atomic hydrogen by protons. Their theory treats the electron motion symmetrically relative to the target and the projectile. The theory, therefore, incorporates CTTC into the resulting cross sections. The theoretical results are in good agreement with the experimental data.

The best overall agreement between the experiment of Shah and Gilbody (1981) and theory appears to be provided by the calculation of Janev and Presnyakov (1980), who used a dipole close-coupling approach. They employed a dipole approximation for the ion–atom interaction and an atomic state close-coupling method. They also used an effective oscillator strength to describe the coupling of the discrete states with the continuum. Their theory, which includes transitions into the continuum through the 2p state but neglects contributions from other intermediate states, is shown in Fig. 24. The agreement with the experimental results of Shah and Gilbody (1981) is excellent.

B. Energy-Loss Differential Cross Sections

Park et al. (1977) report ionization cross sections for $H^+ + H$ collisions that are differential in energy loss (see Fig. 25). The cross sections are indirectly normalized to the Born approximation for excitation of atomic hydrogen to its $n = 2$ level at 200 keV. Normalization of the differential cross sections to the accurate total ionization cross sections of Shah and Gilbody (1981) would lower the experimental results significantly.

Park et al. (1977) compared their experimental differential ionization cross sections with their first Born approximation and Glauber calculations. The comparison in magnitude between theory and experiment is not particularly good for either approximation; however, renormalization of the experimental results would improve the comparison. The binary encounter

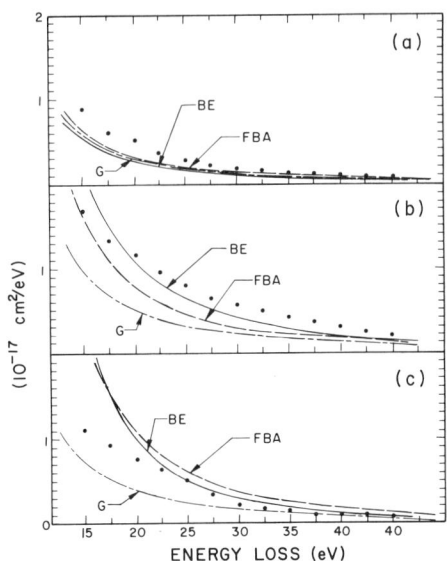

FIG. 25. Differential energy-loss cross sections for ionization of atomic hydrogen by 200-keV (a), 50-keV (b), and 25-keV (c) protons. Experimental data: ●, Park et al. (1977). Theoretical results: ———, BE, Garcia (1969); — —, FBA, Park et al. (1977); — · —, G, Park et al. (1977).

theory of Garcia (1969) is also shown in Fig. 25. Except at 50 keV, the binary encounter and Born results are very similar. At 200 keV, all of the theoretical results are nearly identical. Park et al. (1977) attributed the differences between the theory and experiment to the effects of electron capture to the continuum, which is not adequately described in these first Born approximations and Glauber calculations. However, the only observable effect of CTTC is displayed in the curvature on the 25-keV energy spectrum in the energy-loss region between 15 and 25 eV. The curvature in the cross section for 25-keV differential energy loss is noticeably different than that in the cross sections for higher-energy differential energy loss. The theoretical curves reproduce the shape of the experimental data at 200 keV quite well.

A great deal of work on ionization of helium is available because of the relative simplicity of the experiments. Little of this work is very recent and it will not be discussed here. The subject of ionization of hydrogen targets by heavy ions has recently been reviewed by Gilbody (1981) and will not be repeated here.

The total ionization cross section results display a great deal more spread in both theoretical and experimental values than is observed in total excitation cross sections. The experiments are very difficult, but the recent measurement of Shah and Gilbody (1981) has markedly decreased the uncertainty in the experimental value of the ionization cross section.

Theoretical calculations employing the Born approximation are satisfactory at high impact velocities, but at lower velocities the calculations are very difficult. It appears that heroic efforts are required to provide a sufficient number of accurate basis states to obtain good cross sections for ionization using coupled-channel techniques.

The work on differential cross sections for ionization of atomic hydrogen has just begun. The accuracy of the differential cross sections can be greatly improved and doubly differential cross section measurements (differential in proton energy loss and proton scattering angle) are feasible. However, the problems involved in simultaneously measuring the ejection angle of the electron in proton–atomic hydrogen collisions will not be easily solved.

VII. Elastic Scattering

Conceptually, the simplest collisional process is elastic scattering. The projectile is scattered by the target with no changes in the state of either the target atom or the incident ion. The simplicity of the concept, however, does not result in a simple experiment. The necessity to distinguish elastically scattered ions from both the inelastically scattered ions and the unscattered ions requires high resolution in both energy and angle. Although such measurements are relatively commonplace for very low-energy ions, in the intermediate energy range the requirement for high-energy resolution had until recently restricted efforts to measure total scattering cross sections. Recently Peacher *et al.* (1982) have reported measurements of differential elastic scattering for protons from helium (see Figs. 26–28). The measurements were obtained by using high-resolution energy-loss spectrometry. Differential elastic cross sections were reported for angles between 0.5 and 3 mrad with an angular resolution of 0.12 mrad. The data have a dynamic range covering over three orders of magnitude. Even with high angular resolution the investigators were unable to obtain differential elastic cross sections at scattering angles less than 0.5 mrad because of interference from the unscattered ion beam. Larger-scattering-angle measurements were prevented by a low count rate which rapidly decreased with increasing scattering angle.

In spite of the simplicity of pure elastic scattering, there are few published theoretical calculations in this energy range. Most of the researchers who have published their results, used classical approaches, which are expected to be invalid for small angle scattering. Flannery and McCann (1974b) have

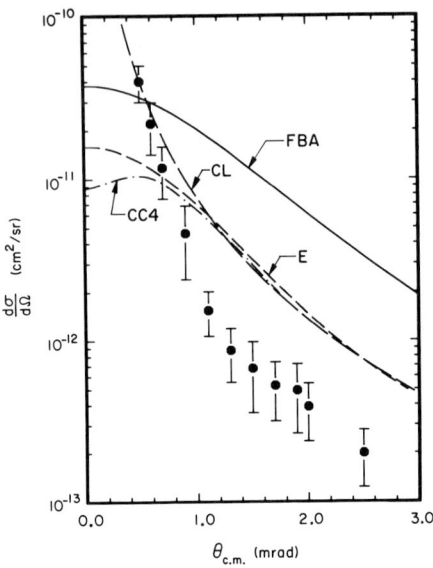

FIG. 26. Elastic scattering differential cross section for 25-keV protons from helium. Experimental data: ●, Peacher et al. (1982). Theoretical results: ———, FBA, Peacher et al. (1982); — —, CL, Peacher et al. (1982); -----, E, Peacher et al. (1982); -·-·, CC4, Flannery and McCann (1974b).

published differential elastic cross sections as part of a four-state close-coupling calculation. Because the elastic channel is usually included in a coupled-state calculation, other coupled-state calculations may have been made but not reported.

Peacher et al. (1982) included their calculations of the differential elastic cross sections using the Born (FBA), eikonal (E), and classical (CL) approximations. They employed an analytical fit to the static potential (Cox and Bonham, 1967). The classical approximation followed the procedures used by Everhart et al. (1955). The data and theoretical results for elastic scattering of protons from helium atoms are shown in Figs. 26–28. The experimentally determined differential elastic cross sections are significantly below all of the theoretical values. With the exception of the first Born approximation results, the theories are in good agreement among themselves at the larger scattering angles.

Peacher et al. (1982) speculated that the source of the discrepancy between the theoretical and experimental results is the failure of the theoretical treatments to adequately account for all of the various scattering channels. They note that the same theories give good agreement at larger angles with experiments which measured total differential scattering (Fitzwilson and

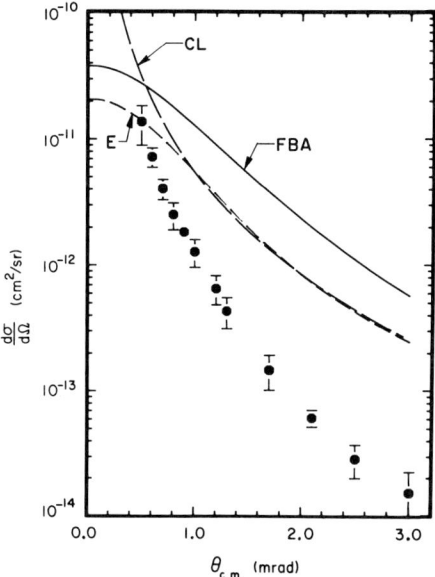

FIG. 27. Elastic scattering differential cross section for 50-keV protons from helium. Experimental data: ●, Peacher et al. (1982). Theoretical results: ——, FBA, Peacher et al. (1982); — —, CL, Peacher et al. (1982); -----, E, Peacher et al. (1982);

Thomas, 1972; Crandall et al., 1973). They also note that the differential elastic cross section for 100-keV protons scattered from helium is smaller than either the electron capture or the ionization cross sections, neither of which was included in the four-state calculations of Flannery and McCann (1974b). Peacher et al. (1982) used their measurements of differential elastic cross sections with earlier measurements of inelastic scattering (Park et al., 1978b) to provide an estimate of the total differential scattering cross section. The estimated total differential scattering cross section is in very good agreement with the classical scattering and, by implication, with the eikonal and the four-state results reported by Flannery and McCann (1974b).

The possibility that the results of the available theoretical calculations really yield total differential scattering cross sections rather than true elastic differential scattering cross sections is quite reasonable, because the static potential used to represent the target does not take into account the effects of other open channels. In the case of the four-state calculation of Flannery and McCann (1974b), only the effect of the $n = 2$ level of helium on the elastic channel is included. Because the $n = 2$ excitation cross section accounts for only about 5% of the total scattering cross section, its effect is small.

Elastic scattering of protons by atomic hydrogen is an even more difficult

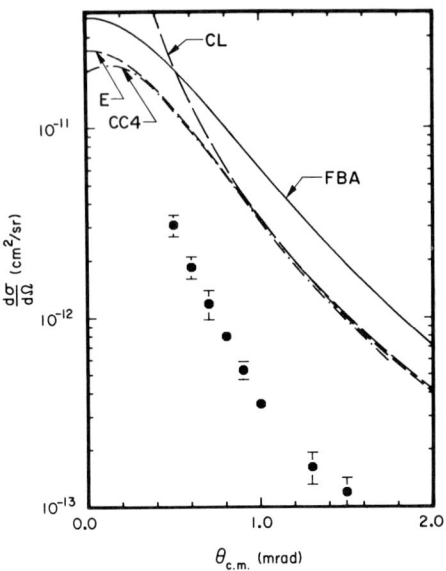

FIG. 28. Elastic scattering differential cross section for 100-keV protons from helium. Experimental data: ●, Peacher et al. (1982). Theoretical results: ———, FBA, Peacher et al. (1982); — —, CL, Peacher et al. (1982); -----, E, Peacher et al. (1982). ----, CC4, Flannery and McCann (1974b).

experimental problem than proton–helium elastic scattering. The only published results are the preliminary results reported by Park (1982). These results are shown in Fig. 29 along with various theoretical calculations. The figure shows the same type of discrepancy between the Born, eikonal, and classical approximation results and the experimental measurements that were observed for proton–helium elastic scattering. In this case, however, a sophisticated multistate calculation is available. Park (1982) reported results obtained by applying a Bessel function transform–Fraunhoffer integral technique, with elastic amplitudes obtained by Shakeshaft (1978). Wadehra and Shakeshaft (1982) later did a more extended calculation for proton–atomic hydrogen collisions using the same elastic amplitudes. This calculation employed a linear combination of 70 basis functions (35 centered around each proton) formed from scaled hydrogenic functions. This large basis set appears to cover the available scattering channels adequately, as is shown in Fig. 29. The results are in good agreement with the experimental results (Park, 1982).

Preliminary results of R. E. Olson (private communication, 1982) for a CTMC calculation were also reported by Park (1982). The CTMC calculation includes all of the major inelastic channels as well as the elastic

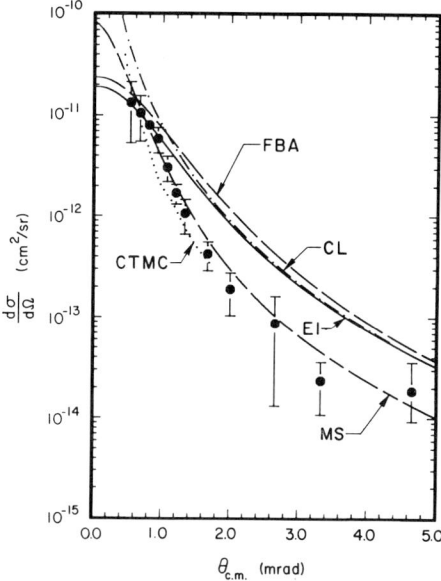

FIG. 29. Differential elastic scattering cross section for 60-keV protons from atomic hydrogen [$H^+ + H(1s) \rightarrow H^+(\theta) + H(1s)$]. Experimental data: ●, Park (1982). Theoretical results: — —, FBA, Park (1982); ----, CL, Park (1982); ———, EI, Park (1982); — - — -, MS, Wadehra and Shakeshaft (1982); ·····, CTMC, R. E. Olson (private communication, 1982).

scattering channel. The CTMC method unambiguously determines the elastic scattering cross section. It shows very good agreement in magnitude, but has a slightly different curve shape than the experimental results.

The comparison of experiment and theory for elastic scattering emphasizes the difficulties faced by the theorists in ion–atom collisions. The cross sections for the possible scattering channels are comparable in magnitude. There is no guide which tells the theorist which states can be safely neglected. Sophisticated calculations, such as those of Shakeshaft (1978) and of R. E. Olson (private communication, 1982), which include or compensate for all of the available scattering channels, appear to give generally good agreement with experiments.

VIII. Conclusions

Both the theoretical efforts and the experiments represent a compromise between the ideal and the possible. Theorists would rather not resort to approximations, which they know will inevitably limit the validity of their calculation in some manner. The best of approximations has only a limited impact energy range where it is valid. The experimentalist would like to deal with infinite resolution, absolute detectors, precise target densities of certain composition, and no hint of systematic errors. These ideal situations are not possible, but the available comparisons between experiment and theory for simple collision systems provide some sound information along with a lot of noise.

Any theory that aspires to reproduce the cross section for a collision process over a wide impact energy range must either include or account for all the alternative processes. The effect of neglecting alternative channels in calculating the elastic scattering cross section, as was demonstrated by Peacher et al. (1982), is an example of how strongly the neglected channels can affect theoretical calculations. The most satisfactory calculations seem to be those which include everything; however, calculations such as the 70-state multistate calculation of Shakeshaft (1978) still display defects in spite of the impressive effort and the complexity of the calculation. When such an effort is required for the simplest case of protons colliding with atomic hydrogen, the effort required to incorporate all of the possible processes for collisions between high-Z projectiles and targets seems overwhelming.

Frequently, arguments of theoretical purity, such as the insistence that an electron capture cross section theory display the proper asymptotic behavior, seem almost irrelevant at intermediate energies; however, a completely satisfying theory will ultimately provide both correct cross sections at intermediate energies and the correct asymptotic behavior.

The need for a more detailed study of the collision process is emphasized throughout this article. State-selected cross sections provide more information than cross sections representing the sum of states. Differential cross sections in turn provide a great deal more information than total cross sections. Several examples have been given in which theoretical total cross sections provided quite satisfactory results, but in which the differential cross sections calculated using the same approximation displayed the wrong angular dependence. In general, because the differential cross sections $d\sigma/d\Omega$ are so sharply peaked in the forward direction, the theoretical total cross sections agree with experimental cross sections if the differential cross section has the right value at the angle θ, where $(\sin \theta) d\sigma/d\Omega$ is a maximum. The differential cross sections can be very wrong at larger scattering angles and still yield the correct value for the total cross section. An example of this behavior is given by the Jackson–Schiff Born calculation of $H^+ + H$ electron capture, which is in good agreement with the experimental total cross section measurements but yields a differential cross section that displays dramatic structure not experimentally observed (Martin *et al.*, 1981b).

A great deal of experimental work remains to be done in this area. Just as the substate total cross sections show marked differences, state-selected differential cross section measurements should show definite state-dependent angular effects. Such measurements for the most basic collision systems are needed in order to provide even more detailed tests of theory. Proton–atomic hydrogen differential cross sections for ionization need to be explored in more detail.

The ultimate experiment in which the energy, momentum, and quantum states of all of the particles before and after the collision are known is technically possible. Such an experiment will not be easy for atomic hydrogen targets, but would probably be possible now if helium targets were used. This type of experiment would provide the ultimate test of theory because all of the uncertainties and averages over final states will have been removed.

The level of recent theoretical and experimental efforts is very high, as can be noted from the fact that most of the data and theory displayed in the figures was published in the 7 years preceding the writing of this article. If the coming years produce the same level of effort, many of the questions which still exist in regard to the interaction of simple ion–atom systems will soon be answered.

ACKNOWLEDGMENTS

The assistance of Mr. Tom Kvale in preparing the figures for this article, of Dr. Jerry Peacher, Dr. Eduard Rille, Mr. Dean Blankenship, Mr. Emmett Redd, and Mr. John Koenig for proofreading, and of Mrs. Ellen Kindle in typing the manuscript is sincerely appreciated. The support of the National Science Foundation for work first reported in this article is gratefully acknowledged.

REFERENCES

Abrines, R., and Percival, I. C. (1966a). *Proc. Phys. Soc., London* **88**, 861.
Abrines, R., and Percival, I. C. (1966b). *Proc. Phys. Soc., London* **88**, 873.
Aldag, J. E., Peacher, J. L., Martin, P. J., Sutcliffe, V., George, J., Redd, E., Kvale, T. J., Blankenship, D. M., and Park, J. T. (1981). *Phys. Rev. A* [3] **23**, 1062.
Andreev, E. P., Ankudinov, V. A., and Bobashev, S. V. (1966). *Sov. Phys.—JETP (Engl. Transl.)* **23**, 375.
Band, Y. B. (1973a). *Phys. Rev. A* [3] **8**, 243.

Band, Y. B. (1973b). *Phys. Rev. A* [3] **8**, 2857.
Band, Y. B. (1973c). *Phys. Rev. A* [3] **8**, 2866.
Bandra, K., and Ghosh, A. A. (1971). *Phys. Rev. Lett.* **26**, 737.
Banks, D., Barnes, K. S., and Wilson, J. McB. (1976). *J. Phys. B* [1] **9**, L141.
Barnett, C. F., and Reynolds, H. K. (1958). *Phys. Rev.* **109**, 355.
Barnett, C. F., Ray, J. A., Ricci, E., Wilker, M. I., McDaniel, E. W., Thomas, E. W., and Gilbody, H. B. (1977). "Atomic Data for Controlled Fusion Research," Rep. No. ORNL-5206 (unpublished).
Bassel, R. H., and Gerjuoy, E. (1960). *Phys. Rev.* **117**, 749.
Basu, D., Mukherjee, S. C., and Sural, D. P. (1978). *Phys. Rep.* **42C**, 145.
Bates, D. R. (1958). *Proc. R. Soc. London, Ser. A* **247**, 294.
Bates, D. R. (1959). *Proc. Phys. Soc., London* **73**, 227.
Bates, D. R. (1961). *Proc. Phys. Soc., London* **77**, 59.
Bates, D. R., and Dalgarno, A. (1953). *Proc. R. Soc. London* **67**, 972.
Bates, D. R., and Griffing, G. (1953). *Proc. Phys. Soc., London, Sect. A* **66**, 961.
Bayfield, J. E. (1969a). *Phys. Rev.* [2] **182**, 115.
Bayfield, J. E. (1969b). *Phys. Rev.* [2] **185**, 105.
Bayfield, J. E. (1969c). *Rev. Sci. Instrum.* **40**, 896.
Belkic, Dz. (1977). *J. Phys. B* [1] **10**, 3491.
Belkic, Dz. (1978). *J. Phys. B* [1] **11**, 3529.
Belkic, Dz. (1979). *J. Phys. B* [1] **12**, 337.
Belkic, Dz., and Gayet, R. (1977a). *J. Phys. B* [1] **10**, 1911.
Belkic, Dz., and Gayet, R. (1977b). *J. Phys. B* [1] **10**, 1923.
Belkic, Dz., and Salin, A. (1978). *J. Phys. B* [1] **11**, 3905.
Belkic, Dz., Gayet, R., and Salin, A. (1979). *Phys. Rep.* **56**, 279.
Bell, K. L., and Kingston, A. E. (1978). *J. Phys. B* [1] **11**, 1259.
Birely, J. H., and McNeal, R. J. (1972). *Phys. Rev. A* [3] **5**, 692.
Bransden, B. H. (1970). "Atomic Collision Theory." Benjamin, New York.
Bransden, B. H. (1979). *Adv. At. Mol. Phys.* **15**, 263.
Bransden, B. H., and Dewangan, D. P. (1979). *J. Phys. B* [1] **12**, 1377.
Bransden, B. H., and Noble, C. J. (1979). *Phys. Lett.* **70A**, 404.
Bransden, B. H., Dewangan, D. P., and Noble, C. J. (1979). *J. Phys. B* [1] **12**, 3563.
Bratton, T. R., Cocke, C. L., and Macdonald, J. R. (1977). *J. Phys. B* [1] **10**, L517.
Brinkman, H. C., and Kramers, H. A. (1930). *Proc. Acad. Sci. Amsterdam* **33**, 973.
Chen, A. C., and Hahn, Y. (1975). *Phys. Rev. A* [3] **12**, 823.
Cheshire, I. M. (1963). *Proc. Phys. Soc., London* **82**, 113.
Cheshire, I. M. (1964). *Proc. Phys. Soc., London* **84**, 89.
Cheshire, I. M., Gallaher, D. F., and Taylor, A. J. (1970). *J. Phys. B* [1] **3**, 813.
Chong, Y. P., and Fite, W. L. (1977). *Phys. Rev. A* [3] **16**, 933.
Coleman, J. P. (1969). *Case Stud. At. Collision Phys.* **11**, 101.
Coleman, J. P., and McDowell, M. R. C. (1965). *Proc. Phys. Soc., London* **85**, 1097.
Coleman, J. P., and Trelease, S. (1971). *J. Phys. B* [1] **4**, L121.
Corigall, B., and Wallace, R. (1971). *J. Phys. B* [1] **4**, 1013.
Cox, H. L., Jr., and Bonham, R. A. (1967). *J. Chem. Phys.* **47**, 1967.
Crandall, D. H., McKnight, R. H., and Jaecks, D. H. (1973). *Phys. Rev. A* [3] **7**, 1261.
Crothers, D. S. F., and Holt, A. R. (1966). *Proc. Phys. Soc., London* **88**, 75.
Datta, S., and Mukherjee, S. C. (1981). *Phys. Rev. A* [3] **23**, 1780.
Datta, S., Mandal, C. R., and Mukherjee, S. C. (1980). *J. Phys. B* [1] **13**, 4791.
Dewangan, D. P. (1977a). *J. Phys. B* [1] **10**, 1083.

Dewangan, D. P. (1977b). *Phys. Lett. A* **62A,** 303.
Eichenauer, D., Grün, N., and Scheid, W. (1981). *J. Phys. B* [1] **14,** 3929.
Eichenauer, D., Grün, N., and Scheid, W. (1982). *J. Phys. B* [1] **15,** L17.
Everhart, E., Stone, G., and Carbone, R. J. (1955). *Phys. Rev.* **99,** 1287.
Fitchard, E. O., Ford, A. L., and Reading, J. C. (1977). *Phys. Rev. A* [3]**16,** 1325.
Fite, W. L., Brackman, R. T., and Snow, W. R. (1958). *Phys. Rev.* **112,** 1161.
Fite, W. L., Stebbings, R. F., Hummer, D. G., and Brackman, R. T. (1960). *Phys. Rev.* **119,** 663.
Fite, W. L., Smith, A. C. H., and Stebbings, R. F. (1962). *Proc. R. Soc. London, Ser. A* **268,** 527.
Fitzwilson, R. L., and Thomas, E. W. (1972). *Phys. Rev. A* [3] **6,** 1054.
Flannery, M. R. (1969). *J. Phys. B* [1] **2,** 1044.
Flannery, M. R., and McCann, K. J. (1974a). *J. Phys. B* [1] **7,** 1349.
Flannery, M. R., and McCann, K. J. (1974b). *J. Phys. B* [1] **7,** 1558.
Franco, V. (1979a). *Phys. Rev. Lett.* **42,** 759.
Franco, V. (1979b). *Phys. Rev. A* [3] **20,** 2297.
Franco, V. (1982). *Phys. Rev. A* [3] **25,** 1358.
Franco, V., and Thomas, B. K. (1971). *Phys. Rev. A* [3] **4,** 945.
Gallaher, D. F., and Wilets, L. (1968). *Phys. Rev.* [2] **169,** 139.
Garcia, J. D. (1969). *Phys. Rev.* [2] **177,** 223.
Garibotti, C. R., and Miraglia, J. E. (1982). *Phys. Rev. A* [3] **25,** 1440.
Geltman, S. (1971). *J. Phys. B* [1] **4,** 1288.
Ghosh, A. S. (1977). *Phys. Rev. Lett.* **38,** 1065.
Gilbody, H. B. (1979). *Adv. At. Mol. Phys.* **15,** 293.
Gilbody, H. B. (1981). *Phys. Scr.* **24,** 712.
Gilbody, H. B., and Ireland, J. V. (1963). *Proc. R. Soc. London, Ser. A* **277,** 137.
Gilbody, H. B., and Ryding, G. (1966). *Proc. R. Soc. London, Ser. A* **291,** 438.
Glembocki, O., and Halpern, A. M. (1977). *Phys. Rev. A* [3] **16,** 981.
Golden, J. E., and McGuire, J. H. (1975). *Phys. Rev. A* [3] **12,** 80.
Grant, T. A., and Shapiro, J. (1965). *Proc. Phys. Soc., London* **86,** 1007.
Gryzinski, M. (1965). *Phys. Rev.* [2] **138,** A336.
Helbig, H. F., and Everhart, E. (1964). *Phys. Rev.* [2] **136,** 674.
Hill, J., Geddes, J., and Gilbody, H. B. (1979a). *J. Phys. B* [1] **12,** L341.
Hill, J., Geddes, J., and Gilbody, H. B. (1979b). *J. Phys. B* [1] **12,** 2875.
Hiskes, J. R. (1965). *Phys. Rev. A* [2] **137A,** 361.
Ho, T. S., Eichler, J., Lieber, M., and Chan, F. T. (1982). *Phys. Rev. A* [3] **25,** 1456.
Holt, A. R., and Moiseiwitsch, B. L. (1968). *J. Phys. B* [1] **1,** 36.
Jackson, J. D., and Schiff, H. (1953). *Phys. Rev.* **89,** 359.
Janev, R. K., and Presnyakov, L. P. (1980). *J. Phys. B* [1] **13,** 4233.
Joachin, C. J., and Vanderpoorten, R. (1973). *J. Phys. B* [1] **6,** 662.
Kingston, A. E., Moiseiwitsch, B. L., and Skinner, B. G. (1960). *Proc. R. Soc. London, Ser. A* **258,** 273.
Kondow, T., Girnius, R. J., Chong, Y. P., and Fite, W. L. (1974). *Phys. Rev. A* [3] **10,** 1167.
Kumar, A., and Roy, B. N. (1979). *J. Phys. B* [1] **12,** 2025.
Lin, C. D. (1978). *J. Phys. B* [1] **11,** L185.
Lin, C. D., and Soong, S. C. (1978). *Phys. Rev. A* [3] **18,** 499.
Lin, C. D., and Tunnell, L. N. (1980). *Phys. Rev. A* [3] **22,** 76.
Lockwood, G. J., and Everhart, E. (1962). *Phys. Rev.* [2] **125,** 567.
Lockwood, G. J., Helbig, H. F., and Everhart, E. (1964). *J. Chem. Phys.* **41,** 3820.
McCarroll, R. (1961). *Proc. R. Soc. London, Ser. A* **264,** 547.

McCarroll, R. (1980). *In* "Atomic and Molecular Collision Theory" (F. A. Gianturco, ed.), p. 165. Plenum, New York.
McClure, G. W. (1966). *Phys. Rev.* [2] **148**, 47.
McDowell, M. R. C. (1961). *Proc. R. Soc. London, Ser. A* **264**, 277.
McDowell, M. R. C., and Coleman, J. P. (1970). "Introduction to the Theory of Ion-Atom Collisions." Am. Elsevier, New York.
McGuire, J. H. (1982). *Phys. Rev. A* [3] **26**, 143.
McKee, J. D. A., Sheridan, J. R., Geddes, J., and Gilbody, H. B. (1977). *J. Phys. B* [1] **10**, 1679.
Mapleton, R. A. (1962). *Phys. Rev.* [2] **126**, 1477.
Mapleton, R. A. (1972). "Theory of Charge Exchange." Wiley (Interscience), New York.
Martin, P. J., Arnett, K., Blankenship, D. M., Kvale, T. J., Peacher, J. L., Redd, E., Sutcliffe, V. C., and Park, J. T., Lin, C. D., and McGuire, J. H. (1981a). *Phys. Rev. A* [3] **23**, 2858.
Martin, P. J., Blankenship, D. M., Kvale, T. J., Redd, E., Peacher, J. L., and Park, J. T. (1981b). *Phys. Rev. A* [3] **23**, 3357.
Moiseiwitsch, B. L., and Perrin, R. (1965). *Proc. Phys. Soc., London* **85**, 51.
Morgan, T. J., Geddes, J., and Gilbody, H. B. (1973). *J. Phys. B* [1] **6**, 2118.
Morgan, T. J., Stone, J., and Mayo, R. (1980). *Phys. Rev. A* [3] **22**, 1460.
Morrison, H. G., and Öpik, U. (1978). *J. Phys. B* [1] **11**, 473.
Morrison, H. G., and Öpik, U. (1979). *J. Phys. B* [1] **12**, L685.
Olson, R. E., and Salop, A. (1977). *Phys. Rev. A* [3] **16**, 531.
Oppenheimer, J. R. (1928). *Phys. Rev.* **31**, 349.
Park, J. T. (1978). *In* "Collision Spectroscopy" (R. G. Cooks, ed.), p. 19. Plenum, New York.
Park, J. T. (1979). *IEEE Trans. Nucl. Sci.* **NS-26**, 1011.
Park, J. T. (1982). *In* "Physics of Electronic and Atomic Collisions" (S. Datz, ed.), p. 109. North-Holland Publ., Amsterdam.
Park, J. T., and Schowengerdt, F. D. (1969). *Phys. Rev.* [2] **185**, 152.
Park, J. T., and Aldag, J. E., and George, J. M. (1975). *Phys. Rev. Lett.* **34**, 1253.
Park, J. T., Aldag, J. E., George, J. M., and Peacher, J. L. (1976). *Phys. Rev. A* [3] **14**, 608.
Park, J. T., Aldag, J. E., George, J. M., Peacher, J. L., and McGuire, J. H. (1977). *Phys. Rev. A* [3] **15**, 508.
Park, J. T., Aldag, J. E., Peacher, J. L., and George, J. M. (1978a). *Phys. Rev. Lett.* **40**, 1646.
Park, J. T., George, J. M., Peacher, J. L., and Aldag, J. E. (1978b). *Phys. Rev. A* [3] **18**, 48.
Park, J. T., Aldag, J. E., Peacher, J. L., and George J. M. (1980). *Phys. Rev. A* [3] **21**, 751.
Peacher, J. L., Kvale, T. J., Redd, E., Martin, P. J., Blankenship, D. M., Rille, E., Sutcliffe, V. C., and Park, J. T. (1982). *Phys. Rev. A* [3] **26**, 2476.
Pradhan, T. (1957). *Phys. Rev.* **105**, 1250.
Rapp, D., and Dinwiddie, D., (1972). *J. Chem. Phys.* **57**, 4919.
Rivarola, R. D., Piacentini, R. D., Salin, A., and Belkic, Dz. (1980). *J. Phys. B* [1] **13**, 2601.
Rogers, S. R., and McGuire, J. H. (1977). *J. Phys. B* [1] **10**, L497.
Roy, A., and Ghosh, A. S. (1979). *J. Phys. B* [1] **12**, 99.
Roy, B. N., and Rai, D. K. (1979). *J. Phys. B* [1] **12**, 2015.
Ryding, G., Wittkower, A. B., and Gilbody, H. B. (1966). *Proc. Phys. Soc., London* **89**, 547.
Ryufuku, H. (1982). *Phys. Rev. A* [3] **25**, 720.
Ryufuku, H., and Watanabe, T. (1978). *Phys. Rev. A* [3] **18**, 2005.
Salin, A. (1969). *J. Phys. B* [1] **2**, 631.
Sellin, I. A. (1982). *In* "Physics of Electronic and Atomic Collisions" (S. Datz, ed.), p. 195. North-Holland Publ., Amsterdam.
Shah, M. B., and Gilbody, H. B. (1981). *J. Phys. B* [1] **14**, 2361.
Shakeshaft, R. (1973). *J. Phys. B* [1] **6**, 2315.
Shakeshaft, R. (1974). *J. Phys. B* [1] **7**, 1734.

Shakeshaft, R. (1976). *Phys. Rev. A* [3] **14,** 1626.
Shakeshaft, R. (1978). *Phys. Rev. A* [3] **18,** 1930.
Shakeshaft, R., and Spruch, L. (1979). *Rev. Mod. Phys.* **51,** 369.
Sil, N. C., Chaudhuri, J., and Ghosh, A. S. (1975). *Phys. Rev. A* [3] **12,** 785.
Stebbings, R. F., Young, R. A., Oxley, C. L., and Ernhardt, H. (1965). *Phys. Rev. A* **138A,** 1312.
Stier, P. M., and Barnett, C. F. (1956). *Phys. Rev.* **103,** 896.
Sur, S. K., Datta, S., and Mukherjee, S. C. (1981). *Phys. Rev. A* [3] **24,** 2465.
Tan, C. K., and Lee, A. R. (1981a). *J. Phys. B* [1] **14,** 2399.
Tan, C. K., and Lee, A. R. (1981b). *J. Phys. B* [1] **14,** 2409.
Theodosiou, C. E. (1980). *Phys. Rev. A* [3] **22,** 2556.
Theodosiou, C. E. (1981). *Phys. Lett. A* **83A,** 254.
Thomas, E. W. (1972). "Excitation in Heavy Particle Collisions." Wiley (Interscience), New York.
Thomas, L. H. (1927). *Proc. R. Soc. London, Ser. A* **114,** 561.
Tripathy, D. N., and Rao, B. K. (1978). *Phys. Rev. A* [3] **17,** 587.
Vainstein, L. A., Presnyakov, L. P., and Sobel'man, I. I. (1964). *Sov. Phys.—JETP (Engl. Transl.)* **18,** 1383.
Van den Bos, J., and de Heer, F. J. (1967). *Physica (Amsterdam)* **34,** 333.
Wadehra, J. M., and Shakeshaft, R. (1982). *Phys. Rev. A* [3] **26,** 1771.
Wadehra, J. M., Shakeshaft, R., and Macek, J. H. (1981). *J. Phys. B* [1] **14,** L767.
Wilets, L., and Gallaher, D. F. (1966). *Phys. Rev.* [2] **147,** 13.
Wilets, L., and Wallace, S. J. (1968). *Phys. Rev.* [2] **169,** 84.
Wittkower, A. B., Ryding, G., and Gilbody, H. B. (1966). *Proc. Phys. Soc., London* **89,** 541.
York, G. W., Jr., Park, J. T., Miskinis, J. J., Crandall, D. H., and Pol, V. (1972). *Rev. Sci. Instrum.* **43,** 230.
Young, R. A., Stebbings, R. F., and McGowan, J. W. (1968). *Phys. Rev.* [2] **171,** 85.

HIGH-RESOLUTION SPECTROSCOPY OF STORED IONS

D. J. WINELAND and WAYNE M. ITANO

Time and Frequency Division
National Bureau of Standards
Boulder, Colorado

R. S. VAN DYCK, JR.

Department of Physics
University of Washington
Seattle, Washington

I. Introduction	136
A. Scope of the Present Article	136
B. The Method	136
II. Ion Storage Techniques	137
A. The Paul (rf) Trap	138
B. The Penning Trap	143
C. The Kingdon Trap	145
D. Ion Creation	147
E. Ion Detection	148
F. Polarization Production/Monitoring	148
G. Other Types of Traps	149
III. Lepton Spectroscopy	149
A. Historical Perspective	149
B. Electron Geonium Experiment	151
C. Positron Geonium Experiment	156
IV. Mass Spectroscopy	159
A. Introduction	159
B. First Mainz Experiment	160
C. Second Mainz Experiment	162
D. University of Washington Experiment	164
V. Atomic and Molecular Ion Spectroscopy	166
A. Optical Atomic Ion Spectroscopy	166
B. Laser Cooling	167
C. Microwave and rf Atomic Ion Spectroscopy	171
D. Application to Frequency Standards	175
E. Molecular Ion Spectroscopy	176
VI. Negative Ion Spectroscopy	176
A. Atoms	177

| B. Molecules 179
| VII. Radiative Lifetime Measurements 180
| References 181

I. Introduction

A. Scope of the Present Article

In this article we attempt to give a review of spectroscopic experiments that have employed the stored ion technique. This review will be biased toward very high-resolution experiments since this is the authors' area of interest. We hope that a description of these high-resolution experiments will illustrate the current state of the art and indicate possible achievements in the future. In this article collision experiments in the usual sense (charge exchange, ion–molecule reactions, etc.) will specifically be excluded—we are primarily interested in photon–ion interactions and will include such topics as spectroscopy of atomic ions, mass spectroscopy using resonant excitation, and measurement of radiative decay times. To make this article more tractable, we do not discuss the interesting spectroscopy that has been done in conjunction with fusion plasma studies or high-energy storage rings. We also omit discussion of the interesting studies of the dynamics of water droplets (Owe Berg and Gaukler, 1969) and of aerosols initiated by C. B. Richardson (private communication, 1982).

The most complete previous review has been given by Dehmelt (1967, 1969) and is still "the" reference for someone starting in this field. Schuessler (1979) has given a review which concentrated on the ion storage exchange collision (ISEC) method in an rf quadruple trap. More recently, Dehmelt (1983) has given a review/introduction as part of a NATO Advanced Study Institute. Todd *et al.* (1976) have given a review of rf traps. In addition, various other partial reviews have also been given (Dehmelt, 1975, 1976, 1981a,b; Minogin, 1982; Neuhauser *et al.* 1981; Toschek and Neuhauser, 1981; Werth, 1982; Wineland *et al.*, 1981b; Wineland, 1983).

Necessarily, there will be some overlap of this review with previous ones; we will tend to rely on these other reviews which have extensively covered some of the techniques such as the ISEC method (Dehmelt, 1969; Schuessler, 1979).

B. The Method

Probably the main advantage of the stored ion techniques is that the ideal of an unperturbed species at rest in space is approached to a high degree.

Specifically, charged particles such as electrons and atomic ions can be stored for long periods of time (essentially indefinitely) without the usual perturbations associated with confinement (for example, the perturbations due to collisions with walls or buffer gases in a traditional optical pumping experiment).

Unfortunately (and necessarily), there is a price to be paid for this property of long storage times with small perturbations—the number of particles that can be stored is typically small (approximately 10^6 or less for a "trap" with centimeter dimensions); the resulting low densities are ultimately governed by the competition between space-charge repulsion and the confining electromagnetic forces obtained under normal laboratory conditions. It should be noted that if we could obtain the high trapping fields necessary to obtain high densities, then we would lose one of the advantages of the technique because, for example, Stark shifts due to confinement would cause problems in very high-resolution work.

As a consequence of the low numbers obtained, many types of experiments may be precluded—for example, spectroscopic experiments on complex molecular ions where only a small fraction of the ions are in a given state. However, in spite of the low numbers obtained, sensitive techniques have been developed to detect simple species such as electrons and atomic ions, so that *single* electrons (Wineland *et al.,* 1973; Van Dyck *et al.,* 1978) and atomic ions (Neuhauser *et al.,* 1980; Wineland and Itano, 1981; Nagourney *et al.,* 1982, 1983; Ruster *et al.,* 1983) can be observed.

In Section II, trapping methods are discussed. In Section III, the electron–positron (g-2) experiments are discussed. These experiments may be the least general of those covered, but a large part of the new developments in the stored ion technique have occurred here. In Section IV applications to mass spectroscopy are discussed. Section V discusses experiments on atomic and molecular ion spectroscopy. Sections VI and VII discuss negative ion experiments and lifetime studies.

II. Ion Storage Techniques

Four types of "traps" have been most commonly used for high-resolution work: the Paul (or rf) trap, the Penning trap, the Kingdon (or electrostatic trap), and the magnetostatic trap (magnetic bottle). Magnetic bottles have not been used extensively for high-resolution work because the trapping relies on the use of inhomogeneous magnetic fields, thus causing inhomogeneities and broadening in field-dependent transitions. An exception to this is the high-resolution electron–positron magnetic moment measurements done by Crane, Rich, and their colleagues (Rich and Wesley, 1972). For the sake of brevity however, magnetic confinement devices will not be discussed further.

A. The Paul (rf) Trap

The Paul or rf trap has the advantage that the trapped ions are bound in a (pseudo) potential well in all directions and no magnetic fields are required for confinement. It has the capability in practice to provide tighter confinement than the Penning trap, but the phenomenon of "rf heating" has been a limitation in some experiments.

1. Theory

The "ideal" Paul trap (Fisher, 1959; Wuerker *et al.*, 1959a) uses three electrodes in a vacuum apparatus as shown in Fig. 1. These electrodes are conjugate hyperboloids of revolution about the z axis, thus allowing a description of the potential in cylindrical coordinates. In general, both a static potential and an alternating potential of frequency Ω are applied between the ring and endcaps, so that

$$\phi(r, z) = \frac{U_0 + V_0 \cos \Omega t}{r_0^2 + 2z_0^2} (r^2 - 2z^2) \tag{1}$$

where r_0 and z_0 are defined in Fig. 1. The electrode surfaces are assumed to be equipotentials of Eq. (1), and we assume $\phi = 0$ at the center of the trap.

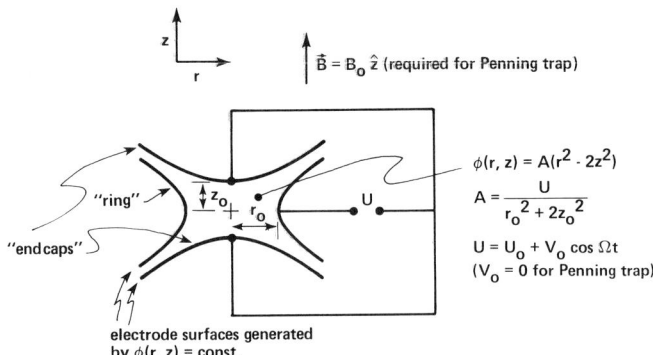

FIG. 1. Schematic representation of the electrode configuration for the "ideal" Paul or Penning trap. Electrode surfaces are figures of revolution about the z axis and are equipotentials of $\phi(r, z) = A(r^2 - 2z^2)$.

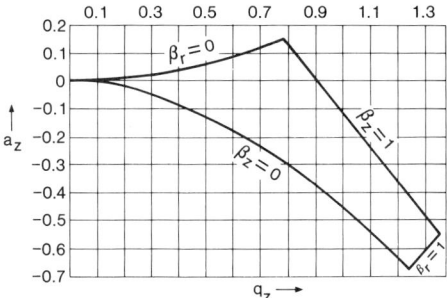

FIG. 2. Theoretical stability diagram for the Paul trap. The region bounded by the solid line represents the values of a_z and q_z giving stable confinement. The validity of this stability diagram has been confirmed in various experiments (see text).

The equations of motion for a positive ion take the form of a Mathieu equation (McLachlan, 1947):

$$\frac{d^2 x_i}{d\tau^2} + (a_i - 2q_i \cos 2\tau)x_i = 0$$

where

$$-a_z = +2a_r = 16qU_0/m\Omega^2(r_0^2 + 2z_0^2)$$
$$+q_z = -2q_r = 8qV_0/m\Omega^2(r_0^2 + 2z_0^2)$$

m is the ion mass, $\tau = \Omega t/2$, and q is the ion charge.

Certain values of a_i and q_i lead to stable bounded motion; these values are bounded by the curve on the stability diagram shown in Fig. 2. Other regions of stability exist (Fisher, 1959; Wuerker et al., 1959a; Dawson, 1980); only the first region of stability, which is shown, is usually used. The perimeter of this curve is determined by the values $\beta_i = 1$ and 0, where $\beta_i^2 = a_i + q_i^2/2$.

A qualitative explanation for the binding due to the rf field in the traps is as follows (Wuerker et al., 1959a; Dehmelt, 1967): For Ω sufficiently large (i.e., $\beta \ll 1$) an ion experiences an rf electric field such that its motion (the "micromotion") is 180° out of phase with respect to the electric force. Because the electric field is inhomogeneous, the force averaged over one period ($T = 2\pi/\Omega$) of the micromotion is in a direction of weaker field amplitude (independent of the sign of the charge), i.e., toward the center of the trap. For Ω sufficiently high, this restoring force gives rise to a pseudopotential (Dehmelt, 1967):

$$\psi(r, z) = \frac{qV_0^2}{m\Omega^2(r_0^2 + 2z_0^2)^2}[(\bar{r})^2 + 4(\bar{z})^2] \qquad (U_0 = 0) \qquad (2)$$

where $\bar{r}$ and $\bar{z}$ are the positions of the ion averaged over T, and q is the ion charge. The resulting "secular" motion (for $\bar{r}$ and $\bar{z}$) is harmonic with frequencies $\bar{\omega}_z = 2\bar{\omega}_r = 2\sqrt{2}qV_0/[\Omega m(r_0^2 + 2z_0^2)]$.

The addition of the static potential U_0 alters the well depth directly so that we must add the potential

$$\phi_{DC} = \frac{U_0}{r_0^2 + 2z_0^2}(r^2 - 2z^2) \tag{3a}$$

to the expression for the pseudopotential. This yields an overall effective (harmonic) potential:

$$\phi(r,z) = \psi + \phi_{DC} = \frac{k_r}{2q}(\bar{r})^2 + \frac{k_z}{2q}(\bar{z})^2$$

$$= \left[\frac{qV_0^2}{m\Omega^2(r_0^2 + 2z_0^2)^2} + \frac{U_0}{r_0^2 + 2z_0^2}\right](\bar{r})^2$$

$$+ \left[\frac{4qV_0^2}{m\Omega^2(r_0^2 + 2z_0^2)^2} - \frac{2U_0}{r_0^2 + 2z_0^2}\right](\bar{z})^2 \tag{3b}$$

A special case is when $U_0 = qV_0^2/m\Omega^2(r_0^2 + 2z_0^2)$, which gives rise to a "spherical" potential where $\phi(r,z) \propto (\bar{r})^2 + (\bar{z})^2$. It is sometimes useful to describe the traps in terms of well depths in the r and z directions; for example, we can define

$$D_r = \phi(r = r_0, 0) - \phi(0, 0)$$

$$D_z = \phi(0, z = z_0) - \phi(0, 0)$$

For the more general potential of Eq. (3b), the secular oscillation frequencies are $\bar{\omega}_r = (k_r/m)^{1/2}$ and $\bar{\omega}_z = (k_z/m)^{1/2}$.

As an example of operating conditions, we take the data from Major and Werth (1978) for ^{199}Hg$^+$ ions. Here $r_0 = \sqrt{2}z_0 = 1.129$ cm, $\Omega/2\pi = 524$ kHz, $U_0 = +8$ V (ring electrode biased positively with respect to the endcaps), and $V_0 = 297$ V. This gives $a_z = -0.024$, $q_z = 0.432$, $\bar{\omega}_r/2\pi = 49$ kHz, $\bar{\omega}_z/2\pi = 69$ kHz. This particular choice of parameters also gives $D_r = D_z = 12$ eV. Very small traps have been used to provide tight confinement for single ion detection. From Neuhauser et al. (1978b), $V_0 \simeq 200$ V and $r_0 = \sqrt{2}z_0 = 354$ μm.

The potential of Eq. (1) is said to be ideal because it has the simplest form which satisfies Laplace's equation and has the desired symmetry (independent of azimuthal angle and reflection symmetry about $z = 0$). For the same symmetry, deviations from the ideal potential introduce higher-order terms; the first of these has spatial dependence $(3r^4 - 24r^2z^2 + 8z^4)$ and introduces anharmonicities in the secular frequencies $\bar{\omega}_z$ and $\bar{\omega}_r$. These deviations can come about due to imperfections in the electrodes such as holes for observation and the fact that the surfaces must be truncated. They are obviously introduced for a trap with simple cylindrical geometry (Benilan and Audoin, 1973; Bonner *et al.*, 1977) or spherical electrodes (O and Schuessler, 1980a), but the region of stability is not substantially altered (Benilan and Audoin, 1973; Mather *et al.*, 1980).

We note that the choice of r_0/z_0 is in principle arbitrary and not inferred from Laplace's equation as has been implied (Dawson and Whetten, 1968; Todd *et al.*, 1976; Bonner *et al.*, 1977). Of course, to yield the ideal potential of Eq. (1), the electrode surfaces must be equipotentials of Eq. (1) and therefore both the ring and endcaps must asymptotically approach the surfaces generated by $z = \pm r/\sqrt{2}$. The choice $r_0^2 = 2z_0^2$ can be made so that the ring and endcap surfaces are equidistant from these asymptotes for large r and z. Knight (1983, also private communication) has also pointed out that the stability diagram is independent of the choice of r_0/z_0 as long as a_i and q_i are determined from the formulas given previously.

The motion of a single ion in the trap is described by the solution of the Mathieu equation. For many ions in the trap, distribution functions can be obtained under various assumptions. Assuming no collisions or space charge, phase space dynamical techniques have been used to calculate distribution functions and give fairly good agreement with observed data (Todd *et al.*, 1980b). For the long-term storage desired in spectroscopic experiments, ion–ion and ion–neutral collisions play an important role. In general, this gives rise to what is called "rf heating"—a process which couples kinetic energy from the micromotion into the secular motion (Church and Demelt, 1969). Elastic collisions with background neutrals will either produce heating if the neutrals are heavy with respect to the ions or an effective cooling (viscous drag) if the neutrals are light (Dehmelt, 1967). This has been examined in detail theoretically (André and Vedel, 1977; Dawson and Lambert, 1975; André *et al.*, 1979; F. Vedel *et al.*, 1981, 1983; M. Vedel *et al.*, 1981), predicting Gaussian density distributions. In addition, resonant charge exchange with the parent neutral may provide a stabilizing influence (Bonner *et al.*, 1976; Dawson, 1980). In many experiments done at high vacuum, it appears that the dominant heating mechanism is due to ion–ion collisions (Dehmelt, 1967). Such heating is not well understood but may arise from the presence of impurity ions or from imperfections in the trap electrodes.

Space charge will have a destabilizing effect on an ion cloud, that is, the wells become more shallow due to the presence of other ions. If the secular motion is "frozen," then the ions form around the center of the trap so that the net electric field an ion experiences inside the cloud is zero. Therefore, following Dehmelt (1967), $\phi + \phi_i = $ constant, where ϕ_i is the potential from the ion cloud. Applying the Laplacian to this equation, we find that the ion cloud has uniform (maximum) density ρ given by

$$\rho_{max} = \frac{1}{4\pi} \nabla^2 \phi = \frac{3qV_0^2}{\pi m \Omega^2 (r_0^2 + 2z_0^2)^2} \quad (4)$$

For the conditions of Major and Werth (1978), this leads to a maximum density of $n_{max} = \rho_{max}/q = 4 \times 10^7$ cm^{-3}.

2. Experiments

The kinematics of ion motion have been verified in the original experiments of Wuerker *et al.* (1959a), where charged aluminum dust particles (ca. 20 μm diameter) were suspended in an rf trap and illuminated so that the paths could be observed. In these beautiful experiments the particles could also be observed to crystallize into lattices when the background gas pressure was increased to cause cooling.

For atomic ions, the validity of the stability diagram has been verified in experiments. Trapping efficiencies have been measured by monitoring the relative intensity of laser-induced fluorescence light from Ba$^+$ ions (Ifflander and Werth, 1977a; Plumelle *et al.*, 1980). Temperature was monitored via the Doppler broadening of the resonance light. Trapping efficiencies have also been measured by observing storage times as a function of trapping conditions (M. Vedel *et al.*, 1981). Temperature also has been measured by the bolometric method (Church and Dehmelt, 1969) discussed below. In addition, the cooling efficiency of buffer gases has also been observed (Vedel, 1976; Dawson, 1980; Plumelle *et al.*, 1980; Schaaf *et al.*, 1981; Ruster *et al.*, 1983). In the experiments of Plumelle *et al.* (1980) using helium buffer gas, trapping times of several days have been observed (Plumelle, 1979; see also Blatt and Werth, 1983). Density measurements have also been made optically (Knight and Prior, 1979; Schaaf *et al.*, 1981), indicating a Gaussian density distribution. The results of these experiments and those of Ifflander and Werth (1977a) and of Church and Dehmelt (1969) showed that the ion temperature was approximately equal to a constant fraction (ca. 0.1) of the well depth. These measurements are also in agreement with the measurement of ion loss based on evaporation (Dehmelt, 1983). Ion–neutral collision studies based on ion creation and loss are given by Todd *et al.* (1976) and can be used to infer ion energies. In the experiments of Knight and Prior (1979) the cloud radius (and therefore the ion energy) was shown to be fairly independent of ion number, indicating that ion–ion rf heating was not important. This was not the case for the very small clouds observed by Neuhauser *et al.* (1980), where a large increase in cloud radius was observed as the ion number increased. In these experiments, however, ion densities were as high as 10^9 cm^{-3}, increasing the importance of ion–ion collisions.

The destabilizing effect of space charge has been observed by Todd *et al.* (1980a) and Neuhauser *et al.* (1978b). However, the densities predicted from the model where the secular motion is cold have not been observed. Usually the densities are one to two orders of magnitude less than this prediction, presumably due to rf heating. Increased densities have been observed by simultaneously storing positive and negative ions (Major and Schermann, 1971), but ion–ion heating would be expected to be important for high densities.

B. The Penning Trap

The Penning trap has the disadvantages of the typically large magnetic fields (larger than about 0.1 T) required and the fact that the motion (magnetron motion) is in an unstable equilibrium in the trap. An important practical advantage is that parasitic heating mechanisms (like rf heating in the Paul trap) are nearly absent. It may also be the clear choice for studies of magnetic field-dependent structure.

1. Theory

The Penning trap (Penning, 1936) uses the same electrode configuration as the rf trap (Fig. 1), but we set V_0 equal to zero in Eq. (3b) such that the charged species see only a static potential well along the z axis given by Eq. (3a). This causes a repulsive potential in the x–y plane which can be overcome by superimposing a static magnetic field along z ($\mathbf{B} = B_0 \hat{z}$). For a single ion in the trap (or neglecting space charge) the equations of motion are (Harrison, 1959; Byrne and Farago, 1965) as follows:

$$\ddot{z} + \omega_z^2 z = 0, \quad \ddot{\mathbf{r}} = \tfrac{1}{2}\omega_z^2 \mathbf{r} - i\omega_c \dot{\mathbf{r}}$$

where $\mathbf{r} = x + iy$, $\omega_z^2 = 4qU_0/m(r_0^2 + 2z_0^2)$, and $\omega_c = qB/mc$. Therefore

$$z = z_0 \cos \omega_z t \text{ and } \mathbf{r} = \mathbf{r}_c e^{-i\omega_c' t} + \mathbf{r}_m e^{-i\omega_m t} \tag{5a}$$

where

$$\begin{Bmatrix} \omega_c' \\ \omega_m \end{Bmatrix} = \tfrac{1}{2}\omega_c \begin{Bmatrix} + \\ - \end{Bmatrix} \left[\left(\frac{\omega_c}{2}\right)^2 - \frac{\omega_z^2}{2} \right]^{1/2} \tag{5b}$$

Some useful expressions are $\omega^2 + \omega_z^2/2 = \omega_c \omega$ ($\omega = \omega_m$ or ω_c'), $\omega_c' + \omega_m = \omega_c$, and $\omega_c' \omega_m = \omega_z^2/2$. Quantum mechanical solutions have also been given (Sokolov and Pavlenko, 1967; Gräff et al., 1968; Van Dyck et al., 1978; Itano and Wineland, 1982); this description is important for the single-electron experiments. Representative values of parameters might be $r_0 = \sqrt{2}z_0 = 0.5$ cm, $B = 2.0$ T, $U_0 = 2$ V. For electrons, $\omega_c/2\pi = \nu_c \cong 55.9$ GHz, $\nu_z = 26.7$ MHz, and $\nu_m = 6.36$ kHz. For $m = 100$ u (atomic mass units) ions, $\nu_c = 305$ kHz, $\nu_z = 62.3$ kHz, and $\nu_m = 6.50$ kHz. It should be noted that the magnetron motion [the $\mathbf{r}_m \exp(-i\omega_m t)$ term in Eq. (5a)] is in unstable equilibrium in the trap. Therefore, if collisions with background neutrals occur, the ions will diffuse out of the trap. For example, when the magnetron velocity is much less than the cyclotron velocity, elastic collisions with heavy neutrals cause r_m to random walk in the $x-y$ plane with step size approximately equal to r_c. In practice however, this is not a limitation because ions can be stored for days in an apparatus at room temperature (Wineland et al., 1978; Drullinger et al., 1980) and electrons for weeks (Dehmelt and Walls, 1968; Ekstrom and Wineland, 1980) in an apparatus at 80 K. Moreover the technique of sideband or radiation pressure cooling (Van Dyck et al., 1978; Itano and Wineland, 1982) can reverse this diffusion process.

Ion–ion collisions can also cause the cloud to spread; however, this spreading is limited because the total canonical angular momentum of the system must be conserved (Wineland and Dehmelt, 1975c; O'Neil and Driscoll, 1979). For very cold clouds (i.e., axial and cyclotron modes at low temperature) this leads to clouds of nearly uniform density (O'Neil and Driscoll, 1979; Prasad and O'Neil, 1979), which for the Penning trap geometry implies that the cloud is a uniformly charged ellipsoid having the potential

$$\phi_i(r, z) = -\tfrac{2}{3}\pi\rho(ar^2 + bz^2) \tag{6}$$

where from Poisson's equation, $2a + b = 3$.

If the voltage U_0 applied to the electrodes becomes too high, then the radial electric field is high enough to overcome the $q\mathbf{v} \times \mathbf{B}/c$ magnetic force and the ions strike the ring electrode in exponentially increasing orbits [the argument of the square root in Eq. (5b) becomes negative.] For singly ionized atoms (or electrons) the voltage where this occurs is given by $V_c \cong 1200B^2(r_0^2 + 2z_0^2)/M$, where V_c is in volts, B is in tesla, M is the mass in u, and dimensions are in centimeters. This same mechanism limits the densities achievable in the Penning trap since space charge also gives radial electric fields. Assuming the ion cloud is a uniformly charged ellipsoid as discussed above, the r motion of an individual ion is now given by Eq. (5a) with

$$\left\{\begin{matrix}\omega_c' \\ \omega_m\end{matrix}\right\} = \frac{\omega_c}{2}\left\{\begin{matrix}+\\-\end{matrix}\right\}\left[\left(\frac{\omega_c}{2}\right)^2 - \frac{\omega_z^2}{2} - \frac{4\pi q\rho a}{3m}\right]^{1/2} \tag{7}$$

where ω_z is the axial frequency for a single ion in the trap. The maximum density allowable (argument of the square root term kept positive) is given by [using $\phi_i(z) = -\phi_T(z)$] $n = \rho/q < B^2/8\pi c^2 m$ or $n < 2.7 \times 10^9 B^2/M$, where B is in tesla and M is the mass in u. For $B = 1$ T, $M = 100$ u, $n < 2.7 \times 10^7$ cm^{-3}.

In the refined work of electron–positron and mass spectroscopy, it is desirable to make the trap as nearly quadratic as possible. Assuming the trap has the desired symmetry, the fourth-order term (proportional to $3r^4 - 24r^2z^2 + 8z^4$) in the potential can be canceled out with correction electrodes (Van Dyck et al., 1976). If the magnetic field is tilted with respect to the trap axis and the potential is imperfect but has the form $[r^2 - \epsilon(x^2 - y^2) - 2z^2]$, then the relation $(\omega'_c)^2 + \omega_z^2 + \omega_m^2 = \omega_c^2$ will still hold (Brown and Gabrielse, 1982; see also Borodkin, 1978; O et al., 1982).

2. Experiments

The main aspects of the theory have been confirmed. This has been done to a very high level in the electron–positron experiments discussed below. For clouds of ions the coupling between modes can be strong (energy transfer times of milliseconds), but the observed frequencies of motion for small clouds are essentially the free space values since only the center-of-mass motion is usually excited (Wineland and Dehmelt, 1975a,c).

Radial transport due to collisions with background neutrals seems to be reasonably well understood (Walls, 1970; McGuire and Fortson, 1974; Jeffries, 1980; deGrassie and Malmberg, 1980; Malmberg and Driscoll, 1980), but transport from ion–ion collisions is not so well understood. This may be caused by plasma oscillation-induced transport. Recent measurements on laser-cooled ion clouds (Bollinger and Wineland, 1983) indicate that they are approximately ellipsoidal in shape and of constant density, in agreement with prediction (Prasad and O'Neil, 1979).

C. The Kingdon Trap

The Kingdon trap (Kingdon, 1923) has the advantage of being simpler than the Paul or Penning trap, requiring only a dc voltage for trapping. Since a potential minimum in free space cannot exist for purely electrostatic fields, the Kingdon trap relies on angular momentum of the ions about a central axis to provide dynamical stability.

Usually, the Kingdon trap has the same symmetry of the Paul and Penning traps: A central wire is surrounded by an outer cylinder as shown in Fig. 3. Perhaps the most desirable potenial takes the form suggested by Prior

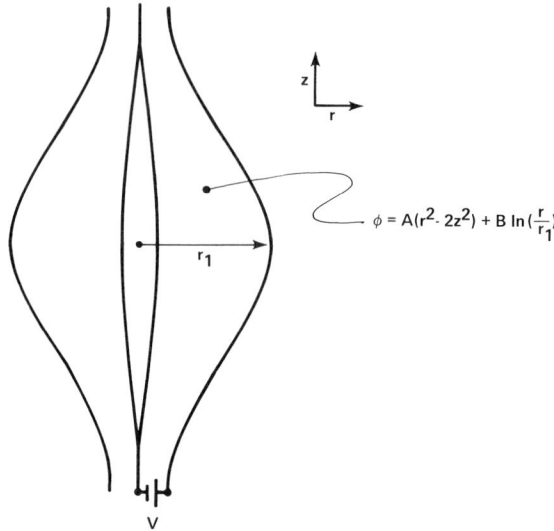

FIG. 3. Schematic representation of the electrode configuration for the "ideal" Kingdon trap. Electrode surfaces are figures of revolution about the z axis and are equipotentials of ϕ. In practice, the central electrode is made as thin as possible to reduce the chance of ions colliding with it.

(Knight, 1981), where

$$\phi = A(r^2 - 2z^2) + B \ln(r/r_1) \quad (8)$$

In this case, the motion is harmonic along z; therefore, detection of ω_z can provide mass analysis. Figure 3 shows a trap which is formed from equipotentials of Eq. (8); in practice, the central electrode would be much thinner but must still conform to an equipotential of Eq. (8) in order to preserve the harmonic well along z. The equations of motion for the $x-y$ plane are not solvable analytically, but the motion is basically composed of precessing orbits about the z axis (Hooverman, 1963). Lewis (1982) has analyzed the case of nearly circular orbits in a cylindrical trap. Storage times in such a trap should be shorter because a single collision with a background neutral is sufficient to cause an ion to collide with the center electrode and be lost. Nevertheless, with a trap whose electrodes approximate equipotentials of Eq. (8), trapping times of about 1 sec were achieved by Knight (1981) at a pressure of 10^{-6} Pa ($\simeq 10^{-8}$ Torr). (The center electrode was 100-μm-diameter wire.)

The case of an axial magnetic field superimposed along the z axis has been investigated theoretically and experimentally (on He$^+$)(Johnson, 1983; Lewis, 1983); in this case storage times should increase dramatically. Ions will eventually be lost by diffusing in toward the center electrode; therefore, storage times should be comparable to those of the Penning trap. However, the frequency of the drift motion about the center of the trap will depend on the distance from the z axis; for the Penning trap this drift (magnetron) frequency is independent of radial position (neglecting nonuniform space charge) and may be part of the reason for the slow evolution of the ion cloud under the influence of ion–ion collisions.

Prior and his colleagues have used the Kingdon traps extensively for spectroscopic studies (Prior, 1972; Prior and Wang, 1977; see also Vane *et al.*, 1981). Lewis (1983) has proposed to study the Aharonov–Bohm effect in a Kingdon trap.

D. Ion Creation

Certainly, the prevalent method for creating ions is from direct ionization of neutrals inside the trap by an externally injected electron beam. However, many times it is desirable to rid the trap of extraneous ions. In the rf trap this can be accomplished by operating the trap in the "mass-selective" mode (i.e., near the bottom of the stability diagram shown in Fig. 2). For the Penning trap, high-mass ions may be excluded by exceeding the "critical" voltage (Section II,B,1). For both traps, strong resonant excitation at the various motional frequencies can be used to drive unwanted ions from the trap.

If an ion is injected into the trap from the outside, it must lose energy inside the trap or it will eventually be lost. Such a scheme utilizing radiative energy loss has been successfully realized for positrons (Dehmelt *et al.*, 1978; Schwinberg *et al.*, 1981c) and various "catching" schemes have been proposed (see, for example, Todd *et al.*, 1976; O and Schuessler, 1981, 1982; Schuessler and O, 1981, and references therein).

Other possibilities for ion creation include photoinduced ion pair formation (Major and Schermann, 1971) and dissociative attachment for negative ion production (Blumberg *et al.*, 1978). Blatt *et al.* (1979) have demonstrated capture of Ba^+ ions from surface ionization followed by slowing down with a high-pressure helium buffer gas. Coutandin and Werth (1982; see also Blatt *et al.*, 1982) have demonstrated a technique whereby ions from an external source are first caught on a Pt ribbon. The neutrals are then evaporated from the ribbon and reionized inside the trap. They report overall efficiency of 10^{-5} and indicate that a significant increase in this number is possible. Knight (1981) reports capture of large numbers of ions from laser-produced plasmas; such a technique might be the best way to produce ions from refractory metals. Charge exchange can also be used, particularly if the parent ion is easier to produce and isolate by direct ionization than the daughter ion (Itano and Wineland, 1981). More recently, highly stripped ions have been trapped as recoils from collisions of neutrals with high-energy, highly stripped ion beams (Vane et al., 1981; Church, 1982). These experiments have so far dealt only with collision processes (electron capture), but the possibility of doing spectroscopy on high-Z ions is very exciting.

E. Ion Detection

Perhaps the most direct way to detect ion number is by ejection-counting methods (Dehmelt and Jefferts, 1962; Jefferts, 1968; Dawson and Whetten, 1968; Gräff et al., 1980; Dawson, 1980). By driving the resonant motion of the ions, the resulting induced currents in the electrodes can be an effective way to monitor ion number (Fisher, 1959; Dehmelt, 1967; Dehmelt and Walls, 1968; Wineland and Dehmelt, 1975a; Blumberg et al., 1978; McIver, 1978a,b; Van Dyck and Schwinberg, 1981a). This can be quite sensitive since a single electron or positron can be detected with good signal to noise ratio. In one case, H_2^+ "test" ions have been used to sample $^3He^+$ ion density by observation of their space-charge-shifted frequencies (Dehmelt, 1967; Schuessler et al., 1969). Both the number and temperature of ions can be monitored with the "bolometric" technique (Dehmelt and Walls, 1968; Church and Dehmelt, 1969; Walls and Dunn, 1974; Wineland and Dehmelt, 1975a; see also Gaboriaud et al., 1981), where one observes the mean square value of the induced currents in the electrodes. Induced charge frequency shifts (Wineland and Dehmelt, 1975a) might also be used to infer ion number. Observation of Doppler-induced (Schuessler, 1971b; Major and Werth, 1973; Major and Duchêne, 1975; Lakkaraju and Schuessler, 1982) or inhomogeneous magnetic field-induced sidebands on rf or microwave transitions can give information on cloud size and temperature. The use of laser-induced fluorescence techniques can be utilized to detect both ion cloud size and number as described previously. In the case of the Penning trap, the cloud density can be measured by observing the magnetron rotation-induced Doppler shift of the ion resonance lines (Bollinger and Wineland, 1983). The sensitivity of these fluorescence methods is indicated by the ability to detect single ions (Neuhauser et al., 1980; Wineland and Itano, 1981; Nagourney et al., 1983; Ruster et al., 1983).

F. Polarization Production/Monitoring

Ion samples have been polarized using spin exchange with polarized neutral beams (Dehmelt, 1969; Gräff et al., 1968, 1969, 1972; Church and

Mokri, 1971; Schuessler, 1979, 1980). Changes of polarization in these experiments have been detected by observing the heating due to spin-dependent inelastic collisions or by detecting spin-dependent ion neutralization by detecting ion number. Changes in polarization of trapped electrons have also been detected by observing changes in polarization of a transmitted atomic beam (Gräff and Holzscheiter, 1980) or by the kinetic energy dependence on polarization of electrons ejected into an inhomogeneous field (Kienow et al., 1974; Gräff et al., 1980). Optical-induced fluorescence techniques discussed below may be the dominant choice in future experiments on atomic ions; we recall, however, that the first optical techniques relied on orientation dependence of ion photodissociation (Dehmelt and Jefferts, 1962). More recently, orientation-dependent photodetachment has been used (Jopson and Larson, 1981).

G. Other Types of Traps

Variations on the types of traps discussed above have been considered. It is possible to use the Paul and Penning traps in a combined mode (Fisher, 1959; O and Schuessler, 1980b), but such traps have not really been used yet. Some rf traps with a "race track" design based on a closed-loop configuration for the Paul mass spectrometer have also been tested (Church, 1969). In addition, six-electrode rf traps with three-phase drive have also been demonstrated (Wuerker et al., 1959b; Haught and Polk, 1966; Zaritskii et al., 1971). For many years, trapped ion cyclotron resonance (ICR) spectrometers (McIver, 1970, 1978a,b; Comisarow, 1981) have been used by chemists; this configuration is basically a Penning trap with rectangular electrodes. (We note also that this is the typical configuration in a sputter ion pump.) Trap arrays (Major, 1977) have been proposed to increase overall trapping efficiency and variations on the Kingdon trap have also been proposed (McIlraith, 1971). Finally we note the possible use of "space charge" traps (Redhead, 1967; Hasted and Awad, 1972; Donets and Pikin, 1976; Hamdan et al., 1978). Such devices have been mainly used in collision studies; for high-resolution spectroscopy the traps described above seem superior.

III. Lepton Spectroscopy

A. Historical Perspective

The only stable leptons that can take advantage of the long containment times possible in the Penning trap are the electron and the positron. Thus, inspired by the early spin exchange, optical pumping experiments at the University of Washington (Dehmelt, 1956, 1958), these particles have subsequently been trapped and studied to high precision in a series of highly successful geonium experiments (Van Dyck, *et al.*, 1977, 1978; Schwinberg *et al.*, 1981a, 1981c, 1983). However, the first truly precise rf spectroscopy experiment, carried out on electrons in such a trap, was conducted by G. Gräff and associates in Bonn and Mainz, West Germany. The Penning trap was placed in the uniform magnetic field region of a standard atomic beam machine (Gräff *et al.*, 1968). A state-selected sodium beam interacts with the trapped electrons, causing the cloud to become polarized; subsequent spin-dependent energy-transfer collisions with the polarized atoms then lead to an observable change in the number of electrons remaining in the trap after a fixed interaction time. By applying a microwave field at the spin precession frequency v_s and then an rf field at the spin-cyclotron difference frequency v_a, the free electron g-factor was measured to a precision of 0.3 ppm (Gräff *et al.*, 1969):

$$g(e^-)/2 = 1.001,159,660(300) \tag{9}$$

This same experiment was subsequently modified to incorporate the non-destructive bolometric detection method discussed in Section II,E, thus yielding the g factor with approximately the same precision (see Church and Mokri, 1971). However, the potential for improvement was limited by relatively large linewidths attributable to relativistic kinematics, second-order Doppler effects, and possible magnetic field inhomogeneity over the large electron cloud. Another possible limitation to this work was the electrostatic cloud shifts associated with the v_a resonance (see Wineland and Dehmelt, 1975c).

About this same time, a different technique for measuring the electron (or positron) g-factor had begun to show great promise at the University of Michigan under the guidance of H. R. Crane (see Wilkinson and Crane, 1963; Rich, 1968a,b). Referred to as the free precession method, the scheme directly observes the relative orientation between the precessing spin and the electron's orbital momentum after a fixed containment time in a weak magnetic mirror trap. This technique was dramatically refined by A. Rich and associates at Michigan (see Wesley and Rich, 1971), achieving a precision of 3 ppb in the electron's g-factor:

$$g(e^-)/2 = 1.001,159,657,700(3,500) \tag{10}$$

At this point, the precession method also found certain limitations such as finite observation time, relativistic mass corrections, possible space-charge shifts, and large containment volumes ($\approx 10^2$ cm^3 with corresponding uncertainty in the magnetic field. This experiment is presently being rebuilt in hopes of achieving a precision of 0.010 ppb in the g-factor.

However, since 1959 H. Dehmelt and colleagues had been studying electron clouds in a Penning trap at very low pressures ($<10^{-9}$ Pa or 10^{-11} Torr), cooled by coupling axially to a resonant tuned circuit. This early work established the basic nondestructive bolometric technique (Dehmelt and Walls, 1968) whereby either excitation of the cyclotron resonance or alternating excitation of spin and g-2 frequencies causes an increase in the monitored axial temperature. This bolometric scheme thus produced a g-factor measurement

$$g(e^-)/2 = 1.001,159,580(80) \tag{11}$$

whose precision appeared to be limited by electrostatic cloud perturbations (Walls, 1970; Walls and Stein, 1973). However, these initial Penning trap studies did point out the advantages of reducing the cloud to its ultimate irreducible size, i.e., a single electron, capable of being continuously observed for several days (Wineland et al., 1973). This now set the stage for the geonium experiments carried out on single electrons (Section III,B) and single positrons (Section III,C) by one of the authors (RSV) in collaboration with P. Schwinberg and H. Dehmelt.

B. Electron Geonium Experiment

1. The Geonium Atom

The basic apparatus (Van Dyck, et al., 1978) is shown schematically in Fig. 4. The electron's driven axial motion at $v_z \approx 60$ MHz induces a measurable voltage across the high-impedance tuned circuit. By keeping a large drive amplitude fixed and the frequency set for off-resonance excitation ($\delta v \approx 10$ kHz), the smallest unit of voltage is found (by loading different bunches of electrons) for which all other signals are integral multiples, thus signifying the presence of a single electron. This can also be verified by comparing respective linewidths. Now isolated from external perturbations within a relatively small volume ($<3 \times 10^{-7}$ cm^3) at ultralow pressures, the trapped electron is bound via the trap's electrode structure and the magnet to the earth as if it were a mestastable pseudoatom. The University of Washington researchers therefore refer to this pseudoatom as "geonium." Further discussion of the geonium state can be found in a series of tutorial lectures (Dehmelt, 1983; Ekstrom and Wineland, 1980).

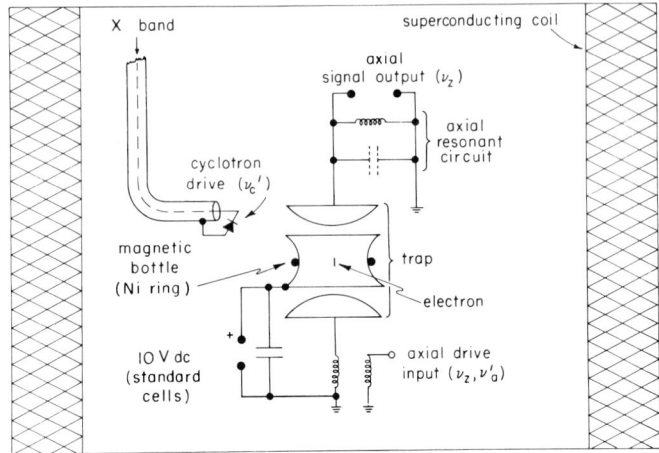

FIG. 4. Schematic of the geonium apparatus. Placed into the bore of a superconducting solenoid and operated at 4.2 K, the basic Penning trap is used to isolate single electrons and detect their driven axial motion via a resonant tuned circuit. Magnetic transitions are observed by coupling respective magnetic moments via the nickel ring to the precise axial resonance frequency. (From Van Dyck *et al.*, 1978.)

2. *Precision Axial Resonance Spectrometer*

Because of the need for holes and slits in the basic electrodes and the unavoidable truncation and machining imperfections of the surfaces, terms higher than quadratic appear in the potential distribution (see Section II). In order to compensate for these effects, guard rings are placed between the endcaps and ring and their potential is varied until the narrowest possible unshifted axial resonances are obtained, i.e., reduced 100-fold relative to those from uncompensated traps (Van Dyck *et al.*, 1976). The resulting single electron linewidth (shown in Fig. 5) is within 10% of the expected damping linewidth and the frequency resolution is approximately 10 ppb (for the signal-to-noise ratio as shown).

This narrow axial resonance is now used in a feedback detection circuit in which a very stable frequency synthesizer supplies the rf drive. Of the two quadrature components shown in Fig. 5, the dispersion-shaped curve yields the error signal, which is integrated to produce a correction that is added back to the ring in order to close the loop. In this way, the axial motion is locked to the synthesizer at v_z and the correction signal is proportional to any shift δv_z. This type of nondestructive synchronous detection of a single electron (or ion) forms the precision axial-resonance spectrometer mode of operation.

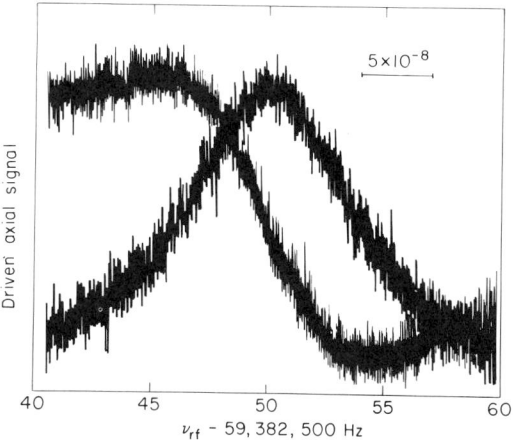

FIG. 5. Graph of the synchronously detected axial resonance. Linewidths of 8 Hz out of 60 MHz are achieved in a well-compensated Penning trap in a cryogenic (4.2 K) environment. Both the absorption and dispersion phases of the signal are given for this resonance using a 0.5-sec time constant. (From Van Dyck et al., 1978.)

3. Magnetic Coupling

The electron's magnetic resonances have been indirectly observed by constructing a shallow magnetic bottle obtained from a small nickel ring located in the midplane of the ring electrode. The uniform axial field B is therefore modified (in cylindrical coordinates) to yield

$$B_z = B'_0 + B_2(z^2 - r^2/2) \tag{12}$$

and

$$B_r = -B_2 rz$$

where B_2 is fixed at approximately 0.012 T/cm² (or 120 G/cm²) when the nickel ring is saturated. The weak interaction with the electron's magnetic moment yields an additional axial restoring force and the following axial frequency shift:

$$\delta v_z = \left(m + n + \frac{1}{2} + \frac{v_m}{v_c} q\right)\delta \tag{13}$$

where $\delta = \mu_B B_2/(2\pi^2 M v_z) \approx 1$ Hz and the integers m, n, and q are respectively spin, cyclotron, and magnetron quantum numbers. Thus, this coupling yields an occasional random jump in the δv_z correction signal and a 1-Hz change in the resulting noise "floor" for each induced spin flip ($\Delta m = \pm 1$).

This same magnetic bottle is also used to generate the perpendicular rf field required to flip the electron's spin. As indicated by the radial field B_r in Eq. (12), an auxiliary axial drive at v'_a combined with the cyclotron motion generates sidebands at $v'_c \pm v'_a$. Since by definition $v'_a \equiv v_s - v'_c$, it follows that v_s is one of the rotating components.

4. Sideband Cooling

In order that the electron may be found reliably at the same radial position in time, it is centered at the top of the radial electric and magnetic hills by using a novel sideband cooling technique first developed at the University of Washington, in which an inhomogeneous electric field is applied at $v_z + v_m$ to the trapping volume (Van Dyck et al., 1978; Wineland, 1979). The electron's magnetron motion in the inhomogeneous field generates a field component at v_z which damps hv_z quanta into the tuned circuit with the extra hv_m quanta being absorbed into the magnetron motion, thereby shrinking r_m. Figure 6 strikingly confirms this process as the magnetic coupling is also used to monitor changes in the magnetron magnetic moment [see Eq. (13)]. Note that r_m grows exponentially when $v_z - v_m$ drive is applied. In addition, this sideband technique allows v_m to be determined and, when compared to $v_z^2/2v'_c$, is found to agree within 100 ppm.

5. Resonance Data

As shown in Fig. 4, a small diode is used both as a frequency multiplier to generate the cyclotron drive and an antenna to launch microwaves into the

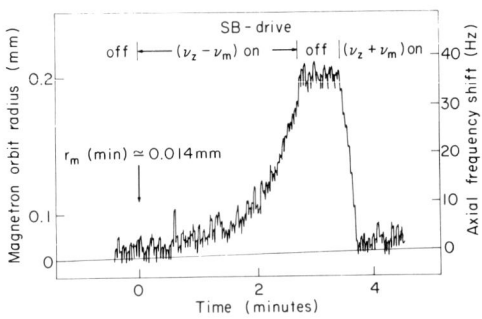

FIG. 6. Graph of changes in the magnetron orbit radius with time. The axial resonance spectrometer technique is used to observe changes in the magnetron radius by driving the axial motion on the $v_z \pm v_m$ sidebands. In this way, hv_m quanta are absorbed (or emitted) in conserving total energy, thereby shrinking (or expanding) the magnetron orbit. (From Van Dyck et al., 1978.)

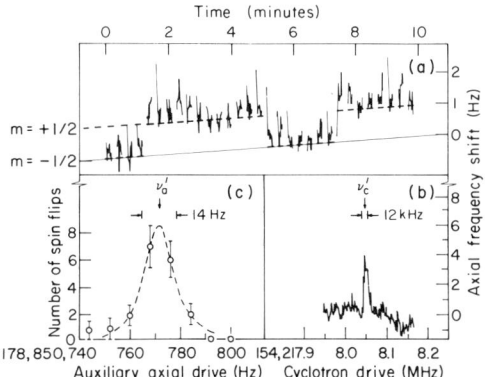

FIG. 7. Graph of electron geonium resonance data. The cyclotron resonance in (b) is measured at 5.5 T by detecting the corresponding axial frequency shifts while using a very low axial drive. The spin–flip data in (a) is obtained by alternating both a strong detection and auxiliary drives; the number of spin flips observed out of say 30 attempts for a fixed-frequency auxiliary drive is plotted in (c).

trap. Figure 7b shows a typical cyclotron resonance obtained at 5.5 T while using the lowest axial drive that still maintains axial lock. The interference between axial drive and thermal noise via the magnetic bottle makes such low drives necessary. Unfortunately, the signal-to-noise ratio for such drives is inadequate to observe the 1-Hz steps in the noise floor. Therefore, an alternating detection/drive scheme was used to obtain the spin flip data shown in Fig. 7a. For a fixed frequency of the auxiliary drive, the number of 1-Hz axial frequency steps produced out of say 30 attempts is counted and plotted versus frequency in order to yield an anomaly resonance typified by that in Fig. 7c; the resolution of this data is ≈ 40 ppb.

6. Results and Conclusions

The procedure for correcting the magnetic resonances for the presence of Penning trap electrodes is nearly invariant to small misalignments or asymmetries (Brown and Gabrielse, 1982). Thus, the electron's g-factor (or anomaly) is given to a precision greatly exceeding 10^{-12} by the following:

$$\frac{g(e^-)}{2} - 1 \equiv a_e = \frac{v'_a - v_z^2/2v'_c}{v'_c + v_z^2/2v'_c} \tag{14}$$

where v'_a, v'_c, and v_z must be the observed resonant frequencies. Presently, the measurement of v'_c has not shown any systematic errors, but v'_a has been found to show a small negative shift ($<4 \times 10^{-8}$ for the full range of anomaly powers used) that depends on the strength of the auxiliary drive. Though at present, no adequate explanation exists for this dependence, an extrapolation to zero amplitude has yielded consistent anomalies for three different magnetic field strengths: 1.8, 3.2, and 5.1 T. Various other small shifts are possible (see Van Dyck et al., 1978; Dehmelt, 1981a, 1983), but at the present level of accuracy, none have been observed to affect the anomaly, either because they are too small or because they affect v'_c and v'_a in the same way. An extrapolated unweighted average of all runs (Van Dyck et al., 1979) yields

$$g(e^-)/2 = 1.001,159,652,200(40) \tag{15}$$

which can be compared with the best theoretical prediction based on quantum electrodynamics (Kinoshita and Lindquist, 1981) and the e/h fine-structure constant α (Williams and Olsen, 1979):

$$g_{thy}/2 = 1.001,159,652,460(147) \tag{16}$$

Conversely, QED theory and the experimental g-factor predicts the fine-structure constant:

$$\alpha^{-1} = 137.035,993(10) \tag{17}$$

which agrees very well with the e/h determination of α:

$$\alpha^{-1} = 137.035,963(15) \tag{18}$$

In order to reach a g-factor precision of 1 part in 10^{12} in the future, a new bottleless Penning trap is being developed [see Dehmelt (1981a) for a possible bottleless detection scheme] which may utilize the relativistic mass shift associated with the cyclotron excitation (Gabrielse and Dehmelt, 1980). A new trap technology is being developed which uses copper electrodes with a split ring for direct production of the spin-flip field, mounted into an all-metal vacuum envelope with indium seals and nonmagnetic ceramic feedthroughs (Gabrielse and Dehmelt, 1983).

C. Positron Geonium Experiment

1. Preparing the Positron Geonium State

The primary technological difference between this and the electron geonium experiment is the continuous positron loading scheme (Schwinberg et al., 1981c) which is completely static and relies upon radiation damping in a separate storage trap (see Fig. 8). A sealed ^{22}Na positron emitter, mounted off-axis (at $\frac{3}{4}r_0$), is biased in such a way that a trappable positron passing

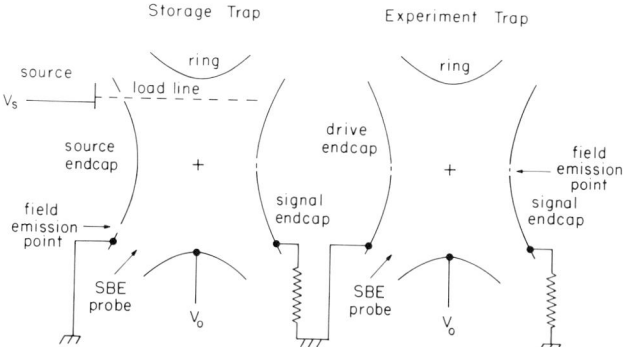

FIG. 8. Schematic of the double-trap configuration. The storage trap contains a sealed ^{22}Na positron emitter located at $\frac{1}{4}r_0$. SBE probes are used in the sideband cooling scheme to radially center the trapped positrons prior to transferring some of them into the highly compensated experiment trap. (From Schwinberg et al., 1981a.)

through the source hole will have most of its energy (≈ 100 keV) in the radial motion and will be temporarily trapped for one magnetron period, before returning to its original entry point. During this time, enough axial energy may be extracted by the damping circuit to permanently trap the positron and to subsequently thermalize its axial motion. In contrast, the relativistic cyclotron energy will damp quickly (a few seconds) by synchrotron radiation. Finally, the trapped positron is centered using the motional sideband technique described in Section III,B,4. A typical loading rate during early liquid helium operation at 5.1 T was 23 e^+/hr with a 0.5 mCi source within a 5-V deep storage trap (Schwinberg et al., 1981c).

The carefully constructed compensated experiment trap (also shown in Fig. 8) was found to be necessary because the large off-axis holes, required for positron loading, prevent the storage trap from being fully compensated. Therefore, some of the centered positrons are moved into the experiment trap by pulsing the two adjacent endcaps down to the approximately common ring potential (-8 to -10 V) for a few microseconds. Once transferred, they are detected by using a large off-resonance axial drive similar to the loading drives described for electrons in Section III,B,1. After determining that more than one have been transferred, the excess are systematically ejected using intense rf pulses at $v_z + v_m$ on the enclosed side band excitation (SBE) probe. The pulse amplitude is carefully adjusted in order to require at least 10 consecutive pulses to eject an e^+ from the cloud. Once a positron is isolated, the axial drive is reduced in order to observe a 4-Hz axial resonance that is used in the precision axial spectrometer mode described in Section III,B,2.

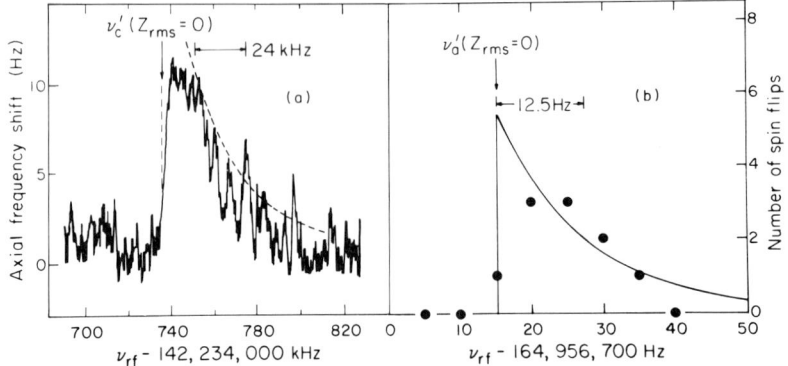

FIG. 9. Graph of the positron geonium resonance data. The cyclotron resonance in (a) has a lineshape characterized by a high-frequency exponential tail and a low-frequency edge (for $z_{rms} = 0$); the dotted curve is an exponential with a 24-kHz decay constant. The corresponding anomaly resonance in (b) is similar to the cyclotron resonance, but with an edge feature convolved with the relaxation rate for the axial resonance. The solid line in (b) represents the ideal lineshape with a 12.5-Hz decay constant. (From Schwinberg et al., 1981a.)

2. Measurements

For this apparatus, the magnetic bottle produces a 1.3-Hz shift per unit change in the magnetic quantum level as described in Section III,B,3. Figure 9a shows a typical positron cyclotron resonance but now with sufficient resolution to observe a linewidth characterized by a high-frequency exponential tail due to the coupled thermal Boltzmann distribution of axial states. The dotted curve represents an exponential with a 24-kHz decay constant. However, this broad linewidth does not limit the resolution since it is the sharp low-frequency edge (for $z_{rms} = 0$) that corresponds to a cyclotron excitation at the bottom of the axial magnetic well. For this example, the field is resolvable to ≈ 10 ppb. Figure 9b shows the corresponding anomaly resonance taken upon alternating the detection and excitation (as explained in Section III,B,5). The characteristic shape is similar to that of the cyclotron resonance, though the edge feature is typically smeared by the relaxation rate of the axial states. Therefore, the precision of the anomaly edge feature, taken as the half linewidth, is ≈ 30 ppb.

3. Results and Conclusions

Time studies of the cyclotron resonances found an effective magnetic field jitter ($\approx \pm 25$ ppb over $\approx 10^2$ sec) which is believed to be due to collisions with background gas varying the minimum radial position in the magnetic bottle. The source of this background was later traced to a small crack in the glass envelope. However, prior to rebuilding the positron apparatus, the only systematic effect observed was again the small negative shift proportional to the strength of the auxiliary axial drive v'_a. From four initial positron runs, a preliminary positron g-factor (Schwinberg et al., 1981a,b)

$$g(e^+)/2 = 1.001,159,652,222(50) \qquad (19)$$

is obtained from a weighted least squares extrapolation using the systematic power dependence observed for the previous electron work. This result agrees well with the electron's g-factor [Eq. (15)] and the theoretical prediction [Eq. (16)]. It is also 20,000 times more precise than the only other direct measurement of the positron g-factor (Gilleland and Rich, 1972) using the Michigan free precession technique. Finally, by combining measured e^- and e^+ g-factors, the comparison

$$g(e^+)/g(e^-) = 1 + (22 \pm 64) \times 10^{-12} \qquad (20)$$

may well represent the most severe symmetry test of charge conjugation–parity reversal–time reversal (CPT) invariance known to date. An excellent discussion of the discrete symmetries using free leptons can be found in a review by Field et al. (1979).

IV. Mass Spectroscopy

A. Introduction

Because of long containment times and small sample volumes, the Penning trap is an ideal device for measuring the mass of stable ions using ion cyclotron resonances. This can be compared to such devices as the resonance rf mass spectrometer of Smith and Wapstra (1975), which has demonstrated a resolution of 1–10 ppb by carefully controlling various systematic effects. In contrast, an ion cooled down in a Penning trap can be confined to such a small volume that both the electric and magnetic fields can be easily controlled by using small electric steering fields (Van Dyck et al., 1980), guard electrodes for compensation, and magnetic shim coils for producing the required field symmetry and uniformity. Finally, refined rf techniques allow small numbers (≤ 10) of light ions (< 10 u) to be observed in such a trap, and weak self-contained sources can be used without appreciably altering the background pressure. Eventually, laser fluorescence techniques may allow simple detection of single ion mass ratios by observing changes in fluorescence as the ion orbits are excited (Wineland et al., 1983).

This section specifically describes direct precision measurements of the proton-to-electron mass ratio (m_p/m_e) using the Penning trap. However, variations of these same techniques could be used to measure any of the light ion masses (such as ^{3}H and ^{3}He) with an *in situ* field calibration obtained using either the electron or proton cyclotron resonance. Ultimately, these techniques may also be used to study more exotic species such as the antiproton (Dehmelt et al., 1979; Torelli, 1980).

Presently, there are three direct measurements of m_p/m_e using the Penning trap to alternately confine both e^- and p^+: two at the University of Mainz and one at the University of Washington.

B. First Mainz Experiment

This experiment by Gärtner and Klempt (1978) has a thermionically produced electron beam which ionizes background hydrogen gas, obtained from a palladium leak, to produce secondary electrons, H$^+$, and H$_2^+$ in the trapping volume. A pressure of 10^{-7} Pa (or 10^{-9} Torr) then yields storage times of ≈ 60 sec for protons and ≈ 60 min for electrons. Typically, a few thousand ions would be trapped and detected via their axial motion interacting with a tuned circuit across the endcaps. By sweeping the dc ring voltage V_0, the axial resonance at v_z can be made to coincide with the frequency of the tuned circuit, thus yielding a detectable decrease in signal as

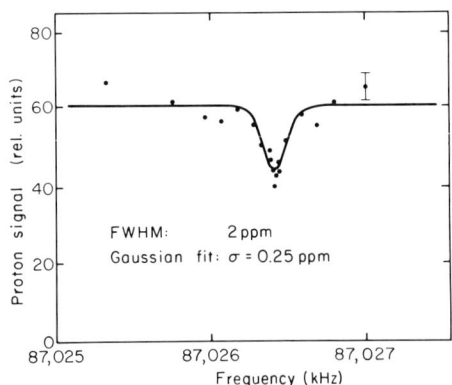

Fig. 10. Graph of the proton cyclotron resonance. The final average number of trapped protons is plotted versus frequency for the case of $V_0 = 20$ V. Resonance is indicated by an $\approx 25\%$ decrease in the number of protons remaining in the trap per cycle after the rf drive is applied, and the solid line is a Gaussian line fit with a 0.25 ppm statistical uncertainty in $v'_c(p^+)$. (From Gärtner and Klempt, 1978.)

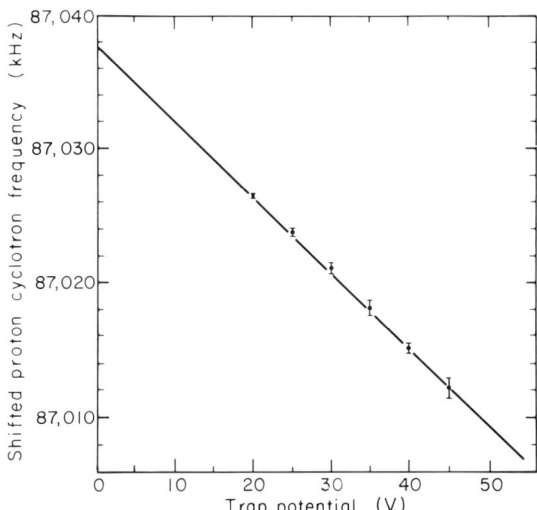

FIG. 11. Graph of the proton cyclotron frequency versus the trapping potential. The bolometrically measured $v'_c(p^+)$ has been corrected for various small systematic effects such as space charge and field inhomogeneity shifts. The straight line represents a linear least squares fit that is extrapolated to zero in order to yield $v_c(p^+)$. (From Gärtner and Klempt, 1978.)

the ions short out the noise coming from the tuned circuit. The width of the resulting notch can be used to estimate the number of trapped ions (Wineland and Dehmelt, 1975a).

A complete detection cycle begins with a pulse of electrons to load the ions, followed by a storage time for cooling the motion, an interaction time when rf fields are applied, a linear sweep of V_0 to detect the ions, and a final electric pulse to clear the trap. Excitation of the cyclotron resonance produces an increase in cloud temperature and an enhanced loss of trapped ions. A typical example of a proton cyclotron resonance is shown in Fig. 10. Each point represents the number of remaining protons per cycle averaged over many complete cycles at each frequency setting. A typical 2-ppm linewidth could be statistically resolved to 0.25 ppm, but the final error is increased by 10 because of such systematic effects as the uncertainty in the electric field strength, inhomogeneity and decay in time of the magnetic field, and space-charge broadening and shifting of the cyclotron resonances in an imperfect Penning trap.

From Section II,B, the trapping electric field shifts the cyclotron frequency as follows:

$$v'_c = v_c + cV_0/[\pi(r_0^2 + 2z_0^2)B_0] \qquad (21)$$

Thus, the extrapolation to zero potential as shown in Fig. 11 yields the unshifted frequency $v_c(p^+)$ with a combined uncertainty of 2.6 ppm. However, because of its much greater frequency (160 GHz), the electron's cyclotron frequency is much less affected by space-charge effects and residual anharmonicity. But in this case, relativistic shifts due to the increased energy of the stored electrons contribute greatly to the 1.3-ppm uncertainty in $v_c(e^-)$; thus, the final mass ratio is

$$m_p/m_e = 1836.15020(530) \tag{22}$$

C. Second Mainz Experiment

This experiment (by Gräff et al., 1980, 1983) has a compensated copper Penning trap with the central ring electrode split parallel to the z axis in order to apply an rf electric field normal to the axial magnetic field. The trap is installed in one end of a 37-cm-long copper drift tube which extends well out of a superconducting solenoid. A channel plate detector is mounted at the receiving end of the drift tube to count ions (or electrons) versus their arrival time at the detector. The entire apparatus is mounted horizontally into the bore of a 6.4-T shimmed superconducting magnet.

Unlike the previous Mainz research in which v_c' is measured versus V_0, this experiment uses a direct induction of the transition at v_c (i.e., $v_c' + v_m$) applied at low electric field strengths. The experimental procedure begins by pulsing thermionic electrons for ≈ 100 msec to produce either trapped electrons or protons. During the following 1 sec, the ring–endcap potential is ramped down to 1 V and (for the proton case) an axial drive is simultaneously applied to sweep out all extraneous ions such as H_2^+, H_3^+, and He^+. The cyclotron frequency v_c is then applied for ≈ 500 msec, after which the trap is cleared by a linear voltage sweep with a superimposed sequence of pulses that define the starting time of the ejected particles and start a set of fast counters which accumulate data to generate the time-of-flight spectrum.

The flight time of the moving charge is determined by its initial kinetic energy along the z axis and by the size of its magnetic moment. In other words, when the rf field is resonant at v_c, the transverse orbit increases, resulting in the greater acceleration of the charge in the inhomogeneous field along the drift tube axis; the corresponding reduction in the proton's average time of flight is shown in Fig. 12. Note that the mean flight time ($\approx 30 \mu sec$) is reduced by several percent and the linewidth (FWHM) is 0.3 ppm. Similar plots were obtained for the electron case, but with a typical average flight time of 3 μsec and a linewidth of 0.4 ppm.

The largest proton systematic effect is shown in Fig. 13, where v_c is plotted versus the number of trapped ions (and extrapolated to zero). However, for

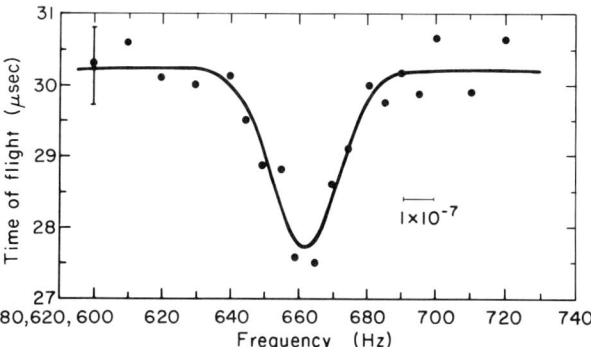

FIG. 12. Graph of the proton cyclotron resonance. The protons' time of flight is plotted as a function of the frequency of the rf drive applied prior to ejection from the Penning trap. The solid curve represents a fit to a Gaussian distribution. (From Gräff et al., 1980.)

the electron the absorbed microwave power produces a sizeable relativistic mass increase, and thus a shift and broadening of the ν_c resonance. Since the lineshape could not be explained theoretically, half the FWHM linewidth is taken as the experimental uncertainty. The experiment was also performed at two different magnetic fields, 5.28 and 5.81 T, and the resulting mass ratios (which agreed within their statistical errors) were averaged to yield

$$m_p/m_e = 1836.15270(110) \tag{23}$$

This result agrees well with the first Mainz experiment and the previous least squares adjusted value (Taylor, 1976; Phillips et al., 1977):

$$m_p/m_e = 1836.15165(68) \tag{24}$$

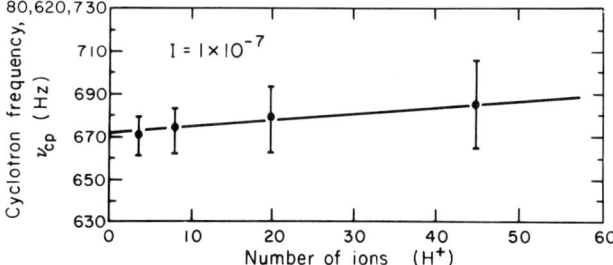

FIG. 13. Graph of the proton cyclotron frequency versus the number of trapped protons. Error bars correspond to $\nu_c(p^+)$ linewidths at FWHM and the relative slope of the dependence is -4.0×10^{-9}/ion. (From Gräff et al., 1980.)

D. UNIVERSITY OF WASHINGTON EXPERIMENT

1. Experimental Method

This experiment by Van Dyck and Schwinberg (1981a, 1983) utilizes a somewhat smaller Penning trap than either Mainz experiment since a smaller size is more favorable (i.e., faster response time) for the synchronous detection scheme discussed in Section III,B,2. The trap contains a field emission point as an electron source, magnetic rings in each endcap for generating the axial magnetic coupling, and a hydrated titanium filament as a hydrogen source.

The fundamental experimental method also differs from the Mainz experiments in that the unperturbed cyclotron frequency v_c is obtained by combining the three actual observed resonant frequencies $\bar{v}'_c$, $\bar{v}_z$, and $\bar{v}_m$ [where the bars allow for small perturbations; see Brown and Gabrielse (1982)]:

$$v_c^2 = \bar{v}_c'^2 + \bar{v}_z^2 + \bar{v}_m^2 \tag{25}$$

This is equivalent to Eq. (5b) in the limit of negligible imperfections and misalignments. The perturbed frequencies are then observed using at least one of three tuned preamplifiers: one on each endcap for axial detection [$v_z(e^-) = 136$ MHz and $v_z(p^+) = 10$ MHz] and one on the ring electrode split into four equal quadrants. This four-part "quadring" design allows the proton's cyclotron resonance to be excited by a balanced drive applied on two opposite quadrants and detected on the other pair externally tuned to the proton's cyclotron resonance at $v_c'(p^+) = 76.4$ MHz.

2. Measurements

With a 2.5-V well depth, the quadring trap could be adequately compensated in order to observe a single electron with a tuned circuit damping linewidth of ≈ 10 Hz (out of 136 MHz). The corresponding cyclotron resonance for this electron has a magnetic bottle-induced linewidth of 20–30 ppb and a low-frequency edge with resolution of 3 ppb (typical of that shown in Fig. 9a). With a well depth of -26 V, small numbers (<40) of protons can also be trapped and synchronously detected to yield very narrow (≤ 0.2 Hz) cyclotron resonances (see Fig. 14). In fact, on one occasion a reproducible linewidth of 0.03 Hz was observed to yield a precision greater than 0.5 ppb. The synchronously detected axial resonances have ≈ 50 times less resolution, but their relative contribution to the accuracy of $v_c(p^+)$ is only 2% via the magnetron (or correction) frequency.

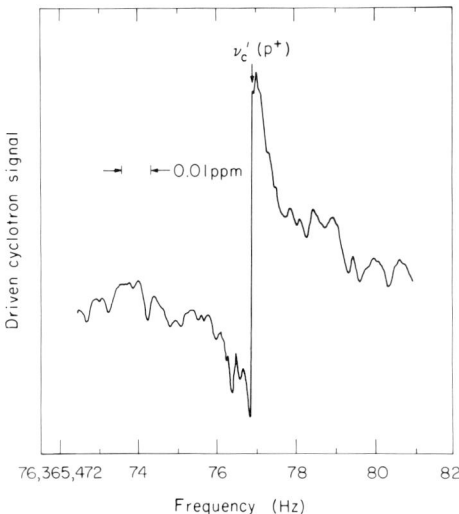

FIG. 14. Graph of the proton cyclotron resonance. This narrow dispersion-shaped curve is the result of direct synchronous detection of the resonance at $v'_c(p^+) = 76,365,476.9$ Hz using the split quadring design in a well-compensated Penning trap (for $V_0 = 54.4$ V). The linewidth, limited primarily by observation time, represents fewer than 40 protons. (From Van Dyck and Schwinberg, 1981a.)

3. Results

In order to eliminate the effects of space-charge fields upon the ion cloud cyclotron resonance (see Wineland and Dehmelt, 1975c), dissimilar ions are selectively removed using strong rf drives at their characteristic axial frequencies. However, this does not eliminate the dependence on the number of trapped ions in an imperfect Penning trap (Van Dyck and Schwinberg, 1983; Liebes and Franken, 1959). Since $v_c(e^-)$ was conveniently measured in preliminary work using small electron clouds (< 50 e^-) and did not exhibit a number dependence, the mass ratio $m_p/m_e = v_c(e^-)/v_c(p^+)$, plotted in Fig. 15, reflects only the number dependence of $v_c(p^+)$. A linear fit of this data finds a relative slope of -1.2×10^{-10}/ion. This can be compared with -4.0×10^{-9}/ion found in the second Mainz experiment and $+1.1 \times 10^{-8}$/ion produced by an earlier uncompensated quadring trap.

The intercept of the linear fit in Fig. 15 corresponds to the preliminary mass ratio $m_p/m_e = 1836.15300(7)$. However, an analysis of bottle-related position dependence of the magnetic field suggests that a systematic error as large as 0.1 ppm may exist if the two charge types do not have the same average position in space. Thus, a preliminary mass ratio (Van Dyck and

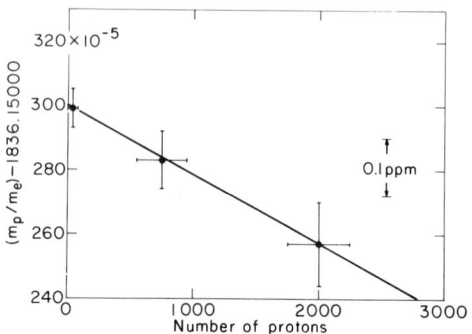

FIG. 15. Graph of the mass ratio versus the proton number. This dependence follows from the sensitivity of $v_c(p^+)$ to the number of trapped protons. The vertical and horizontal error bars are determined primarily by the $v_c(e^-)$ linewidth and the axial proton number calibration, respectively; the relative slope of the dependence is -1.2×10^{-10}/ion. (From Van Dyck and Schwinberg, 1981a.)

Schwinberg, 1981b) of

$$m_p/m_e = 1836.15300(25) \qquad (26)$$

reflects the location uncertainty in the magnetic field. This systematic effect can be carefully checked in future experiments by directly varying the magnetic bottle strength B_2.

V. Atomic and Molecular Ion Spectroscopy

A. Optical Atomic Ion Spectroscopy

Relatively little optical spectroscopy has been performed on stored atomic ions, even though laser-induced fluorescence (LIF) provides a very sensitive detection method. Probably, this is due to the fact that few ionic resonance lines lie in the visible region of the spectrum, where tunable lasers are available.

Fluorescence of Ba$^+$ ions stored in rf traps has been observed, using a hollow cathode lamp (Duchêne et al., 1977) or a dye laser (Iffländer and Werth, 1977a) as a light source. The hyperfine splitting of the 493-nm D_1 line of ^{137}Ba$^+$ was observed by Blatt et al. (1979). The absolute wavelength (Drullinger et al., 1980; Nagourney and Dehmelt, 1981a,b) and isotope and hyperfine splittings (Drullinger et al., 1980) of the 280-nm D_2 line of Mg$^+$ have been measured, using frequency-doubled dye lasers. A saturation dip in the 493-nm Ba$^+$ line was observed by Iffländer and Werth (1977b), showing that Doppler-free optical spectroscopy of ions is possible. Prior and Knight (1980) have observed hysteresis and line-narrowing effects due to hyperfine optical pumping in the LIF spectrum of metastable Li$^+$ in an rf trap. As discussed in Section II, optical excitation of ions has been used as a probe to measure ion cloud spatial distributions (Knight and Prior, 1979; Schaaf *et al.*, 1981), ion trapping efficiencies (Iffländer and Werth, 1977a; Plumelle *et al.*, 1980; Blatt *et al.*, 1979), lifetimes of metastable levels (Schneider and Werth, 1979; Knight and Prior, 1980; Plumelle *et al.*, 1980), and decay branching ratios (Osipowicz and Werth, 1981).

B. Laser Cooling

Laser cooling (also called optical sideband cooling or radiation-pressure cooling) was first proposed for trapped ions by Wineland and Dehmelt (1975b) and for free atoms by Hänsch and Schawlow (1975). This is a method by which a beam of light can be used to damp the velocity of an atom or ion. Thus, it is useful for reducing the second-order Doppler (time dilation) shift and first-order Doppler broadening of resonance lines. Most theoretical treatments of laser cooling of stored ions assume a harmonic trapping potential, a case which is approximated by the rf trap (Neuhauser *et al.*, 1980; Wineland and Itano, 1979; Javanainen and Stenholm, 1980, 1981a,b; Javanainen, 1980, 1981a,b; André *et al.*, 1981; Itano and Wineland, 1982). Itano and Wineland (1982) treat laser cooling in a Penning trap.

A harmonically bound ion can be cooled by a monochromatic laser beam tuned slightly lower in frequency than a strong resonance transition. For the usual experimental case, in which the natural linewidth of the transition is much greater than the ion's frequencies of motion, the process can be described as follows: When the velocity of the ion is directed against the laser beam, the light frequency in the ion's frame is Doppler-shifted closer to resonance, so that light scattering takes place at an increased rate. Since the photons are emitted in random directions, the net effect is to slow the ion down, due to the absorption of photon momentum. When the velocity of the ion is directed along the laser beam, the light frequency is shifted farther from resonance, so that it scatters photons at a reduced rate. The net effect, over a motional cycle, is to damp the ion's velocity. If the laser is tuned above resonance, it causes heating. The minimum number of scattering events required to cool an ion substantially, starting from 300 K, is on the order of 10^4, which is the ratio of the ion's momentum to the photon's momentum. When the transition linewidth is larger than the motional frequencies the lowest temperature that can be achieved is on the order of

$\frac{1}{2}\hbar\gamma/k_B$, where γ is the natural linewidth in angular frequency units. Typically, this is on the order of 0.1 to 1 mK. This limit is the result of a balance between the average damping force just described and heating due to fluctuations in the force, caused by the discreteness of the momentum transfers in the photon scattering events. Other cases have been treated theoretically. If the natural linewidth of the transition is less than the frequencies of motion, laser cooling is analogous to motional side-band cooling of the magnetron mode, as described for electrons in Section III.

Laser cooling of the cyclotron and axial modes of an ion in a Penning trap works in the same way as in a harmonic trap, for the case where the natural linewidth is much greater than any of the frequencies of motion. However, cooling of the magnetron mode requires that the momentum transfers take place preferentially in the same direction as the magnetron velocity (Wineland *et al.*, 1978). This can be arranged by focusing the laser beam so that it is more intense on the side of the trap axis on which the magnetron velocity is directed along the direction of light propagation. Simultaneous cooling of all three modes is possible with the proper frequency detuning, orientation, and intensity profile of the laser beam (Itano and Wineland, 1982). The minimum temperature is again about $\frac{1}{2}\hbar\gamma/k_B$.

Laser cooling of stored ions has been achieved by groups at Heidelberg and Boulder and more recently at Seattle (Nagourney *et al.*, 1983). The Heidelberg group has cooled Ba$^+$ in an rf trap using a cw dye laser tuned to the 493-nm D$_1$ resonance line (Neuhauser *et al.*, 1978a, b, 1980). Ions excited to the 6 $^2P_{1/2}$ level can decay to the 5 $^2D_{3/2}$ metastable level as well as to the ground state. In order to provide continuous laser cooling, a second cw dye laser at 650 nm was used to transfer ions in the 5 $^2D_{3/2}$ level back to the 6 $^2P_{1/2}$ level, from which they could decay back to the ground state. Small numbers of ions (down to one ion) were observed visually and photographically, through a microscope, by LIF. Figure 16 is a photograph of a single ion from these experiments. The image of a single ion, after deconvolution of instrumental effects, was determined to be about 0.2 μm in diameter, implying a temperature on the order of 36 mK. The 6 $^2S_{1/2}$–6 $^2P_{1/2}$–5 $^2D_{3/2}$ stimulated Raman resonance on a single ion has been observed by scanning the 650-nm laser, while keeping the 493-nm laser fixed (Neuhauser *et al.*, 1981; Dehmelt, 1982). This resonance is potentially very sharp, because the lifetime of the metastable level is 17.5 sec (Schneider and Werth, 1979). The rf trap used for the single-ion experiments was extremely small, with a separation between endcap electrodes $2z_0 \simeq 0.5$ mm, in order to provide strong confinement. This makes it easier to satisfy the Dicke criterion (confinement to dimensions less than the wavelength), which makes the first-order Doppler broadening disappear (Dicke, 1953).

The Boulder group has cooled Mg$^+$ (Wineland *et al.*, 1978; Drullinger *et*

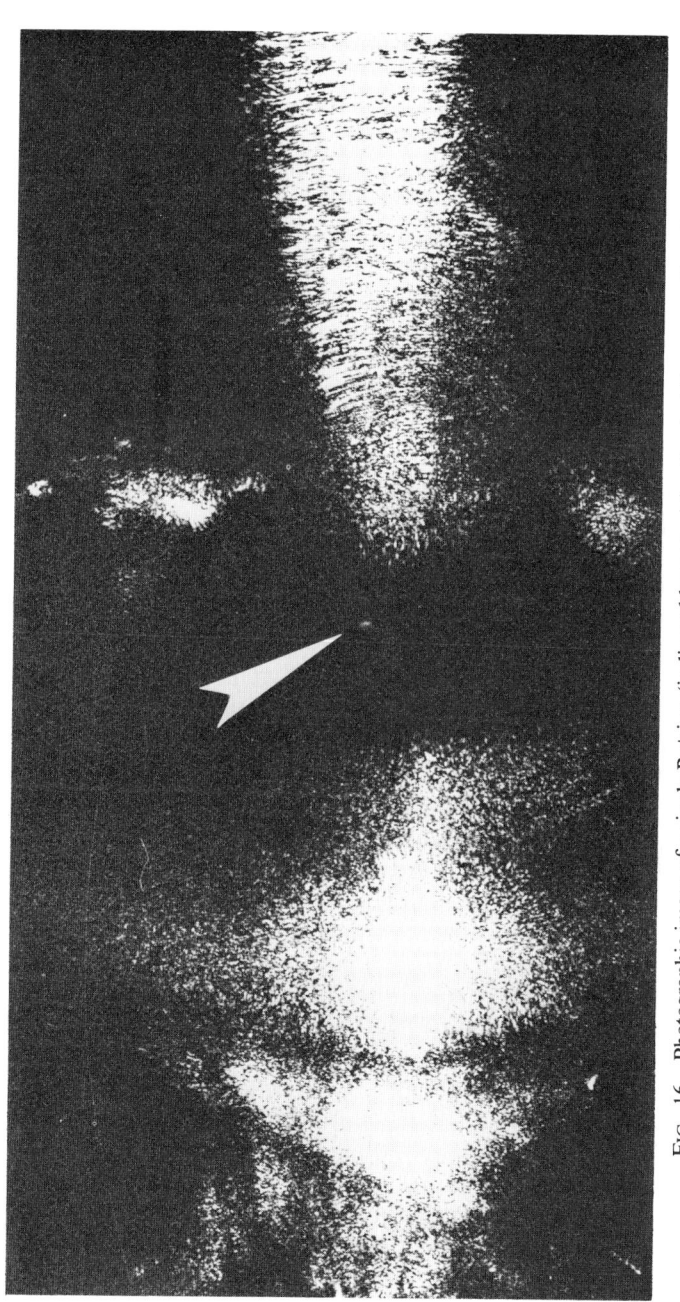

FIG. 16. Photographic image of a single Ba⁺ ion (indicated by arrow) localized at the center of an rf trap.

al., 1980; Wineland and Itano, 1981) and Be$^+$ (Wineland and Itano, 1982), stored in Penning traps. The D_2 resonance lines at 280 nm and 313 nm for Mg$^+$ and Be$^+$, respectively, were used to provide cooling. The UV radiation was generated by frequency doubling the outputs of cw dye lasers in nonlinear crystals. The UV power was typically on the order of 10 μW. These ions can each be cooled using only one laser frequency, unlike Ba$^+$, because of the absence of intermediate metastable levels. The primary advantage of the Penning trap over the rf trap is the absence of rf heating, so that a cloud of ions can be cooled, even with a low-power source. (For a single ion, rf heating is not as severe a problem.) For some spectroscopic applications, the magnetic field required for operation of the Penning trap may be a disadvantage. Also, because of practical limitations on the magnitude of the magnetic field which can be applied, satisfying the Dicke criterion with a cooled ion is more difficult than with an rf trap.

The first observation of laser cooling of Mg$^+$ ions was made by monitoring the power of the currents induced by the thermal motions of the ions at the axial frequency (the bolometric technique discussed in Section II). For a fixed number of ions, this power is proportional to the temperature. The temperature was observed to decrease when the ions were illuminated by light tuned below the resonance frequency and to increase when the light was tuned above resonance, clearly demonstrating the laser cooling and heating effects (Wineland et al., 1978). In later experiments, the resonance fluorescence photons emitted by the Mg$^+$ ions were counted by a photomultiplier tube. The measurement of the Doppler widths of the resonance curves, obtained by scanning the laser, indicated that the temperature of the ion cloud was below 0.5 K (Drullinger et al., 1980).

It is easier to achieve very low temperatures with small numbers of ions, because space charge increases the radial electric field, which increases the magnetron velocity. In the limit of a single ion, this effect is absent. In order to achieve the lowest possible temperatures, experiments were performed on single ions (Wineland and Itano, 1981). In Fig. 17, the detected fluorescence intensity from a small cloud of ^{24}Mg$^+$ ions is shown as a function of time. The three step-decreases are due to the loss of individual ^{24}Mg$^+$ ions by charge exchange with ^{25}Mg atoms coming from an oven. The last plateau above background is due to a single, isolated ion. Optical Doppler width measurements showed that the combined cyclotron and magnetron temperature was 50 ± 30 mK. The axial motion was not cooled as efficiently, because the direction of propagation of the light beam was nearly perpendicular to the z axis. The axial temperature was estimated to be about 600 mK from a measurement of the extent of the axial excursions. Preliminary, unpublished results on clouds of Be$^+$ ions indicate that cyclotron–magne-

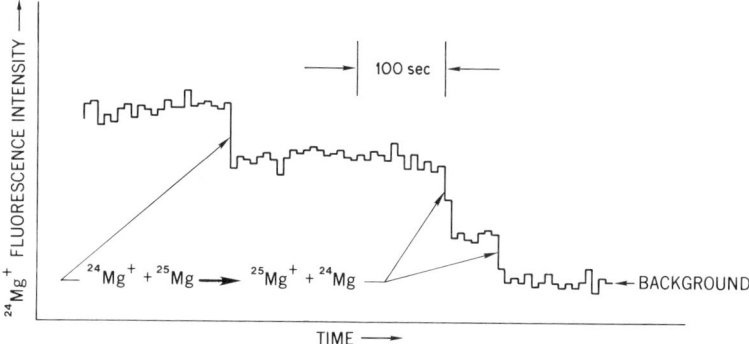

FIG. 17. Graph of the fluorescence from a small cloud of ^{24}Mg$^+$ ions. The three large steps are due to the loss of individual ions. The last plateau above background is the fluorescence from a single ion. (From Wineland and Itano, 1981.)

tron temperatures less than 100 mK have been achieved (Bollinger and Wineland, 1983).

C. Microwave and rf Atomic Ion Spectroscopy

Observation of microwave or rf resonance transitions by absorption of the radiation is not generally feasible because of the small number of stored ions, so more sensitive detection methods have to be devised. The general scheme of all such methods includes three steps: (1) creation of a population difference between two states, (2) transfer of population from the greater to the lesser populated state by resonant radiation, and (3) detection of the population transfer.

In the ion storage exchange collision (ISEC) method, collisions with polarized atoms are used to carry out steps (1) and (3). Dehmelt and Major (1962) detected the rf Zeeman resonance of ^{4}He$^+$ ions stored in an rf trap. The ions were polarized by spin-exchange collisions with a beam of Cs atoms, which were polarized by irradiation with circularly polarized resonance light. A change of the He$^+$ polarization, induced by resonant rf, was detected by a change of the number of ions remaining in the trap, due to spin-dependent charge-exchange collisions with the same atomic beam. The method was later used to observe hyperfine transitions in ground-state ^{3}He$^+$, with linewidths as small as 10 Hz (Fortson et al., 1966; Major and Dehmelt, 1968; Schuessler et al., 1969). Auxiliary rf fields were required in order to make transitions which did not change the average electron spin polarization detectable. The zero-field hyperfine separation $\Delta\nu_1$ was determined to be 8,665,649,867(10) Hz.

Prior and Wang (1977) have measured the zero-field hyperfine splitting in the metastable 2s level of $^3\text{He}^+$, Δv_2, obtaining a value of 1,083,354,980.7(8.8) Hz. The ions were stored in a Kingdon trap. The population difference was created by using an auxiliary microwave field to drive ions in one hyperfine state to the $2\,^2P_{1/2}$ level, which decayed immediately. The hyperfine transition was then driven, and ions which made the transition were detected by counting the Lyman α decay photons when the auxiliary microwave field was applied again. The quantity $D_{21} = (8\Delta v_2 - \Delta v_1)$, obtained from these two ion storage experiments, is of interest as a test of quantum electrodynamic calculations, because it is largely free of nuclear structure corrections.

Optical-pumping double-resonance methods, in which population differences are created by illumination with resonance light and transitions are detected by changes in the absorption or fluorescence of the light, have been used previously on many neutral and charged atomic systems. Major and Werth (1973, 1978) were the first to apply this method to ion storage spectroscopy, in a measurement of the ground-state hyperfine structure of $^{199}\text{Hg}^+$ ions in an rf trap. A $^{202}\text{Hg}^+$ 194-nm D_1 resonance lamp was used to optically pump the $^{199}\text{Hg}^+$ ions into the $F = 0$ hyperfine state, because its spectrum strongly overlapped the $^{199}\text{Hg}^+$ D_1 resonance transition from the $F = 1$ state. The hyperfine resonance was detected by an increase in the fluorescence intensity. Work on this system has continued, because of its possible use as a frequency standard (Schuessler, 1971a; McGuire et al., 1978; Jardino and Desaintfuscien, 1980; Jardino et al., 1981a,b; Cutler et al., 1982; Maleki, 1982). The most recent determination of the zero-field frequency is $\Delta v = 40,507,347,996.9(0.3)$ Hz (Cutler et al., 1982).

Recently, optical pumping experiments on stored ions have been performed using tunable lasers as light sources. The ground-state hyperfine splittings of $^{137}\text{Ba}^+$ (Blatt and Werth, 1982), $^{135}\text{Ba}^+$ (Becker et al., 1981), and $^{171}\text{Yb}^+$ (Blatt et al., 1982) have been measured, using pulsed dye lasers and rf traps. The results were $\Delta v(^{137}\text{Ba}^+) = 8,037,741,667.69(0.37)$ Hz, $\Delta v(^{135}\text{Ba}^+) = 7,183,340,234.35(0.47)$ Hz, and $\Delta v(^{171}\text{Yb}^+) = 12,642,812,124.2(1.4)$ Hz. Microwave resonances as narrow as 60 mHz were observed in $^{171}\text{Yb}^+$ (see Fig. 18). This has a line Q (resonance frequency divided by FWHM) of 2×10^{11}. In some cases, optical pumping out of the absorbing ground state prevents use of the double-resonance method. This problem may be overcome, however, with the use of collisional relaxation (Blatt and Werth, 1982; Ruster et al., 1983).

Microwave and rf transitions in laser-cooled Mg^+ (Wineland et al., 1980; Itano and Wineland, 1981) and Be^+ (Wineland and Itano, 1982) stored in Penning traps have been observed by optical-pumping, double-resonance techniques. Laser cooling greatly reduces the second-order Doppler shift,

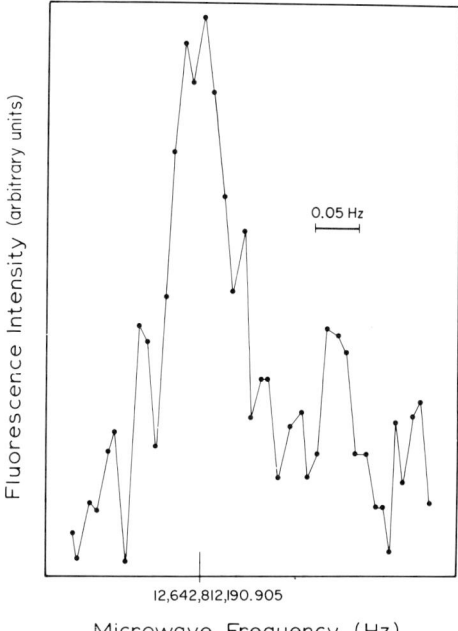

FIG. 18. Graph of the hyperfine resonance of trapped ^{171}Yb$^+$ ions. The line Q is 2×10^{11}. (From Blatt *et al.*, 1982.)

which is a major source of uncertainty in other high-resolution stored-ion spectroscopic measurements. The optical pumping process is somewhat unusual in that population is transferred *into* the ground-state sublevel which is coupled most strongly to the excited state by the laser. A transition from this sublevel to another causes a temporary decrease in the fluorescence level. Since the number of photons *not* scattered during the time it takes the laser to pump an ion back into its original sublevel can be much greater than one, transitions can be detected with almost 100% efficiency (Wineland *et al.*, 1980). Electronic spin reorientation transitions in ^{24}Mg$^+$ and both electronic and nuclear spin reorientation transitions in ^{25}Mg$^+$ were observed. Auxiliary rf fields were required in order to make some of the transitions observable. At a magnetic field near 1.24 T, the first derivative of the $(M_I = -\frac{3}{2}, M_J = \frac{1}{2})$ to $(M_I = -\frac{1}{2}, M_J = \frac{1}{2})$ transition goes to zero. The resonance shown in Fig. 19 was obtained at that field and has a linewidth of only 12 mHz. The oscillatory lineshape results from the use of the Ramsey interference technique (Ramsey, 1956), applied in the time domain. Two coherent rf pulses, 1.02 sec long and separated by 41.4 sec, were applied. By fitting resonance frequencies to the Breit–Rabi formula, values were ob-

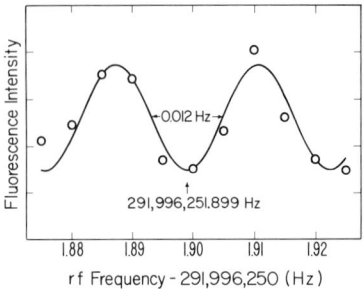

FIG. 19. Graph of the hyperfine resonance of trapped ^{25}Mg$^+$ ions. The oscillatory lineshape results from the use of the Ramsey resonance method, implemented by applying two coherent rf pulses 1.02 sec long, separated by 41.4 sec. The solid curve is a theoretical lineshape. (From Itano and Wineland, 1981.)

tained for the hyperfine constant [$A = -596,254,376(54)$ Hz] and the nuclear to electronic g-factor ratio ($g_I/g_J = 9.299,484(75) \times 10^{-5}$). The uncertainties could be greatly reduced by observing another field-independent transition. Similar spectroscopy was performed on ^{9}Be$^+$. Two field-independent transitions have been observed: ($M_I = \frac{3}{2}$, $M_J = -\frac{1}{2}$) to ($M_I = \frac{1}{2}$, $M_J = -\frac{1}{2}$,) at 0.68 T and ($M_I = -\frac{3}{2}$, $M_J = \frac{1}{2}$) to ($M_I = -\frac{1}{2}$, $M_J = \frac{1}{2}$) at 0.82 T. The magnetic field was calibrated by observing the cyclotron resonance of electrons which were trapped alternately in the same apparatus as the ions. Preliminary values for the ground-state constants are $A = -625,008,837.048(10)$ Hz, $g_I/g_J = 2.134,779,853(2) \times 10^{-4}$, and $g_J = 2.002,263(6)$. A ^{9}Be$^+$ Ramsey resonance curve is shown in Fig. 20. The rf pulses were 2 sec long and were separated by 4 sec.

Gräff (1982) has proposed a method of observing parity- and time reversal-violating interactions by driving nuclear spin reorientation transitions in stored atomic ions, using rf electric and magnetic fields of the same fre-

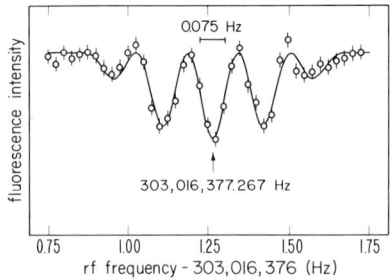

FIG. 20. Graph of the hyperfine resonance of trapped ^{9}Be$^+$ ions, obtained by the Ramsey method. Two 2-sec rf pulses, separated by 4 sec, were used. The solid curve is a least squares fit to the theoretical lineshape.

quency but different relative phases. The method is particularly sensitive to the nuclear spin-dependent part of the parity-violating weak neutral current interaction, which has not yet been observed in atomic systems. Relative to atomic beam methods, the stored ion method has the advantage of longer coherent interaction times and the disadvantage of lower signal-to-noise ratio.

D. Application to Frequency Standards

The use of microwave or optical transitions of stored atomic ions in frequency standards has the combined advantages of long coherent interaction times (hence narrow linewidths) and small perturbations. The main disadvantage is the low signal-to-noise ratio due to the small number of ions that can be stored.

Work on frequency standards based on the 40.5-GHz hyperfine transition of ^{199}Hg$^+$ ions has already been mentioned. The use of a field-independent transition in laser-cooled ^{201}Hg$^+$ ions stored in a Penning trap has been proposed for a primary microwave frequency standard (Wineland et al., 1981a). For microwave frequency standards, it may be particularly desirable to use a large number of ions in order to increase the signal-to-noise ratio. Therefore, the largest systematic frequency shift may be due to the second-order Doppler shift. For an rf trap this will be caused by the kinetic energy in the micromotion; for a Penning trap this will be caused by the kinetic energy in the magnetron motion (Wineland, 1983).

Optical frequency standards have the basic advantage of higher Q for fixed coherent interaction time. Dehmelt has proposed optical frequency standards based on forbidden transitions of single, laser-cooled group IIIA ions (Tl$^+$, In$^+$, Ga$^+$, Al$^+$, or B$^+$) stored in small rf traps (Dehmelt, 1982). Penning traps or Penning/rf trap combinations might also be used (Wineland and Itano, 1982b). The $6\,^2S_{1/2}$–$6\,^2P_{1/2}$–$5\,^2D_{3/2}$ Raman transition in Ba$^+$ could be used as a reference to generate a stable infrared difference frequency in a nonlinear crystal (Neuhauser et al., 1981; Dehmelt, 1982). Also in Ba$^+$, the $5\,^2D_{5/2}$ to $5\,^2D_{3/2}$ 12-μm transition (Dehmelt et al., 1982) and the quadrupole-allowed $6\,^2S_{1/2}$ to $5\,^2D_{5/2}$ 1.8-μm transition (Dehmelt, 1981b) have been proposed as standards. Other high-Q optical transitions in Sr$^+$ (Dehmelt and Walther, 1975), Pb$^+$, I$^+$, and Bi$^+$ (Strumia, 1978) have been suggested for stored ion frequency standards. The two-photon (Bender et al., 1976; Wineland et al., 1981a) or single-photon quadrupole (Wineland et al., 1981b) $5d^{10}\,6s\,^2S_{1/2}$ to $5d^9\,6s^2\,^2D_{5/2}$ transition in Hg$^+$ has also been suggested. Two-photon transitions have the advantage of being first-order Doppler free even for a cloud of many ions, where it is impossible to satisfy the Dicke criterion at optical wavelengths. They have the disadvantage that the large fields required to drive the transition cause ac Stark shifts.

E. Molecular Ion Spectroscopy

As the species under investigation becomes more complex, fewer of the trapped ions reside in the particular states of interest. Therefore, the fact that molecular spectroscopy is possible at all in the ion traps where so few ions are present is indeed remarkable. Nevertheless, the traps do provide some advantages over other methods for molecular ion spectroscopy. First, the environment in a high vacuum can provide cleaner operating conditions for the ions, thereby suppressing collision-induced chemical reactions. Because of the long storage times, the state distributions of ions can be allowed to relax, thus increasing the number of ions in a particular ground state. In addition, the traps (particularly the rf trap) can be operated in a mass-selective mode in order to reduce the effects of background ions. The use of LIF can also be used to "tag" certain ions for chemical studies.

The first successful molecular ion experiments measured high-resolution rf spectra of H_2^+ ions (Dehmelt and Jefferts, 1962; Jefferts, 1968, 1969; Richardson et al., 1968; Menasian and Dehmelt, 1973). Transitions were detected by using the orientation dependence of photodissociation. These measurements were particularly interesting because they could be compared to high accuracy with theory. More recently, the group of Mahan has measured the spectra of heavier ions using LIF techniques (Grieman et al., 1980, 1981a,b; see also Danon et al., 1982). Dunbar and Beauchamp (Dunbar and Kramer, 1973; Freiser and Beauchamp, 1974; Dunbar et al., 1977) have also used LIF for photodissociation studies.

VI. Negative Ion Spectroscopy

Negative ions have been observed to have only one, or at most a few, bound electronic states, in contrast to neutral or positively charged atoms or molecules, which have an infinite number. Therefore, line emission or absorption spectroscopy is not generally feasible. Most spectroscopy of negative ions is based on photodetachment (absorption of a photon with loss of an electron) or, for molecules, on photodissociation (absorption of a photon with fragmentation of the molecule). For more general reviews, including results of other experimental methods, such as ion beams and drift tubes, see the article by Hotop and Lineberger (1975) on atomic negative ions, the articles by Corderman and Lineberger (1979) and by Janousek and Brauman (1979) on molecular negative ions, and the book by Massey (1976).

A. Atoms

The only spectroscopic studies of negative atomic ions to use ion storage techniques are those of Larson and co-workers. Perhaps this is because photodetachment cross sections and electron affinities can be determined using ion beams, either by using a tunable light source and observing the threshold wavelengths or by using a fixed-wavelength source and measuring the energies of the detached electrons. However, the long observation times available with trapped ions can be useful in high-resolution studies. Also,

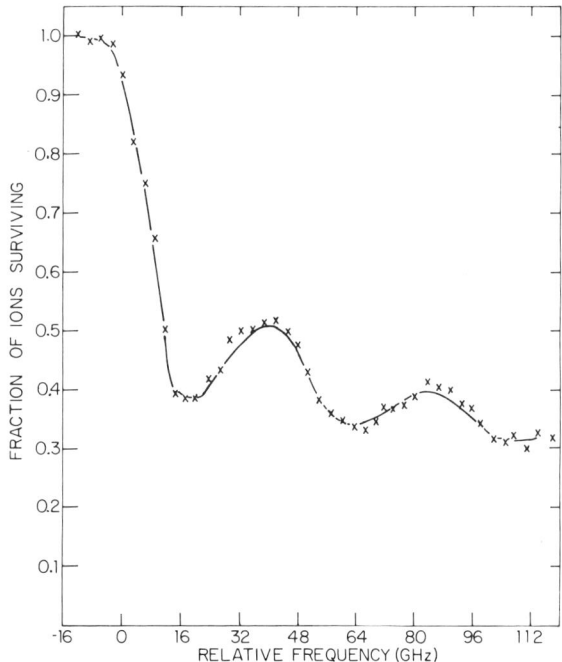

FIG. 21. Photodetachment data near the $S^-(^2P_{3/2}) \rightarrow S(^3P_2)$ threshold. The fraction of ions surviving laser illumination is plotted as a function of the light frequency (with an arbitrary zero). The light had π polarization and the magnetic field was 1.57 T. The solid curve is a theoretical prediction, with three parameters adjusted to give agreement with the data. (From Blumberg *et al.*, 1979.)

effects due to high magnetic fields can be conveniently studied in a Penning trap.

Near-threshold photodetachment of S⁻ was observed in a Penning trap in a magnetic field of the order of 1 T by Blumberg et al. (1978). The relative number of ions surviving after irradiation by a tunable cw dye laser was measured by resonantly exciting their axial motion and detecting the image current at the ring electrode at the second harmonic. The cross section was found to have an oscillatory dependence on light frequency, with a spacing between peaks approximately equal to the electron cyclotron frequency. These peaks correspond to thresholds for excitation of the detached electron to quantized cyclotron levels. The light polarization and frequency dependences of the cross sections are well described by a theory which ignores the final-state interaction between the detached electron and the neutral atom, but which includes the Zeeman splittings of the initial ionic state and the final atomic state and the broadening due to the Doppler effect and the motional electric field (Blumberg et al., 1979) (see Fig. 21). The effect of including the final-state interaction in lowest order is to reduce the cross section near the thresholds (Larson and Stoneman, 1982).

Different magnetic sublevels have different cross sections for photode-

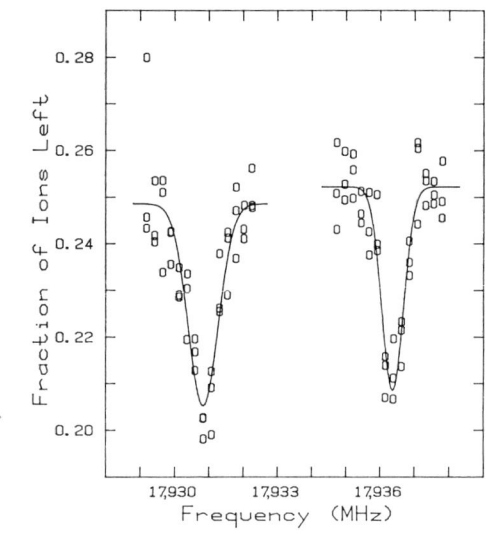

FIG. 22. Microwave Zeeman resonances between magnetic sublevels of the $^2P_{3/2}$ level of S⁻ at a magnetic field of about 0.96 T, observed by state-dependent photodetachment. The lower frequency transition is $m_J = -\frac{1}{2}$ to $m_J = -\frac{3}{2}$ and the higher frequency one is $m_J = +\frac{1}{2}$ to $m_J = +\frac{3}{2}$. (From Jopson and Larson, 1981.)

tachment by polarized light, a fact which can be used to detect microwave Zeeman transitions (Jopson and Larson, 1980). The magnetic moments of S$^-$ (Jopson and Larson, 1981) and O$^-$ (Larson and Jopson, 1981), and the hyperfine constants of ^{33}S$^-$ (Jopson et al., 1981) have been measured by this technique (see Fig. 22). In these experiments, the magnetic field is calibrated by driving the cyclotron resonance of electrons trapped alternately in the same apparatus.

B. MOLECULES

Larson and Stoneman (1982) have observed photodetachment of SeH$^-$ near threshold, with a resolution of about 0.2 cm^{-1}. The rotational structure is well resolved. The oscillatory structure at the electron cyclotron frequency continues for many cycles, in contrast to the atomic case (see Fig. 23). This

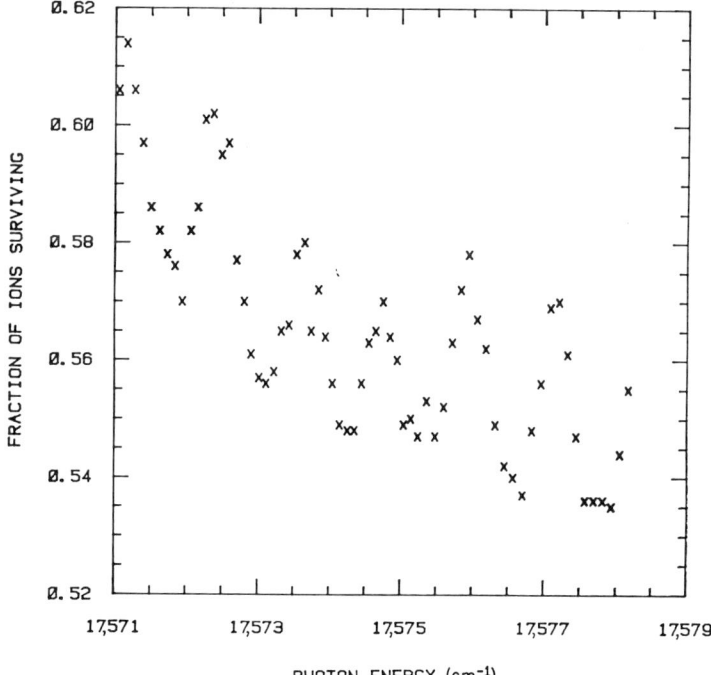

FIG. 23. Photodetachment data for SeH$^-$ in the vicinity of a rotational ($J \rightarrow J'$) threshold. The magnetic field was 1.3 T. The oscillatory structure has a period at or very close to the electron cyclotron frequency.

may be due to the final-state interaction between the electron and the molecular electric dipole moment.

Photodetachment or photodissociation of many different negative molecular ions has been observed in ICR spectrometers (Smyth et al., 1971; Eyler, 1974; Dunbar and Hutchinson, 1974). Most of this work has been done by Brauman and co-workers. The basic advantages of ICR techniques over beam techniques are the large variety of ions that can be produced by dissociative attachment, followed by ion–molecule reactions, and the long trapping times, which allow many of the excited states to decay before interrogation, thus simplifying the spectra. The quantities obtained are the electron affinities and some spectroscopic constants (vibrational, spin–orbit, etc.). Much of this work has been reviewed by Corderman and Lineberger (1979) and by Janousek and Brauman (1979). More recently, ICR techniques have been used to observe infrared multiphoton photodetachment (Rosenfeld et al., 1979) and narrow resonances near threshold which may be due to Rydberg-like states bound by the electric dipole moment of the neutral molecule (Jackson et al., 1979, 1981).

VII. Radiative Lifetime Measurements

For longer-lived states, the ion traps may be particularly well suited to make measurements of radiative lifetimes. Clearly this is a result of the relatively perturbation-free environment and the long storage times possible. This is indicated by the measurements of the lifetimes of the $5\,^2D_{3/2}$ state in Ba^+, $\tau \simeq 17.5$ sec (Schneider and Werth, 1979), the $5\,^2D_{5/2}$ state in Ba^+, $\tau \simeq 47$ sec (Plumelle et al., 1980), and the $2\,^3S_1$, state in Li^+, $\tau \simeq 58.6$ sec (Knight and Prior, 1980; see also Osipowicz and Werth, 1981). Of particular interest for atomic physics are the lifetime measurements of Prior and colleagues at Berkeley on simple atomic ions since these can be compared with various theories. Lifetimes of $2\,^1S_0\,Li^+(\tau \simeq 503\,\mu sec)$ in a Penning trap (Prior and Shugart, 1971), 2s He^+ ($\tau \simeq 1.92$ msec) in a Kingdon trap (Prior, 1972), and $2\,^3S_1\,Li^+$ ($\tau \simeq 58.6$ sec) in an rf trap (Knight and Prior, 1980) have been measured. More recently, lifetimes of atomic and molecular ions of atmospheric and astrophysical interest have been determined. For example, the lifetimes of the $^5S_2$ state of N^+ (Knight 1982) and the $^3P_1$ metastable level of Si^{2+} (Kwong et al., 1983), and lifetimes in N_2^+ (Mahan and O'Keefe, 1981a,b) and CO^+ and CH^+ (Mahan and O'Keefe; 1981a,b; Danon et al., 1982) have been measured.

ACKNOWLEDGMENTS

The authors wish to acknowledge many of the researchers in this field who have contributed information in the form of preprints, reprints, and private communications. We especially thank R. D. Knight, G. H. Dunn, J. J. Bollinger, and H. G. Dehmelt for their suggestions and comments on portions of the manuscript. Two of the authors (D.J.W.) and (W.M.I.) acknowledge the support of the Air Force Office of Scientific Research and the Office of Naval Research; the third author (R.S.V.) acknowledges the support of the National Science Foundation.

REFERENCES

André, J., and Vedel, F. (1977). *J. Phys. (Orsay, Fr.)* **38**, 1381.
André, J., Vedel, F., and Vedel, M. (1979). *J. Phys. (Orsay, Fr.)* **40**, L-633.
André, J., Teboul, A., Vedel, F., and Vedel, M. (1981). *J. Phys. (Orsay, Fr.)* **42**, C8-315.
Becker, W., Blatt, R., and Werth, G. (1981). *J. Phys. (Orsay, Fr.)* **42**, C8-339.
Bender, P. L., Hall, J. L., Garstang, R. H., Pichanick, F. M. J., Smith, W. W., Barger, R. L., and West, J. B. (1976). *Bull. Am. Phys. Soc.* [2] **21**, 599.
Benilan, M. N., and Audoin, C. (1973). *Int. J. Mass Spectrom. Ion Phys.* **11**, 421.
Blatt, R., and Werth, G. (1982). *Phys. Rev. A*[3] **25**, 1476.
Blatt, R., and Werth, G. (1983). *In* "Precision Measurements and Fundamental Constants. II" (B. N. Taylor and W. D. Phillip, eds.). *NBS Spec. Publ. (U.S.)* **617** (in press).
Blatt, R., Schmeling, U., and Werth, G. (1979). *Appl. Phys.* **20**, 295.
Blatt, R., Schnatz, H., and Werth, G. (1982). *Phys. Rev. Lett.* **48**, 1601.
Blumberg, W. A. M., Jopson, R. M., and Larson, D. J. (1978). *Phys. Rev. Lett.* **40**, 1320.
Blumberg, W. A. M., Itano, W. M., and Larson, D. J. (1979). *Phys. Rev. A* [3] **19**, 139.
Bollinger, J. J., and Wineland, D. J. (1983). To be published.
Bonner, R. F., March, R. E., and Durup, J. (1976). *Int. J. Mass Spectrom. Ion Phys.* **22**, 17.
Bonner, R. F., Fulford, J. E., March, R. E., and Hamilton, G. F. (1977). *Int. J. Mass Spectrom. Ion Phys.* **24**, 255.
Borodkin, A. S. (1978). *Zh. Tekh. Fiz.* **48**, 889; *Sov. Phys.—Tech. Phys. (Engl. Transl.)* **23**, 520.
Brown, L. S., and Gabrielse, G. (1982). *Phys. Rev. A* [3] **25**, 2423.
Bryne, J., and Farago, P. S. (1965). *Proc. Phys. Soc. London* **86**, 801.
Church, D. A. (1969). *J. Appl. Phys.* **40**, 3127.
Church, D. A. (1982). *In* "Physics of Electronic and Atomic Collisions" (S. Datz, ed.), p. 533. North-Holland Publ., Amsterdam.
Church, D. A., and Dehmelt, H. G. (1969). *J. Appl. Phys.* **40**, 3421.
Church, D. A., and Mokri, B. (1971). *Z. Phys.* **244**, 6.
Comisarow, M. B. (1981). *Int. J. Mass Spectrom. Ion Phys.* **37**, 251.
Corderman, R. R., and Lineberger, W. C. (1979). *Annu. Rev. Phys. Chem.* **30**, 347.
Coutandin, J., and Werth, G. (1982). *Appl. Phys. B* **29**, 89.
Cutler, L. S., Giffard, R. P., and McGuire, M. D. (1982). *NASA Conf. Publ.* **2220**, 563.
Danon, J., Mauclaire, G., Govers, T. R., and Marx, R. (1982). *J. Chem. Phys.* **76**, 1255.
Dawson, P. H. (1980). *Adv. Electron. Electron Phys.* Suppl. **13B**, 173.
Dawson, P. H., and Lambert, C. (1975). *Int. J. Mass. Spectrom. Ion Phys.* **16**, 269.
Dawson, P. H., and Whetten, N. R. (1968). *J. Vac. Sci. Technol.* **5**, 1.

deGrassie, J. S., and Malmberg, J. M. (1980). *Phys. Fluids* **23**, 63.
Dehmelt, H. G. (1956). *Phys. Rev.* **103**, 1125.
Dehmelt, H. G. (1958). *Phys. Rev.* **109**, 381.
Dehmelt, H. G. (1967). *Adv. At. Mol. Phys.* **3**, 53.
Dehmelt, H. G. (1969). *Adv. At. Mol. Phys.* **5**, 109.
Dehmelt, H. G. (1975). *In* "The Physics of Electronic and Atomic Collisions" (J. S. Risley and R. Geballe, eds.), p. 857. Univ. of Washington Press, Seattle.
Dehmelt, H. G. (1976). *In* "Atomic Masses and Fundamental Constants" (J. H. Sanders and A. H. Wapstra, eds.), 5th, p. 499. Plenum, New York.
Dehmelt, H. G. (1981a). *At. Phys.* **7**, 337.
Dehmelt, H. G. (1981b). *J. Phys. (Orsay, Fr.)* **42**, C8-299.
Dehmelt, H. (1982). *IEEE Trans. Instrum. Meas.* **IM-31**, 83.
Dehmelt, H. G. (1983). *In* "Advances in Laser Spectroscopy" (F.T. Arecchi, F. Strumia, and H. Walther, eds.). Plenum, New York.
Dehmelt, H. G., and Jefferts, K. B. (1962). *Phys. Rev.* **125**, 1318.
Dehmelt, H. G., and Major, F. G. (1962). *Phys. Rev. Lett.* **8**, 213.
Dehmelt, H. G., and Walls, F. L. (1968). *Phys. Rev. Lett.* **21**, 127.
Dehmelt, H. G., and Walther, H. (1975). *Bull. Am. Phys. Soc.* [2] **20**, 61.
Dehmelt, H. G., Schwinberg, P. B., and Van Dyck, R. S., Jr., (1978). *Int. J. Mass Spectrom. Ion Phys.* **26**, 107.
Dehmelt, H. G., Van Dyck, R. S., Jr., Schwinberg, P., and Gabrielse, G. (1979). *Bull. Am. Phys. Soc.* [2] **24**, 757.
Dehmelt, H., Nagourney, W., and Janik, G. (1982). *Bull. Am. Phys. Soc.* [2] **27**, 402.
Dicke, R. H. (1953). *Phys. Rev.* **89**, 472.
Donets, E. D., and Pikin, A. I. (1976). *Zh. Eksp. Teor. Fiz.* **70**, 2025; *Sov. Phys.—JETP (Engl. Transl.)* **43**, 1057.
Drullinger, R. E., Wineland, D. J., and Bergquist, J. C. (1980). *Appl. Phys.* **22**, 365.
Duchêne, J. L., Audoin, C., and Schermann, J. P. (1977). *Metrologia* **13**, 157.
Dunbar, R. C., and Hutchinson, B. B. (1974). *J. Am. Chem. Soc.* **96**, 1154.
Dunbar, R. C., and Kramer, J. M. (1973). *J. Chem. Phys.* **58**, 1266.
Dunbar, R. C., Fu, E. W., and Olah, G. A. (1977). *J. Am. Chem. Soc.* **99**, 7502.
Ekstrom, P., and Wineland, D. (1980). *Sci. Am.* **243**, No. 2, 104.
Eyler, J. R. (1974). *Rev. Sci. Instrum.* **45**, 1154.
Field, J. H., Picasso, E., and Combley, F. (1979). *Usp. Fiz. Nauk* **127**, 553; *Sov. Phys.—Usp. (Engl. Transl.)* **22**, 199.
Fisher, E. (1959). *Z. Phys.* **156**, 1.
Fortson, E. N., Major, F. G., and Dehmelt, H. (1966). *Phys. Rev. Lett.* **16**, 221.
Freiser, B. J., and Beauchamp, J. L. (1974). *J. Am. Chem. Soc.* **96**, 6260.
Gaboriaud, M. N., Desaintfuscien, M., and Major, F. G. (1981). *Int. J. Mass Spectrom. Ion Phys.* **41**, 109.
Gabrielse, G., and Dehmelt, H. G. (1980). *Bull. Am. Phys. Soc.* [2] **25**, 1149.
Gabrielse, G., and Dehmelt, H. G. (1983). *In* "Precision Measurements and Fundamental Constants. II" (B. N. Taylor and W. D. Phillips, eds.). *NBS Spec. Publ. (U.S.)* **617** (in press).
Gärtner G., and Klempt, E. (1978) *Z. Phys. A* **287**, 1.
Gilleland J. R., and Rich, A. (1972). *Phys. Rev. A* [3] **5**, 38.
Gräff, G. (1982). *Z. Phys. A* **305**, 107.
Gräff, G., and Holzscheiter, M. (1980). *Phys. Lett. A* **79A**, 380.
Gräff, G., and Major, F. G., Roeder, R. W. H., and Werth, G. (1968). *Phys. Rev. Lett.* **21**, 340.
Gräff, G., Klempt, E., and Werth, G. (1969). *Z. Phys.* **222**, 201.
Gräff, G., Huber, K., Kalinowsky, H., and Wolf, H. (1972). *Phys. Lett. A* **41A**, 277.

Gräff, G., Kalinowsky, H., and Traut, J. (1980). *Z. Phys.* **297**, 35.
Gräff, G., Kalinowsky, H., and Traut, J. (1983). *In* "Precision Measurements and Fundamental Constants. II" (B. N. Taylor and W. D. Phillips, eds.). *NBS Spec. Publ. (U.S.)* **617** (in press).
Grieman, F. J., Mahan, B. H., and O'Keefe, A. (1980). *J. Chem. Phys.* **72**, 4246.
Grieman, F. J., Mahan, B. H., and O'Keefe, A. (1981a). *J. Chem. Phys.* **74**, 857.
Grieman, F. J., Mahan, B. H., O'Keefe, A., and Winn, J. S. (1981b). *Faraday Discuss. Chem. Soc.* **71**, 191.
Hamdan, M., Birkinshaw, K., and Hasted, J. B. (1978). *J. Phys. B* **11**, 331.
Hänsch, T. W., and Schawlow, A. L. (1975). *Opt. Commun.* **13**, 68.
Harrison, E. R. (1959). *Am. J. Phys.* **27**, 314.
Hasted, J. B., and Awad, G. L. (1972). *J. Phys. B* **5**, 1719.
Haught, A. F., and Polk, D. H. (1966). *Phys. Fluids* **9**, 2047.
Hooverman, R. M. (1963). *J. Appl. Phys.* **34**, 3505.
Hotop, H., and Lineberger, W. C. (1975). *J. Phys. Chem. Ref. Data* **4**, 539.
Iffländer, R., and Werth, G. (1977a). *Metrologia* **13**, 167.
Iffländer, R., and Werth, G. (1977b). *Opt. Commun.* **21**, 411.
Itano, W. M., and Wineland, D. J. (1981). *Phys. Rev. A* [3] **24**, 1364.
Itano, W. M., and Wineland, D. J. (1982). *Phys. Rev. A* [3] **25**, 35.
Jackson, R. L., Zimmerman, A. H., and Brauman, J. I. (1979). *J. Chem. Phys.* **71**, 2088.
Jackson, R. L., Hiberty, P. C., and Brauman, J. I. (1981). *J. Chem. Phys.* **74**, 3705.
Janousek, B. K., and Brauman, J. I. (1979). *In* "Gas Phase Ion Chemistry" (M. T. Bowers, Vol. 2, ed.), pp. 53–86. Academic Press New York.
Jardino, M., and Desaintfuscien, M. (1980). *IEEE Trans. Instrum. Meas.* **IM-29**, 163.
Jardino, M., Desaintfuscien, M., Barillet, R., Viennet, J. Petit, P., and Audoin, C. (1981a). *Appl. Phys.* **24**, 107.
Jardino, M., Desaintfuscien, M., and Plumelle, F. (1981b). *J. Phys. (Orsay, Fr.)* **42**, C8-327.
Javanainen, J. (1980). *Appl. Phys.* **23**, 175.
Javanainen, J. (1981a). *J. Phys. B* **14**, 2519.
Javanainen, J. (1981b). *J. Phys. B* **14**, 4191.
Javanainen, J. and Stenholm, S. (1980). *Appl. Phys.* **21**, 283.
Javanainen, J. and Stenholm, S. (1981a). *Appl. Phys.* **24**, 71.
Javanainen, J. and Stenholm, S. (1981b). *Appl. Phys.* **24**, 151.
Jefferts, K. B. (1968). *Phys. Rev. Lett.* **20**, 39.
Jefferts, K. B. (1969). *Phys. Rev. Lett.* **23**, 1476.
Jeffries, J. B. (1980). Ph. D. thesis, University of Colorado, Boulder.
Johnson, C. E. (1983). *Bull. Am. Phys. Soc.* [2] **28**, 796.
Jopson, R. M., and Larson, D. J. (1980). *Opt. Lett.* **5**, 531.
Jopson, R. M., and Larson, D. J. (1981). *Phys. Rev. Lett.* **47**, 789.
Jopson, R. M., and Trainham, R., and Larson, D. J. (1981). *Bull. Am. Phys. Soc.* [2] **26**, 1306.
Kienow, E., Klempt, E., Lange, F., and Neubecker, K. (1974). *Phys. Lett. A* **46A**, 441.
Kingdon, K. H. (1923). *Phys. Rev.* **21**, 408.
Kinoshita, T., and Lindquist, W. B. (1981). *Phys. Rev. Lett.* **47**, 1573.
Knight, R. D. (1981). *Appl. Phys. Lett.* **38**, 221.
Knight, R. D. (1982). *Phys. Rev. Lett.* **48**, 792.
Knight, R. D. (1983). *Int. J. Mass Spectrom. Ion Phys.* (*In press*).
Knight, R. D., and Prior, M. H. (1979). *J. Appl. Phys.* **50**, 3044.
Knight, R. D., and Prior, M. H. (1980). *Phys. Rev. A* [3] **21**, 179.
Kwong, H. S., Johnson, B. C., Smith, P. L., and Parkinson, W. H. (1983). *Phys. Rev. A* [3] **27**, 3040.
Lakkaraju, H. S., and Schuessler, H. A. (1982). *J. Appl. Phys.* **53**, 3967.

Larson, D. J., and Jopson, R. M. (1981). *Springer Ser. Opt. Sci.* **30**, 369.
Larson, D. J., and Stoneman, R. (1982). *J. Phys. (Orsay, Fr.)* **43**, C2-285.
Lewis, R. R. (1982). *J. Appl. Phys.* **53**, 3975.
Lewis, R. R. (1983) *Phys. Rev. A (in press)*.
Liebes, S., Jr., and Franken, P. (1959). *Phys. Rev.* **116**, 633.
McGuire, M. D., and Fortson, E. N. (1974). *Phys. Rev. Lett.* **33**, 737.
McGuire, M. D., Petsch, R., and Werth, G. (1978). *Phys. Rev. A* [3] **17**, 1999.
McIlraith, A. H. (1971). *J. Vac. Sci. Technol.* **9**, 209.
McIver, R. T., Jr. (1970). *Rev. Sci. Instrum.* **41**, 555.
McIver, R. T., Jr. (1978a). *Lect. Notes Chem.* **7**, 97.
McIver, R. T., Jr. (1978b). *Rev. Sci. Instrum.* **49**, 111.
McLachlan, N. W. (1947). "Theory and Application of Mathieu Functions." Oxford Univ. Press (Clarendon), London and New York.
Mahan, B. H., and O'Keefe, A. (1981a). *J. Chem. Phys.* **74**, 5606.
Mahan, B. H., and O'Keefe, A (1981b). *Astrophys. J.* **248**, 1209.
Major, F. G. (1977). *J. Phys. (Orsay, Fr.)* **38**, L-221.
Major, F. G., and Dehmelt, H. G. (1968). *Phys. Rev.* [2] **170**, 91.
Major, F. G., and Duchêne, J. L. (1975). *J. Phys. (Orsay, Fr.)* **36**, 953.
Major, F. G., and Schermann, J. P. (1971). *Bull. Am. Phys. Soc.* [2] **16**, 838.
Major, F. G., and Werth, G. (1973). *Phys. Rev. Lett.* **30**, 1155.
Major, F. G., and Werth, G. (1978). *Appl. Phys.* **15**, 201.
Maleki, L. (1981). *NASA Conf. Publ.* **2220**, 593.
Malmberg, J. H., and Driscoll, C. F. (1980). *Phys. Rev. Lett.* **44**, 654.
Massey, H. S. W. (1976). "Negative Ions," 3rd ed. Cambridge Univ. Press, London and New York.
Mather, R. E., Waldren, R. M., Todd, J. F. J., and March, R. E. (1980). *Int. J. Mass Spectrom. Ion Phys.* **33**, 201.
Menasian, S. C., and Dehmelt, H. G. (1973). *Bull. Am. Phys. Soc.* [2] **18**, 408.
Minogin, V. G. (1982). *Usp. Fiz. Nauk* 137, 173 [*Sov. Phys. Usp.* **25**, 359—*Engl. Transl.*]
Nagourney, W., and Dehmelt, H. G. (1981a). *Bull. Am. Phys. Soc.* [2] **26**, 797.
Nagourney, W., and Dehmelt, H. (1981b). *Bull. Am. Phys. Soc.* [2] **26**, 805.
Nagourney, W., Janik, G., and Dehmelt, H. (1982). *Bull. Am. Phys. Soc.* [2] **27**, 839.
Nagourney, W., Janik, G., and Dehmelt, H. (1983). *Proc. Natl. Acad. Sci. U.S.A.* **80**, 643.
Neuhauser, W., Hohenstatt, M., Toschek, P., and Dehmelt, H. G. (1978a). *Phys. Rev. Lett.* **41**, 233.
Neuhauser, W., Hohenstatt, M., Toschek, P., and Dehmelt, H. (1978b). *Appl. Phys.* **17**, 123.
Neuhauser, W., Hohenstatt, M., Toschek, P., and Dehmelt, H. (1980). *Phys. Rev. A* [3] **22**, 1137.
Neuhauser, W., Hohenstatt, M., Toschek, P., and Dehmelt, H. (1981). *In* "Spectral Line Shapes" (B. Wende, ed.), Vol. 5, p. 1045. De Gruyter, Hawthorne, New York.
O, C. S., and Schuessler, H. A. (1980a). *Int. J. Mass Spectrom. Ion Phys.* **35**, 305.
O, C. S., and Schuessler, H. A. (1980b). *J. Appl. Phys.* **52**, 2601.
O, C. S., and Schuessler, H. A. (1981). *Int. J. Mass Spectrom. Ion Phys.* **39**, 95.
O, C. S., and Schuessler, H. A. (1982). *Rev. Phys. Appl.* **17**, 83.
O, C. S., and Gärtner, G. F., and Schuessler, H. A. (1982). *Appl. Phys. Lett.* **41**, 33.
O'Neil, T. M., and Driscoll, C. F. (1979). *Phys. Fluids* **22**, 266.
Osipowicz, A., and Werth, G. (1981). *Opt. Commun.* **36**, 359.
Owe Berg, T. G., and Gaukler, T. (1969) *Am. J. Phys.* **37**, 1013.
Penning, F. M. (1936). *Physica* **3**, 873.
Phillips, W. D., Cooke, W. E., and Kleppner, D. (1977). *Metrologia* **13**, 179.

Plumelle, F. (1979). Ph. D. thesis, Université de Paris-Sud, Orsay.
Plumelle, F., Desaintfuscien, M., Duchêne, J. J., and Audoin, C. (1980). *Opt. Commun.* **34**, 71.
Prasad, S. A., and O'Neil, T. M. (1979). *Phys. Fluids* **22**, 278.
Prior, M. H. (1972). *Phys. Rev. Lett.* **29**, 611.
Prior, M. H., and Knight, R. D. (1980). *Opt. Commun.* **35**, 54.
Prior, M. H., and Shugart, H. A. (1971). *Phys. Rev. Lett.* **27**, 902.
Prior, M. H., and Wang, E. C. (1977). *Phys. Rev. A* [3] **16**, 6.
Ramsey, N. F. (1956). "Molecular Beams." Oxford Univ. Press, London and New York.
Redhead, P. A. (1967). *Can. J. Phys.* **45**, 1791.
Rich, A. (1968a). *Phys. Rev. Lett.* **20**, 967.
Rich, A. (1968b). *Phys. Rev. Lett.* **21**, 1221.
Rich, A., and Wesley, J. C. (1972). *Rev. Mod. Phys.* **44**, 250.
Richardson, C. B., Jefferts, K. B., and Dehmelt, H. G. (1968). *Phys. Rev.* [2] **165**, 80.
Rosenfeld, R. N., Jasinski, J. M., and Brauman, J. I. (1979). *J. Chem. Phys.* **71**, 1030.
Ruster, W., Bonn, J., Peuser, P., and Trautmann, N. (1983). *Appl. Phys.* **B30**, 83.
Schaaf, H., Schmeling, U., and Werth, G. (1981). *Appl. Phys.* **25**, 249.
Schneider, R., and Werth, G. (1979). *Z. Phys. A* **293**, 103.
Schuessler, H. A. (1971a). *Metrologia* **7**, 103.
Schuessler, H. A. (1971b). *Appl. Phys. Lett.* **18**, 117.
Schuessler, H. A. (1979). *In* "Physics of Atoms and Molecules, Progress in Atomic Spectroscopy" (W. Hanle and H. Kleinpoppen, eds.), p. 999. Plenum, New York.
Schuessler, H. A. (1980). *In* "Physics of Atoms and Molecules, Coherence and Correlation in Atomic Collisions " (H. Kleinpoppen and J. F. Williams, eds.), p. 423. Plenum, New York.
Schuessler, H. A., and O, C. S. (1981). *Nucl. Instrum. Methods* **186**, 219.
Schuessler, H. A., Fortson, E. N., and Dehmelt, H. G. (1969). *Phys. Rev.* [2] **187**, 5.
Schwinberg, P. B., Van Dyck, R. S., Jr., and Dehmelt, H. G. (1981a). *Phys. Rev. Lett.* **47**, 1679.
Schwinberg, P. B., Van Dyck, R. S., Jr., and Dehmelt, H. G. (1981b). *Bull. Am. Phys. Soc.* [2] **26**, 597.
Schwinberg, P. B., Van Dyck, R. S., Jr., and Dehmelt, H. G. (1981c). *Phys. Lett. A* **81A**, 119.
Schwinberg, P. B., Van Dyck, R. S., Jr., and Dehmelt, H. G. (1983). *In* "Precision Measurements and Fundamental Constants. II" (B. N. Taylor and W. D. Phillips, eds.). *NBS Spec. Publ. (U.S.)* **617** (in press).
Smith, L. G., and Wapstra, A. H. (1975). *Phys. Rev. C* [3] **11**, 1392.
Smyth, K. C., McIver, R. T., Jr., Brauman, J. I., and Wallace, R. W. (1971). *J. Chem. Phys.* **54**, 2758.
Sokolov, A. A., and Pavlenko, Yu. G. (1967). *Opt. Spectrosc.* **22**, 1.
Strumia, F. (1978). *Proc. 32nd Ann. Symp. Freq. Control* p. 444.
Taylor, B. N. (1976). *Metrologia* **12**, 81.
Todd, J. F. J., Lawson, G., and Bonner, R. F. (1976). *In* "Quadrupole Mass Spectroscopy and its Applications" (P. H. Dawson, ed.), Chapter VIII. Elsevier, Amsterdam.
Todd, J. F. J., Waldren, R. M., and Mather, R. E. (1980a). *Int. J. Mass Spectrom. Ion Phys.* **34**, 325.
Todd, J. F. J., Waldren, R. M., Freer, D. A., and Turner, R. B. (1980b). *Int. J. Mass Spectrom. Ion Phys.* **75**, 107.
Torelli, G. (1980). *Proc. Eur. Symp. Nucleon, Anti Nucleon Interact., 5th, 1980* p. 43.
Toschek, P. E., and Neuhauser, W. (1981). *At. Phys.* **7**, 529.
Van Dyck, R. S., Jr., and Schwinberg, P. B. (1981a). *Phys. Rev. Lett.* **47**, 395.
Van Dyck, R. S., Jr., and Schwinberg, P. B. (1981b). *Bull. Am. Phys. Soc.* [2] **26**, 796.
Van Dyck, R. S., Jr., and Schwinberg, P. B. (1983). *In* "Precision Measurements and Funda-

mental Constants. II" (B. N. Taylor and W. D. Phillips, eds.). *NBS Spec. Publ. (U.S.)* **617** (in press).
Van Dyck, R. S., Jr., Wineland, D. J., Ekström, P. A., and Dehmelt, H. G. (1976). *Appl. Phys. Lett.* **28,** 446.
Van Dyck, R. S., Jr., Schwinberg, P. B., and Dehmelt, H. G. (1977). *Phys. Rev. Lett.* **38,** 310.
Van Dyck, R. S., Jr., Schwinberg, P. B., and Dehmelt, H. G. (1978). *In* "New Frontiers in High-Energy Physics" (B. M. Kursunoglu, A. Perlmutter, and L. F. Scott, eds.), p. 159. Plenum, New York.
Van Dyck, R. S., Jr., Schwinberg, P. B., and Dehmelt, H. G. (1979). *Bull. Am. Phys. Soc.* [2] **24,** 758.
Van Dyck, R. S., Jr., Schwinberg, P. B., and Bailey, S. H. (1980). *In* "Atomic Masses and Fundamental Constants" (J. A. Nolen, Jr. and W. Benenson, eds.), 6th, p. 173. Plenum, New York.
Vane, C. R., Prior, M. H., and Marrus, R. (1981). *Phys. Rev. Lett.* **46,** 107.
Vedel, F., André, J., and Vedel, M. (1981). *J. Phys. (Orsay, Fr.)* **42,** 1611.
Vedel, F., André, J., Vedel, M., and Brincourt, G. (1983). *Phys. Rev. A* [3] **27,** 2321.
Vedel, M. (1976). *J. Phys. (Orsay, Fr.)* **37,** L-339.
Vedel, M., André, J., Chaillat-Negrel, S., and Vedel, F. (1981). *J. Phys. (Orsay, Fr.)* **42,** 541.
Walls, F. L. (1970). Ph. D. thesis, University of Washington, Seattle.
Walls, F. L., and Dunn, G. H. (1974). *Phys. Today* **27,** No. 8, 30.
Walls, F. L., and Stein, T. S. (1973). *Phys. Rev. Lett.* **31,** 975.
Werth, G. (1982). *Acta Phys. Polonica* **A61,** 213.
Wesley, J. C., and Rich, A. (1971). *Phys. Rev. A* [3] **4,** 1341.
Wilkinson, D. T., and Crane, H. R. (1963). *Phys. Rev.* **130,** 852.
Williams, E. R., and Olsen, P. T. (1979). *Phys. Rev. Lett.* **42,** 1575.
Wineland, D. J. (1979). *J. Appl. Phys.* **50,** 2528.
Wineland, D. J. (1983). *In* "Precision Measurements and Fundamental Constants. II" (B. N. Taylor and W. D. Phillips, eds.). *NBS Spec. Publ. (U.S.)* **617** (in press).
Wineland, D. J., and Dehmelt, H. G. (1975a). *J. Appl. Phys.* **46,** 919.
Wineland, D. J., and Dehmelt, H. G. (1975b). *Bull. Am. Phys. Soc.* [2] **20,** 637.
Wineland, D. J., and Dehmelt, H. G. (1975c). *Int. J. Mass Spectrom. Ion Phys.* **16,** 338; erratum: **19,** 251 (1976).
Wineland, D. J., and Itano, W. M. (1979). *Phys. Rev. A* [3] **20,** 1521.
Wineland, D. J., and Itano, W. M. (1981). *Phys. Lett. A* **82A,** 75.
Wineland, D. J., and Itano, W. M. (1982a). *Bull. Am. Phys. Soc.* [2] **27,** 471.
Wineland, D. J., and Itano, W. M. (1982b). *Bull. Am. Phys. Soc.* [2] **27,** 864.
Wineland, D. J., Ekstrom, P., and Dehmelt, H. (1973). *Phys. Rev. Lett.* **31,** 1279.
Wineland, D. J., Drullinger, R. E., and Walls, F. L. (1978). *Phys. Rev. Lett.* **40,** 1639.
Wineland, D. J., Bergquist, J. C., Itano, W. M., and Drullinger, R. E. (1980). *Opt. Lett.* **5,** 245.
Wineland, D. J., Itano, W. M., Bergquist, J. C., and Walls, F. L. (1981a). *Proc. 35th Annu. Symp. Freq. Control* p. 602 (copies available from Electronic Industries Assoc., 2001 Eye St., NW, Washington, DC 20006).
Wineland, D. J., Bergquist, J. C., Drullinger, R. E., Hemmati, H., Itano, W. M., and Walls, F. L. (1981b). *J. Phys. (Orsay, Fr.)* **42,** C8-307.
Wineland, D. J., Bollinger, J. J., and Itano, W. M. (1983). *Phys. Rev. Lett.* **50,** 628; erratum: **50,** 1333 (1983).
Wuerker, R. F., Shelton, H., and Langmuir, R. V. (1959a). *J. Appl. Phys.* **30,** 342.
Wuerker, R. F., Goldenberg, H. M., and Langmuir, R. V. (1959b). *J. Appl. Phys.* **30,** 441.
Zaritskii, A. A., Zukharov, S. O., and Kryukov, P. G. (1971). *Sov. Phys.—Tech. Phys. (Engl. Transl.)* **16,** 174.

SPIN-DEPENDENT PHENOMENA IN INELASTIC ELECTRON–ATOM COLLISIONS*

K. BLUM

Institut für Theoretische Physik I
Universität Münster
Münster, Federal Republic of Germany

H. KLEINPOPPEN

Atomic Physics Laboratory
University of Stirling
Stirling, Scotland

I. Introduction	188
II. Exchange Effects in Inelastic Collisions between Electrons and Light Atoms	189
III. Excitation of Intermediate and Heavy Atoms: Spin Polarization and Asymmetry Studies	192
A. General Theory	192
B. The Number of Independent Parameters. The Difference between Polarizing and Analyzing Power	198
C. Results for Mercury Excitation	201
IV. Influence of Spin-Dependent Interactions on Coherence Parameters .	204
A. Parametrization of Atomic Density Matrix	204
B. Physical Importance of Parameters $\cos \epsilon$ and $\cos \Delta$	207
C. Influence of Coherence Parameters on Polarization of Emitted Light	210
D. Results and Analysis of Numerical Calculations	214
V. Excitation of Heavy Atoms by Polarized Electrons: Stokes Parameter Analysis	225
A. Symmetry Considerations	225
B. Probing the Spin–Orbit Interaction by Means of Integrated Stokes Parameters	229
C. Classification of Resonances Close to Threshold	234
VI. Electron–Photon Coincidence Experiments with Polarized Electrons	236

* The material presented in Sections III–VI is part of the Habilitation thesis of K. Blum, accepted by the Physics Department of the University of Münster.

VII. Ionization 241
   A. Theory of Measurement for Ionization Processes with
       Polarized Electrons and Polarized Atoms 242
   B. Results of Spin-Dependent Ionization Asymmetries. 243
   Appendix A 259
   Appendix B. 260
   References 261

I. Introduction

Spin-dependent electron–atom scattering processes have attracted attention since shortly after Pauli's introduction of the concept of the spinning electron in quantum mechanics. Mott (1928) first developed the theory of production and detection of spin-polarized electrons by including spin–orbit interactions in the scattering process. Massey and Mohr (1941) extended this theory by introducing screening effects in the Mott scattering. Various authors (Kessler, 1976; Bederson, 1969a,b, 1973; Celotta and Pierce, 1980; Kleinpoppen, 1977; Lubell, 1980) summarized the progress of theoretical and experimental investigations of spin effects in electron–atom collisions. In this review we concentrate on recent results on spin-dependent effects in inelastic electron–atom collisions; elastic scattering is not to be included.

In Section II electron-exchange effects in inelastic collisions with light atoms are briefly dealt with. In Sections III–VI various aspects of spin-dependent effects in collisions with intermediate and heavy atoms are considered. In Section III a review is given on recent work on spin polarization and asymmetry measurements. The general theory is developed and applied to various experimental conditions.

In Sections IV–VI some new spin-sensitive parameters are introduced which contain independent information. What can be learned from a determination of these parameters is discussed and first experimental and numerical results are given. Of particular interest is that some of the new parameters can be linked to *specific* details of the spin-coupling mechanism.

The developments in Sections III–VI are based on the concept of the reduced density matrix, which allows a discussion of the various cases from a unified point of view. Finally, in Section VII spin-dependent effects in ionization processes cure dealt with. The spin-dependent ionization asymmetry in electron-impact ionization can be linked to the analysis for triplet and singlet ionization cross sections, which are of particular interest in connection with ionization threshold laws.

II. Exchange Effects in Inelastic Collisions between Electrons and Light Atoms

In general, spin-polarization effects can be caused by electron exchange, by explicit spin-dependent forces, or by a combination of both effects. By the term "explicit spin-dependent" we mean interactions which correspond to a particular term in the Hamiltonian. This is in contrast to effects which are due to the Pauli principle and which are caused by electron exchange only.

The most important explicit spin-dependent term in the Hamiltonian corresponds to the spin–orbit interaction and throughout this article we will mainly consider this interaction. Discussions of other effects, corresponding to specific terms in the Hamiltonian, have been given, for example, by Farago (1976) and by Walker (1976).

In this section we consider excitation of light atoms by electron impact where all explicit spin-dependent forces can be neglected during the collision. More explicitly, we neglect, first, the spin–orbit interaction between the continuum electron and the atomic core, and second, the fine-structure interaction inside the atom.

It is useful to consider the latter approximation in more detail. In the excited atomic states the orbital angular momentum **L** and the spin **S** couple under the influence of the fine-structure interaction and precess around the total angular momentum **J** of the atom. This precession takes place in a time $t_{LS} \sim 1/\Delta E_{LS}$, where ΔE_{LS} denotes the fine-structure splitting of the relevant level. If the collision time t_c is much shorter than the spin–orbit precession time, then the spin vector will not have time to precess appreciably during the collision, and **L** and **S** can be considered to be uncoupled during the excitation. The state of the excited atoms immediately after the collision can then adequately be described in the LS-coupling scheme. The assumption that $t_c \ll t_{LS}$ means that the atoms can be considered as instantaneously excited with respect to the much longer spin–orbit precession time.

Throughout this article we use the term "LS-coupling" if the total spin of the combined system atom plus electron and its z component are conserved during the collision.

We consider excitation of atoms from an initial state with quantum numbers $\gamma_0 = n_0 s_0 M_{S_0}$ and $L_0 = 0$ to a final state characterized by $\gamma_1 = n_1 L M S_1 M_{S_1}$. Here, S_0, M_{S_0} and S_1, M_{S_1} denote initial and final atomic spins and their z components, respectively, LM denotes the orbital angular momentum and its z component of the excited atoms, and n_0 and n_1 denote all other quantum numbers necessary for a complete characterization of the initial and final states. Incident and scattered electrons are specified by their respective momenta $\mathbf{p}_0$ and $\mathbf{p}_1$ and their spin components m_0 and m_1.

A transition between states

$$\Gamma_0 \equiv \gamma_0 p_0 m_0 \to \Gamma_1 \equiv \gamma_1 p_1 m_1 \quad (1.1)$$

is then completely characterized by the corresponding scattering amplitude $f(\Gamma_1, \Gamma_0)$, defined as the matrix element of the transition operator **T**:

$$f(\Gamma_1, \Gamma_0) = \langle \Gamma_1 | \mathbf{T} | \Gamma_0 \rangle \quad (1.2)$$

We normalize according to

$$|f(\Gamma_1, \Gamma_0)|^2 = \sigma(\Gamma_1, \Gamma_0) \quad (1.3)$$

where $\sigma(\Gamma_1, \Gamma_0)$ is the differential cross section for the transition (1.1). In the following we assume that $\mathbf{p}_0, \mathbf{p}_1, S_0, S_1$, and L are fixed by the experimental conditions and we suppress the dependence of the amplitudes on these variables.

In the LS-coupling limit the amplitudes can be written in the form

$$f(\Gamma_1, \Gamma_0) = \sum_{SM_S} (S_1 M_{S_1}, \tfrac{1}{2} m_1 | SM_S)(S_0 M_0, \tfrac{1}{2} m_0 | SM_S) f(M)^{(S)} \quad (1.4)$$

(for details see Appendix A). Here, $S = S_0 \pm \tfrac{1}{2} = S_1 \pm \tfrac{1}{2}$ denotes the total (channel) spin and $M_S = M_0 + m_0 = M_{S_1} + m_1$ its z component. $(\cdots|\cdots)$ is a standard Clebsch–Gordan coefficient.

All information on the dynamics is contained in the amplitudes $f(M)^{(S)}$, which depend only on the magnetic quantum number M and the total spin S (apart from the dependence on the fixed variables). Hence, the set of all transitions (1.1) (with $\mathbf{p}_0, \mathbf{p}_1, S_0, S_1$, and L sharp) is completely characterized in terms of $2(2L + 1)$ independent (complex) amplitudes if both S_0 and S_1 are different from zero.

The number of independent amplitudes is further reduced by reflection invariance. $\mathbf{p}_0$ and $\mathbf{p}_1$ define a plane, the scattering plane, and the interaction must be invariant under reflection in this plane. From this requirement we obtain the symmetry condition (see Appendix B for details):

$$f(M)^{(S)} = (-1)^M f(M)^{(S)} \quad (1.5)$$

For example, in a $^1S_0 \to {}^3P_1$ transition the total spin is fixed ($S = \tfrac{1}{2}$) and the excitation is completely characterized in terms of three real parameters, for example, the magnitudes $|f(1)|$ and $|f(0)|$ and the relative phase between these amplitudes.

Instead of using a description in terms of the total spin and its z component, one can also use "direct" and "exchange" amplitudes, commonly denoted by $f(M)$ and $g(M)$, respectively. For example, for elastic scattering of electrons on spin-½ atoms f and g can be distinguished experimentally by performing scattering experiments with polarized particles, for example,

$$e(\uparrow) + A(\downarrow) \rightarrow e(\uparrow) + A(\downarrow) \qquad (1.6)$$

is a direct process characterized by f,

$$e(\uparrow) + A(\downarrow) \rightarrow e(\downarrow) + A(\uparrow)$$

is an exchange process characterized by g,

$$e(\uparrow) + nA(\uparrow) \rightarrow e(\uparrow) + A(\uparrow)$$

is pure triplet scattering described by the triplet amplitude $f^{(1)} = f - g$. Similarly, the singlet amplitude is given by $f^{(0)} = f + g$ (see, e.g., Kleinpoppen, 1971).

For elastic scattering the independent parameters $|f|$, $|g|$, and their relative phase can be extracted from measurements with polarized particles. This was first discussed by Bederson (1969a,b), who introduced the term "complete experiment" for measurements where all independent parameters are determined.

A formal theory for scattering processes between electrons and light atoms has been developed by Burke and Schey (1962), Rubin et al. (1969), and Kleinpoppen (1971) where the observables are related to the scattering amplitudes. We refer the reader to these articles for further details.

In elastic scattering processes the amplitudes depend on the magnetic quantum number M. A complete experiment in the sense defined by Bederson requires then an electron–photon coincidence experiment with polarized particles (Rubin et al., 1969; Kleinpoppen, 1971) (see also Section VI).

Inelastic collisions with a spin analysis of atomic systems by low-energy electrons were first studied by Rubin et al. (1969) and Goldstein et al. (1972). The authors studied the 4s → 4p transition in potassium and measured the ratio R of the spin flip to the total differential cross section. The first numerical calculations, obtained in the close-coupling approximation, were performed by Karule and Peterkop (1965), by Burke and Taylor (1969), and by Moores and Norcross (1972).

New techniques have been developed for producing beams of polarized electrons and atoms (Hils et al., 1981; Baum et al., 1980; Pierce and Celotta, 1982) which allow the study of spin effects in inelastic collisions in greater detail. Baum et al. (1982) measured the spin asymmetry for the 2s → 2p transition in lithium where both electrons and atoms were polarized. Some results are shown in Fig. 1.

A theoretical analysis of this and similar transitions has been given by Khalid et al. (1982). These authors derived formulas for the extraction of partial differential cross sections for elastic scattering and for excitation of

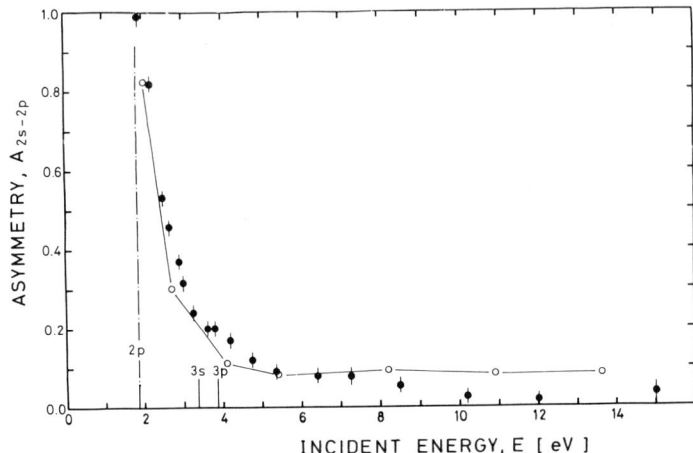

FIG. 1. Total spin asymmetry for the 2s → 2p transition in Li. Experimental data points (●) from Schröder (1982) compared to theoretical predictions: (○) Burke and Taylor (1969) and Kennedy et al. (1977).

the ^{2}P states of light alkalis. Results of numerical calculations for various cases are also reported, obtained from reactance matrices calculated in the close-coupling approximation.

Some examples are shown in Fig. 2 (3s → 3p excitation of Na) and in Fig. 3 (2s → 2p excitation of Li). The figures show a pronounced structure which is contrary to the rather smooth behavior of differential cross sections. The further investigation of these effects seems to be a fruitful area for confronting theory with experiment.

III. Excitation of Intermediate and Heavy Atoms: Spin Polarization and Asymmetry Studies

A. GENERAL THEORY

1. The Reduced Density Matrix of Scattered Electrons

In this and the following sections we consider excitation of intermediate and heavy atoms by polarized or unpolarized electrons where explicit spin-dependent effects cannot be neglected during the excitation.

We will characterize initial and final atomic states in terms of quantum numbers $n_0 J_0 M_0$ and $n_1 J M$, respectively, where J and M denote the total

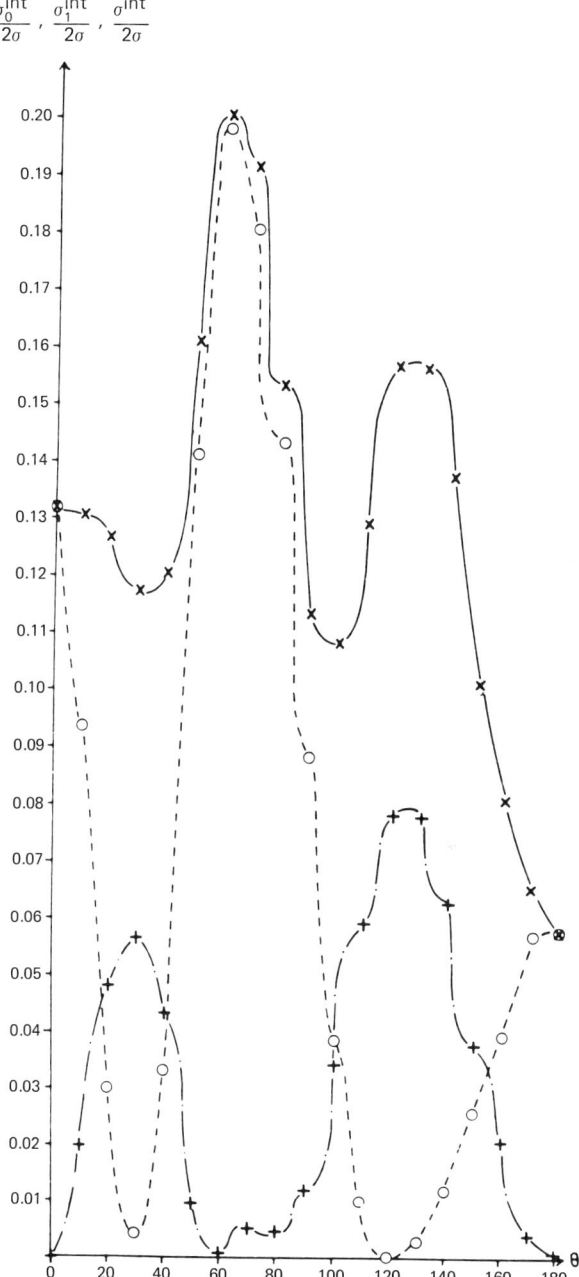

FIG. 2. Relative differential interference cross sections (including sublevel cross sections for $m_l = 0$ and ± 1) for the $2s \to 2p$ transition in Li at $E = 4.0$ eV (theoretical predictions by Khalid et al., 1982): (O) $\sigma_0^{int}/2\sigma$, (+) $\sigma_1^{int}/2\sigma$, ($\times$) $\sigma^{int}/2\sigma$.

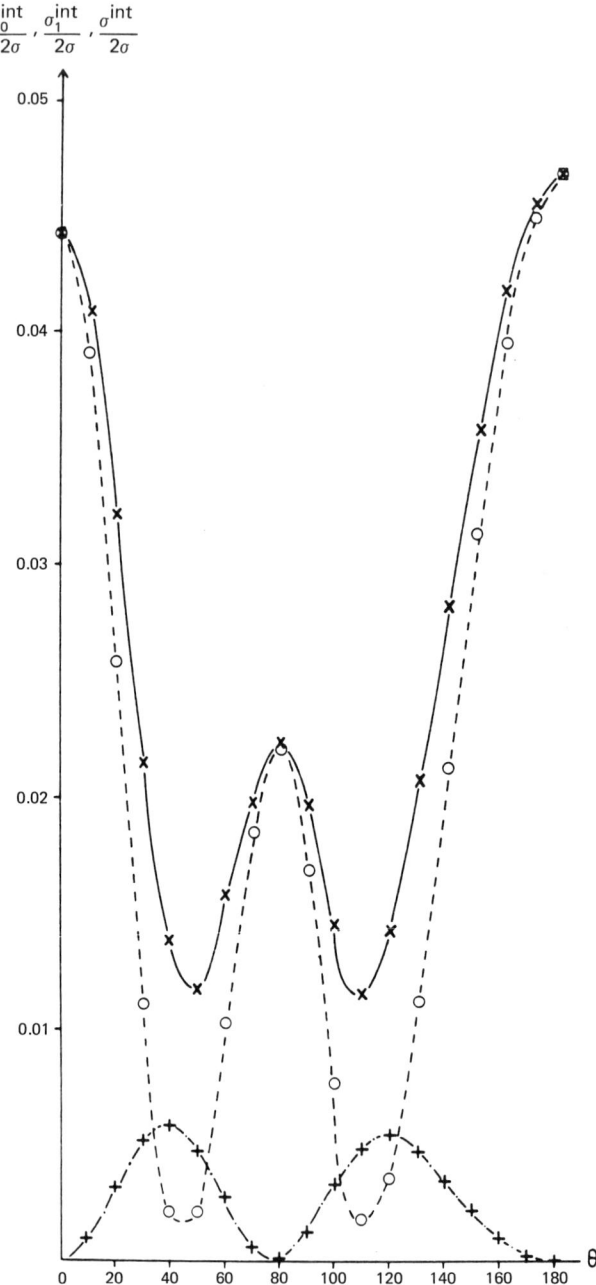

FIG. 3. Same as in Fig. 2 for the 3s → 3p transition in Na at $E = 2.2$ eV.

atomic angular momentum ($\mathbf{J} = \mathbf{J} + \mathbf{S}$) and its z component, respectively A transition

$$\Gamma_0 = n_0 J_0 M_0 p_0 m_0 \to \Gamma_1 \equiv n_1 J M p_1 m_1 \qquad (2.1)$$

is then specified in terms of the corresponding scattering amplitude

$$f(\Gamma_1, \Gamma_0) = \langle \Gamma_1 | \mathbf{T} | \Gamma_0 \rangle \qquad (2.2)$$

[see Eq. (1.20)]. As in Section II, we apply the normalization

$$|f(\Gamma_1, \Gamma_0)|^2 = \sigma(\Gamma_1, \Gamma_0) \qquad (2.3)$$

where the sum runs over the variables $J_0 M_0 m_0 J M m_1$.

The number of independent amplitudes is reduced by the requirement that the interaction dynamics must be invariant against reflection in the scattering plane, which yields the condition

$$f(JMm_1, J_0 M_0 m_0)$$
$$= (-1)^{J+M+m_1+L-J_0-M_0-m_0} f(J-M-m_1, J_0-M_0-m_0) \qquad (2.4)$$

where it has been assumed that the level under consideration has a definite orbital angular momentum L (see Appendix B for details).

Throughout this section it is always assumed that the atoms are initially unpolarized and that the excited atoms are not detected. Observables are the differential cross section and the spin polarization and asymmetry of the scattered electrons.

An analysis of spin effects in inelastic collisions between electrons and two-electron atoms has been given by Hanne (1976), by Drukarev and Obedkov (1979), and by Obedkov (1980). These authors generalized a theory developed by Burke and Mitchel (1974) for elastic scattering processes. Various aspects have also been considered by Farago (1971).

Here, we apply a slightly different procedure. All information which can be extracted from experiments of the type described above is contained in the *reduced density matrix* of the scattered electrons. In this section we construct this quantity in a systematic way. In Section IV the atomic density matrix is considered for experiments without any spin selection, and in Section VI more complex collisions are analyzed. The density matrix method allows one to discuss these various cases from a unified point of view.

The initial spin state of the electrons may be arbitrary and described in terms of the corresponding density matrix ρ_0 with elements $\langle m'_0 | \rho_0 | m_0 \rangle$. Assuming unpolarized atoms and a sharp value of J_0, the initial state of the combined system of atoms and electrons is specified by the density matrix

$$\rho_{\text{in}} = (2J_0 + 1)^{-1} \mathbf{1} \times \rho_0 \qquad (2.5)$$

where **1** is the $(2J_0 + 1)$-dimensional unit matrix in the atomic space and the cross denotes the direct product. We will always normalize according to

$$\mathrm{tr}\, \rho_{\mathrm{in}} = 1 \tag{2.6}$$

After the scattering, the combined system is characterized by the density matrix ρ_{out}, which is related to ρ_{in} by the relation

$$\rho_{\mathrm{out}} = \mathbf{T} \rho_{\mathrm{in}} \mathbf{T}^\dagger \tag{2.7}$$

where **T** is the transition operator (see, for example, Blum, 1981a, Appendix E). Taking matrix elements, we obtain

$$\langle J'M'm_1' | \rho_{\mathrm{out}} | JMm_1 \rangle = (2J_0 + 1)^{-1} \sum_{M_0' m_0' m_0} \langle J'M'm_1' | \mathbf{T} | J_0 M_0 m_0' \rangle$$
$$\times \langle m_0' | \rho_0 | m_0 \rangle \langle J_0 M_0 m_0 | \mathbf{T}^\dagger | JMm_1 \rangle \tag{2.8}$$

where the dependence of the T-matrix elements on the momenta has been suppressed. This expression is our starting point for the construction of the various reduced density matrices in this and the following sections.

In this section we are only interested in the spin polarization of the scattered electrons with the final atoms undetected. The corresponding information is contained in the reduced density matrix ρ_1 of the electrons, which follows from Eq. (2.8) by summing over all elements diagonal in all unobserved variables (see Blum, 1981a, Section 3.2, for details):

$$\langle m_1' | \rho_1 | m_1 \rangle = \sum_{JM} \langle JMm_1' | \rho_{\mathrm{out}} | JMm_1 \rangle \tag{2.9}$$

It should be noted that Eq. (2.9) is *incoherent* in all atomic variables J and M. This is a consequence of the *experimental conditions*, that is, of the fact that only the scattered electrons are observed (for a detailed discussion see, for example, Blum, 1981a, Chapter 3). This result holds even if the fine-structure levels overlap.

From Eqs. (2.2), (2.8), and (2.9) we obtain

$$\langle m_1' | \rho_1 | m_1 \rangle = (2J_0 + 1)^{-1} \sum_{\substack{JMM_0 \\ m_0' m_0}} f(JM'm_1', J_0 M_0 m_0')$$
$$\times f(JMm_1, J_0 M_0 m_0)^* \langle m_0' | \rho_0 | m_0 \rangle \tag{2.10}$$

This expression relates the spin-density matrix of the scattered electrons to the initial density matrix ρ_0 and the scattering amplitudes.

As is well known, the spin-density matrix can be written in the form

$$\rho = \frac{1}{2} \begin{pmatrix} 1 + P_z & P_x + iP_y \\ P_x + iP_y & 1 - P_z \end{pmatrix} \tag{2.11}$$

where P_x, P_y, and P_z are the components of the corresponding polarization vector.

Using Eqs. (2.10) and (2.11), expressions for the components P_i can easily be obtained. The resulting expressions are simplified by applying the symmetry condition (2.4). It should be noted that, contrary to the elastic case, time-reversal invariance does not reduce the number of independent amplitudes. Invariance of the interaction under time reversal allows one only to relate the amplitudes of the excitation process to those of the corresponding superelastic scattering.

In the remainder of this article we are mainly concerned with experiments where a single fine-structure level is separated in the *electronic* channel. The relevant electronic density matrix is then given by Eq. (2.10) with the sum over J omitted. In these cases the spin–orbit interaction inside the target atoms is apparently so strong that the spin–orbit precession time is smaller than the collision time. This is contrary to the other extreme case of LS-coupling discussed in Section II (see also the following section).

2. The LS-Coupling Limit. Discussion of Fine-Structure Effects

With regard to some recent discussions it is useful to consider the LS-coupling limit. As discussed in Section II, this means that all explicit spin-dependent effects during the collision are neglected so that the total spin and its z component are conserved. In this section we consider the case of unpolarized incident electrons $\langle m'_0 | \rho_0 | m_0 \rangle = \frac{1}{2} \delta_{m'_0 m_0}$ and excitation from a $^1S_0$ ground state ($J_0 = 0$).

By using the expansion

$$|JM\rangle = \sum_{M_L M_{S_1}} (LM_L S_1 M_{S_1} | JM) |LM_L S_1 M_{S_1}\rangle$$

and inserting it into Eq. (2.10), we obtain

$$\langle M'_1 | \rho_1 | m_1 \rangle = [2(2J_0 + 1)]^{-1} \sum_{\substack{m_0 \\ M'_L M_L \\ M'_{S_1} M_{S_1}}} \left[\sum_{JM} (LM'_L S_1 M'_{S_1} | JM)(LM_L S_1 M_{S_1} | JM) \right]$$

$$\times f(LM'_L S_1 M'_{S_1} M'_1; m_0) f(LM_L S_1 M_{S_1} M_1; m_0)^* \quad (2.12a)$$

$$= [2(2J_0 + 1)]^{-1} \sum_{\substack{m_0 \\ M_L M_{S_1}}} f(LM_L S_1 M_{S_1} m'_1, m_0)$$

$$\times f(LM_L S_1 M_{S_1} m_1, m_0)^* \quad (2.12b)$$

where the sum over J and M has been performed by applying the orthogonality relations of the Clebsch–Gordan coefficients. It should be noted that the remaining sum is incoherent in M_L and particularly in M_{S_1}, that is, only terms with $M'_L = M_L$ and $M'_{S_1} = M_{S_1}$ contribute to the sum (2.12b) (see Section IV,A for a general discussion of coherence).

From spin conservation it follows that the scattering amplitudes vanish unless the conditions $M_{S_1} + m'_1 = M_0$ and $M_{S_1} + m_1 = m_0$ are simultaneously satisfied. This is only possible if $m'_1 = m_1$. Consequently, all off-diagonal elements of ρ_1 vanish. Using reflection invariance [Eq. (2.4)], it can be shown that the diagonal elements of ρ_1 are equal. We have obtained the well-known result that the scattered electrons are unpolarized in the LS-coupling limit if the initial polarization vector is zero (Blum and Kleinpoppen, 1975; Hanne, 1976). Electron-exchange effects alone cannot produce a polarization of the scattered electrons.

We have discussed these derivations in some detail with regard to recent suggestions by Hanne (1981). In order to see the essential point of his argument, let us consider the following extreme case. Assume that LS-coupling is a reasonable assumption but that it is nevertheless possible to separate one single fine-structure level with definite J in the experiment (which requires an extremely sharp energy resolution of the electron beams). The corresponding electronic density matrix is then given by Eq. (2.12a) with the sum over J omitted. Since J is fixed, it is not possible to reduce Eq. (2.12a) to the incoherent sum (2.12b). In other words, there is now no general argument which gives the restrictions $M'_{S_1} = M_{S_1}$ and $m'_1 = m_1$. Hence, the off-diagonal elements of ρ_1 are not necessarily zero.

This means that the scattered electrons can be expected to be polarized, *even if all explicit spin-dependent interactions are very small,* provided it is possible to separate a single fine-structure level in the electronic channel. That such an effect may exist was first suggested by Hanne (1981). We refer the reader to this article for a detailed discussion of various theoretical and experimental aspects.

B. The Number of Independent Parameters. The Difference between Polarizing and Analyzing Power

In this section we consider in more detail what information can be obtained by scattering polarized electrons on unpolarized atoms. As shown in Section III,A, this information is contained in the reduced density matrix ρ_1 of the scattered electrons. Rewriting Eq. (2.10), we have

$$\langle m'_1|\rho_1|m_1\rangle = (2J_0 + 1)^{-1} \sum_{\substack{JMJ_0M_0 \\ m'_0 m_0}} \langle JM, \mathbf{p}_1 m'_1|\mathbf{T}|J_0 M_0, \mathbf{p}_0 m'_0\rangle$$
$$\times \langle JM, \mathbf{p}_1 m_1|\mathbf{T}|J_0 M_0, \mathbf{p}_0 m_0\rangle^* \langle m'_0|\rho_0|m_0\rangle \quad (2.13)$$

All information on the scattering dynamics is contained in the **T**-matrix elements; the elements of ρ_0 specify the initial conditions. We now determine how many independent parameters can be determined from the experiments under discussion.

In order to achieve this, we first introduce a convenient notation. We introduce a set of parameters by the following definition:

$$\langle m_1' m_0' m_1 m_0 \rangle = (2J_0 + 1)^{-1} \sum_{JMJ_0M_0} \langle JM, \mathbf{p}_1 m_1' | \mathbf{T} | J_0 M_0, \mathbf{p}_0 m_0' \rangle$$
$$\times \langle JM, \mathbf{p}_1 m_1 | \mathbf{T} | J_0 M_0, \mathbf{p}_0 m_0 \rangle^* \qquad (2.14)$$

Substitution of Eq. (2.14) into Eq. (2.13) yields

$$\langle m_1' | \rho_1 | m_1 \rangle = \sum_{m_0' m_0} \langle m_1' m_0' m_1 m_0 \rangle \langle m_0' | \rho_0 | m_0 \rangle \qquad (2.15)$$

This expression shows that the information obtainable from the experiments under discussion is given by the quantities $\langle m_1' m_0' m_1 m_0 \rangle$. These parameters can be measured by determining the elements of ρ_1 under given initial conditions, that is, by measuring the polarization vector of the scattered electrons as a function of the initial polarization.

With $m = \pm \tfrac{1}{2}$ there are 16 (complex) parameters. Not all of these are independent. From the definition (2.14) follows immediately the condition

$$\langle m_1' m_0' m_1 m_0 \rangle = \langle m_1 m_0 m_1' m_0' \rangle^* \qquad (2.16)$$

and from reflection invariance in the scattering plane [Eq. (B3)] it follows that

$$\langle m_1' m_0' m_1 m_0 \rangle = (-1)^{m_1' + m_1 - m_0' - m_0} \langle -m_1' -m_0' -m_1 -m_0 \rangle \qquad (2.17)$$

By applying the conditions (2.16) and (2.17), relations between the parameters (2.14) can be deduced. For example

$$\langle \tfrac{1}{2} -\tfrac{1}{2} -\tfrac{1}{2} \tfrac{1}{2} \rangle = \langle -\tfrac{1}{2} \tfrac{1}{2} \tfrac{1}{2} -\tfrac{1}{2} \rangle^*$$
$$= \langle \tfrac{1}{2} -\tfrac{1}{2} -\tfrac{1}{2} \tfrac{1}{2} \rangle^*$$

from which it follows that $\langle \tfrac{1}{2} -\tfrac{1}{2} -\tfrac{1}{2} \tfrac{1}{2} \rangle$ is real. In deriving this result, we have first applied Eq. (2.16) and then Eq. (2.17).

In this way it can be shown that *there are in general eight real independent parameters which characterize the scattering of polarized electrons on unpolarized atoms.*

A possible choice is the following set of four real parameters: $\langle \tfrac{1}{2} \tfrac{1}{2} \tfrac{1}{2} \tfrac{1}{2} \rangle$, $\langle \tfrac{1}{2} -\tfrac{1}{2} \tfrac{1}{2} -\tfrac{1}{2} \rangle$, $\langle \tfrac{1}{2} \tfrac{1}{2} -\tfrac{1}{2} -\tfrac{1}{2} \rangle$, and $\langle \tfrac{1}{2} -\tfrac{1}{2} -\tfrac{1}{2} \tfrac{1}{2} \rangle$, and the real and imaginary parts of the two complex parameters $\langle \tfrac{1}{2} \tfrac{1}{2} \tfrac{1}{2} -\tfrac{1}{2} \rangle$ and $\langle \tfrac{1}{2} \tfrac{1}{2} -\tfrac{1}{2} \tfrac{1}{2} \rangle$.

Some special cases are of interest. The important case of elastic scattering on spin-$\tfrac{1}{2}$ atoms ($J = J_0 = \tfrac{1}{2}$) has been discussed by Burke and Mitchel

(1974). Let us briefly consider scattering on spinless atoms ($J = J_0 = 0$). In this case the parameters (2.14) factorize as follows:

$$\langle m_1' m_0' m_1 m_0 \rangle = \langle m_1' m_0' \rangle \langle m_1 m_0 \rangle$$
$$= \langle \mathbf{p}_1 m_1' | T | \mathbf{p}_0 m_0' \rangle \langle \mathbf{p}_1 m_1 | T | \mathbf{p}_0 m_0 \rangle \quad (2.18)$$

Reflection invariance can then be applied to each factor separately, which reduces the number of independent parameters.

It is therefore convenient to use the direct amplitude $\langle \mathbf{p}_1 \tfrac{1}{2} | T | \mathbf{p}_0 \tfrac{1}{2} \rangle$ and the spin-flip amplitude $\langle \mathbf{p}_1 \tfrac{1}{2} | T | \mathbf{p}_0 -\tfrac{1}{2} \rangle$ as independent parameters. One phase can be choosen arbitrarily and there are therefore three independent real parameters as is well known from potential scattering.

If the excitation process can be described in the LS-coupling scheme, then ρ_1 depends on the dynamical parameters

$$\sum_{M_L} \langle LM_L, \mathbf{p}_1 | T^{(S)} | 0, \mathbf{p}_0 \rangle \langle LM_L, \mathbf{p}_1 | T^{(S)} | 0, \mathbf{p}_0 \rangle$$

which follows from the formulas given in Appendix A (with $L_0 = 0$). If both S_0 and S_1 are different from zero, then the scattering process under discussion is completely characterized by four real parameters, two absolute squares $\sum_{M_L} |\langle LM_M, \mathbf{p}_1 | T^{(S)} | 0, \mathbf{p}_0 \rangle|^2$ (with $S = S_0 \pm \tfrac{1}{2} = S_1 \pm \tfrac{1}{2}$) and the real and imaginary parts of the interference term with $S' \neq S$:

$$\sum_{M_L} \langle LM_L, \mathbf{p}_1 | T^{(S')} | 0, \mathbf{p}_0 \rangle \langle LM_L, \mathbf{p}_1 | T^{(S)} | 0, \mathbf{p}_0 \rangle$$

In elastic scattering (with $L = L_0 = 0$) this number is reduced to three: the magnitudes of the amplitudes $\langle \mathbf{p}_1 | T^{(S)} | \mathbf{p}_0 \rangle$ and their relative phase (Burke and Schey, 1962).

The components P_i' of the polarization vector of the scattered electrons ($i = x, y, z$) are obtained from the relations

$$\sigma P_i' = \operatorname{tr} \rho_1 \sigma_i = \sum_{m_1' m_1} \langle m_1' | \rho_1 | m_1 \rangle \langle m_1 | \sigma_i | m_1' \rangle \quad (2.19)$$

where σ is the differential cross section and σ_i denotes the Pauli matrices (with $i = x, y, z$). By using the representation (2.11) for ρ_0 and by applying the relations (2.15)–(2.17), it is straightforward to express the final polarization P_i^* in terms of the independent parameters $\langle m_1' m_0' m_1 m_0 \rangle$ and the initial polarization P_i. For example, if the initial electrons are unpolarized, we obtain for P' (the "polarizing power") the $P_x' = P_z' = 0$ and

$$\sigma_{un} P_y' = 2 \operatorname{Im} \langle \tfrac{1}{2} \tfrac{1}{2} - \tfrac{1}{2} \tfrac{1}{2} \rangle \quad (2.20)$$

where σ_{un} is the differential cross section for excitation with unpolarized electrons.

Let us now consider excitation by electrons initially polarized in the y direction. The differential cross section is given by the relation

$$\sigma = \sum_{m_1} \langle m_1 | \rho_1 | m_1 \rangle$$

$$= \sum_{m_1 m_0' m_0} \langle m_1 m_0' m_1 m_0 \rangle \langle m_0' | \rho_0 | m_0 \rangle$$

$$= \sigma_{\text{un}} + 2P_y \, \text{Im} \langle \tfrac{1}{2} \tfrac{1}{2} \tfrac{1}{2} -\tfrac{1}{2} \rangle \quad (2.21)$$

where Eq. (2.11) with $P_x = P_z = 0$ has been used. The scattering asymmetry A (the "analyzing power") is given by the last term in Eq. (2.21):

$$\sigma_{\text{un}} A = 2 P_y \, \text{Im} \langle \tfrac{1}{2} \tfrac{1}{2} \tfrac{1}{2} -\tfrac{1}{2} \rangle \quad (2.22)$$

The difference between P and A is of some interest with respect to recent discussions. By comparing Eqs. (2.20) and (2.22), it can be seen that the equality $P = A$ requires the symmetry condition

$$\langle \tfrac{1}{2} \tfrac{1}{2} -\tfrac{1}{2} \tfrac{1}{2} \rangle = \langle \tfrac{1}{2} \tfrac{1}{2} \tfrac{1}{2} -\tfrac{1}{2} \rangle \quad (2.23)$$

In elastic scattering processes this condition, and hence $P = A$, holds as a consequence of time-reversal invariance. In inelastic scattering the parameters $\langle \tfrac{1}{2} \tfrac{1}{2} -\tfrac{1}{2} \tfrac{1}{2} \rangle$ and $\langle \tfrac{1}{2} \tfrac{1}{2} \tfrac{1}{2} -\tfrac{1}{2} \rangle$ are in general not related by time reversal. There are however, some limiting cases where one expects Eq. (2.23) and, hence $P = A$, to hold which have been discussed in nuclear physics [see, for example, Amado (1982), and the references therein].

Another case has been considered recently in atomic physics. By analzying a particular dynamical model (the distorted wave Born approximation), it has been shown that Eq. (2.23) holds as a consequence of special dynamical properties and expresses then a *dynamical* symmetry property of the model under discussion (Bartschat and Blum, 1982b). Since this investigation is closely related to the properties of some new spin-sensitive parameters, which are introduced in Section IV, we will postpone further discussion of $P-A$ until Section IV,D,2.

C. Results for Mercury Excitation

The experimental investigation of spin-orbit phenomena in inelastic collisions has been mainly limited to excitation processes of mercury atoms. Eitel and Kessler (1971) measured differential cross sections and polarizations for the $6\ ^1S_0 \rightarrow 6\ ^1P_1$ transition. Theories for this case have been developed by Madison and Shelton (1973), by Bonham (1974), and by Yamazaki *et al.* (1977). These theories are in good agreement with the experimental results.

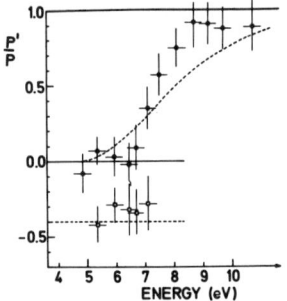

FIG. 4. Ratio P'/P measured in the forward direction as a function of the incident energy for the transition $6\,^1S_0 \rightarrow 6\,^3P_1$ (●) and $6\,^1S_0 \rightarrow {}^3P_2$ (○).

A particularly interesting feature is the strong similarity that exists between the results of Eitel and Kessler and those for elastic electron–mercury scattering for energies down to 30 eV. The reason for this similarity and the underlying theoretical model is most conveniently discussed in terms of some new spin-sensitive parameters which are introduced in Section IV. We therefore refer the reader to a more detailed discussion in Section IV,D,2.

The excitation of the $6\,^3P_1$ state of mercury from its $6\,^1S_0$ ground state was first studied by Hanne and Kessler (1976). They determined the ratio P'/P of the final to the initial transversal polarization of electrons scattered in the forward direction after excitation of the $6\,^3P_1$ state.

An interesting feature of the results is that $P'/P \approx 1$ for energies $E \gtrsim 9$ eV (see Fig. 4). Using the intermediate coupling scheme, this fact was interpreted as an indication that only the ^1P admixture is noticeably excited in forward direction if $E \gtrsim 9$ eV (Hanne and Kessler, 1976; Hanne, 1980; Hanne et al., 1981). We refer the reader to these articles for a detailed discussion of the results and their theoretical implications.

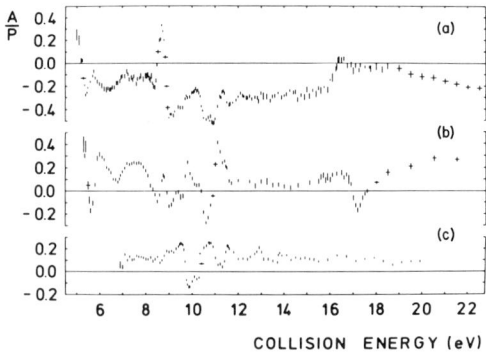

FIG. 5. Measured A/P versus incident energy: (a) Hg $6\,^3P_1$, (b) Hg $6\,^3P_2$, (c) Hg $6\,^1P_1$.

Scattering asymmetries A have been studied by Bartschat et al. (1981b). These authors measured the energy dependence of A in inelastic collisions where the 6 $^3P_1$, 6 $^3P_2$, and 6 $^1P_1$ states of mercury have been excited by polarized electrons. The results are shown in Fig. 5 for a scattering angle of 90°. P denotes the (transverse) polarization of the incident electrons. The figures show pronounced structures with large values of A/P, particularly at energies close to the threshold. These structures can be compared with those found in intensity measurements, which are attributed to the formation and decay of autoionizing states of the Hg$^-$ ion. We shall return to these effects in Section V,C.

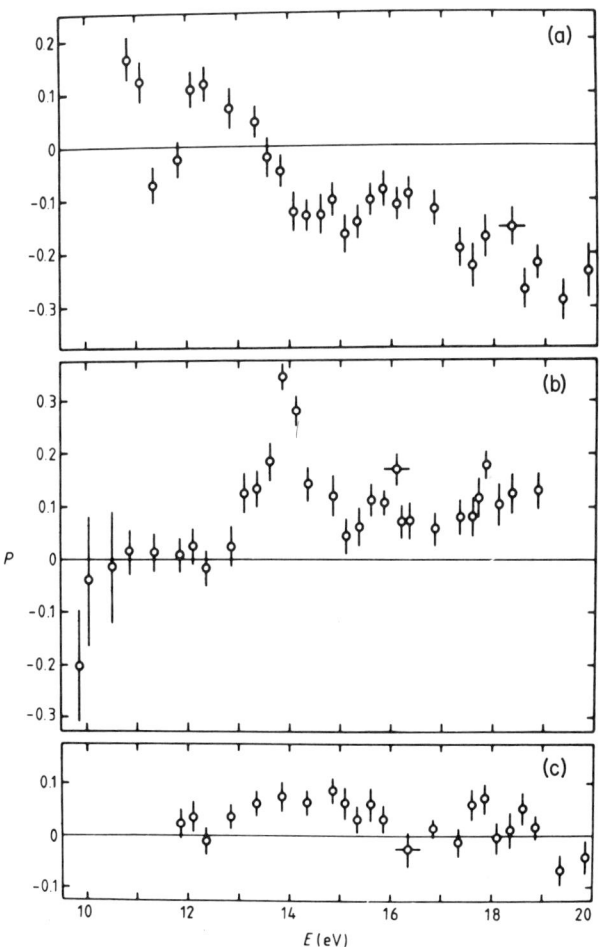

FIG. 6. Polarization P of the scattered electrons versus collision energy: (a) 6 $^3P_{0,1}$, (b) 6 $^3P_2$, (c) 6 $^1P_1$. The scattering angle is 60°.

The same problem was considered by Franz et al. (1982). Here, the 6 ^3P_J states of mercury were excited by unpolarized electrons and the spin polarization P of the scattered electrons was determined. The results are shown in Fig. 6 and compared with the measurements of Eitel and Kessler (1971) for the 6 $^1P_1$ excitation.

No numerical results for the 6 $^3P_1$ excitation of mercury are available at present. Several methods of including relativistic effects in the theory have been developed [see Scott and Burke, (1980), and the references therein]. However, there is still a need to develop further theoretical models which are valid for inelastic collisions, particularly at low energies.

IV. Influence of Spin-Dependent Interactions on Coherence Parameters

A. Parametrization of Atomic Density Matrix

So far, spin-dependent phenomena have been discussed in terms of the spin polarization and asymmetry of the scattered electrons. In the remainder of this article we expand this general framework and introduce new "spin-sensitive" parameters. A discussion is undertaken with regard to can be learned from a determination of these quantities.

Of particular interest is that some of the new parameters can be linked to *specific* details of the spin-coupling mechanisms. By measuring these parameters, one can test whether very specific assumptions of a given theoretical model are valid — without the necessity of performing detailed numerical calculations. We will give examples in Sections IV,D and V,B. These results show the usefulness of collecting data for many independent parameters. Of principal interest is the question whether such connections between observables and specific dynamical effects can be established in more complex cases.

In this section electron–photon coincidence experiments *without any spin selection* are discussed. More definitely, we concentrate on experiments of the following type. We consider electron–atom collisions where the atoms are excited from an ^1S ground state to a $^1P_1$ or $^3P_1$ state. The initial electrons are unpolarized; the spin of the scattered electrons is unobserved. The excited atoms will subsequently decay by photoemission and the scattered electrons and emitted photons are detected in coincidence. Schematically

$$e + A(^1S) \to e' + A(^1P_1, {}^3P_1)$$
$$\hookrightarrow A(^1S) + \gamma \quad (3.1)$$

It is always be assumed that collision and decay can be considered as independent processes. This requires that the atomic mean lifetime be sufficiently long so that the projectile electrons have long left the atom before a noticeable number of these will have decayed.

The essential point for the understanding of electron–photon coincidence experiments is that the observation is restricted to radiation emitted only by those atoms which "scattered" the electrons with a given energy in a given direction defined by the detector. Thus, a certain subensemble of atoms is selected in the experiment. It is the quantum mechanical state of this subensemble which will be considered throughout this section (Blum and Kleinpoppen, 1979).

The excitation process is described in a coordinate system where the Z axis is parallel to the initial electron momentum $\mathbf{P}_0$ and where the X-Z plane is the scattering plane spanned by $\mathbf{P}_0$ and the momentum $\mathbf{P}_1$ of the scattered electrons ("collision system").

The excited atoms of interest are characterized in terms of the corresponding reduced density matrix ρ. We obtain the elements of this matrix by putting $m_1' = m_1$ in Eq. (2.8) and summing over all m_1. This gives the expression

$$\langle M'|\rho|M\rangle = \frac{1}{2} \sum_{m_1 m_0} \langle M'\mathbf{p}_1 m_1|\mathbf{T}|0, \mathbf{p}_0 m_0\rangle \langle M\mathbf{p}_1 m_1|\mathbf{T}|0, \mathbf{p}_0 m_0\rangle^* \quad (3.2)$$

[see, for example, Section 3.5 in Blum (1981a)]. Here M' and M denote the Z components of the total atomic angular momentum $J = 1$ and $|0\rangle$ denotes the atomic ground state.

We use notation similar to that used in Section III:

$$f(M m_1 m_0) = \langle M\mathbf{p}_1 m_1|\mathbf{T}|0, \mathbf{p}_0 m_0\rangle \quad (3.3)$$

for the scattering amplitudes. We normalize according to

$$|f(M m_1 m_0)|^2 = \sigma(M m_1 m_0) \quad (3.4a)$$

where $\sigma(M m_1 m_0)$ denotes the (partial) differential cross section for the transition $|0\rangle|\mathbf{p}_0 m_0\rangle \to |M\rangle|\mathbf{p}_1 m_1\rangle$. The diagonal elements of the density matrix (3.2) are then the various differential cross sections

$$\langle M|\rho|M\rangle = \sigma(M) \quad (3.4b)$$

where $\sigma(M)$ denotes the differential cross section for the excitation of the atomic sublevel $|M\rangle$ independent of the electronic spins. The trace of ρ is then given by the (total) differential cross section:

$$\text{tr } \rho = \sum_M \sigma(M) = \sigma \tag{3.4c}$$

In the following we use the notation

$$\langle M'|\rho|M\rangle = \langle f(M')f(M)^*\rangle \tag{3.5}$$

for the density matrix elements where the angular brackets indicate that averages are taken over all spin components.

The nine (complex) elements of the matrix (3.2) are not independent. With $J = L = 1$ and $J_0 = 0$, we obtain from Eq. (2.4) the symmetry condition:

$$f(Mm_1m_0) = (-1)^{M+m_1+m_0} f(-M-m_1-m_0) \tag{3.6}$$

From condition (3.6) and the hermiticity of the density matrix it follows that ρ is completely determined by five independent real parameters. For example, $\sigma(1)$, $\sigma(0)$, $\langle f(1)f(-1)^*\rangle$, and the magnitude and phase of $\langle f(0)f(1)^*\rangle$ (note that $\langle f(1)f(-1)^*\rangle$ is real because of Eq. (3.6) and the hermiticity of ρ). It is convenient to use the following set of parameters:

$$\lambda = \frac{\sigma(0)}{\sigma}, \qquad \cos\chi = \frac{\text{Re}[f(0)f(1)^*]}{|\langle f(0)f(1)^*\rangle|}$$

$$\cos\Delta = \frac{|\langle f(0)f(1)^*\rangle|}{\sigma(0)\sigma(1)}, \qquad \cos\epsilon = \frac{-\langle f(1)f(-1)^*\rangle}{\sigma(1)} \tag{3.7}$$

(Blum et al., 1980; da Paixao et al., 1980). The density matrix (3.2) is completely characterized by the parameters in Eqs. (3.7) and the differential cross section in Eq. (3.4c).

The parameters in Eqs. (3.7) [in addition to the differential cross section in Eq. (3.4c)] contain the maximal possible information which can be extracted from experiments of the type (3.1). The parameters σ and λ contain all the information on the population of substates $|\pm 1\rangle$ and $|0\rangle$.

The other parameters in Eqs. (3.7) depend on nondiagonal elements of the atomic density matrix given in Eq. (3.2). The off-diagonal elements of ρ are combinations of the scattering amplitudes with different quantum numbers M. We may say that these elements "measure" the interference which—immediately after the excitation—exists between the various sublevels. Following the usual custom, we say that the sublevels with different M have been coherently excited (or that the atomic state is a *coherent superposition* of these substates) if the corresponding off-diagonal element of ρ is different from zero (Cohen-Tannouidji, 1962). Coherence in this sense means that the atoms do not have a well-defined value of M.

We say that the state of the excited atoms is an *incoherent superposition* of substates $|JM\rangle$ if ρ is diagonal in this representation. As a model we may think of an incoherently excited system as an ensemble of atoms where any atom is in exactly one state $|JM\rangle$ and different atoms may be in different substates. As a consequence, such an ensemble is completely characterized if we give the number of atoms populating the various substates. The matrix ρ is then diagonal and its diagonal elements contain all relevant information on the system.

We briefly compare these results with the information which can be obtained from measurements of the spin polarization of the scattered electrons (without detecting the emitted photons). In these experiments one determines the elements of the electronic density matrix. Assuming unpolarized electrons, we obtain from Eq. (2.10) that

$$\langle m_1'|\rho|m_1\rangle = \frac{1}{2} \sum_{Mm_0} f(Mm_1'm_0)f(Mm_1m_0)^* \tag{3.8}$$

As discussed in Section III,A,1, this matrix contains only terms with $M' = M$ so that no information on the coherence between the excited atomic states can be obtained. On the other hand, the matrix (3.2) gives no information on terms with $m_1' \neq m_1$. We will return to this point in Section VI.

Finally, we briefly consider experiments where the emitted light is detected but the scattered electrons are not. The atomic ensemble of interest consists of *all* atoms in the excited state. The corresponding density matrix follows from Eq. (3.2) by integration over all electron scattering angles. The calculation shows that the resulting density matrix is diagonal. Any interference between states with difference values of M vanishes as a consequence of the integration.

This result follows more directly from symmetry considerations. The excitation process is axially symmetric with respect to the incoming beam axis if the scattered electrons are not observed. An immediate consequence of this symmetry is that all nondiagonal elements of ρ vanish [see, for example, Section 4.5 in Blum (1981a)]. Therefore, any experimental determination of coherence parameters requires that the excitation geometry not contain a symmetry axis as was first pointed out by Macek and Jaecks (1971).

B. Physical Importance of Parameters $\cos \epsilon$ and $\cos \Delta$

The parametrization (3.7) is valid for $^1P_1$ and $^3P_1$ excitation or a combination of both states (see Section IV,D). In this section we concentrate on *^1P excitation* and consider the coherence properties of the excited states. The parameter χ has been extensively discussed in the literature. Here we will be concerned with $\cos \epsilon$ and $\cos \Delta$.

In order to see the physical importance of $\cos \epsilon$ and $\cos \Delta$, we assume that the collision can adequately be described in the LS-coupling scheme (see Section II,A). In this case it follows for ^1P excitation that the spin components cannot change ($m_1 = m_0$) and that the amplitudes are independent of the electronic spin (see Appendix A). The angular brackets in eqs. (3.5) and (3.7) are then superfluous and it follows from Eqs. (3.7) that

$$\cos \epsilon = \cos \Delta = 1 \qquad (3.9)$$

where the symmetry condition (3.6) has been used.

Hence, $\cos \epsilon$ and $\cos \Delta$ are equal to one if the excitation process can be described in the LS-coupling scheme and this result holds independently of excitation energy and scattering angle. Electron-exchange processes alone cannot produce values of these parameters different from one.

The parameters $\cos \epsilon$ and $\cos \Delta$ contain therefore direct information on spin–orbit effects during the scattering process. From a measurement of these parameters one can immediately deduce whether or not LS-coupling is violated during the collision—without comparing it with numerical calculations. This result has some interesting consequences for the interpretations of superelastic scattering experiments, which will be discussed in the following section. It is interesting that such direct information can be obtained from the experiments without any spin selection.

Another aspect of these results may be of interest for the theoretical treatment. We are concerned with the influence of a weak force (the spin–orbit interaction) which acts in the presence of a much stronger interaction (the Coulomb force). The influence of the weak forces can often be neglected in calculating cross sections; however, it can be expected that they will influence the phases of the scattering amplitudes and therefore the coherence parameters. A comparison between calculated and measured values of the nondiagonal elements of ρ should therefore be a very sensitive test of the validity of the theoretical model. We return to this point in Section IV,D.

The importance of the parameters $\cos \epsilon$ and $\cos \Delta$ has also been discussed recently by Hermann and Hertel (1982).

It is interesting to consider our results from a more general quantum mechanical point of view. In the case of spin-independent amplitudes it follows from Eq. (3.2) that

$$\text{tr}(\rho^2) = (\text{tr } \rho)^2 = \sigma^2 \qquad (3.10)$$

and the density matrix can be represented by the projector

$$\rho = |\psi\rangle\langle\psi| \tag{3.11a}$$

with

$$|\psi\rangle = \sum_M f(M)|M\rangle \tag{3.11b}$$

This means that the state of the atomic ensemble of interest is completely characterized in terms of a single state vector $|\psi\rangle$. All atoms of the ensemble are therefore in one and the same state $|\psi\rangle$ ("pure" quantum mechanical state).

From Eqs. (3.6) and (3.11b) it follows that $|\psi\rangle$ is completely characterized in terms of three parameters: the magnitudes $|f(1)| = |f(-1)|$ and $|f(0)|$ (or by σ and λ) and the relative phase between $f(1)$ and $f(0)$, which is equal to χ in the case under discussion. In this case an electron–photon coincidence experiment allows the complete determination of the scattering amplitudes, whereas in general only the parameters $\langle f(M') f(M)^* \rangle$ can be obtained which are averaged over all spin components.

The parameters λ and χ were introduced by Eminyan et al. (1974) in order to discuss the excitation of helium atoms where all spin–orbit effects can be neglected.

Equation (3.11b) describes a state without a well-defined quantum number M and, by definition, describes a coherent superposition state. In addition, it represents atoms in identical states. In the recent literature such states are sometimes called "completely coherent" superpositions in order to distinguish this case from the more general one of coherent superposition (Blum and Kleinpoppen, 1979; Hermann and Hertel, 1982). For many experiments this distinction is without importance and in the literature the term "coherent" is usually applied to both situations. The distinction is important, however, for experiments of the type discussed here. In the case of complete coherence, additional constraints [Eq. (3.10), for example] reduce the number of independent density matrix elements and, hence, the number of measurements necessary for a complete determination of ρ.

In conclusion, it has been shown that the density matrix description given by Eqs. (3.2) and (3.7) is necessary in order to discuss all relevant coherence aspects of the excitation. The most obvious consequence of explicit spin-dependent interactions is quantum kinematical in nature: the number of independent parameters is increased. In the literature this consequence was at first overlooked, leading to an incorrect parametrization of experimental results.

Departures of the parameters $\cos \epsilon$ and $\cos \Delta$ from unity are only possible if the collision is influenced by explicit spin-dependent forces. Conversely, if the conditions $\cos \epsilon = 1$ and $\cos \Delta = 1$ are simultaneously fulfilled, then it follows from Eqs. (3.2) and (3.7) that the condition (3.10) is satisfied. The state of the atomic subensemble of interest can then be represented by a single state vector.

C. Influence of Coherence Parameters on Polarization of Emitted Light

1. General Discussion

In this section we briefly consider the determination of the atomic density matrix. The differential cross section given by Eq. (3.4c) can be obtained by applying standard methods of collision physics. The determination of the four parameters of Eqs. (3.7) requires a measurement of the angular distribution and the polarization of the emitted light observed in coincidence with the scattered electrons.

We discuss the polarization properties of the emitted light in the "helicity system" of the emitted photons which is spanned by three unit vectors $\hat{n}$, $\hat{e}_1$, and $\hat{e}_2$. The vector $\hat{n}$ denotes the direction of observation of the emitted radiation (quantization axis in the helicity system) with polar angles θ and ϕ in the collision system (see Fig. 7). The vector $\hat{e}_1$ is chosen to lie in the plane formed by $\hat{n}$ and the Z axis and to point in the direction of increasing θ. The vector $\hat{e}_2$ is then chosen to be perpendicular to both $\hat{n}$ and $\hat{e}_1$, and to point in the direction of increasing ϕ. The vector $\hat{e}_1$ has then polar angles $(\theta + 90°, \phi)$ in the collision system and $\hat{e}_2$ has the polar angles $(90°, \phi + 90°)$.

The intensity emitted in the direction $\hat{n} = (\theta, \phi)$ will be denoted by $I(\theta, \phi) \equiv I$. The polarization properties of the radiation are conveniently char-

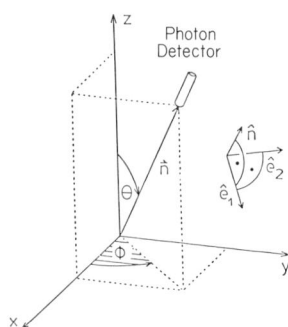

FIG. 7. Coordinate system used in the description of the coincidence experiment.

acterized in terms of the Stokes parameters η_1, η_2, and η_3. Here, η_2 denotes the degree of circular polarization:

$$\eta_2 = (I_+ - I_-)/(I_+ + I_-) \tag{3.12a}$$

Applying the optical convention, I_+ denotes left-handed circularly polarized radiation and I_- refers to right-handed polarized radiation. The parameters η_1 and η_3 are related to the linear polarization. $I(\beta)$ may denote the intensity transmitted by a Nicol prism which is oriented at an angle β with respect to the $\hat{e}_1$ axis (in the $\hat{e}_1$-$\hat{e}_2$ plane). The Stokes parameter η_3 is then defined by the expression

$$\eta_3 = [I(0) - I(90)]/I \tag{3.12b}$$

and the parameter η_1 by

$$\eta_1 = [I(45) - I(135)]/I \tag{3.12c}$$

The Stokes parameters are functions of the parameters given in Eqs. (3.7) which characterize the excited atoms of interest. This relaxation is expressed by the following set of equations (Blum and Kleinpoppen, 1982):

$$I = \frac{A\sigma}{3\gamma} \left[\frac{1-\lambda}{2} (2 - \sin^2\theta) - \frac{1-\lambda}{2} \cos\epsilon \sin^2\theta \cos 2\phi \right.$$
$$\left. + \lambda \sin^2\theta + [\lambda(1-\lambda)]^{1/2} \cos\Delta \cos\chi \sin 2\theta \cos\phi \right] \tag{3.13a}$$

$$I\eta_3 = \frac{A\sigma}{3\gamma} \left[\frac{1-\lambda}{2} \cos\epsilon(2 - \sin^2\theta) \cos 2\phi - \frac{1-\lambda}{2} \cos\epsilon \sin^2\theta \cos 2\phi \right.$$
$$\left. + \lambda \sin^2\theta + [\lambda(1-\lambda)]^{1/2} \cos\Delta \cos\chi \sin 2\theta \cos\phi \right] \tag{3.13b}$$

$$I\eta_1 = -\frac{A\sigma}{3\gamma} [(1-\lambda) \cos\epsilon \cos\theta \sin 2\phi$$
$$+ 2[\lambda(1-\lambda)] \cos\Delta \cos\chi \sin\theta \sin\phi] \tag{3.13c}$$

$$I\eta_2 = -\frac{A\sigma}{3\gamma} 2[\lambda(1-\lambda)]^{1/2} \cos\Delta \sin\chi \sin\theta \sin\phi \tag{3.13d}$$

where γ is the decay constant and A is a spectroscopic factor (Blum and Kleinpoppen, 1980).

Equations (3.13) afford several possibilities for determining the parameters given in Eqs. (3.7). For example, I can be measured for three different pairs of angles θ, ϕ, and $I\eta_2$ can be measured for one set of angles θ, ϕ. Alternatively, all four parameters of Eqs. (3.13) can be measured in the same direction θ, and ϕ. In any case, a complete determination of the set of parameters given in Eqs. (3.7) requires a measurement of the circular polarization.

Experimental determinations of the parameters of Eqs. (3.7) was first performed by Kleinpoppen and collaborators for the 6 $^3P_1$ excitation of mercury (Zaidi et al., 1980, 1981) and by Hippler et al. (1982) for the excitation of argon by protons.

If the photons are detected in the scattering plane ($\phi = 0$), then $\eta_1 = \eta_2 = 0$. If the light is then passed through a linear polarizer, oriented at an angle β to the scattering plane, then the transmitted intensity is given by the following expression [see Eq. (1.2.29) in Blum (1981a)]:

$$I(\beta) = \tfrac{1}{2}(1 + \eta_3 \cos 2\beta) \qquad (3.14)$$

In the case $\cos \epsilon = 1$ it follows from Eqs. (3.13a) and (3.13b) that $I\eta_3 = I$, with $\eta_3 = 1$. Equation (3.14) is then reduced to the expression

$$I(\beta) = \tfrac{1}{2}(1 + \cos 2\beta) \qquad (3.15)$$

Hence, the intensity $I(\beta)$ must be zero for $\beta = \pi/2$ and $\beta = 3\pi/2$ if $\cos \epsilon = 1$.

2. Scattering from Laser-Excited Atoms

The results obtained so far can also be applied to processes where electrons are "superelastically" scattered on laser-excited atoms:

$$\gamma + A \to A^*, \quad A^* + e \to A + e' \qquad (3.16)$$

Experiments of this type were first performed by Hertel and his collaborators (Hertel and Stolle, 1978).

The process (3.16) is inverse to the process (3.1) if excitation geometry and atomic states are appropriately chosen. In both cases the same atomic density matrix can be determined (Hermann and Hertel, 1980).

Here, we are interested in the "superelastic" cross section σ_s (that is, the differential cross section of the electrons scattered by the excited atoms). In the case that the collision can be described in the LS-coupling scheme, it has been shown that

$$\sigma(\beta)_s \sim 1 + \cos 2\beta \qquad (3.17)$$

(Macek and Hertel, 1974; Register, 1981). The relation (3.17) holds under the assumption that the atoms were excited by linearly polarized laser light incident in the scattering plane. β is the angle between the polarization vector and the scattering plane.

Equation (3.17) corresponds to Eq. (3.15). The superelastic cross section should vanish for $\beta = \pi/2, 3\pi/2$. Investigating the 6 $^1P \to 6$ ^1S transition in barium, Register, Trajmar, and collaborators found minima of $\sigma(\beta)_s$ but the minimum is not zero (Register et al., 1980).

The experimental results obtained by Register et al. are shown in Fig. 8 for an incident electron energy of 100 eV and an electron scattering angle $\theta_e = 5°$. On the ordinate $\sigma(\beta)_s$ is given in arbitrary units. The solid line represents a fit to the experimental results.

The most striking and obvious difference between the results shown in Fig. 8 and the earlier Na data of Hertel and Stoll (1974) is the nonzero minimum. In order to explain this feature, Register et al. applied the parametrization (3.7) to this problem. They showed that $\cos \epsilon \neq 1$ additional terms are present in Eq. (3.17) which prevent the vanishing of σ_s. The corresponding equations are similar to Eq. (3.14) [and to Eq. (3.15) in the LS-coupling case].

Hence, the nonvanishing of the superelastic cross section was interpreted by Register et al. (1980) as being due to a spin–orbit effect in the collision. This result shows the usefulness of the parametrization (3.7) for investigations of spin-dependent effects in superelastic scattering processes. A measurement of $\cos \epsilon$ and $\cos \Delta$ gives direct information on whether or not LS-coupling is violated in the collision. Furthermore, Register et al. found that the parameter $\cos \Delta$ is nearly zero:

$$\cos \Delta \approx 0 \tag{3.18}$$

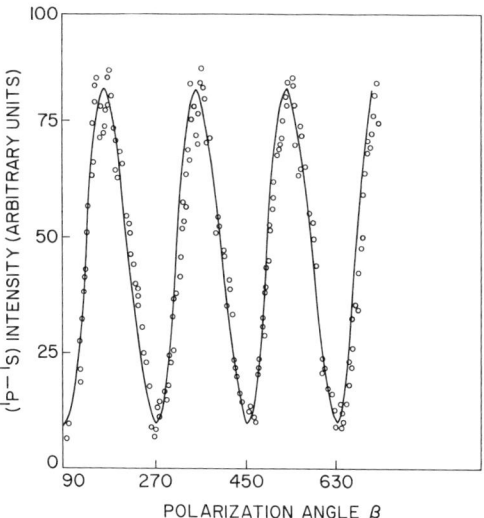

FIG. 8. Superelastic scattering intensity in barium as a function of laser polarization angle; impact energy $E_0 = 100$ eV; electron scattering angle $\theta = 5°$.

This result is in striking contrast to the behavior of cos ϵ. The parameter cos ϵ is clearly different from one but the deviation is small:

$$\cos \epsilon \approx 1 \tag{3.19}$$

Hence, spin effects introduce a large amount of incoherence in cos Δ but not in cos ϵ. This striking difference is discussed further in Section IV,D,2.

D. Results and Analysis of Numerical Calculations

1. Results for Ne and Ar

First calculations of the complete density matrix given by Eq. (3.2) have been performed for various Ne and Ar states. The main goal of these investigations was to consider the dependence of the parameters of Eqs. (3.7) on spin-dependent forces in some detail, using a simple model.

The calculations are based on the so-called "first-order many-body theory" developed by Csanak et al. (1971). This method was first applied to scattering processes between electrons and light atoms where all explicit spin-dependent interactions during the collisions can be neglected.

This model has now been generalized. Spin–orbit effects have been included by making the following simple assumptions:

(1) The coupling between the spin variables and the orbital angular momentum of the continuum electron in the field of the atomic core can be neglected. This assumption is reasonable for intermediate atoms and intermediate energies.

(2) The spin–orbit coupling inside the excited target states is taken into account by representing the excited state by the superposition

$$|^1P_1\rangle' = a|^1P\rangle + b|^3P_1\rangle \tag{3.20}$$

where $|^1P\rangle$ and $|^3P_1\rangle$ are Russell-Saunders states (intermediate coupling scheme). Similarly, we write

$$|^3P_1\rangle' = b|^1P\rangle - a|^3P_1\rangle \tag{3.20a}$$

The values for a and b are obtained from spectroscopic data (Cowan, 1968). In the following we mainly consider excitation of the state (3.20).

The scattering amplitudes are then written in the following form:

$$f(Mm_1m_0)' = af(Mm_1m_0)^{(1)} + bf(Mm_1m_0)^{(3)} \tag{3.21}$$

where f' denotes the amplitude for the excitation of the state (3.20), and $f^{(1)}$ and $f^{(3)}$ denote the amplitudes related to the singlet and triplet parts, respectively. Using the amplitudes of Eq. (3.21) the parameters of Eqs. (3.7) have been calculated.

If only the $|^1P\rangle$ part of the state (3.20) has been excited, then it follows that $\cos \epsilon = \cos \Delta = 1$ according to Section IV,B. Deviations from unity indicate that the $|^3P_1\rangle$ part also contributes to the scattering.

Using the relation (3.21) and the angular momentum coupling rules valid for the Russell–Saunders states, one obtains the inequality

$$-\lambda/(1-\lambda) \leq \cos \epsilon \leq 1 \qquad (3.22)$$

For $\lambda \leq 0.5$ this gives a nontrivial lower limit of $\cos \epsilon$. The relation (3.22) holds for any scattering angle and energy. [The same relation holds also for the state (3.20a).]

The parameter $\cos \epsilon$ is equal to $-\lambda/(1-\lambda)$ if only the $|^3P_1\rangle$ part of the states (3.20) and (3.20a) contributes (note that in this case we have $0 \leq \lambda \leq 0.5$, as can be shown by applying the usual coupling rules valid for Russel–Saunders states).

The inequality (3.22) can be used in order to test the described model. If one finds experimental deviations then additional spin-dependent effects must be taken into account. We give an example at the end of Section IV,D,2.

The theory described here has been applied to the excitation of the 3 $^1P_1$ state of Ne (Machado et al., 1982) and to the excitation of the 4 $^1P_1$ state of Ar (Blum et al., 1980; da Paixao et al., 1980, 1982). The coefficients a and b have been determined by R. Cowan (private communication, 1981) for neon:

$$a = 0.964, \quad b = 0.266 \qquad (3.23a)$$

and for argon:

$$a = 0.893, \quad b = -0.450 \qquad (3.23b)$$

In the following we give some of the results. Figure 9 shows results for the differential cross section obtained by Padial et al. (1981). The calculations have been performed (i) in the LS-coupling scheme ($b = 0$) and (ii) by using the values for a and b given in Eq. (3.23b). As shown by the figure, both curves are compatible with the results obtained by Csanak from an analysis of the measurements of Lewis et al. (1975).

The influence of the $|^3P_1\rangle$ admixture to the $^1P_1$ excitation of Ne is vanishingly small (Csanak, 1981). The same result holds for the parameter λ as has been shown by Machado et al. (1982) (see Table I).

Figure 10 shows results for the ϵ parameter of the Ne 3 $^1P_1$ excitation. Clearly, the $|^3P_1\rangle$ admixture cannot be neglected when calculating this

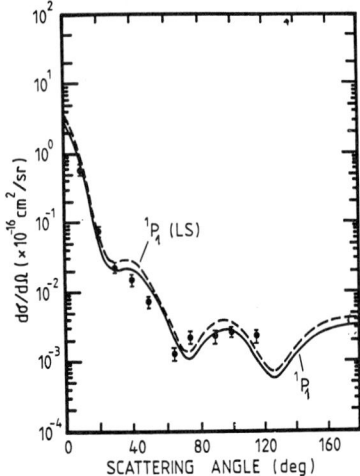

Fig. 9. Differential cross section for excitation of the $^1P_1$ state of argon at $E_0 = 50$ eV, with and without spin–orbit coupling. (From Padial et al., 1981.)

parameter. Similar results have been obtained for ϵ and Δ in the case of the 4 $^1P_1$ state of argon (see Figs. 11–13).

The results show that, in the cases under discussion, spin-dependent effects can safely be ignored when calculating the diagonal elements of the density matrix. These effects cannot be neglected, however, for the ϵ and Δ

TABLE I

Calculated Values of λ for $E = 20$ and 50 eV and Various Electron Scattering Angles θ[a]

	\multicolumn{2}{c}{λ}		
	$E = 20$ eV		$E = 50$ eV	
θ (deg)	LS[b]	Text[c]	LS	Text
30	0.69	0.68	0.04	0.04
60	0.03	0.04	0.99	0.98
90	0.35	0.34	0.87	0.87
120	0.59	0.59	0.23	0.23
150	0.88	0.86	0.40	0.39

[a] From Machado et al. (1982).
[b] LS-coupling.
[c] The model described in the text.

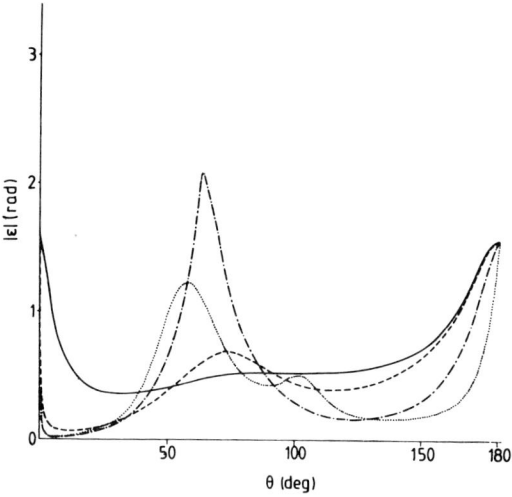

FIG. 10. $|\epsilon|$ as a function of the scattering angle for the Ne $^3P_1$ excitation for various energies E (eV): (———) 20, (----) 30, (·····) 40, (-·-·-) 50. (From Machado et al., 1982.)

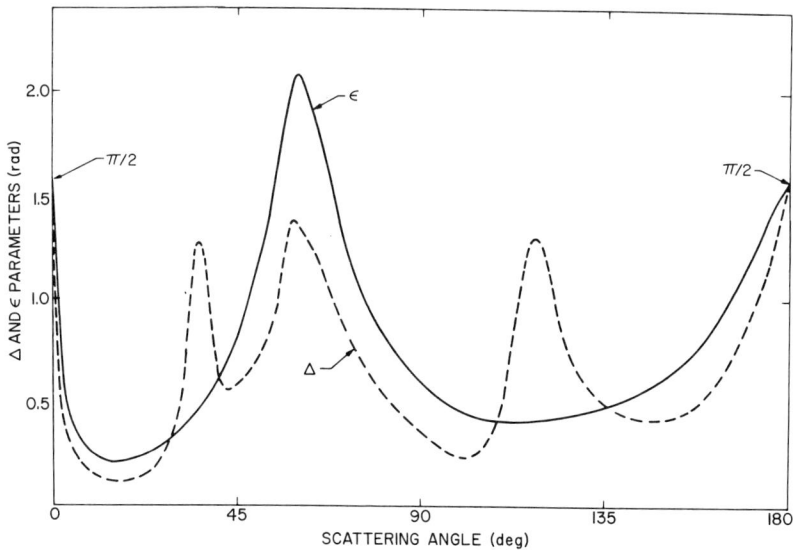

FIG. 11. ϵ and Δ parameters for the excitation of the $^1P_1$ state of argon as a function of the scattering angle. (From da Paixao et al., 1980.)

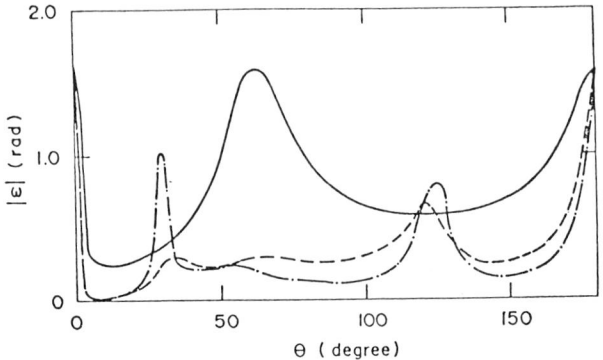

FIG. 12. $|\epsilon|$ as a function of the scattering angle for the Ar $^1P_1$ excitation for various energies E (eV): (———) 16, (----) 50 (-·-·-) 80.4. (From da Paixao et al. 1982.)

parameters. Both parameters can therefore be used as a sensitive test for the assumptions made on the spin–orbit interactions in the theoretical model.

The parameters ϵ and Δ show a pronounced structure. Csanak (1981) has shown that, in general, maxima of ϵ are correlated with minima of $\sigma(1)$. Similarly, the maxima of Δ are related to the minima of $\sigma(0)$ on $\sigma(1)$. These correlations are similar to the relations which exist between the spin polarization and the corresponding differential cross sections (Kessler, 1976).

Finally, we consider the behavior of ϵ and Δ for electron scattering angles $\theta \to 0°$ and $\theta \to 180°$. In the forward and backward directions the excitation process is axially symmetric with respect to the incident beam axis. As a consequence, the density matrix (3.2) must be diagonal [see, for example, Section 4.5 in Blum (1981a)]. That is, the numerators of cos ϵ and cos Δ

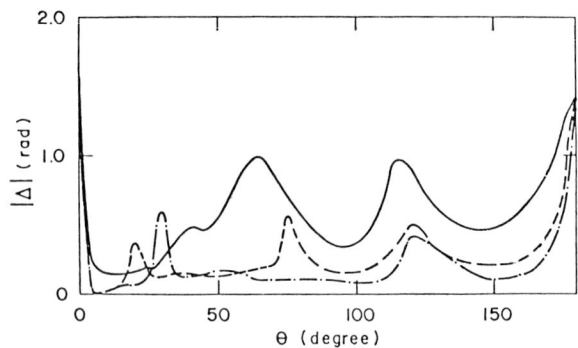

FIG. 13. $|\Delta|$ parameter as a function of the scattering angle for the Ar $^1P_1$ excitation for various energies E (eV): (———) 16, (----) 50, (-·-·-) 80.4. (From da Paixao et al., 1982.)

vanish according to Eqs. (3.7). If $\sigma(1)$ and $\sigma(0)$ are different from zero for these directions, then it follows that

$$\cos \epsilon \to 0, \quad \cos \Delta \to 0 \qquad (3.24)$$

for $\theta \to 0°$ and $\theta \to 180°$. (Note, however, that very small errors in the numerical calculations of $\sigma(1)$ and $\sigma(0)$ will heavily influence the behavior of $\cos \epsilon$ and $\cos \Delta$ for very small and very large angles.)

The coherence parameter χ for the $4\,^1P_1$ excitation of Ar has also been calculated. The calculations have been performed (1) in the LS-coupling limit, (2) by using the coefficients (3.23). The results show distinct differences (da Paixao et al., 1980). A determination of this parameter therefore gives additional information on spin–orbit effects in scattering processes. However, $\cos \chi$ does not have the convenient properties of $\cos \epsilon$ and $\cos \Delta$, which are equal to unity in the LS limit.

Further numerical results for the parameters of Eqs. (3.7) have been obtained by da Paixao et al. (1982).

2. Results for Hg Analysis of the DWBA

a. Numerical results obtained in the DWBA. The excitation of the $6\,^1P_1$ state of Hg has been investigated theoretically by Madison and Shelton (1973). These authors calculated the differential cross section and the spin polarization of the scattered electrons for various energies, assuming that the incident particles are unpolarized.

The calculations were performed in the "distorted-wave Born approximation" (DWBA). The coupling of the spin and orbital angular momentum of the continuum electron in the field of the atomic core was taken into account. However, any fine-structure effect inside the excited atoms has been neglected in the model under discussion. The numerical results, obtained by Madison and Shelton, are in good agreement with the experiments of Eitel and Kessler (1971) at higher energies ($E \gtrsim 80$ eV).

We have calculated the parameters $\cos \Delta$ and $\Delta - 1$, using the results for the scattering amplitudes obtained by Madison and Shelton (Bartschat and Blum, 1982b). The results are shown in Fig. 14 for an energy of 180 eV. The parameters $\cos \Delta$ and $\Delta - 1$ are shown and the spin polarization of the scattered electrons has been reproduced. No results for very small and large scattering angles have been given since the numerical values, obtained in the described model, are not very accurate in these regions.

The figure shows that deviations of $\cos \Delta$ from unity (respectively, of Δ from zero) are only important in those regions where also the spin polarization is largest—in particular, for scattering angles between 80° and 120°.

More interesting and completely unexpected results have been obtained

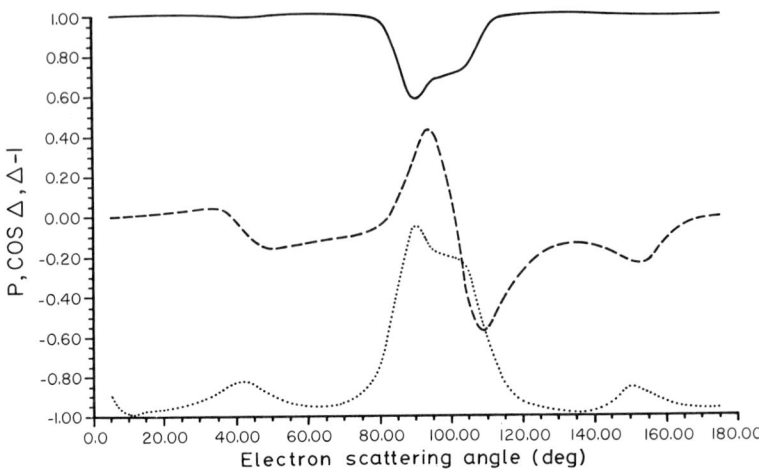

FIG. 14. Spin polarization P, cos Δ, and Δ − 1 as functions of the electron scattering angle for energy of 180 eV: (———) cos Δ; (·····) Δ − 1; (-----) averaged P. (From Bartschat and Blum, 1982b.)

for cos ϵ. For $E = 125$ eV and $E = 180$ eV we have found cos $\epsilon \approx 1$ (cos $\epsilon >$ 0.99) independent of the scattering angle (except very small and very large angles). Although spin–orbit effects have been explicitly taken into account, cos ϵ behaves in the LS-coupling limit.

This result is also surprising from the following point of view. The parameters of Eqs. (3.7) have been defined for experiments without any spin selection. If spin–orbit effects are present, then, because of the incoherence in the spin states, the coherence between all *atomic* states should be reduced compared to the LS-coupling case [see, for example, Fano (1957) or Chapter 3 in Blum (1981a)]. As a consequence, the values of *both* cos Δ and cos ϵ should be smaller than unity.

b. An approximate symmetry property of the DWBA amplitudes. In order to understand these results, we have analyzed the model of Madison and Shelton in more detail. The particular approximation, important for the subsequent discussions, is the following. At the end of their article Madison and Shelton showed numerically that, for higher energies, the final-state distorted wave can be replaced by a plane wave without greatly disturbing the results for cross sections and polarizations. Electron-exchange effects can be neglected in the energy region considered ($E > 100$ eV).

In this approximation the direct DWBA amplitude is given by the expression

$$f(Mm_1m_0) = \langle M, \mathbf{p}_1 m_1 | V_c | 0, \phi(m_0)^+ \rangle \tag{3.25}$$

Here, V_c is the Coulomb potential between the incident and the outer atomic electrons. The term $|\phi(m_0)^+\rangle$ is a scattering state of the continuum electron defined as a solution of the integral equation

$$|\phi(m_0)^+\rangle = |\mathbf{p}_0 m_0\rangle + (E - H_0 - U + i\delta)^{-1} U |\phi(m_0)^+\rangle \quad (3.26)$$

Here, H_0 denotes the Hamiltonian of the free electron and the potential U contains the Coulomb and spin–orbit interaction between the continuum electron and the atomic core. (More precisely, U is the atomic ground-state potential) (Madison and Shelton, 1973).

Equations (3.25) and (3.26) can be interpreted in terms of the following physical picture where the collision is interpreted as a two-step process: The incident electron is elastically scattered in the field of the atomic core [Eq. (3.26)] followed by an inelastic scattering caused by the Coulomb interaction with the electrons in the outer shell. Only the first process is then influenced by spin–orbit forces and is responsible for the polarization effects. This model has been introduced and discussed by Massey and Burhop (1969) and by Eitel and Kessler (1971) in order to understand the similarity between elastic and inelastic cross sections and polarizations.

If $|\phi(m_0)^+\rangle$ is replaced by the plane wave $|\mathbf{p}_0 m_0\rangle$ in Eq. (3.26), then one obtains the amplitude in the first Born approximation. In this approximation, however, the experimental results cannot be reproduced (Madison and Shelton, 1973).

The amplitude (3.25) possesses an interesting symmetry property. Expanding the distorted waves into plane waves, using the transformation properties of the states under reflections in planes through $\mathbf{p}_0$, and taking the spin independent of V_c into account, we obtain the following symmetry relation (Bartschat and Blum, 1982b):

$$f(M m_1 m_0) = (-1)^{m_1 - m_0} f(M - m_1 - m_0) \quad (3.27a)$$

Following the discussion by Madison and Shelton, one can expect that the full DWBA amplitude approaches that given by Eq. (3.25) with increasing energy. Hence, in this sense, Eq. (3.27a) represents an approximate symmetry property of the full DWBA amplitude which should be better satisfied the higher the energy is.

As an example, we give in Table II numerical results for the scattering amplitudes for 180 eV, calculated by Madison and Shelton in the DWBA model but without using the additional approximation of Eq. (3.25).

Care should be taken by applying the relation (3.27) to forward and backward scattering. In these cases the exact selection rule

$$M + m_1 = m_0$$

holds. Therefore, the amplitudes $f(1, \frac{1}{2}, \frac{1}{2})$, $f(-1, -\frac{1}{2}, -\frac{1}{2})$, and $f(1, \frac{1}{2}, -\frac{1}{2})$

TABLE II

Scattering Amplitudes for $E = 180$ eV[a,b]

Angle (deg)	$f(1, \frac{1}{2}, \frac{1}{2})$		$f(1, -\frac{1}{2}, -\frac{1}{2})$	
	Re	Im	Re	Im
30	1.14	0.247	1.13	0.236
60	0.850	−0.299	0.843	−0.312
90	−0.058	0.049	−0.064	0.041
120	0.203	−0.367	0.196	−0.372
150	−0.043	−0.068	−0.047	−0.070

Angle (deg)	$f(1, -\frac{1}{2}, \frac{1}{2})$		$f(1, \frac{1}{2}, -\frac{1}{2})$	
	Re	Im	Re	Im
30	−0.047	−0.053	0.043	0.010
60	−0.095	−0.036	0.093	0.012
90	−0.087	−0.003	0.085	−0.008
120	−0.054	0.009	0.054	−0.016
150	−0.016	−0.001	0.019	−0.006

[a] Re, real part; Im, imaginary part.
[b] From Madison and Shelton (1973).

must vanish for these angles, whereas $f(1, -\frac{1}{2}, \frac{1}{2})$ may be different from zero. It can therefore be expected that for very small and very large scattering angles Eq. (3.27a) will in general be better satisfied by the nonflip than by the spin-flip amplitudes.

Finally, it should be noted that, for fixed M, Eq. (3.27a) closely resembles the relation $f(m_1 m_0) = (-1)^{m_1 - m_0} f(-m_1 - m_0)$, which holds exactly for the elastic scattering amplitudes.

c. Consequences for coherence parameters. Under reflection in the scattering plane the exact symmetry property of the amplitudes is given by Eq. (3.6). From the relations (3.6) and (3.27a) we obtain

$$f(M m_1 m_0) = (-1)^M f(-M m_1 m_0) \qquad (3.27)$$

This equation relates the amplitudes for $M = 1$ and $M = -1$ excitation, and this relation is independent of m_1 and m_0. In other words, there exists *a definite phase relation between $f(1 m_1 m_0)$ and $f(-1 m_1 m_0)$ which is independent of the spin components.* One may say that the atomic states $|M = 1\rangle$ and $|M = -1\rangle$ are completely coherently excited.

These results explain the observed behavior of $\cos \epsilon$ discussed above.

From Eqs. (3.7) and (3.27) it follows immediately that

$$\cos \epsilon = 1 \tag{3.28}$$

The incoherence in the spin states has no consequences for $\cos \epsilon$. No similar argument follows from Eq. (3.27) for $\cos \Delta$. The two parameters can therefore behave completely differently.

Let us return to the experimental results of Register and Trajmar for the superelastic scattering of electrons on barium. As discussed in Section III,C,2, the authors found $\cos \Delta$ is nearly zero. The parameter $\cos \epsilon$ is clearly different from unity but the deviation is small. Hence, the states $|M = 1\rangle$ and $|M = -1\rangle$ are nearly completely coherently excited, whereas there exists practically no coherence between $|M = \pm 1\rangle$ and $|M = 0\rangle$. As discussed above, such a difference is surprising.

Our results suggest a possible explanation: The scattering process may be described in terms of the two-step model (3.25) and (3.26) in a reasonable approximation so that Eq. (3.28) is nearly satisfied. In this case it should be expected that Eq. (3.28) is better satisfied the higher the energy is. Experiments are in progress to measure $\cos \epsilon$ at higher energies and larger scattering angles in order to test these conclusions (S. Trajmar, private communication, 1982).

As shown by Eq. (3.28), the determination of $\cos \epsilon$ offers *a direct experimental test* of the validity of the approximations (3.25) and (3.26) without the necessity of performing detailed numerical calculations. If one finds experimentally noticeable deviations from $\cos \epsilon = 1$, then the process under investigation certainly cannot be described in terms of the two-step model.

Therefore, a measurement of $\cos \epsilon$ allows *one* particular dynamical assumption to be tested: the validity of Eqs. (3.25) and (3.26). On the other hand, a comparison between experimental and numerical results allows one to test the theoretical model only as a whole, including *all* approximations made on the scattering process, the atomic wave functions, etc.

d. Consequences for the scattering asymmetry. Finally, we compare the spin polarization P of the scattered electrons and the asymmetry A. It is well known that for elastic scattering $P = A$ holds as a consequence of time-reversal invariance. For inelastic scattering, however, time-reversal invariance does not reduce the number of independent parameters and, in general, P and A will be different from each other.

However, if Eqs. (3.25) and (3.26) are good approximations for the scattering amplitudes, then it follows that $P = A$ as a consequence of the symmetry relation (3.27a) (Bartschat and Blum, 1982b).

For example, the partial polarization connected with the $M = 1$ sublevel is given by

$$P(M=1) = -\text{Im}[f(1, \tfrac{1}{2}, \tfrac{1}{2})f(1, -\tfrac{1}{2}, \tfrac{1}{2})^*$$
$$+ f(1, \tfrac{1}{2}, -\tfrac{1}{2})f(1, -\tfrac{1}{2}, -\tfrac{1}{2})^*]/\sigma \qquad (3.29a)$$

and the asymmetry by the expression

$$A(M=1) = \text{Im}[f(1, \tfrac{1}{2}, \tfrac{1}{2})f(1, \tfrac{1}{2}, -\tfrac{1}{2})^*$$
$$+ f(1, -\tfrac{1}{2}, \tfrac{1}{2})f(1, -\tfrac{1}{2}, -\tfrac{1}{2})^*]/\sigma \qquad (3.29b)$$

If Eq. (3.27a) is satisfied, then it follows immediately that $P(M=1) = A(M=1)$. Similar relations hold for $M=-1$ and $M=0$. For the total parameters

$$P = \sum_M P(M), \quad A = \sum_M A(M)$$

it follows then that

$$P = A \qquad (3.29)$$

Hence, in the framework of the DWBA, P and A should approach each other with increasing energy when excitation of atomic states with $J=1$ is considered.

e. Discussion of experimental results. The new parameters $\cos \epsilon$ and $\cos \Delta$ have been measured for the excitation of the 6 $^3P_1$ state of Hg (Zaidi *et al.*, 1980, 1981). The authors have determined the parameters of Eqs. (3.7) for two energies (5.5 and 6.5 eV) and two scattering angles (50° and 70°) (see Table III).

The experimental results are of interest in connection with theoretical ideas developed in order to explain electron scattering from heavy atoms. It has been suggested that spin effects in *inelastic* scattering processes at low energies can be described in terms of the theoretical model, described in Section IV,D,1. Spin–orbit effects between atom and continuum electron should only be important if the collision is influenced by Feshbach resonances.

So far, no numerical calculations for the Hg excitation in the resonance region have been performed. In this situation it is interesting that a direct test of the proposed model is possible by applying the inequality (3.22).

When the experimental values of Zaidi *et al.* are inserted in the inequality (3.22), it follows that both values for 70° violate (3.22) (taking the experimental error bars into account). The excitation of the 6 $^3P_1$ state of Hg is influenced by a Feshbach resonance at 5.5 eV. At 6.5 eV, however, no compound state of the negative Hg ion has been found to date. Hence, the suggested theoretical model cannot be valid in general. In addition to the assumptions included in Section III,D,1, further spin-dependent interac-

TABLE III

COINCIDENT PHOTON STOKES PARAMETERS FOR THE $6\,^3P_1 \rightarrow 6\,^1S_0$ TRANSITION OF MERCURY[a,b]

Incident electron energy (eV)	Scattering angle (deg)	Normalized Stokes parameters			
		$\phi_\gamma = \pi/2$		$\phi_\gamma = 3\pi/4$	
		P_1	P_2	P_3	P_1
5.5	50	−0.42 ± 0.08	0.34 ± 0.08	0.37 ± 0.07	−0.17 ± 0.09
	70	−0.13 ± 0.08	0.52 ± 0.08	0.27 ± 0.04	−0.39 ± 0.10
6.5	50	−0.26 ± 0.09	0.39 ± 0.09	0.44 ± 0.04	−0.24 ± 0.09
	70	−0.13 ± 0.06	0.45 ± 0.09	0.42 ± 0.07	−0.45 ± 0.10

Incident electron energy (eV)	Scattering angle (deg)	λ	$\bar{\chi}$ (rad)	$|\epsilon|$ (rad)	$|\Delta|$ (rad)
5.5	50	0.26 ± 0.03	0.83 ± 0.15	0.77 ± 0.62	1.04 ± 0.11
	70	0.18 ± 0.03	0.48 ± 0.08	2.02 ± 0.16	1.25 ± 0.06
6.5	50	0.23 ± 0.03	0.85 ± 0.12	1.55 ± 0.26	1.12 ± 0.06
	70	0.16 ± 0.03	0.75 ± 0.13	2.10 ± 0.15	1.26 ± 0.06

[a] Quoted experimental uncertainties represent one standard deviation.
[b] From Zaidi et al. (1981).

tions must be taken into account in order to explain the experiments by Zaidi et al. (1980, 1981).

V. Excitation of Heavy Atoms by Polarized Electrons: Stokes Parameter Analysis

A. SYMMETRY CONSIDERATIONS

When atoms are excited by unpolarized electrons and the light emitted in the subsequent decay is observed without detecting the scattered electrons, it is well known that the polarization properties of the emitted radiation depend on two parameters, the total cross section and the atomic alignment (Percival and Seaton, 1957).

In this section we discuss excitation of heavy atoms by polarized electrons, taking into account spin–orbit effects during the collision. It will always be assumed that *the scattered electrons are not observed*. By using simple symmetry arguments, we first be discuss how the initial polarization influences the polarization of the emitted light and how this can be described in terms of additional orientation and aligment parameters (Bartschat, 1981; Bartschat and Blum, 1982a). The more formal aspects of the theory can be found in a recent publication (Bartschat et al., 1981a).

In the following section we consider what information on the spin-dependent interactions can be extracted from such measurements. Our main case of interest is the excitation of the 6 $^3P_1$ state of Hg and the subsequent decay of the 6 $^1S_0$ state. Experimental results for this case and the measuring apparatus have been described by Bartschat et al. (1982).

An atomic transition is excited by an unpolarized electron beam directed along the z axis. Radiation subsequently emitted during the decay will have an intensity I which varies with the angle between the z axis and the direction of observation. When viewed in y direction, the intensities are $I(0)$ and $I(90)$ for radiation with the electric vector oscillating along the z and x axis, respectively. Since the geometry of the experiment has cylindrical symmetry rather than complete spherical symmetry, the z and x axes are not equivalent, and $I(0)$ and $I(90)$ are not necessarily equal. Consequently, defining the standard linear polarization fraction by

$$\eta_3 = [I(0) - I(90)]/[I(0) + I(90)] \quad (4.1a)$$

In general, η_3 will be different from zero (Percival and Seaton, 1957).

Consider now the two axes 1 and 2 at angles of 45° and 135° to the incident beam axis (Fig. 15). If the light is observed in the y direction and if $I(45)$ and $I(135)$ denote the intensities for radiation with the electric vector parallel to the lines 1 and 2, respectively, then the polarization fraction

$$\eta_1 = [I(45) - I(135)]/[I(45) + I(135)] \quad (4.1b)$$

must be zero. This follows from the axial symmetry of the excitation process

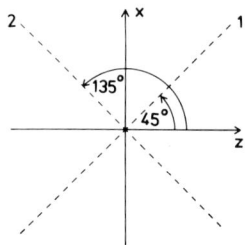

FIG. 15. Excitation of atoms by unpolarized electrons and symmetry consequences for the polarization of the emitted radiation. The photon detector is set up in the y direction perpendicular to the drawing plane.

with respect to the incident beam axis. Under a rotation of 180° around z the two lines are interchanged and are therefore equivalent. That is, because of the symmetry, light with the electric vector oscillating along line 1 must be emitted with the same probability as light with the field vector oscillating along line 2, and $I(45) = I(135)$.

Consider now the case where the incoming electrons are polarized with polarization vector **P** parallel to the beam axis z (see Fig. 16, where **P** is represented by a double arrow). In this case the axial symmetry is retained and η_1 must vanish. Since **P** is an axial vector, a sense of rotation is defined in the $x-y$ plane. It can therefore be expected that a nonvanishing atomic orientation $\langle J_z \rangle$ will be induced during the collision by electron exchange or spin–orbit effects. Circular polarization of the emitted light is an observable consequence of the atomic orientation.

The relation between a longitudinal spin polarization of the incident electrons and the circular polarization η_2 was first considered by Farago and Wykes (1969) and by Wykes (1971). Wykes (1971) discussed the special case of excitation of a $^3S_1$ state from a $^1S_0$ ground state and the subsequent decay to a ^3P_J state. Assuming that LS-coupling holds during the excitation process, Wykes showed that the degree of circular polarization η_2, observed in a certain direction, is proportional to the component of **P** in this direction. This result allows an optical determination of the spin polarization of electrons.

Such a transfer of longitudinal electron polarization to a circular polarization of the emitted light (via an atomic orientation) has been demonstrated by Eminyan and Lampel (1980). The authors studied the excitation of zinc ($4s^2\ ^1S \rightarrow 4s5s\ ^3S$) by longitudinally polarized electrons and the decay into the $4s4p\ ^3P_J$ states. From their measurements of η_2 Eminyan and Lampel determined the polarization of the incident electrons obtained from a GaAs source.

In the rest of this section we always assume that the incident electrons are *transversally* polarized with polarization vector **P** parallel to the y axis and

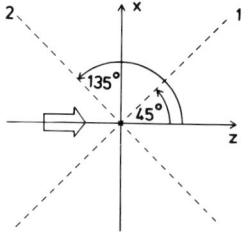

FIG. 16. Symmetry properties of the excitation process with longitudinally polarized electrons.

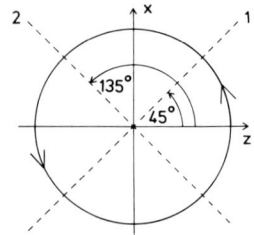

FIG. 17. Excitation by transversely polarized electrons.

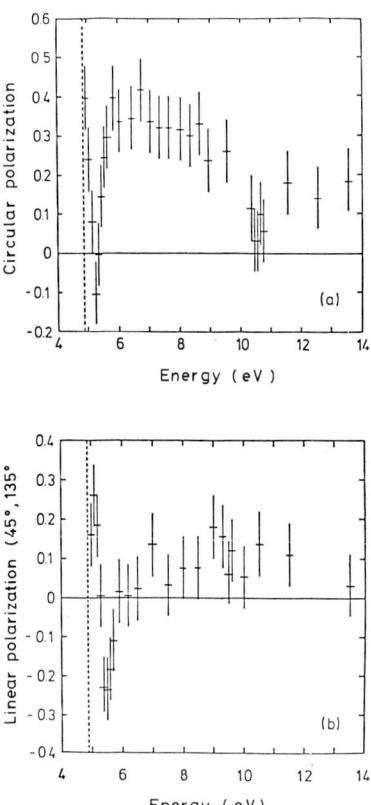

FIG. 18. Measurements of the Stokes parameters η_2 (a) and η_1 (b) for the 6 $^3P_1 \to$ 6 $^1S_0$ transition in mercury. The photon detector was set up in the direction of the initial (transverse) electron polarization; the scattered electrons were not observed. The dashed line denotes the excitation threshold, (From Wolcke et al., 1982.)

we always assume that the emitted light is observed in the y direction. With **P** parallel to the y axis the collision geometry loses its axis of symmetry. The polarization **P** defines a sense of rotation in the $x-z$ plane and the geometry of the experiment possesses only reflection symmetry in this plane (see Fig. 17). The two lines 1 and 2 are then not equivalent, $I(45)$ and $I(135)$ are not necessarily equal, and η_1 can be expected to be different from zero in general.

Furthermore, similar to the case of longitudinally polarized electrons discussed above, it can be expected that a nonvanishing atomic orientation $\langle J_y \rangle$ perpendicular to the $x-y$ plane will be induced during the collision and that the circular polarization fraction

$$\eta_2 = [I_+ - I_-]/[I_+ + I_-] \tag{4.1c}$$

will be different from zero. (Here, I_+ and I_- denote the intensity of light with positive and negative helicity, respectively).

For unpolarized electrons no sense of rotation is defined, the mean atomic angular momentum $\langle \mathbf{J} \rangle = 0$, and η_2 vanishes.

Experimental results for η_1 and η_2 are shown in Fig. 18. These results show that, in addition to η_3, the emitted light has (*i*) a circular polarization and (*ii*) a linear polarization with respect to two axes oriented at angles of 45° and 135° to the incident beam axis. This is a situation quite different from the case of excitation by unpolarized electrons where the only nonvanishing polarization component is η_3.

B. Probing the Spin–Orbit Interaction by Means of Integrated Stokes Parameters

In this section we show that under the conditions described above (transversal polarization of the initial electron, no observation of the scattered electrons) the Stokes parameters η_1 and η_2 contain direct information on explicit spin-dependent forces acting during the collision.

In order to show this, it is convenient to express the Stokes parameters in terms of the atomic orientation and alignment. The number of independent parameters characterizing the excited atomic state is restricted by symmetry requirements (Fano and Macek, 1973). For transversely polarized electrons there are in general five independent parameters because of the reflection invariance in the scattering plane (see, for example, Bartschat *et al.*, 1981a). Using the language of state multipoles and assuming that states with sharp angular momentum J have been excited, these parameters are as follows [see, for example, Chapter 4 of Blum (1981a) for details]: (1) a monopole $\langle T(J)_{00}^+ \rangle$ which is proportional to the total cross section, (2) an orientation parameter $\langle T(J)_{11}^+ \rangle$ which is proportional to the mean value $\langle J_y \rangle$ of the atomic angular momentum, and (3) three alignment parameters $\langle T(J)_{20}^+\rangle$, $\langle T(J)_{21}^+\rangle$, and $\langle T(J)_{22}^+\rangle$ which are proportional to the expectation values of bilinear combinations $\langle 3J_z^2 - J^2\rangle$, $\langle J_x J_z + J_z J_x\rangle$, and $\langle J_z^2 - J_z^2\rangle$, respectively, of the angular momentum operators. Since the scattered electrons are not observed, these parameters are integrated over all scattering angles. As a consequence of this integration, it follows that $\langle T(J)_{22}^+\rangle = 0$ (Bartschat et al., 1981a). We refer to the parameters $\langle T(J)_{KQ}^+\rangle$ as "integrated state multipoles" in order to distinguish these from the parameters $\langle T(J)_{KQ}^+\rangle$ introduced to describe electron–photon coincidence experiments.

The four independent integrated state multipoles are related to the Stokes parameters by the equations

$$I = C(2/3\sqrt{3})\langle T(J)_{00}^+\rangle - (1/3\sqrt{6})\langle T(J)_{20}^+\rangle) \tag{4.2a}$$

$$\eta_3 = -(C/I)(1/\sqrt{6})\langle T(J)_{20}^+\rangle \tag{4.2b}$$

$$\eta_1 = (C/I)\tfrac{2}{3}\langle T(J)_{21}^+\rangle \tag{4.2c}$$

$$\eta_2 = -i(C/I)\tfrac{2}{3}\langle T(J)_{11}^+\rangle \tag{4.2d}$$

Here, C is a numerical constant. Equations (4.2) follow by specializing Eqs. (13) of Bartschat et al. (1981a).

The parameters $\langle T(J)_{00}^+\rangle$ and $\langle T(J)_{20}^+\rangle$ are not influenced by the initial polarization and have therefore the same values as in scattering processes with unpolarized electrons. A measurement of η_1 and η_2 allows the determination of the parameters $\langle T(J)_{21}^+\rangle$ and $\langle T(J)_{11}^+\rangle$, respectively. These quantities characterize the departure from cyclindrical symmetry due to the initial polarization. Therefore, both parameters depend on how the excitation is influenced by the polarization vector **P**. Consequently, these parameters will contain valuable information on spin-dependent forces acting during the collision, as will be demonstrated in the following (Blum, 1982; Bartschat and Blum, 1982a).

Transfer of polarization between the projectile electrons and the atoms can be caused by electron-exchange processes, by spin–orbit coupling effects, or by a combination of both. It is interesting to try to disentangle these effects. In order to achieve this, let us assume that during the scattering spin and orbital angular momenta are completely decoupled and that the excitation can therefore adequately be described in the LS-coupling scheme. In this case a polarization transfer from the incident electrons to the atoms is only possible by electron exchange. Immediately after the excitation the atoms will have obtained a certain spin polarization $\langle S_y\rangle$ (which is proportional to a parameter $\langle T(S)_{11}^+\rangle$ in the multipole language) which is parallel to the initial electron polarization (see, for example, Blum and Kleinpoppen,

1975). For example, in a singlet–triplet transition with atomic spin $S = 1$ we have $\langle T(S)_{11}^+ \rangle = (i/3)P$, where P is the magnitude of the incident electron polarization. The orbital system alone is described in terms of two parameters $\langle T(L)_{00}^+ \rangle$ and $\langle T(L)_{20}^+ \rangle$ after integration over all scattering angles.

After the excitation the atomic spin S and the orbital angular momentum L will couple to the total angular momentum J of the atom. In the coupled representation the relevant state multipoles are given by the following relation [see, for example, Eq. (B6) of Blum (1981a)]:

$$\langle T(J)_{KQ}^+ \rangle = \sum_{K_1} [3(2K_1 + 1)]^{1/2}((2J+1)(K_1 0, 11|KQ)$$

$$\times \begin{Bmatrix} K_L & 1 & K \\ L & S_1 & J \\ L & S_1 & J \end{Bmatrix} \langle T(L)_{K0}^+ \rangle \langle T(S_1)_{11}^+ \rangle$$

with $K_1 = 0, 2$. It has been assumed that the excited atoms have sharp L and sharp spin S_1 (this assumption is relaxed below).

From the symmetry properties of the Clebsch–Gordan coefficient $(K0, 11|KQ)$ follows $Q = 1$ and from those of the $9j$ symbol

$$\begin{Bmatrix} K_1 & 1 & K \\ L & S_1 & J \\ L & S_1 & J \end{Bmatrix}$$

follows that $(-1)^{K+K_1+1} = 1$ [this is a consequence of the fact that the $9j$ symbol has two identical rows and is therefore invariant under an interchange of these rows; see, for example, Eq. (6.4.5) of Edmonds (1975)]. Hence, for $K = 2$, we obtain

$$\langle T(J)_{21}^+ \rangle = 0 \qquad (4.4)$$

Assuming that photons emitted from an atomic state with definite J are observed, it follows from Eqs. (4.2c) and (4.4) that η_1 vanishes.

Consider now the orientation parameter $\langle T(J)_{11}^+ \rangle$ for the case $J = L = S = 1$. Monopole and alignment parameters are then given by [see Eqs. (4.6.10) and (4.6.11) in Blum (1981a)]

$$\langle T(L)_{00}^+ \rangle = Q/\sqrt{3} \qquad (4.5a)$$

$$\langle T(L)_{20}^+ \rangle = (2/\sqrt{6})[Q(1) - Q(0)] \qquad (4.5b)$$

where $Q(M)$ denotes the total cross section for excitation of the atomic state with $L = 1$ and magnetic quantum number M, and $Q = \Sigma_M Q(M)$. Substituting numerical values for the Clebsch–Gordan coefficient and the $9j$

symbol into Eq. (4.3) and using Eqs. (4.5), we obtain

$$\langle T(J)_{11}^+ \rangle = (i/6)Q(1)P_y \tag{4.5c}$$

that is, the atomic orientation is parallel to the initial electron polarization vector when LS-coupling holds during the scattering process.

This result can be understood as a consequence of angular momentum conservation. After the collision the component of the total angular momentum in the y direction is $\langle J_y \rangle \sim \langle S_y \rangle$. Since the atom is a closed system, this component cannot be changed by spin–orbit coupling inside the atom.

In conclusion, a nonvanishing value of $\langle T(J)_{21}^+ \rangle$ and the Stokes parameter η_1 is only possible if spin–orbit effects *during the collision* cannot be neglected. Electron-exchange processes alone cannot produce a finite η_1. A comparison of theoretical and numerical values of η_1 should therefore be a sensitive test of those parts of the theoretical models which describe the spin–orbit coupling during the excitation. Similarly, $\langle T(J)_{11}^+ \rangle$ and η_2 have the same sign as P_y if the collision is dominated by electron-exchange processes. A change in sign of η_2 can only be caused by strong spin–orbit coupling effects during the scattering. When η_2 is measured as a function of the incident energy, spin–orbit effects can be expected to be strong, particularly in those energy regions where η_2 and P_y have opposite signs.

These results can be generalized in the following way (Blum, 1983). In order to describe spin-dependent effects in excitation processes of heavy atoms, one has two effects to take into account (see, e.g., Hanne, 1981).

1. Fine-Structure Interaction Inside the Target Atoms

The fine-structure interaction inside the target atoms is taken into account by applying the intermediate coupling scheme. We represent, for example, the $6\ ^3P_1$ state of Hg by the following superposition:

$$\begin{aligned}|6\ ^3P_1, M\rangle &= a\overline{|6\ ^3P_1, M\rangle} + b\overline{|6\ ^1P_1, M\rangle} \\ &= \sum_S a(S)|(LS)JM\rangle\end{aligned} \tag{4.6a}$$

Here, the states $\overline{|\cdots\rangle} = |(LS)JM\rangle$ are Russel–Saunders states with definite orbital angular momentum L and spin S and with $a = 0.987$ and $b = -0.16$. The **T**-matrix elements are then given by the following expression:

$$\langle JM, \mathbf{p}_1 m_1 | \mathbf{T} | 0, \mathbf{p}_0 m_0 \rangle = \sum_S a(S) \langle (LS)JM, \mathbf{p}_1 m_1 | \mathbf{T} | 0, \mathbf{p}_0 m_0 \rangle \tag{4.6b}$$

(This effect has been neglected in the treatment given above.)

2. Spin–Orbit Interaction during Excitation

The spin–orbit interaction during excitation is taken into account by including relativistic terms in the relevant interaction potentials (see, for example, Madison and Shelton, 1973; Scott and Burke, 1982).

It is of interest to try and separate the effects of the fine-structure interaction and the spin–orbit interaction in order to find out how strongly the excitation process is influenced by the spin–orbit interaction, particularly at lower energies.

Since no numerical results are available for the Hg excitation at low energies, one must perform a theoretical analysis of the experimental data. We will therefore look for an observable which is particularly sensitive to the spin–orbit coupling. In order to do this, we will apply the following model *where only the fine-structure interaction inside the atoms is taken into account.* The **T**-matrix elements on the right-hand side of Eq. (4.6b) are calculated by neglecting any further explicit spin-dependent term in the Hamiltonian.

This model has been used by various authors as a basis for numerical calculations using additional approximations. McConnell and Moisewitsch (1968) calculated oscillator strengths and cross sections at higher energies, using the Born–Ochkur approximation and additional assumptions on the radial parts of the atomic wave functions. Moisewitsch (1976) used the Born–Ochkur approximation and calculated the spin polarization of electrons scattered in the forward direction after exciting the 6 $^3P_1$ state of Hg. Moisewitsch found good agreement with the experimental results of Hanne (1976). These results have been discussed further by Bonham (1974).

It is of interest that a measurement of the Stokes parameter η_1 allows a direct test of the assumptions made above on the spin coupling: *η_1 vanishes identically if the effects of spin–orbit interaction can be neglected* (Blum, 1983).

Hence, the integrated Stokes parameter given by Eq. (4.2c) depends sensitively on *one* specific dynamical effect: the spin–orbit interaction between projectile and target during the collision. It is this interaction which gives nonvanishing values to η_1.

Therefore, a measurement of η_1 gives direct information on the model under discussion without the necessity of performing detailed numerical calculations. If one finds experimentally that—under the conditions stated above—*the integrated Stokes parameter η_1 is noticeably different from zero, then the spin–orbit coupling effect cannot be neglected during the collision.*

It should be noted that a test of the type described here has the following advantage. By comparing numerical and experimental results for various observables, one can only test the theoretical model as a whole with *all* approximations made on the scattering dynamics and the atomic wave functions. A measurement of η_1 allows one to test *specifically* the assumption made on the spin-coupling mechanism.

The experimental data obtained by Wolcke *et al.* (1982) are shown in Fig. 18. In the vicinity of resonances, strong spin–orbit coupling effects can be expected. It is therefore not too surprising to find nonvanishing values of η_1 just above the threshold. It is of interest that η_1 becomes very small for energies $E \geqslant 6$ eV. For $E \geqslant 7.3$ eV the experimental results are influenced by cascades. However, arguments have been put forward that these effects can be expected to be small (Hanne, 1981).

Finally, we briefly comment on experiments where atoms are excited by transversely polarized electrons and where the emitted light is detected in coincidence with the electrons scattered in the forward direction. The main result holds also for this case: If the parameter η_1 is found experimentally to be noticeably different from zero, then the effects of spin–orbit interaction cannot be neglected and must be taken into account in numerical calculations.

Such an experiment is in progress in Münster (by Wolcke, Hanne, and Kessler). The results will be of interest as an independent test of theoretical models (Moisewitsch, 1976).

C. Classification of Resonances Close to Threshold

The classification of mercury resonances in electron mercury scattering close to the 6 $^3P_1$ threshold has been studied by many authors since their first observation by Kuyatt *et al.* (1965). Fano and Cooper (1965) suggested that these resonances should have a 6s6p^2 configuration and used the Russel–Saunders scheme for a classification. Later, various groups studied these structures trying to measure the exact energy position as well as to identify the corresponding Hg$^-$ states (Zapesochny and Spenik, 1966; Düwecke *et al.*, 1973; Ottley and Kleinpoppen, 1975; Burrow and Micheda, 1975; Heddle, 1975, 1978; Albert *et al.*, 1977).

Several classification schemes have been suggested which differ particularly in the assignment of quantum numbers to the structure at 4.92 eV just above the 6 $^3P_1$ threshold.

Furthermore, not all of the compound states which can be expected according to Fano and Copper (1965) or Heddle (1978) have been fully confirmed experimentally.

Additional information on these questions can be extracted from measurements of the integrated Stokes parameters. For excitation by unpolarized electrons this has been discussed by Baranger and Gerjuoy (1958) and by Ottley and and Kleinpoppen (1975) for the linear polarization fraction η_3. More recently, it has been shown that independent information can be obtained from an analysis of experimental results of the new parameters η_1 and η_2 (Wolcke et al., 1982). Assuming that close to threshold the nonresonant background can be neglected, it can be shown that the integrated Stokes parameters depend only on geometric factors (6j and 9j symbols) and can easily be calculated for various sets of quantum numbers.

The results, specialized to $^3P_1$ excitation, are given in Table IV (for details, see Bartschat, 1981; Wolcke et al., 1983). The circular light polarization η_2 for longitudinally polarized initial electrons has been calculated (first column in Table IV) and for transversely polarized elecltrons (second column in Table IV). In both cases it has been assumed that the emitted light is observed in the direction of the initial electron polarization P.

The depolarization of the emitted light by hyperfine structure interaction is taken into account [see the appendix in the article by Wolcke et al. (1982) for details]. The final state of Hg has been described in the intermediate coupling scheme.

The linear light polarization η_3 has also been calculated for unpolarized initial electrons and for observation of the emitted light perpendicular to the incoming electron beam (third column of Table IV). The Stokes parameter η_1 vanishes if the collision is dominated by a single resonance.

A comparison between Table IV and the experimental results shown in Fig. 19 gives an independent test of the classification schemes suggested in the literature for the resonance at 4.92 eV immediately above the 6 $^3P_1$ threshold. The main conclusion is that the results for the circular polarization are compatible with the assignment of a total angular momentum $\bar{J} = \frac{5}{2}$ to the threshold resonance. This is in accordance with the conclusions of Albert et al. (1977) and of Heddle (1978). The assignment of $\bar{J} = \frac{3}{2}$, sug

TABLE IV

Results for the Light Polarizations $\eta_2(\theta = 0)/P_z$, $\eta_2(\theta = \Phi = \pi/2)/P_y$, and $\eta_3(\theta = \Phi = \pi/2)$ for the $[(6s6p^2)\ ^3P]^{2S+1}\bar{L}_{\bar{J}}$ Resonance of Hg

Classification of the resonance	$\eta_2(\theta = 0)/P_z$	$\eta_2(\theta = \Phi = \pi/2)/P_y$	$\eta_3(\theta = \Phi = \pi/2)$
$^4P_{1/2}$	+0.44	+0.44	0
$^4P_{3/2}$	0.13	−0.43	−0.58
$^4P_{5/2}$	0.39	0.69	0.41
$^2D_{3/2}$	0.17	−0.47	−0.40
$^2D_{5/2}$	0.39	0.69	0.41

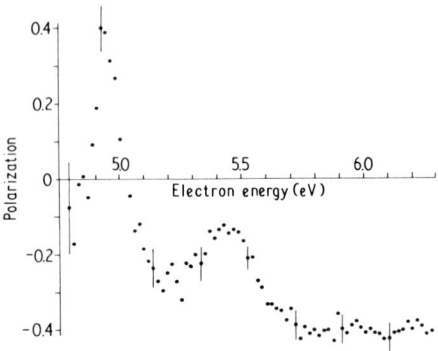

FIG. 19. Measurement of the linear polarization η_3 for the 6 $^3P_1 \rightarrow\ ^1S_0$ transition in mercury for excitation with unpolarized electrons. (From Ottley and Kleinpoppen, 1975.)

gested by Ottley and Kleinpoppen (1975) and by Heddle (1975), gives the wrong sign for η_2. The same conclusions are obtained by comparing the theoretical values for η_3 with the measurements by Ottley and Kleinpoppen. A more detailed numerical analysis, however, is required in order to check the assumptions made in the calculations.

Where elastic scattering calculations are in reasonable agreement with experiments (Walker, 1975; Sin Fai Lam and Baylis, 1981), the situation above the 6s6s $^3P_0$ threshold is much less clear. First numerical results for excitation cross sections have been presented by Scott *et al.* (1983). These calculations are based on the *R*-matrix method and electron-exchange, relativistic, and channel coupling effects have been included for the first time. The authors have discussed the influence of various resonances on the cross section and clarified in particular the role of the so-called "missing" resonance at 5.2 eV.

VI. Electron–Photon Coincidence Experiments with Polarized Electrons

Considering the results given in Sections IV and V, one can expect that a combination of spin polarization techniques with the methods of coincidence experiments will give additional information (Hanne, 1980). Such an experiment is in progress in Münster (excitation of the 6 $^3P_1$ state of mercury by polarized electrons and observation of scattered electrons and emitted photons in coincidence). A formal theory has been developed recently

(Bartschat *et al.*, 1981a), which is briefly be discussed in the following. Some results of numerical calculation are given as an illustration.

An important aspect of the formal theory of the measurement is to show how to disentangle the P-independent and the P-dependent parts. That is, how to separate those parts which are independent of the initial spin polarization P (and which can be measured in experiments with unpolarized particles) from those parts which contain new information. This aspect is of central importance for the planning and performance of the experiments and we will discuss this point in the following.

The most efficient way of obtaining the desired results is in terms of tensor operators. In order to do this, first we have to consider how to construct the tensor operators: whether eigenstates of orbital angular momentum and spin should be used or eigenstates of the total atomic angular momentum $\mathbf{J}$ ($\mathbf{J} = \mathbf{L} + \mathbf{S}$). In order to decide which basis is most useful, we consider the physical assumptions with regard to the excitation process.

The disccisions given in Section II were based on the assumption that the collision can be described in the LS-coupling scheme and that the spin–orbit coupling inside the atom influences only the time evolution of the excited states between excitation and decay. This conception cannot be applied to experiments where atomic states with sharp J are separated in the *electronic* channel [as is the case, for example, in the experiments by Malcolm and McConkey (1979), by Zaidi *et al.* (1980), or in progress at Münster].

In these cases the spin–orbit interaction inside the target atoms is apparently so strong that the following physical picture should be applied. The fine-structure levels are "impulsively" excited during the collision (the electrons "see" the fine-structure separation) so that immediately after the excitation the atoms are found in states with definite J. Among these states one is selected experimentally and separated in the electronic channel from the other states with different J.

The atomic ensemble of interest consists then of all atoms with the given value of J which "scattered" the electrons in the given direction (see the discussion at the beginning of Section III,A). This ensemble is conveniently described in terms of state multipoles $\langle T(J)_{KQ}^+ \rangle$, where the corresponding tensor operators $T(J)_{KQ}$ are constructed from the states $|JM\rangle$ [see, for example, Chapter 4 in Blum (1981a)]. As coordinate system we will use the "collision system" introduced in Section IV.

It is assumed that the atoms are excited from an ^1S ground state. More general cases have been considered by Bartschat *et al.* (1981a).

The scattering amplitude for a transition

$$|0\rangle|\mathbf{p}_0 m_0\rangle \rightarrow |M\rangle|\mathbf{p}_1 m_1\rangle \tag{5.1}$$

will be denoted by $f(M_1 m_0)$ using the notation of Section IV, where $|0\rangle$ is again the atomic ground state.

First, let us consider excitation by unpolarized electrons. As discussed in Section IV,A, the collision is invariant under reflection in the scattering plane. From this condition and the hermiticity of the density matrix follows the relation

$$\langle T(J)^+_{KQ}\rangle = (-1)^K \langle T(J)^+_{KQ}\rangle^* \tag{5.2}$$

for the state multipoles [see, for example, Chapter 4 in Blum (1981a)]. The asterisk denotes the complex conjugate. Equation (5.2) shows that the components of the alignment tensor ($K = 2$) are real, the components $\langle T(J)^+_{1\pm 1}\rangle$ of the orientation vector are imaginary, and $\langle T(J)_{10}\rangle = 0$.

Hence, for $J = 1$, we are left with five independent parameters:

$$\langle T^+_{00}\rangle, \quad \text{Im}\langle T^+_{11}\rangle, \quad \text{Re}\langle T^+_{22}\rangle, \quad \text{Re}\langle T^+_{21}\rangle, \quad \langle T^+_{20}\rangle \tag{5.3}$$

which can be expressed in terms of the parameters of Eqs. (3.7).

If the atoms are excited by polarized electrons, then additional terms are present which contain new information. The state multipoles are then given by the following expressions [see Bartschat et al. (1981a) for details]:

$$\langle T(J)^+_{KQ}\rangle = \sum_{MM'} (-1)^{J-M}(JM'J - M|KQ)[\langle f(M')f(M)^*\rangle_{un} \tag{5.4a}$$
$$+ \langle f(M')f(M)^*\rangle_p]$$
$$\equiv \langle T(J)^+_{KQ}\rangle_{un} + \langle T(J)^+_{KQ}\rangle_p \tag{5.4}$$

Here, the term

$$\langle f(M')f(M)^*\rangle_{un} = \tfrac{1}{2} \sum_{m_0 m_1} f(M'm_1 m_0) f(MM_1 m_0)^* \tag{5.5a}$$

is independent of the initial polarization and is equal to the term given in Eq. (3.5). The second term in Eq. (5.4a) is given by the following expression:

$$\langle f(M')f(M)^*\rangle_p$$
$$= \sum_{\substack{m_0 m'_0 \\ m_1 q}} f(M'm_1 m'_0) f(MM_1 m_0)^*(-1)^{1/2-m_0}(\tfrac{1}{2}m_0, \tfrac{1}{2} - m_0|1q\rangle\langle T(\tfrac{1}{2})^+_{1q}\rangle \tag{5.5b}$$

where the spin tensor $\langle T(\tfrac{1}{2})^+_{1q}\rangle$ is proportional to the corresponding spherical component of the polarization vector of the incident electrons:

$$\langle T(\tfrac{1}{2})^+_{1q}\rangle = (\tfrac{1}{2})^{1/2} P^*_q \tag{5.6}$$

[with $P_{\pm 1} = \mp (\tfrac{1}{2})^{1/2}(P_x + iP_y)$ and $P_0 = P_z$].

The statistical tensors $\langle T(J)^+_{KQ}\rangle_{un}$ are the same as in experiments with unpolarized electrons [as shown by Eqs. (5.4) and (5.5a)]. The multiples

$\langle T(J)^+_{KQ}\rangle_p$ contain the additional information which can be obtained from measurements with unpolarized electrons.

It is important to distinguish between the following two cases. If the initial polarization vector is perpendicular to the scattering plane, then the reflection invariance with respect to this plane is retained (since **P** transforms as an axial vector). With the help of Eqs. (3.6), (5.4), and (5.5), it can be shown that the relevant state multipoles satisfy condition (5.2). The excited atomic ensemble of interest is then characterized in terms of the five parameters (5.3) (for $J = 1$). These tensors now contain a term $\langle T(J)^+_{KQ}\rangle_p$ which depends on the initial polarization.

Reflection invariance is violated if the polarization vector **P** has a component in the scattering plane. Then Eq. (5.2) cannot be applied, the tensor components with $Q \neq 0$ become complex quantities, and—in addition to the usual components (5.3)—there are four new independent parameters:

$$\text{Im}\langle T^+_{22}\rangle, \quad \text{Im}\langle T^+_{21}\rangle, \quad \text{Re}\langle T^+_{11}\rangle, \quad \langle T^+_{10}\rangle \qquad (5.7)$$

Let us briefly consider how to separate experimentally the components (5.3) and (5.7). The tensors can be determined by measuring angular dependence and polarization of the emitted light in coincidence with the scattered electrons. As an example we give here an expression for the angular distribution of light emitted in a $^3P_1 \rightarrow {}^1S_0$ transition (for example, in mercury) (Bartschat et al., 1981a):

$$\begin{aligned}I(\theta, \phi) = &\ W(\tfrac{4}{3})^{1/2}\langle T^+_{00}\rangle + W[\text{Re}\langle T^+_{22}\rangle \sin^2\theta \cos 2\phi \\ &- \text{Re}\langle T^+_{21}\rangle \sin 2\theta \cos \phi \\ &+ (\tfrac{1}{6})^{1/2}\langle T^+_{20}\rangle(3\cos^2\theta - 1)] \\ &- W(\text{Im}\langle T^+_{22}\rangle \sin^2\theta \sin 2\phi - \text{Im}\langle T^+_{21}\rangle, \sin 2\theta \sin \phi \quad (5.8)\end{aligned}$$

Here, W is a spectroscopic factor which depends on the reduced matrix elements; θ and ϕ are the polar angles of the direction of observation in the collision system (see Section IV,C).

Equation (5.8) shows that any tensor component gives rise to a characteristic angular dependence of the emitted light. Note in particular that the ϕ dependence is different for the "old" parameters (5.3) and the "new" parameters (5.7). This difference in the azimuth dependence is related to the different transformation properties of the tensor components under reflections as discussed above. Both sets of parameters can therefore be separated by measuring the intensities emitted in the directions (θ, ϕ) and $(\theta, -\phi)$ and summing and subtracting, respectively, the results. The sum $I(\theta, \phi) + I(\theta, -\phi)$ depends only on the parameters (5.3); the difference $I(\theta, \phi) - I(\theta, -\phi)$ depends only on the components (5.7) containing the new information.

It should be noted that a determination of the orientation vector requires a measurement of the circular polarization as discussed in Section IV,C.

A separation of the parts $\langle T^+_{KQ}\rangle_{un}$ and $\langle T^+_{KQ}\rangle_p$ can be obtained in the following way. First, the Stokes parameters are measured with a given initial polarization vector **P** and the parameters $\langle T^+_{KQ}\rangle$ are determined. Then the measurements are repeated with the inverse polarization $-\mathbf{P}$. From the first set of measurements one obtains

$$\langle T^+_{KQ}\rangle^{(1)} = \langle T^+_{KQ}\rangle_{un} + \langle T^+_{KQ}\rangle_p \tag{5.8a}$$

and from the second set of experiments

$$\langle T^+_{KQ}\rangle^{(2)} = \langle T^+_{KQ}\rangle_{un} + \langle T^+_{KQ}\rangle_{-p}$$
$$= \langle T^+_{KQ}\rangle_{un} - \langle T^+_{KQ}\rangle_p \tag{5.8b}$$

Here, it has been noted that the P-dependent terms change their sign if **P** is replaced by $-\mathbf{P}$ as shown by Eq. (5.5b). By adding and subtracting Eqs. (5.8a) and (5.8b), the "unpolarized" terms can be separated from the parameters $\langle T^+_{KQ}\rangle_p$ of interest.

By performing numerical calculations, one obtains the scattering amplitudes from which the various tensors can be calculated by use of Eqs. (5.4) and (5.5). The comparison of these theoretical values with experimental results then allows a very detailed test of the theoretical model.

Explicit expressions for the tensors as functions of the scattering amplitudes have been given by Bartschat *et al.* (1981a) for the case of $J = 1$ excitation. With the help of the symmetry condition (3.6) it can be shown that the excitation process is completely characterized in terms of 11 independent parameters. The general analysis given by Bartschat *et al.* shows that, in this case, the experiments under discussion allow a *complete* determination of the scattering amplitudes — no spin analysis of the scattered electrons is necessary. This is a "complete experiment" in the sense defined by Bederson (1969a,b) (see Section II).

As an illustration of the formal theory we give some numerical results for the atomic orientation for the 6 $^1P_1$ excitation of Hg.

The components of the orientation vector are proportional to the mean values of the angular momentum $\langle J_Y \rangle$ transferred to the atoms during the collision. Because of reflection invariance for the case under discussion, we have $\langle T^+_{10}\rangle = 0$ and $\mathrm{Re}\langle T^+_{1\pm1}\rangle = 0$. It can be shown that the following relations hold [see, for example, Eq. (4.3.15a) in Blum (1981a)]:

$$\mathrm{Im}\langle T^+_{11}\rangle = -(\sigma/2)\langle J_Y\rangle$$

where σ denotes the differential cross section. The orientation parameter

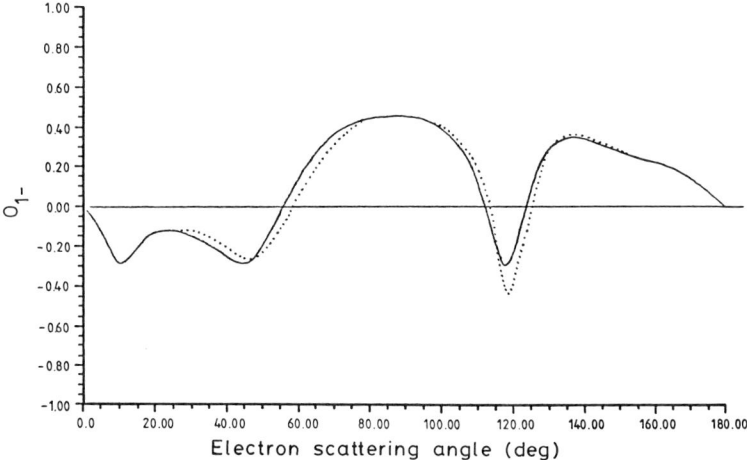

FIG. 20. Orientation parameter O_{1-} for excitation of the 6 $^1P_1$ state of mercury with unpolarized (——) and with polarized (·····) electrons.

O_{1-}, introduced by Fano and Macek (1973), is given by

$$O_{1-} = \tfrac{1}{2}\langle J_Y \rangle$$

In Fig. 20 we give numerical results for O_{1-} for incident energies of $E = 180$ eV, respectively, where the results of Madison and Shelton have been used for the scattering amplitudes (see Section IV,D). The calculations show that the influence of the polarization **P** on the orientation is very small. We have obtained similar results for the alignment parameters for $E = 180$ and 125 eV. It is interesting that the data in Fig. 20 show a more pronounced structure than is the case for ^1P excitation of He states (Steph and Golden, 1980).

VII. Ionization

The first successful collision experiments with both polarized electrons and polarized atoms were reported at the Satellite Meeting on Electron Atom Collisions of the 10th ICPEAC Conference at Paris in 1977 (Lubell, 1977a; Hils, 1977). In this connection it is interesting to note historically the fact that it appeared to be "easier" to the experimentalist to detect ionization products rather than scattered electrons in the first attempts of experiments with polarized electrons and polarized atoms. The results of these experiments provided a new kind of information on ionization, the so-called spin-dependent ionization asymmetry, which can also be related to ratios of singlet to triplet or of interference to total ionization cross sections. We will first describe the theory of the measurement of such experiments.

A. Theory of Measurement for Ionization Processes with Polarized Electrons and Polarized Atoms

We restrict ourselves to electron atom scattering processes where only Coulomb direct or exchange interactions take place. This matches the present situation in which the experimental data from ionization studies with polarized electrons and polarized atoms are restricted to light one-electron atoms. The analysis of heavier atoms would require additional terms in order to include spin–orbit interactions, which will be neglected in our theoretical analysis. We therefore apply the previously reported spin analysis without spin–orbit terms in order to describe the ionization reactions with spin–polarized electrons $\{e(\uparrow), e(\downarrow)\}$ and spin-polarized atoms $\{A(\uparrow), A(\downarrow)\}$:

(1) $e(\uparrow) + A(\downarrow) \rightarrow A^+ + e(\uparrow) + e(\downarrow)$ (amplitude f)

$\rightarrow A^+ + e(\downarrow) + e(\uparrow)$ (amplitude g)

(2) $e(\uparrow) + A(\uparrow) \rightarrow A^+ + e(\uparrow) + e(\uparrow)$ (amplitude f-g)

The two ionization processes are described by the direct amplitude (f) and the exchange amplitude (g). The cross sections for the two reactions can be written as

$$\sigma^{\uparrow\downarrow} = |f|^2 + |g|^2 = \sigma + \text{Re}(f^*g) \qquad (6.1)$$

$$\sigma^{\uparrow\uparrow} = |f|^2 + |g|^2 - 2\,\text{Re}(f^*g) = \sigma - \text{Re}(f^*g) \qquad (6.2)$$

with

$$\sigma = \sigma(E, k_1, k_2) = \tfrac{1}{2}|f|^2 + \tfrac{1}{2}|g|^2 + \tfrac{1}{2}|f - g|^2 \qquad (6.3)$$

as differential cross sections depending on the electron impact energy E and wave vectors k_1 and k_2 of the two outgoing electrons after the ionization process. By integrating over all directions, we obtain the relevant total ionization cross sections:

$$Q^{\uparrow\downarrow} = \int\!\!\int (\sigma + \text{Re}(f^*g))\, d\vec{k}_1\, d\vec{k}_2 = Q + Q_{\text{int}} \qquad (6.4)$$

$$Q^{\uparrow\uparrow} = \int\int \underbrace{(\sigma - \text{Re}(f^*g))\, dk_1\, dk_2}_{= Q_{\text{int}}} = Q - Q_{\text{int}} \qquad (6.5)$$

with the total ionization cross section Q and its interference term Q_{int}. The difference between the two equations determines $2Q_{\text{int}} = Q^{\uparrow\downarrow} - Q^{\uparrow\uparrow}$ or

$$\frac{Q_{\text{int}}}{Q} = \frac{Q^{\uparrow\downarrow} - Q^{\uparrow\uparrow}}{Q^{\uparrow\downarrow} + Q^{\uparrow\uparrow}} = \frac{Q^{\uparrow\downarrow} - Q^{\uparrow\uparrow}}{2Q} = A \qquad (6.6)$$

is the spin-dependent ionization asymmetry. Alternatively, the asymmetry A can be expressed in terms of the singlet (Q_s) and triplet (Q_t) ionization cross sections through the following relations:

$$Q_s = Q^{\uparrow\downarrow} + 2Q_{\text{int}}, \qquad Q_t = Q^{\uparrow\uparrow}$$
$$Q = \tfrac{1}{4}Q_s + \tfrac{3}{4}Q_t \qquad (6.7)$$

$$A = \frac{Q_s - Q_t}{4Q} = \frac{1 - Q_t/Q_s}{1 + 3Q_t/Q_s} \qquad (6.8)$$

B. Results of Spin-Dependent Ionization Asymmetries

1. Atomic Hydrogen Data

Figure 21 shows a schematic diagram of the experimental layout applied in the Yale experiment for polarized electrons scattered by polarized atomic hydrogen (Alguard *et al.*, 1977). The four main parts of the experiment are the interaction chamber with electron, atom, and ion detectors, the Mott analyzer for measuring the electron spin, the polarized atomic hydrogen source, and the polarized electron source. The polarized electron beam crosses the polarized atomic hydrogen beam.

In this experiment the production of polarized electrons was obtained by means of the Fano effect (Fano, 1969) (Fig. 21). Circularly polarized UV light from a 1000-W Hg–Xe arc lamp photoionizes an atomic Cs beam in a region maintained at -1 kV electrostatic potential. The highly polarized photoelectrons are extracted from the ionization region by a potential gradient of ~ 1 V/cm and formed into a beam by electo-optical focusing elements. A bending magnet can deflect the 1-keV polarized electron beam either into the interaction region or the Mott detector. In the Mott analyzer the electron beam is first transferred from longitudinal to transversal spin polarization in a Wien filter. After acceleration of the electron beam to 100 keV, the transversal spin polarization is measured with the Mott analyzer and typically has a value $P_e = 0.63 \pm 0.03$.

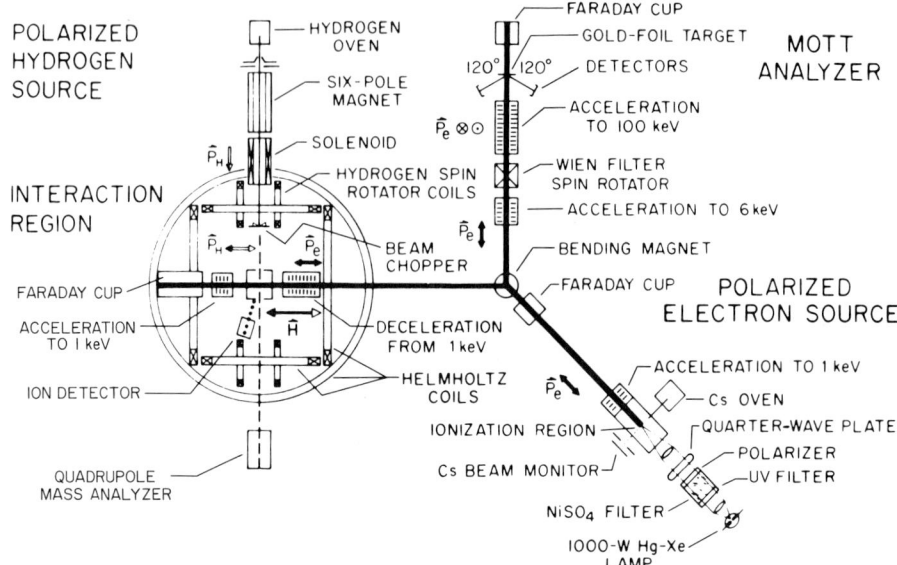

FIG. 21. Scheme of the experiment for ionization of polarized atomic hydrogen by polarized electron impact. The apparatus is divided into parts for the polarized hydrogen source, the polarized electron source, the Mott analyzer, and the interaction chamber. (From Alguard et al., 1977.)

The atomic hydrogen beam is produced in a tungsten oven (typical temperature 2800 K) by thermal dissociation of molecular hydrogen. After collimation, the hydrogen beam passes a six-pole magnet. The special focusing properties of the six-pole magnet have been analyzed based upon an optical model (Hughes et al., 1972). The six-pole magnet applied in this experiment had a selectivity of $S = 99\%$ for selecting the $m_j = \frac{1}{2}$ magnetic substate of the hydrogen ground state. This high field polarization is reduced to $P_H = \frac{1}{2}S \approx 50\%$ in a near-zero magnetic field due to the hyperfine structure interaction in the hydrogen ground state. To avoid nonadiabtic Majorana flips which would depolarize the hydrogen beam, the atoms emerging from the six-pole magnet are first spin-oriented in a ~ 200-G axial field of a small solenoid; following this the spin is adiabatically rotated into the orientation of the polarized electrons by further sets of Helmholtz pairs. The small magnetic field of the interaction region of ~ 200 mG defines the quantization axis for the spins of the electrons and atoms.

By applying the apparatus as illustrated in Fig. 21, the ionization asymmetry factor A for atomic hydrogen was measured.

Figure 22 shows results for the asymmetry factor of atomic hydrogen as a

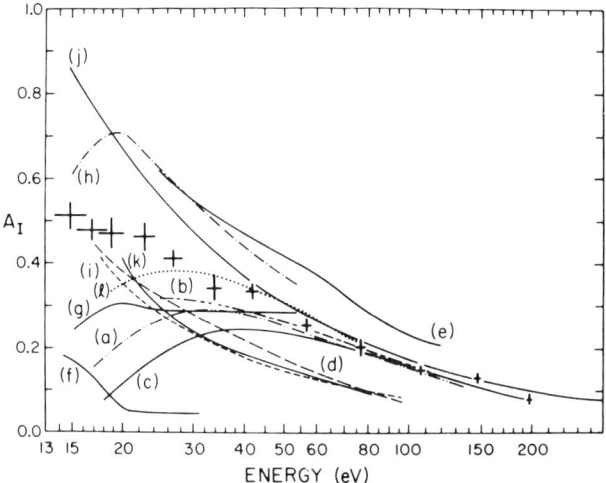

FIG. 22. Experimental and theoretical data for the ionization asymmetry factor $A = Q_{int}/Q$ for atomic hydrogen by electron impact. The experimental data were obtained with the apparatus described in Fig. 21. For comparison, data of the following theoretical approximations are given: Born exchange (BE) calculations from (a) Rudge and Seaton (1965), (b) Peterkop (1961a, 1962), (c) Geltman *et al.*, (1963), and (d) Goldin and McGuire (1974); (e) BE calculation with maximum interference taken from Peterkop (1961a, 1962); (f) BE calculation with angle-dependent potential from Rudge and Schwartz (1966); (g, h) special average exchange calculations allowing for maximum interference according to Rudge and Seaton (1965); (i) Glauber exchange (Goldin and McGuire, 1974), (j) modified Born–Oppenheimer (Ochkur, 1964), and (k) close-coupling calculations (Gallaher, 1974); (*l*) Be calculations taken from Rudge (1978).

function of the electron energy. The errors include one standard deviation and systematic uncertainties from the measurements of P_e ($\pm 4\%$), P_H ($\pm 2\%$), the degree of dissociation (1%), and the collinearity between the small magnetic field of the interaction region and the electron beam. The energy uncertainty of the electrons amounted to several electron volts.

Various theoretical approximations have been applied for predictions of the asymmetry factor A. It is surprising that no theoretical approximation predicts the experimental data below 50 eV. Of course, Born and Glauber approximations are not expected to reproduce reliable data at low energies. However, the disagreement between the close-coupling data and the experimental data is notable and surprising. The work of Klar and Schlecht (1976) predicted a threshold value of $A = 1$. Of course, the hydrogen data are too far away from threshold to draw any conclusions about the validity of this threshold prediction (for a further discussion on threshold arguments see Section VII,B,3).

2. Lithium Data

The first results on the ionization asymmetry of lithium were reported by Baum et al. (1981) from Bielefeld University. In their experiment they applied the following experimental method (Fig. 23). A beam of polarized ^{6}Li (95% enriched isotope contribution) atoms is crossed by a polarized beam of electrons; the ions from the impact ionization process are extracted by a small electric field, accelerated, and focused onto a channeltron multiplier. The ionization asymmetry was measured as the relative difference between the ionization count rate for parallel and antiparallel spin positions of electrons and atoms.

The production of polarized electrons was achieved by cooling ferromagnetic EuS on a tungsten tip below the Curie temperature. Electrons can tunnel through the potential barrier from the internal W–EuS interface. The ferromagnetic splitting of the W–EuS conducting band acts as an electron spin filter for the internal field-emission process. The EuS is evaporated onto a clean oriented $\langle 112 \rangle$ tungsten tip and subsequently annealed to produce the characteristic electron emission which has a high transverse polarization (for cooling temperatures below 10 K). Typically the effective transverse polarization of the electrons is about 60% and its current

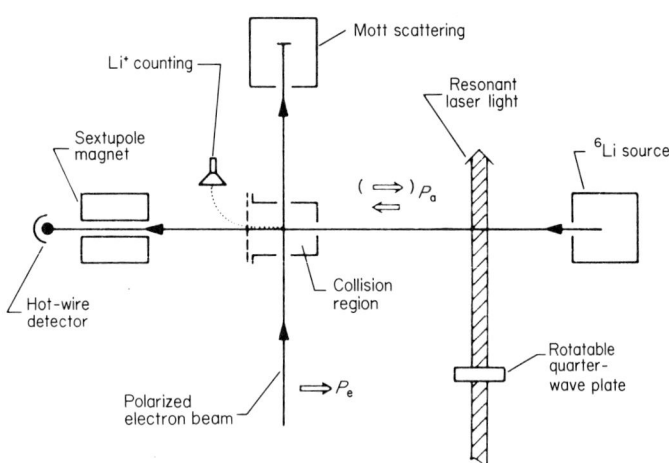

FIG. 23. Schematic diagram for the measurement of the lithium ionization asymmetry. The Li^6 atomic beam is first crossed by a resonant laser beam and afterward is crossed by a polarized electron beam. The polarized electron beam is Mott analyzed in the Mott scattering chamber; after passing a sectupole magnet, the polarized atomic beam is detected by a hot-wire detector. (From Baum et al., 1981.)

is about 10 nA. The ^{6}Li atomic beam was polarized as follows (Baum *et al.*, 1980). The atomic beam is exposed to the resonant photon radiation from a single-mode dye laser inducing optical transitions from $2\,^2S_{1/2}$ to $2\,P_{3/2}$. A complete electron spin polarization of the atomic beam can be obtained by transferring all atoms in the various magnetic m_F substates of the two hyperfine structure states $F = \frac{3}{2}$ and $F = \frac{1}{2}$ into the $F = \frac{3}{2}$, $m_F = +\frac{3}{2}$ (or $-\frac{3}{2}$) substate. This can be achieved by optical pumping with circularly polarized photons inducing transitions to the hyperfine states $F = \frac{5}{2}$ and $F = \frac{3}{2}$ of the first excited $2\,^2P_{3/2}$ state. Simultaneous optical pumping of the two hyperfine ground states is achieved by applying an acousto-optical modulator which generates two laser beams differing in the frequency separation between the two states (corresponding to 228 MHz). A small magnetic field of 7 G is applied parallel to the light beam in the pumping region. The atomic spin direction is then adiabatically transferred into the longitudinal mode by means of a small guiding magnetic field of 100 mG in the center of the collision chamber. A six-pole magnet is applied to analyze the atomic polarization; the atomic polarization achieved by this method is typically $P_{Li} = 70\%$.[1]

Figure 24 displays results of the ionization asymmetry of ^{6}Li. The experimental data near threshold have been obtained with an energy resolution of 0.3 eV. Theoretical data are given based upon the binary encounter model of Vriens (1966) and the theory of Peach (1966) in Born exchange as well as in the Ochkur approximation. Apart from the higher energy region, none of these theories approaches the experimental data. The binary encounter theory appears to qualitatively describe the shape or trend of the measured data. As in the case of the hydrogen data (Section VII,B,1), the Li data near threshold show little indication for a trend of the asymmetry factor toward $A = 1$ at threshold.

3. Sodium and Potassium Data

Data for the ionization asymmetry of sodium and potassium were reported by a Stirling group (Hils, 1977; Hils and Kleinpoppen, 1978; Hils *et al.*, 1980, 1981). In their first experiment (Hils, 1977; Hils and Kleinpoppen, 1978; Hils *et al.*, 1980) with potassium they used a six-pole magnet to polarize the atoms; because of the ground-state hyperfine splitting (nuclear spin $I = \frac{3}{2}$), the upper limit of the atomic spin polarization in a low magnetic

---

[1] The fact that 100% polarization was not achieved in this experiment has been attributed to the effect of coherent population trapping. This can be of importance in optical pumping in which two frequencies couple to an upper common level, as is the case in this experiment.

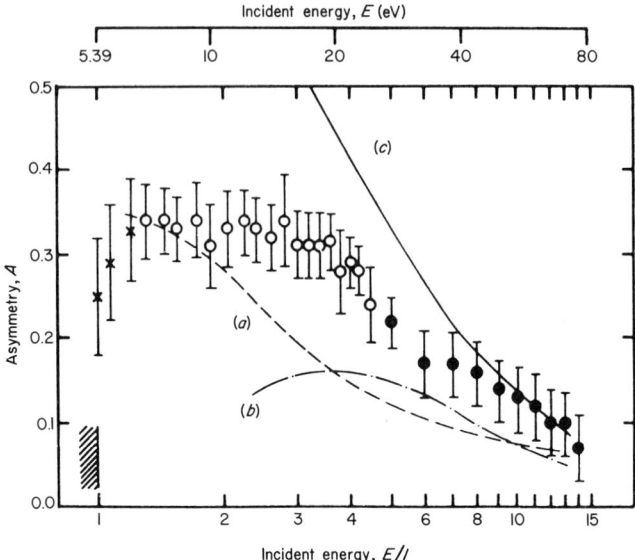

FIG. 24. Ionization asymmetry factor A of ^{6}Li plotted against the electron energy E in units of the ionization energy $I = 5.39$ eV. (×) Experimental data with an energy resolution $\Delta E_{\text{FWAM}} = 0.3$ eV; (○) for $\Delta E \approx 1$ eV; (●) measurements averaged over 5 eV range. The lines represent various theoretical approximations: (a) binary encounter formulas of Vriens et al. (1966), (b) Born exchange theory (Peach, 1966), (c) Ochkur approximation (Peach, 1966). (From Baum et al., 1981.)

field is $P_K = 1/(2I + 1) = 25\%$. By taking into account the finite selectivity of the six-pole magnet, the polarization of the potassium atoms was $P_K \approx 20\%$. Polarized electrons were obtained by spin–orbit interaction in the scattering of electrons on a mercury beam. In a compromise with regard to intensity and polarization of the electrons scattered by mercury atoms, typical values obtained were 10^{-10} A for the current and a degree of polarization of 20%. Nevertheless, it was possible to measure the ionization asymmetry of potassium (Fig. 25) even with the low product $P_e P_K = 0.04$.

In an attempt to improve the polarization product $P_e P_A$, Hils et al. (1981) combined two experimental techniques for preparing atoms in selected states with a high degree of atomic polarization. First, they applied state selection by an inhomogeneous magnetic field (six-pole magnet) and, second, optical pumping with laser light. In this way two techniques described in Sections VII,B,1 and 2 are brought together whereby the unwanted effect of coherent population trapping (see footnote on page 247) is eliminated. The scheme of the experimental apparatus used is shown in Fig. 26. The sodium atomic beam effuses from a two-stage oven, passes a hexapole

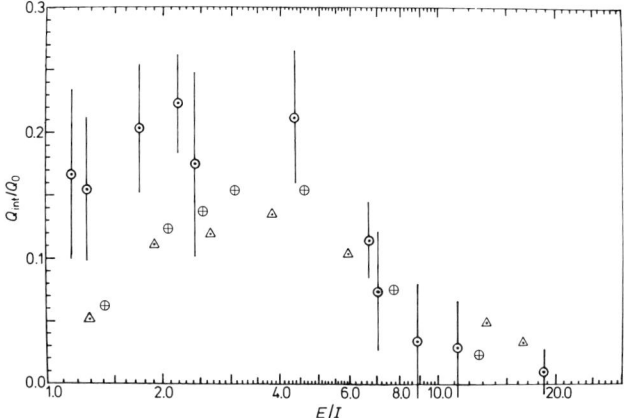

FIG. 25. The ionization asymmetry factor Q_{int}/Q_0 of potassium (Hils and Kleinpoppen, 1978; Hils et al., 1980) as a function of the electron impact energy E (in units of the ionization energy I). Experimental results: (⊙) K. Born approximation by Peach (1965, 1966); (⊕) Li; (△) Na.

magnet, three guiding magnetic fields, and an analyzer magnet, and finally is detected by a hot-wire detector. After leaving the hexapole, the atoms are optically pumped by the circularly polarized light from a continuous dye laser. Figure 27 shows the hyperfine splittings of the ground and first excited 3 $^2P_{3/2}$ state, whereas Fig. 28 displays the electron spin polarization of the various hyperfine ground states as a function of the magnetic field. Obviously aiming for complete electron spin polarization (and automatically for complete nuclear spin polarization) would require that either the magnetic substate $m_F = +2$ or $m_F = -2$ be exclusively populated. After passing the hexapole magnet, the sodium atoms are preferentially populated in the magnetic substates with $F = 2$ and $m_F = 2, 1, 0,$ and -1. A most efficient way of achieving an exclusive population of the magnetic substates $m_F = \pm 2$ can be obtained by optically pumping the transition between the hyperfine ground state $\bar{F} = 2$ and the $F = 3$ hyperfine state in 3 $^2P_{3/2}$ with circularly polarized light. The transition to this excited $^2P_{3/2}$ state is most suitable for the optical pumping since it can only decay back to the $\bar{F} = 2$ ground state from which the pumping process started. It is important to study in detail the influence of the laser intensities in connection with the optical pumping process. Figures 29 and 30 display atomic beam intensities as a function of the laser frequency for the spin-up component after the Rabi type analyzer magnet (the Rabi analyzer completely separates the two electron spin components). At the highest laser intensities (60 mW/cm²) the hyperfine structure components in the atomic beam signal are broad and

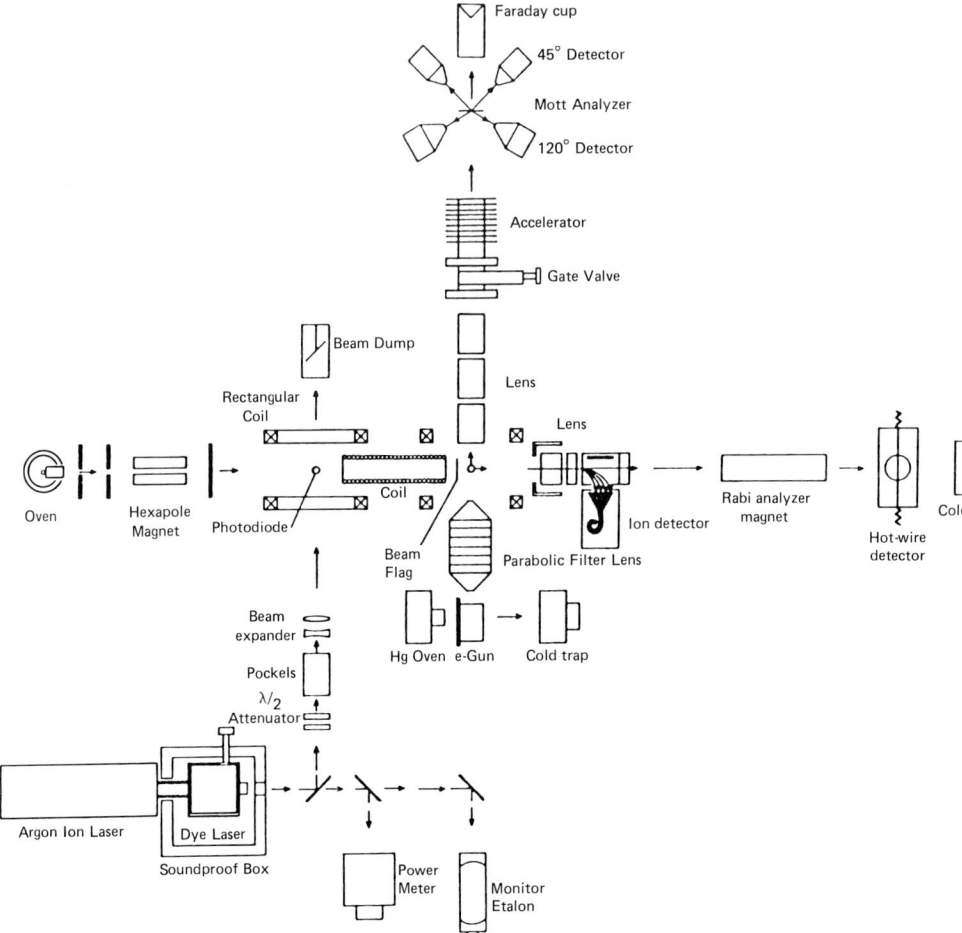

FIG. 26. Scheme of the polarized electron–polarized sodium scattering apparatus for measuring the sodium ionization asymmetry. The polarized electrons are produced by electron scattering on a mercury atomic beam; the Mott detector determines the degree of electron spin polarization. The atomic polarization is produced by combining the methods of state selection by a hexapole magnet with that of resonant optical pumping by the circularly polished D_2 line of sodium. The atomic beam passes a hexapole magnet for polarization analysis and is detected by a hot-wire device.

partially overlap; at reduced intensity all three transitions from $3\ ^2S_{1/2}$, $\bar{F} = 2$ to $3\ ^2P_{3/2}$, $F = 1, 2$, and 3 are well resolved and disturbances of the pumping process by simultaneous excitation of several transitions are avoided. By pumping with σ^+ light, transitions to $F = 1$ and 2 depopulate the spin-up component, whereas transitions to $F = 3$ increase the population of the spin-up component (Fig. 29). On the other hand, pumping with

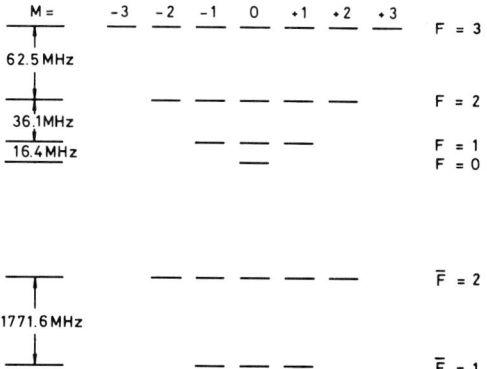

FIG. 27. Hyperfine splitting and magnetic sublevels of the ²³Na ground state and its first excited 3 ²P₃/₂ state.

σ^- light depopulates the spin-up component of the atomic beam signal for all three transitions (Fig. 30).

The degree of population of the sodium atoms can be determined by measuring the relative spin-up and spin-down components after the Rabi magnet for optical pumping with right- or left-handed circularly polarized light. From this there follows an expression

$$P_{Na} = P_\infty(1 - X - Y) \qquad (6.9)$$

where $P_\infty = \tfrac{3}{8}s + \tfrac{5}{8}$ for ideal optical pumping and s is the hexapole selection

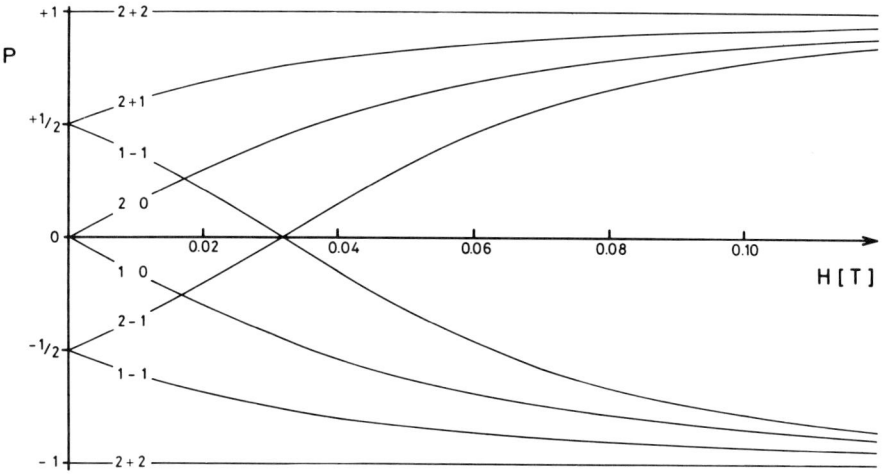

FIG. 28. Electron spin polarization of the various hyperfine ground states as a function of the magnetic field applied.

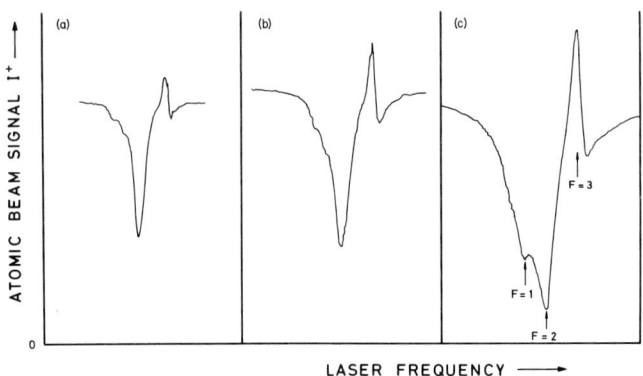

FIG. 29. Atomic beam signal of the Rabi analyzer after optical pumping with σ^+ polarized light. Laser intensities (mW/cm²): (a) 5, (b) 10, (c) 60.

parameter; X is the relative spin-up component of the atomic beam by pumping with σ^- light, Y is the fraction of atoms lost during optical pumping from the $\bar{F} = 2$ magnetic substates to the $\bar{F} = 1$ substates. In the experiment of Hils et al. (1981) $S \approx 70\%$ and $X + Y = 11\%$, which gives a total polarization of $P_{\text{Na}} \approx 80\%$.

In the sodium experiment a time-of-flight (TOF) technique was used to improve the signal-to-noise ratio, which is an important consideration for measurements in the vicinity of the 5.12-eV ionization threshold. Most of the ion background counts originate from the Penning ionization of sodium atoms by metastable mercury $^3P_2$ atoms which escape from the polarized electron source region. Sodium ions formed via the Penning process can

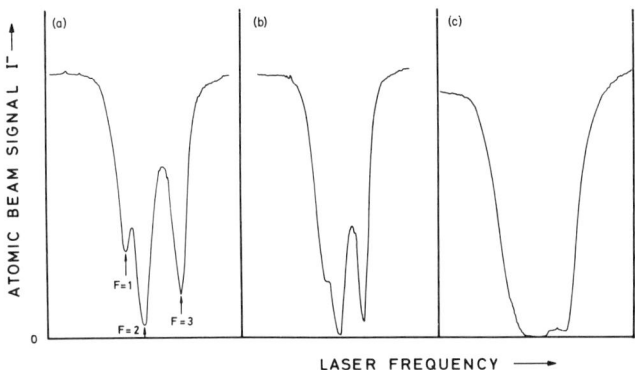

FIG. 30. The same signal as in Fig. 29, but for σ^- polarized laser light.

clearly be distinguished from sodium ions formed by electron impact by their time of arrival in the ion detector. The average time of flight of the metastable $^3P_2$ atoms is roughly 1 msec for our geometry, whereas electrons from the polarized electron source arrive at the point of ionization within approximately 100 nsec (see Fig. 31). Accordingly, the polarized electron source is operated in a pulsed mode whereby a square voltage 0.48 msec in duration is applied to the grid of the Pierce electron gun with a repetition time of 3.15 msec. TOF ion spectra are accumulated in two separate multichannel analyzers (MCAs) in the multiscaling mode. The dwell time chosen was 0.1 msec/channel. By a suitable choice of gating pulses applied to the MCAs, both the ($\sigma^{\uparrow\downarrow} - \sigma^{\uparrow\uparrow}$) difference and ($\sigma^{\uparrow\downarrow} + \sigma^{\uparrow\uparrow}$) sum TOF spectra were obtained simultaneously. Figure 32 shows a block diagram of the timing electronics together with the timing sequence of the gating pulses. As already mentioned, one of the MCSs measures the antiparallel/parallel difference directly. This has the advantage that all background contributions cancel automatically. For electron energies above the 10.4-eV mercury ionization threshold the antiparallel/parallel sum obtained with the second MCA had to be corrected for background ions formed by electron impact of the background gas. The background ions consisted mainly of mercury ions with a much smaller contribution from genuine background gas ions. The mercury ions are due to the ionization of mercury atoms escaping from the polarized electron source ($n_{Hg} \cong n_{gas} \cong 10^{10}$ cm^{-1}). No attempt was made to discriminate by TOF methods against background ions produced by electron impact because of the much faster MCA sweep rates required (microseconds), which would pose problems with one of the MCAs in use. At the same time dc methods proved to be quite adequate in eliminating background contributions, the reason being that the background level varies only very slowly with time, which is mainly due to the fact that the Hg oven has a large mass (several kilograms).

The procedure for data acquisition was the following. At the beginning and end of each TOF spectrum accumulated by the two MCAs ($T = 500$ sec), signal and background contributions were obtained with counting times of typically 50 sec. For electron energies above 10.4 eV, these consisted of measurements with the sodium beam on and off as well as with the electron beam on and off. The sodium beam could be blocked with a beam flag, whereas the electron beam was blocked with the filter lens by the application of a retarding voltage to its central electrode.

All results were printed out on a teletype and then the cycle repeated itself. For a given electron energy, data was taken until the statistical error in the asymmetry A_M was typically 10%. From the background measurements, averages were obtained and these provided the corrections applied to the individual TOF sum spectra to obtain the true sodium ion sum. The

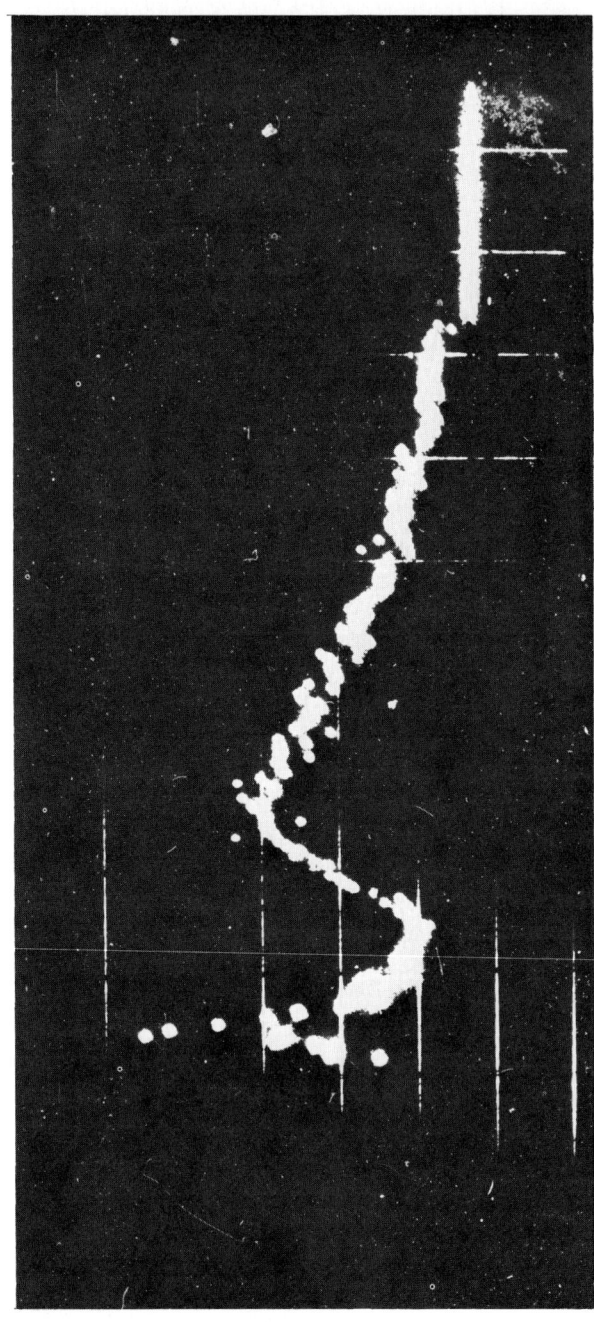

FIG. 31. Time-of-flight spectrum of ions arriving at the ion detector as described in Fig. 26. The sharp peak is due to ions produced by electron impact, whereas broad Boltzmann distribution is due to Penning ionization of sodium by metastable $^3P_2$ atoms. Integrating over the first peak represents the signal from electron impact ionization which is well separated from the unwanted Penning background signal.

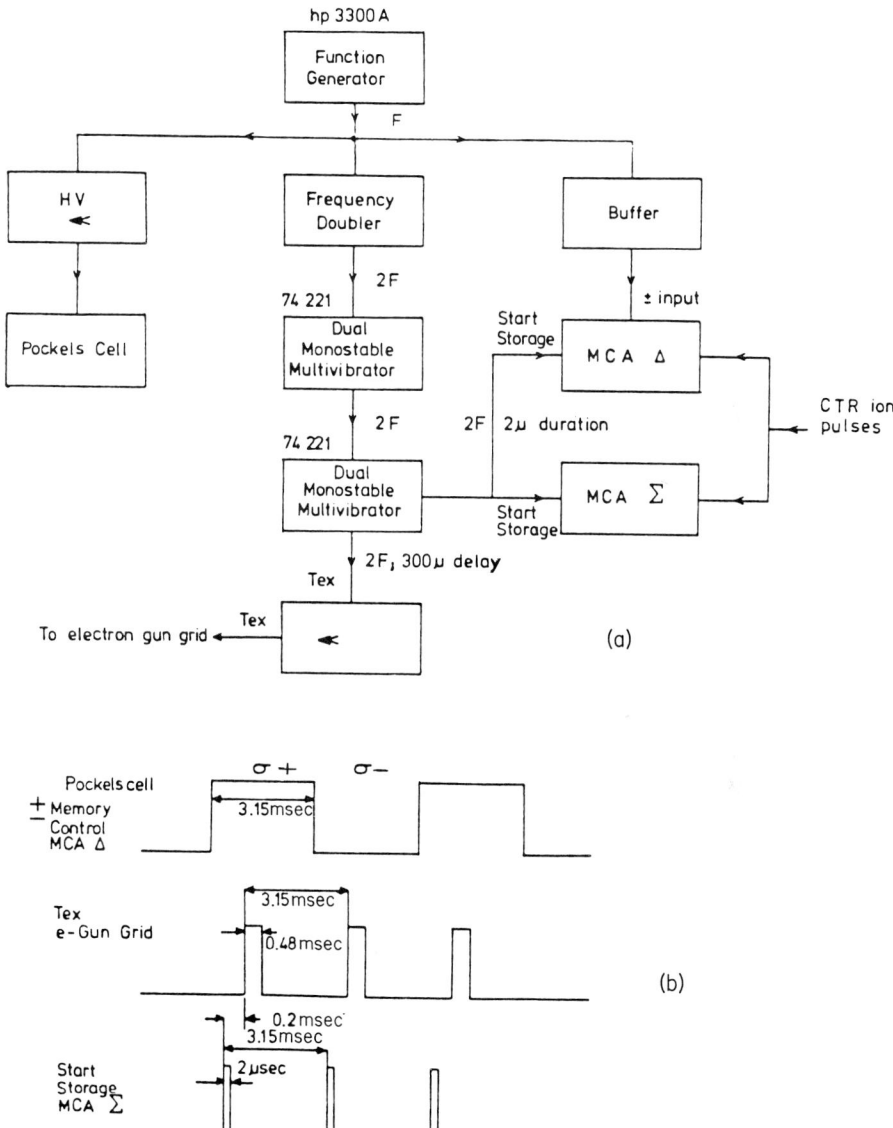

FIG. 32. Block diagram of gating pulses (a) and electronics (b) for the sodium ionization experiment. (From Hils et al., 1981.)

measured asymmetry parameter A_M was finally converted into values for A by the relation

$$A = A_M/P_e P_{Na}$$

Figure 33 shows the results for the ionization asymmetry factor A of sodium together with Born approximation calculations by Peach (1965, 1966). From threshold to approximately three times the ionization energy, the measured asymmetry A shows very little variation with energy. There is an indication that A is increasing as the threshold energy is approached. Above three times the ionization energy, the observed asymmetry A is decreasing.

At electron energies of more than four times the threshold energy, the agreement between the observed and theoretical asymmetry values from the Born approximation is satisfactory. The energy required to remove a 2p electron is roughly 38 eV and therefore our results are not influenced by inner-shell transitions.

At low energies, theory and experiment show a marked discrepancy. It is

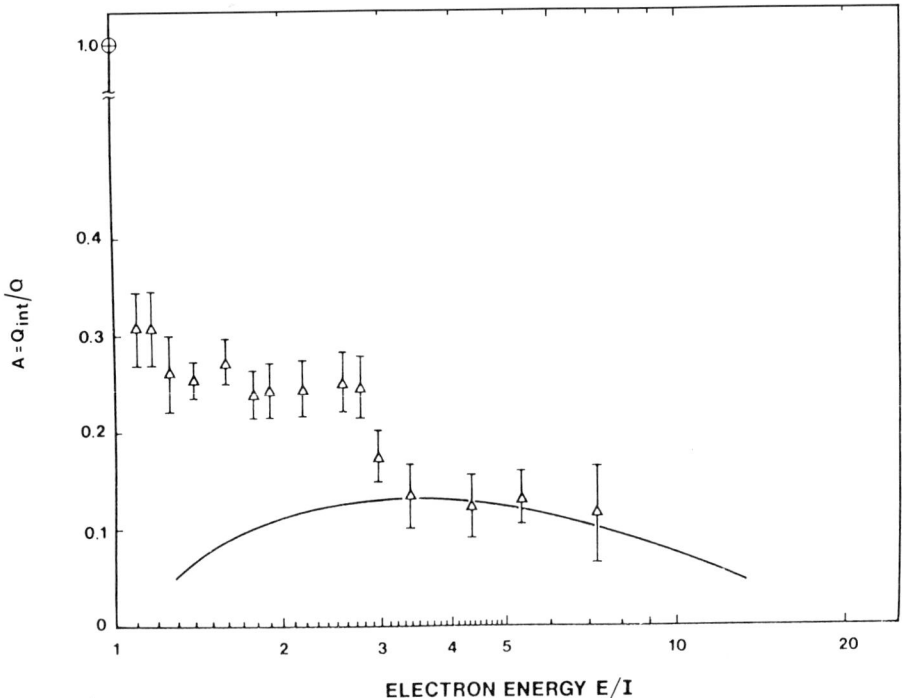

FIG. 33. Experimental data for ionization asymmetry factor A of sodium (Hils et al., 1981) compared with Born approximation calculations by Peach (1965, 1966).

not surprising that the Born approximation fails at the lower energies since the wave function describing the final state includes no correlations and the Born approximation is not expected to be valid for these energies. However, the comparison of the experimental data with the theoretical predictions of Wannier (1953), of Rau (1971), and of Klar and Schlecht (1976) are of much greater interest. This theory predicts that at the threshold limit, the ionization proceeds entirely via the singlet channel, and the asymmetry parameter A therefore takes on the value $A = 1$. The former argument derives from the fact that singlet and triplet S state functions obey different threshold laws $\sigma \propto E^n$, where $n(^3S) = 3.88$ and $n(^1S) = 1.127$. It is interesting in this connection to consider the ratios Q_t/Q_s, which can also be extracted from

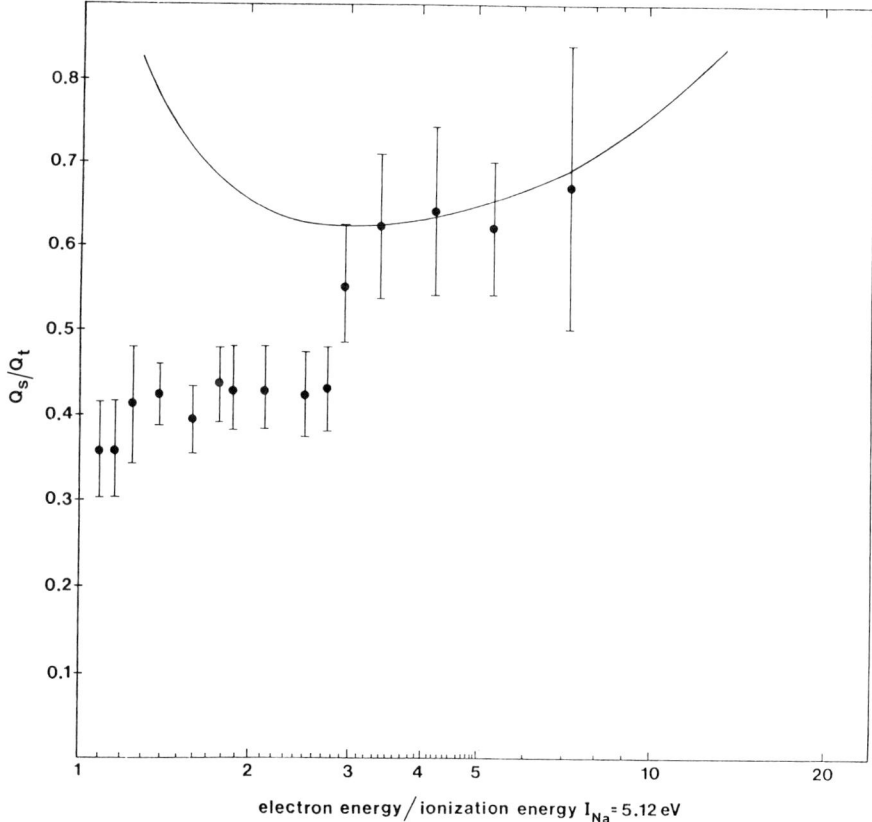

FIG. 34. Experimental data for ratios of triplet to singlet ionization cross sections (Hils *et al.*, 1981) determined from the data of Fig. 33 by using Eq. (6.8) and compared to the Born approximation (Peach, 1965, 1966).

the asymmetry measurement Eq. (4). Figure 34 displays such data, whereas in Fig. 35 data for log Q_t/Q_s are compared to the threshold law $Q_t/Q_s \propto E^{2.756}$ following the theory of Klar and Schlecht (1976). It is obvious that the present accuracy of the data in Fig. 35 should be improved in the future for a more critical test on the spin dependence of the threshold law. However, as can be seen in Fig. 35, the data for the four ratios at lowest energy are not incompatible with the threshold law $\sigma_t/\sigma_s \propto E^{2.756}$. On the other hand, the considerable difference between the observed asymmetry near threshold ($A \approx 0.3$) and unity may lead to the following conclusions:

(1) The range of validity of the threshold laws is $E < 0.5$ eV. One would expect that in the case of many-electron atoms such as sodium, the polarizability of the core electrons with their long-range effect will restrict the range of validity compared to hydrogen, for example.

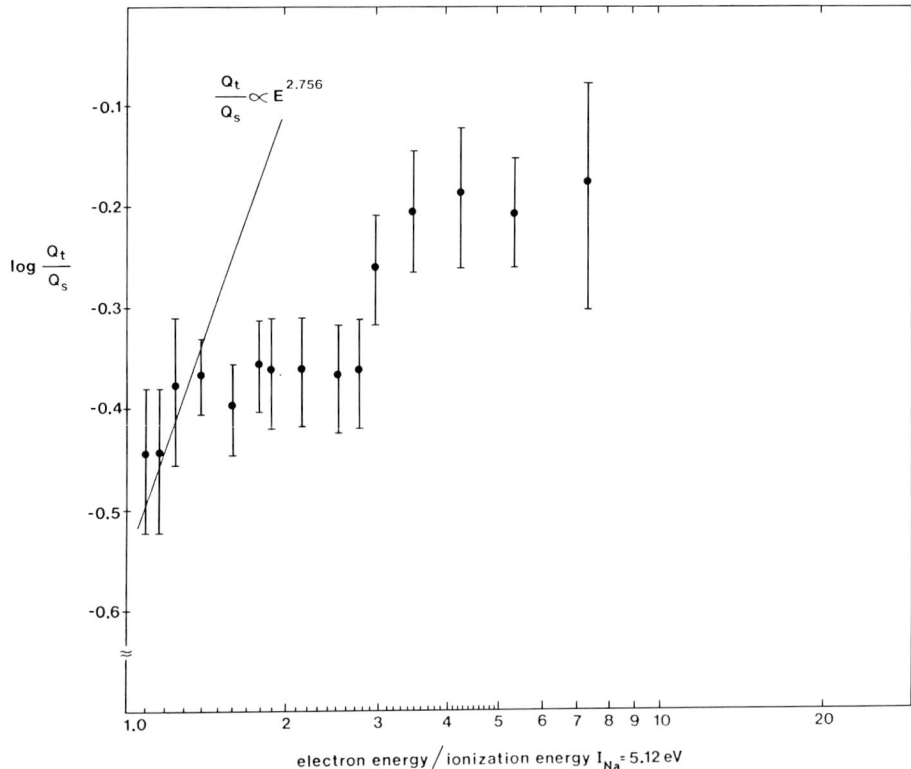

FIG. 35. Double logarithmic scale for Q_t/Q_s data of Fig. 34 compared to the threshold law $Q_t/Q_s \propto E^{2.756}$ following the theory of Klar and Schlecht (1976).

(2) Singlet and triplet wave functions obey the same threshold laws, in which case A would become energy independent near threshold and A could take on values different from unity.

(3) Green and Rau (1982) have pointed out that the ^1P_e state wave function for escaping electrons should also be governed by the exponent $n = 3.88$ in the threshold law. Apart from the ^3S_e and ^1P_e states, all other states should follow the Wannier law (Wannier, 1953) with the exponent $n = 1.127$. Of course, a contribution from the ^1P_e state would affect the threshold law for the ratio of triplet to singlet contributions in the ionization process. Any ^3P contribution would also result in $A < 1$ near threshold as obviously was observed in all the measurements of ionization asymmetries reported so far. Greene and Rau concluded that even $L = 2$ states may have appreciable contributions near threshold.

Finally, we note that all recent measurements of spin-dependent ionization asymmetries described in Sections VII,B,1–3 resulted in data which are substantially lower than unity near threshold. More accurate measurements with higher energy resolution are required for more crucial tests on the spin-dependent threshold laws of electron-impact ionization.

Appendix A

In this appendix we derive some useful expressions for the scattering amplitudes. It will be assumed that any explicit spin-dependent interaction can be neglected during the scattering process so that the collision can adequately be described in the LS-coupling scheme (as defined in Section II).

In this case we can describe the transition between atomic states $|S_0 M_{S_0}\rangle \rightarrow |LMS_1 M_1\rangle$ by electron impact in terms of the amplitudes $\langle LMS_1 M_{S_1}, \mathbf{p}_1 m_1 | \mathbf{T} | S_0 M_{S_0}, \mathbf{p}_0 m_0 \rangle$, assuming that the atoms are initially in their ground state with orbital angular momentum equal to zero. By coupling initial and final states to states of the total angular momentum S and its z component M_S, we obtain

$$\langle LM_L S_1 M_{S_1}, \mathbf{p}_1 m_1 | \mathbf{T} | S_0 M_{S_0}, \mathbf{p}_0 m_0 \rangle$$
$$= \sum_{SM_S S' M'_S} (S_1 M_{S_1} \tfrac{1}{2} m_1 | S' M'_S)(SM_S \tfrac{1}{2} m_0 | SM_S)$$
$$\times \langle LM_L, \mathbf{p}_1, S' M'_S | \mathbf{T} | \mathbf{p}_0, SM_S \rangle \quad \text{(A1)}$$

where $(\cdots|\cdots)$ is a standard Clebsch–Gordan coefficient.

Under the conditions stated above, the total spin and its z component are conserved. The **T**-matrix elements vanish, therefore, unless the conditions $S' = S$ and $M'_S = M_S$ are simultaneously satisfied. Furthermore, the dynamics must be invariant under rotations in the combined spin space. This is only possible if the **T** matrix elements are independent of M_S. Hence,

$$\langle LM_L S_1 M_{S_1}, \mathbf{p}_1 m_1 | \mathbf{T} | S_0 M_{S_0}, \mathbf{p}_0 m_0 \rangle$$
$$= \sum_{SM_S} (S_1 M_{S_1} \tfrac{1}{2} m_1 | SM_S)(S_0 M_{S_0} \tfrac{1}{2} m_0 | SM_S) \langle LM_L, \mathbf{p}_1 | \mathbf{T}^{(S)} | 0, \mathbf{p}_0 \rangle \quad (A2)$$

where the **T**-matrix element depends only on the total spin S but is independent of all spin components.

Spin conservation gives two values to the total spin (the "channel" spin W): $S = S_0 \pm \tfrac{1}{2} = S_1 \pm \tfrac{1}{2}$. If $S_0 = 0$ (or $S_1 = 0$), there is only one allowed spin channel with $S = \tfrac{1}{2}$ and Eq. (A2) reduces to ($S_0 = 0$)

$$\langle LM_L S_1 M_{S_1}, \mathbf{p}_1 m_1 | \mathbf{T} | \mathbf{p}_0 m_0 \rangle = (S_1 M_{S_1} \tfrac{1}{2} m_1 | \tfrac{1}{2} m_0) \langle LM_L, \mathbf{p}_1 | \mathbf{T} | 0, \mathbf{p}_0 \rangle \quad (A3)$$

Appendix B

In this appendix some consequences of reflection invariance in the scattering plane ($x-z$ plane) will be derived. We first consider atomic states with total angular momentum J and component M. Denoting the relevant reflection operator by **R**, the transformation properties of the atomic states are given by the following relation [see, for example, Taylor (1972) for details]:

$$\mathbf{R} |JM\rangle = (-1)^{L+J-M} |J-M\rangle \quad (B1a)$$

where it has been assumed that the atoms have a definite orbital angular momentum L (and hence a definite parity). The electronic states transform according to

$$\mathbf{R} |\mathbf{p} m\rangle = (-1)^{1/2-m} |\mathbf{p} -m\rangle \quad (B1b)$$

We now require that the scattering dynamics be invariant under reflection in the scattering plane. This requirement is expressed mathematically by the following relation:

$$\mathbf{T} = \mathbf{R}^\dagger \mathbf{T} \mathbf{R} \quad (B2)$$

It should be noted that Eq. (B2) contains the essential physical condition; Eq. (B1) expresses merely the transform properties of the states.

From Eq. (B2) there follows

$$\langle JM, \mathbf{p}_1 m_1 | \mathbf{T} | J_0 M_0, \mathbf{p}_0 m_0 \rangle$$
$$= \langle JM, \mathbf{p}_1 m_1 | \mathbf{R}\dagger \mathbf{T} \mathbf{R} | J_0 M_0, \mathbf{p}_0 m_0 \rangle$$
$$= (-1)^{L+J-M-L_0-J_0-M_0+m_1-m_0}$$
$$\times \langle J-M, \mathbf{p}_1 -m_1 | \mathbf{T} | J_0 -M_0, \mathbf{p}_0 -m_0 \rangle \quad (B3)$$

where $\mathbf{R}\dagger$ has been applied to the bra vector and $\mathbf{R}$ to the ket vector, and where L_0 denotes the an orbital angular momentum of the initial atomic states.

Similarly, a condition for LS-coupled states can be derived by using the transformation properties

$$\mathbf{R} | L M_L S M_S, \mathbf{p} m \rangle = (-1)^{L-M_L+S-M_S+1/2-m} | L-M_L S-M_S, \mathbf{p} -m \rangle \quad (B4)$$

By using Eq. (B2), inverting Eq. (A2), and applying the symmetry properties of the Clebsch–Gordan coefficients, the following symmetry relation can be obtained:

$$\langle L M_L \mathbf{p}_1 | \mathbf{T}^{(S)} | 0, \mathbf{p}_0 \rangle = (-1)^{M_L} \langle L-M_L, \mathbf{p}_1 | \mathbf{T}^{(S)} | 0, \mathbf{p}_0 \rangle \quad (B5)$$

ACKNOWLEDGMENTS

It is a pleasure to thank our colleagues at Bielefeld, Münster, and Stirling for many helpful discussions, in particular, K. Bartschat, G. F. Hanne, W. Jitschin, J. Kessler, and I. McGregor. One of us (H. K.) also gratefully acknowledges support by the German Research Council (DFG) for visits to Bielefeld University.

REFERENCES

Albert, K., Christian, C., Heindorff, T., Reichert, E., and Schön, S. (1977). *J. Phys.* B **10**, 3733.
Alguard, M. J., Hughes, V. W., Lubell, M. S., and Wainwright, P. F. (1977). *Phys. Rev. Lett.* **39**, 334.
Amado, R. D. (1982). *Phys. Rev.* C **26**, 270.
Baranger, E., and Gerjuoy, E. (1958). *Proc. Phys. Soc. London* **72**, 326.
Bartschat, K. (1981). Diplomarbeit, Universität Münster.
Bartschat, K., and Blum, K. (1982a). *Z. Phys.* A **304**, 85.
Bartschat, K., and Blum, K. (1982b). *J. Phys.* B **15**, 2747.
Bartschat, K., Blum, K., Hanne, G. F., and Kessler, J. (1981a). *J. Phys.* B **14**, 3761.
Bartschat, K., Hanne, G. F., Wolcke, A., and Kessler, J. (1981b). *Phys. Rev. Lett.* **47**, 997.
Bartschat, K., Hanne, G. F., and Wolcke, A. (1982). *Z. Phys.* A **304**, 90.
Baum, G., Caldwell, C. D., and Schröder, W. (1980). *Appl. Phys.* **25**, 39.

Baum, G., Kisker, E., Raith, W., Schröder, W., Sillman, U., and Zenses, D. (1981). *J. Phys. B* **14**, 4377.
Baum, G., Raith, W., Schröder, W., and Tanaka, H. (1982). *Verh. Dtsch. Phys. Ges.* **4**, 324.
Bederson, B. (1969a). *Comments At. Mol. Phys.* **1**, 71; Bederson, B. (1969b). *Comment At. Mol. Phys.* **2**, 65.
Bederson, B. (1973). *At. Phys.* **3**, 398.
Blum, K. (1980). *Book Invit. Pap., ICPEAC, 11th, 1979* p. 579.
Blum, K. (1981a). "Density Matrix Theory and Applications." Plenum, New York.
Blum, K. (1981b). *Invit. Pap., Int. Workshop Spin Polariz.*, unpublished.
Blum, K. (1982). Habilitationsschrift, Universität Münster.
Blum, K. (1983). *Nat. Adv. Study Inst.* to be published.
Blum, K., and Kleinpoppen, H. (1975). *Int. J. Quantum Chem.* **S9**, 415.
Blum, K., and Kleinpoppen, H. (1976). *Int. J. Quantum. Chem.* **S10**, 231.
Blum, K., and Kleinpoppen, H. (1979). *Phys. Rep.* **52**, 203.
Blum, K., and Kleinpoppen, H. (1982). *Book Invit. Pap. Int. Workshop Electron-Atom Mol. Collisions, 1983* p. 1.
Blum, K., da Paixao, F., and Csanak, G. (1980). *J. Phys. B* **13**, L257.
Bonham, R. A. (1974). *Electron Spectrosc.* **3**, 85.
Burke, P. G. (1977). *"Potential Scattering."* Plenum, New York.
Burke, P. G., and Mitchel, J. (1974). *J. Phys. B* **7**, 214.
Burke, P. G., and Robb, W. D. (1975). *Adv. At. Mol. Phys.* **11**, 143.
Burke, P. G., and Schey, H. M. (1962). *Phys. Rev.* **126**, 163.
Burke, P. G., and Taylor, A. J. (1969). *J. Phys. B* **2**, 44.
Burrow, P. D., and Michejda, J. A. (1975). *Int. Symp. Stirling Abstr.*, p. 50.
Celotta, R. J., and Pierce, D. T. (1980). *Adv. At. Mol. Phys.* **16**, 102.
Cohen-Tannouidji, C. (1962). *Ann. Phys. (Paris)* [13] **7**, 423.
Cowan, R. (1968). *J. Opt. Soc. Am.* **58**, 808.
Cowan, R. (1981): private Communication
Csanak, G. (1981). *Invit. Pap., Int. Workshop Spin Polariz.* unpublished.
Csanak, G., Taylor, H. S., and Yaris, R. (1971). *Adv. At. Mol. Phys.* **7**, 287.
da Paixao, F., Padial, N. T., Csanak, G., and Blum, K. (1980). *Phys. Rev. Lett.* **45**, 1164.
da Paixao, F., Padial, N. T., and Csanak, G. (1982). Private communication.
Düwecke, M., Kircher, N., Reichert, E., Staudt, E. (1973). *J. Phys. B* **6**, L208.
Drukarev, G. F., and Obedkov, V. D. (1979). *Usp. Fiz. Nauk* **127**, 621.
Edmonds, A. R. (1957). "Angular Momentum in Quantum Mechanics." Princeton Univ. Press, Princeton, New Jersey.
Eitel, W., and Kessler, J. (1971). *Phys. Rev. Lett.* **24**, 1972.
Eminyan, M., and Lampel, G. (1980). *Phys. Rev. Lett.* **45**, 1171.
Eminyan, M., McAdam, K., Slevin, J., and Kleinpoppen, H. (1974). *J. Phys. B* **7**, 1519.
Fano, U. (1957). *Rev. Mod. Phys.* **29**, 74.
Fano, U. (1969). *Phys. Rev.* **178**, 131.
Fano, U. (1978). *Phys. Rev.* **178**, 131.
Fano, U., and Cooper, J. (1965). *Phys. Rev. A* **138**, 400.
Fano, U., and Macek, J. (1973). *Rev. Mod. Phys.* **45**, 553.
Farago, P. S. (1971). *Rep. Prog. Phys.* **34**, 1055.
Farago, P. S. (1976). "Electronic and Atomic Interactions." Plenum, New York.
Farago, P. S., and Wykes, J. S. (1969). *J. Phys. B* **2**, 747.
Franz, K., Hanne, G. F., Hiddemann, K. H., and Kessler, J. (1982). *J. Phys. B* **15**, L115.
Gallaher, D. F. (1974). *J. Phys. B* **7**, 362.
Geltmann, S., Rudge, M. R. M., and Seaton, M. J. (1963). *Proc. Phys. Soc. London* **81**, 315.
Goldin, J. F., and McGuire, J. H. (1974). *Phys. Rev. Lett* **32**, 1218.

Goldstein, M., Kasdan, A., and Bederson, B. (1972). *Phys. Rev. A* **5**, 660.
Greene, C. H., and Rau, A. R. P. (1982). *Phys. Rev. Lett.* **48**, 553.
Hanne, G. F. (1976). *J. Phys. B* **9**, 805.
Hanne, G. F. (1980). *In* "Coherence and Correlation in Atomic Physics." (H. Kleinpoppen and J. Williams, eds.), p. 593. Plenum, New York.
Hanne, G. F. (1981). Habilitationsschrift, Universität Münster.
Hanne, G. F., and Kessler, J. (1976). *J. Phys. B* **9**, 791.
Hanne, G. F., Wemhoff, H., Wolcke, A., and Kessler, J. (1981). *J. Phys. B* **14**, L507.
Heddle, D. (1975). *J. Phys. B* **8**, L23.
Heddle, D. (1978). *J. Phys. B* **11**, L711.
Hermann, H., and Hertel, I. V. (1980). *In* "Coherence and Correlation in Atomic Physics" (H. Kleinpoppen and J. Williams, eds.), p. 625. Plenum, New York.
Hermann, H., and Hertel, I. V. (1982). *Comments At. Mol. Phys.* **12**, 61.
Hertel, I. V., and Stoll, W. (1974). *J. Phys. B* **7**, 570.
Hertel, I. V., and Stoll, W. (1978). *Adv. At. Mol. Phys.* **13**, 162.
Hils, D. (1977). *Paris Satellite Meet., 10th ICPEAC*, unpublished.
Hils, D., and Kleinpoppen, H. (1978). *J. Phys. B* **11**, L283.
Hils, D., Rubin, K., and Kleinpoppen, H. (1980). *In* "Coherence and Correlation in Atomic Physics (H. Kleinpoppen and J. Williams, eds.), p. 689. Plenum, New York.
Hils, D., Jitschin, W., and Kleinpoppen, H. (1981). *Appl. Phys.* **25**, 39.
Hippler, R., Malunat, G., Faust, M., Lutz, H. O., and Kleinpoppen, H.(1982). *Z. Phys.* **304**,63.
Hughes, V. W., Long, R. L., Lubell, M. S., Posner, M., and Raith, W. (1972). *Phys. Rev. A* **5**, 195.
Jost, K., and Kessler, J. (1965). *Z. Phys.* **195**, 1.
Karule, E., and Peterkop, R. (1965). *At. Collisions* **3**, 31.
Kennedy, J. V., Meyerscough, V. P., and McDowell, M. R. C. (1977). *J. Phys. B* **10**, 3759.
Kessler, J. (1969). *Rev. Mod. Phys.* **41**, 1.
Kessler, J. (1976). "Polarised Electrons." Springer-Verlag, Berlin and New York.
Khalid, W., and Kleinpoppen, H. (1982). *Z. Phys. A* **311**, 57.
Khalid, W., Kleinpoppen, H., Jakubowicz, H., and Moores, D. L. (1982). To be published.
Klar, H., and Schlecht, W. (1976). *J. Phys. B* **9**, 1699.
Kleinpoppen, H. (1971). *Phys. Rev. A* **3**, 2015.
Kleinpoppen, H. (1977). *Adv. Quantum. Chem.* **10**, 77.
Kleinpoppen, H., and Williams, J., eds. (1980). "Coherence and Correlation in Atomic Physics." Plenum, New York.
Kuyatt, C. E., Simpson, J. A. (1965). *Phys. Rev.* **138**, 385.
Lewis, B., Weigold, E., and Teubner, J. O. (1975). *J. Phys. B* **8**, 212.
Lubell, M. S. (1977a). *Paris Satellite Meet., 10th, ICPEAC*, unpublished.
Lubell, M. S. (1977b). *At. Phys.* **5**, 325.
Lubell, M. S. (1980). *In* "Coherence and Correlation in Atomic Collisions" (H. Kleinpoppen and J. F. Williams, eds.), p. 663. (Plenum, New York.
McConkey, J. W. (1980). *Book Invit. Pap. 11th, ICPEAC, 1979* p. 225.
McConnell, J. C., and Moisewitsch, B. L. (1968). *J. Phys. B* **2**, 406.
Macek, J. (1976). *In* "Electron and Photon Interactions with Atoms" (H. Kleinpoppen and M. R. C. McDowell eds.), p. 485. Plenum, New York.
Macek, J., and Hertel, I. V. (1974). *J. Phys. B* **7**, 2173.
Macek, J., and Jaecks, D. (1971). *Phys. Rev. A* **4**, 1288.
McGregor, I., Hils, D., Hippler, R., Malik, N., Williams, J., Zaidi, A., and Kleinpoppen, H. (1982). *J. Phys. B* **12**, L411.
Machado, L., Leal, E., and Csanak, G. (1982). *J. Phys. B* **15**, 1773.
Madison, D. H., and Shelton, W. N. (1973). *Phys. Rev. A* **7**, 499, 514.

Malcolm, I. C., and McConkey, J. W. (1979). *J. Phys. B* **12**, 511.
Massey, H. S. W., and Burhop, E. H. S. (1969). "Electronic and Ionic Impact Phenomena." Oxford Univ. Press (Clarendon), London and New York.
Massey, H. S. W., and Mohr, C. B. O. (1941). *Proc. Phys. Soc., London, Sect. A* **177**, 341.
Moisewitsch, B. L. (1976). *J. Phys. B* **9**, L245.
Moores, D. L., and Norcross, D. W. (1972). *J. Phys. B* **5**, 1482.
Mott, N. F. (1929). *Proc. Soc. London, Ser. A* **124**, 425.
Obedkov, V. D. (1980). *Book Invit. Pap., 11th, ICPEAC, 1979:* p. 219.
Ochkur, V. I. (1964). *Zh. Eksp. Teor. Fiz,* **41**, 1938.
Ottley, T. W., and Kleinpoppen, H. (1975). *J. Phys. B* **8**, 621.
Padial, N. T., Meneses, G., da Paixao, F., and Csanak, G. (1981). *Phys. Rev. A* **23**, 2194.
Peach, G. (1965). *Proc. Phys. Soc., London* **85**, 709.
Peach, G. (1966). *Proc. Phys. Soc., London* **87**, 381.
Percival, I., and Seaton, M. (1957). *Philos. Trans. R. Soc. London, Ser. A* **252**, 113.
Peterkop, R. (1961). *Zh. Eksp. Teor. Fiz.* **41**, 1746 (*JETP* **14**, 1377—*Engl. Transl*).
Pierce, D., and Celotta, R. J. (1982). *Adv. Electron. Electron Phys.* **56**, p. 214.
Raith, W. (1969). *At. Phys.* **1**, p. 389.
Rau, A. R. P. (1971). *Phys. Rev. A* **4**, 207.
Register, D. (1981). *Invit. Pap., Int. Workshop Spin Polariz.*, unpublished.
Register, D., Trajmar, S., Jensen, S., and Pol, R. (1980). *In* "Coherence and Correlation in Atomic Physics" (H. Kleinpoppen and J. Williams, eds.), p. 641. Plenum, New York.
Reichert, E. (1963). *Z. Phys.* **173**, 1.
Rodberg, J., and Thaler, R. (1967). "Introduction to the Quantum Theory of Scattering." Academic Press, New York.
Rubin, K., Bederson, B., Goldstein, M., and Collins, R. (1969). *Phys. Rev.* **182**, 201.
Rudge, M. R. M., and Seaton M. J. (1965). *Proc. R. Soc. London A* **283**, 262.
Rudge, M. R. M., and Schwartz S. B. (1966). *Proc. Phys. Soc. London* **88**, 563.
Rudge, M. R. M. (1978). *J. Phys. B* **11**, L149.
Schröder, W. (1982). Dissertation Universität Bielefeld.
Schultz, G. J. (1962). *Rev. Mod. Phys.* **45**, 378.
Scott N. S., Burke P. G. (1980). *J. Phys. B* **13**, 4299.
Scott, N. S., Burke, P. G., Bartschat, K. (1983). *J. Phys. B* (in press).
Sin Fai Lam, L. T., Baylis, W. E. (1981). *J. Phys. B* **14**, 559.
Steffen, R. M., and Alder, K. (1975). *In* "Electromagnetic Int. in Nuclear Spectroscopy" (W. D. Hamilton, ed.), p. 505. North-Holland Publ., Amsterdam.
Steidl, H., Reichert, E., and Deichsel, D. (1965). *Phys. Lett.* **17**, 31.
Steph, N., and Golden, D. (1980). *Phys. Rev. A* **21**, 1848.
Taylor, J. R. (1972). "Scattering Theory." Wiley, New York.
Ugbabe, A., Teubner, P., Weigold, E., and Arriola, H. (1977). *J. Phys. B* **10**, 71.
Vriens, L. (1966). *Proc. Phys. Soc. London* **89**, 13.
Walker, D. W. (1975). *J. Phys. B* **8**, L161.
Walker, D. W. (1976). *In* "Electron and Photon Interactions with Atoms" (H. Kleinpoppen and M. R. C. McDowell, eds.), p. 203. Plenum, New York.
Wannier, G. H. (1953). *Phys. Rev.* **90**, 817.
Wolcke, A., Bartschat, K., Blum, K., Borgmann, H., Hanne, G. F., and Kessler, J. (1983). *J. Phys. B* **16**, 639 (cited and corrected p. 235).
Wykes, J. S. (1971). *J. Phys. B* **4**, L91.
Yamazaki, Y., Shimizu, R., Ueda, K., and Hashimoto, H. (1977). *J. Phys. B* **10**, L731.
Zaidi, A., McGregor, I., and Kleinpoppen, H. (1980). *Phys. Rev. Lett.* **45**, 1168.
Zaidi, A., Khalid, S., McGregor, I., and Kleinpoppen, H. (1981). *J. Phys. B* **14**, L503.
Zapesochny, I. P., and Spenik, O. (1966). *Sov. Phys.—JETP* (*Engl. Transl.*) **23**, 592.

THE REDUCED POTENTIAL CURVE METHOD FOR DIATOMIC MOLECULES AND ITS APPLICATIONS

F. JENČ

Department of Theoretical Physics
University of Marburg
Marburg, Federal Republic of Germany

I. Introduction	266
II. Methods for Construction of Potential Curves	267
A. Theoretical Potential Curves	267
B. Rydberg–Klein–Rees–Vanderslice Method for Construction of Potential Curve	269
III. Reduced (Internuclear) Potential of Diatomic Molecules	271
A. Definition of Reduced Quantities and Their Properties	271
B. Reduced Ground-State RKRV Potentials	275
C. Sensitivity of RPC Geometry to Changes in Values of Molecular Constants	284
D. Reduced Theoretical Potential Curves	284
E. Mathematical Foundations of RPC Method	290
F. Excited States	291
IV. Applications of RPC Method	294
A. Construction of Internuclear Potentials of Diatomic Molecules in a Sufficiently Wide Range of Internuclear Distance	294
B. Estimation of Molecular Constants	296
C. Classification of "Empirical" Potential Functions	297
D. Detection of Errors in Experimental Values of Molecular Constants and in Analysis of Spectrum. Detection of Perturbations; Detection of Nonvalidity of Some Commonly Used Formulas	298
V. Comments on Some Misunderstandings about the RPC Method	302
VI. Conclusions	305
References	306

I. Introduction

In absence of nonadiabatic perturbations, the physical system of a diatomic molecule[1] is completely determined through its interatomic (internuclear) potential. Hence the determination of the interatomic potentials of diatomic molecules (eventually in numerical form) or of the graphs of these potentials (the potential curves) is of fundamental importance in spectroscopy and in many practical applications.

On the other hand, the values of molecular constants of diatomics, the geometry of the potential curves, and the location of the curves in the $r-U$ diagram (r is the internuclear distance, U is the potential energy) are so diversified that one certainly would like to find a unifying scheme in which a systematical study of diatomic potentials could be pursued.

The reduced potential curve (RPC) method has been developed for both of these purposes. Indeed, the RPC method makes possible a systematic study of internuclear potentials of diatomic molecules in a unified scheme, using universal "reduced" quantities, where the potential curves of different diatomic molecules may be directly compared and the regularities in the behavior of the potentials of various diatomics conveniently visualized.

Moreover, the RPC method has the following practical applications:

(1) Construction of reliable interatomic potentials of diatomic molecules in a sufficiently large range of internuclear distance:

(a) from experimentally determined spectral lines of molecules with adjacent values of atomic numbers;
(b) from *approximate ab initio* calculated theoretical potentials.

(2) Estimation of spectroscopic molecular constants, in particular r_e (the equilibrium internuclear distance), D_e (the depth of the minimum of the potential curve), and k_e (the force constant).

(3) Classification of "empirical" potential functions proposed to approximate the internuclear potentials—e.g., also the potential functions proposed for the description of the interatomic interaction of rare gases.

(4) Detection of errors in the experimentally determined values of spectroscopic constants and in the analysis of the spectrum, in general as well as in the use of spectroscopic formulas.

(5) Detection of perturbations.

---

[1] The neutral molecule as well as the molecular ion. Only singly ionized molecules were studied here.

II. Methods for Construction of Potential Curves

A. Theoretical Potential Curves

In the Born–Oppenheimer approximation, the internuclear potential of diatomic molecule is defined by Eq. (1):

$$U(r) = E_{el}(r) + Z_1 Z_2 / r \tag{1}$$

where $E_{el}(r)$ denotes the adiabatic electronic energy as a function of the internuclear distance of the clamped nuclei; Z_1 and Z_2 are the atomic numbers of the two atoms (Fig. 1).

Here, spin effects and relativistic effects may be neglected for not too heavy molecules. At any rate, these corrections are relatively small with respect to the degree of accuracy that may be obtained presently in calculations of molecular constants of heavier (not too heavy) molecules. We refer here to the "zero-order adiabatic approximation" of clamped nuclei (Born and Oppenheimer, 1927), which was meant by the term "adiabatic approximation" in earlier papers of the author. For presently used standard terminology and an up-to-date discussion of the adiabatic approximation, see the survey of Kołos (1970). We shall use this terminology in the following text.

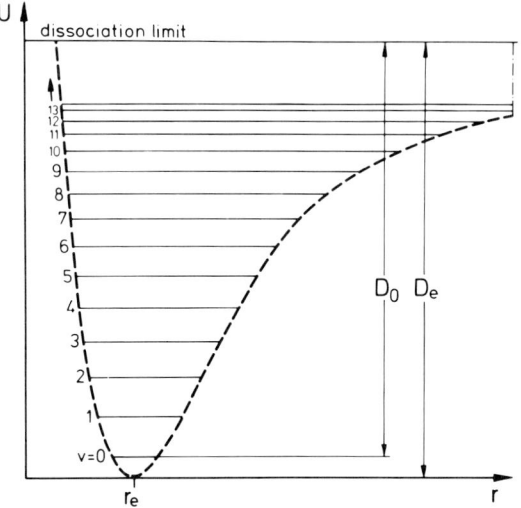

FIG. 1. Typical example of a potential curve for a diatomic molecule. The horizontal lines indicate the levels of vibrational energy; r_e is the equilibrium internuclear distance; D_0 is the dissociation energy; D_e is the dissociation energy + zero-point energy.

Let us recall that the "adiabatic," i.e., diagonal corrections for nuclear motion are generally very small for internuclear distances of interest, particularly for heavier molecules and ground states.

It seems at present impossible to deduce a manageable analytical expression for the function $E_{el}(r)$ from the first principles of the theory (even modern mathematical methods are of little help here). The eigenvalue problem corresponding to $E_{el}(r)$ may be rigorously solved only for the one-electron problem (e.g., H_2^+). Therefore, the important question is the following: Can one calculate molecular (and atomic) electronic energy for a sufficient number of different values of the internuclear distance ("*ab initio*" calculations) to obtain a satisfactory approximation of the potential curve (in numerical form, say)?

For many-electron problems, variational methods seem to be most efficient; the calculations are, however, so complex that only rough approximations to the potential curve, and the molecular constants related to its geometry, may be obtained for heavier molecules even if rather sophisticated calculations are performed.

Rather than the calculated potential curve ever being "parallel" to the true potential curve, it has a different form (so that a "shift" in D_e does not suffice). This is illustrated on the example of a concrete calculation (Jenč, 1963b) in Fig. 2.

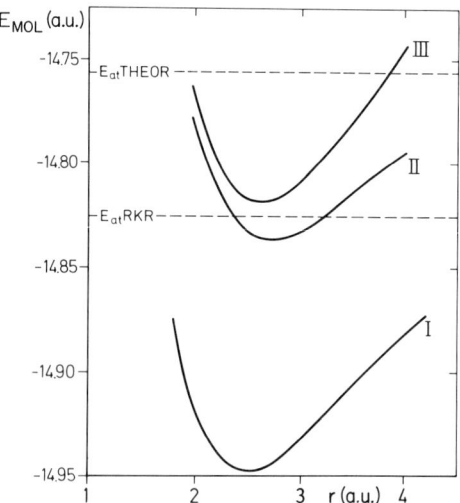

FIG. 2. Example of an "*ab initio*" calculation of the potential curve of a diatomic molecule (ground state of BeH$^+$): (I) RKRV potential curve; (II) CI–MO (SCF–LCAO)-calculated potential curve; (III) MO (SCF–LCAO)-calculated potential curve (incorrect asymptotic behavior). (From Jenč, 1963b.)

Convergence problems play the central role here. In the configuration interaction (CI) method, the convergence could be substantially improved through variation of nonlinear parameters (which, of course, makes the calculation rather entangled). A massive use of modern computers and the employment of suitable function bases for the test functions have brought about a great improvement of the results of *ab initio* calculations nowadays (forcing the number of configurations up to the order of 10^3–10^4 with one-electron functions of Gaussian type, for instance). Nevertheless, the errors in the calculated values of molecular constants of heavy molecules are still too large and there seem to exist some practical limits to the accuracy of such calculations. The James–Coolidge-type method [considerably developed by Kołos and Wolniewicz (1965)] is so complicated that significant results have as yet been obtained only for the lightest hydrogen molecule.

Interesting results have been obtained for heavy molecules also using the "method of effective core potentials" (see, e.g., the series of articles by M. Krauss *et al.*). However, one still has to reckon with errors in the molecular constants up to at least 20% (one also has to analyze carefully the applicability of the method in special cases).

All in all, we may assume that, at present, *ab initio* calculations may be expected to yield only a rather modest approximation for the potential curves of heavy molecules.

B. Rydberg–Klein–Rees–Vanderslice Method for Construction of Potential Curve

On the basis of the work of Rydberg (1931, 1933), Klein (1932) proposed a method for the determination of internuclear potentials of diatomic molecules (in numerical form) directly from the measured spectral lines. Vibrational as well as rotational spectra are needed for this construction of the potential. The experimental data must be sufficiently accurate, the method being rather sensitive to errors in the analysis of the rotational spectrum. A priori, this method is not applicable in a *close* neighborhood of the dissociation limit owing to certain singularities in the formulas.

The Rydberg–Klein method was later more satisfactorily formulated by Rees (1947a,b) in an analytical form that is suitable for numerical calculations. Vanderslice and others (1959) have employed an improved modification of the method in many calculations [for literature, cf. Jenč and Plíva (1963)].

The Rydberg–Klein–Rees–Vanderslice (RKRV) method has the following advantage: In contradistinction to *ab initio* calculations, the complexity of the RKRV calculations does *not* increase with increasing number of electrons in the molecule. (Slight modifications of the method have been made by several authors; in general, the method is denoted as the RKR method in the literature.)

In principle, the RKRV method has the character of a semiclassical approximation: One calculates the turning points of an oscillator. Therefore, one certainly must ask about the goodness of this approximation. It is not easy to obtain a clear general estimate of the errors involved from the analytical formulation. However, the RKRV method has been subjected to a series of critical tests and its reliability is supported essentially by the following points:

(1) The RKRV potential curve for the ground state of the H_2 molecule coincides with the theoretical potential curve calculated by Kołos and Wolniewicz (1965) using a sophisticated James–Coolidge-type method (all molecular spectroscopic constants approximated practically to 100% of the experimental value). Since, as is well known, there are no significant perturbations in the ground state of H_2, the RKRV curve may represent, in principle, the adiabatic potential (moreover, Kołos and Wolniewicz have proved the goodness of the Born–Oppenheimer approximation in calculating the mean value of the nuclear kinetic energy, which is shown to be very small). This coincidence of both curves is most important, since possible errors in the semiclassical RKRV approximation should be largest for the *lightest* molecule.

(2) The reliability of the RKRV potential is further supported by the coincidence of the RKRV potential curves of isotopic molecules, e.g., $^{14}N_2$ and $^{14-15}N_2$ (Benesch et al., 1965) (i.e., at least for molecules with a normal isotope shift).

(3) A direct proof of the reliability of the RKRV potentials may be found in the work of Zare (1963, 1964), of Zare and Cashion (1963), and of Zare et al. (1965), who calculated numerically the eigenvalues of the radial Schrödinger equation for these potentials with high degree of accuracy (for cases where a sufficient number of spectral lines have been measured and the gap between adjacent energy levels is not too large). It is interesting to note that not only do the calculated energy levels coincide with experimentally determined values within the limit of experimental error, but also the B_v constants could be evaluated in a very good approximation with the use of the wave functions obtained with the help of this calculation. Quite sophisticated programs have been worked out recently by the co-workers of Zare and may be used as standard checking procedures for testing the calculated potentials in such cases where a correspondingly large number of points may be obtained from the measured spectra (Zare et al., 1973).

Moreover, the RKRV method has been successfully employed in a number of practical applications. At present, the RKRV method seems to be the most reliable method for the determination of internuclear potentials of diatomic molecules. The value of the dissociation energy D_0 (or depth of the minimum, D_e) is *not* needed in the RKRV calculation; an accurate determination of this quantity usually poses serious problems. (Note that the value of D_e is needed in practically all so-called "empirical" potential functions.)

The most serious practical problems encountered in the use of the RKRV method are the following:

(1) There still exists only a rather limited supply of sufficient and sufficiently reliable experimental data on diatomic molecules. In some cases, only a few spectral lines have been measured.

(2) For heavy molecules, the RKRV curve can be constructed only in a relatively narrow region of internuclear distance, even though relatively many spectral lines have been measured: The energy difference between adjacent vibrational levels is, for heavy molecules, only a very small fraction of the dissociation energy (of the depth of the minimum). One often cannot get over the value $U + D_e = 0.2 D_e$.

(3) As has already been mentioned, reliable RKRV potentials can only be calculated if sufficiently complete and accurate spectral data are available from the lowest level on. In some cases where some doubts on the accuracy of the RKRV method could arise, one also could attempt to employ the method developed recently by Vidal and Scheingraber (1977). This method, however, should still be more deeply analyzed and tested by reliable checks. This method probably should be employed in such cases where well-founded doubts on the applicability of the Dunham-type approximation arise.

(4) Quite generally speaking, the RKRV method — like any other method based directly on the spectral data, e.g., Vidal's method — can in practice only give the picture of the potential well, since the spectral lines cannot be measured up to the dissociation limit (moreover, the errors in the calculation would become considerable in the neighborhood of the dissociation limit). Hence, it does not yield a description of the "van der Waals part" of the potential.

III. Reduced (Internuclear) Potentials of Diatomic Molecules

A. Definition of Reduced Quantities and Their Properties

In Eq. (1), the numbers $E_{el}(r)$ are eigenvalues of the electronic Schrödinger operators $H_{el}(Z_1, Z_2, n, r)$, n being the number of electrons. Hence $U(r)$ is a function of the three parameters Z_1, Z_2, and n. The structure of the Hamiltonians is quite analogous in the various functional spaces (of $3n$ variables at a time); hence, one could conjecture the existence of a universal potential function containing the three parameters Z_1, Z_2, and n. Since all other molecular constants may be deduced from the potential function (i.e., are closely related to the potential function), one may surmise that conversely (at least three) molecular constants α_i could be used as parameters for the construction of a universal function $U(\alpha_i, r)$ that could describe interatomic potentials of all diatomic molecules. That also was, in principle, the foundation of the idea of so-called "empirical" potential functions. However, so far, none of the "empirical" potential functions proposed has been proved to yield a generally acceptable approximation to the adiabatic protential [this fact may also be shown in evaluating molecular constants corresponding to such "empirical" potential functions; cf., for instance, Varshni (1957)]. Indeed, in general, there exist considerable discrepancies between the RKRV potentials determined from the measured spectra and the "empirical" potential functions (Jenč and Plíva, 1963). It appears evident that such functions may sometimes be successfully used as a rough approximation for internuclear potentials of a rather limited group of molecules; however, they fail completely for molecules of a different type.

Pursuing further this idea along somewhat different lines, one could conjecture that the same graph of the adiabatic potential curve may be obtained for all diatomic molecules by using, instead of the variables r and U, some dimensionless "reduced" quantities ρ and u, say, the definition of which could include, at any rate, at least the molecular constants D_e and r_e (possibly also other molecular constants). This ideas is akin to the idea of the reduced state equation of gases and is based on the assumption that the *essential structure* of the vibrational problem is, in principle, the same in all diatomics.

The existence of such a "reduced" picture in which the graphs of all diatomic-potential curves practically coincide would, eventually, imply the existence of an (possibly rather complicated) analytical expression for the potential function in the sense of the idea of "empirical" potential functions. This was the pioneer idea of Puppi (1946) and of Frost and Musulin[2] (1954). However, they were not able to obtain any positive results confirming their hypothesis and abandoned the subject. Indeed, postulating the existence of a universal reduced potential curve valid for all diatomic systems certainly seems *too strong a hypothesis.*

On the other hand, the following physically reasonable hypothesis—confirmed by extensive research—has proved correct: The essential structure of

---

[2] These authors did not employ the RKR potentials.

the vibrational problem is not identical in all diatomic molecules and depends in a defined way on the atomic numbers. This dependence is, however, rather weak in the sense that the potential curves in reduced form practically coincide for diatomics possessing adjacent values of both atomic numbers (i.e., where Z_1 of molecule A differs at most by a few units from Z_1 of molecule B, and analogously for Z_2; there will, of course, be only a slight difference in n).

The whole scheme should then exhibit certain *regularities* in the geometry of the reduced potential curves with respect to the values of the parameters Z_1, Z_2 (and n). Irregularities in such a scheme should a priori be expected for excited states, where nonadiabatic perturbations may appear. On the other hand, such perturbations are rare in ground states of diatomics (and, moreover, the ground states have, in most cases, the same symmetry $^1\Sigma$ and represent the lower bounds for the eigenvalues of the electronic Schrödinger operators in subspaces of this particular symmetry). Therefore, ground states are here of special interest.

To check on the ideas proposed, one may employ the RKR potentials for the construction of the reduced potential curves. With "reduced" quantities of the type $u = U/D_e$, $\rho = r/r_e$, one does not obtain any positive results.

Important results appear, however, with the following definition of the "reduced" internuclear distance and the "reduced" potential energy (Jenč and Plíva, 1963):

$$u = U/D_e \tag{2}$$

$$\rho = \frac{r - [1 - \exp(-r/\rho_{ij})]\rho_{ij}}{r_e - [1 - \exp(-r/\rho_{ij})]\rho_{ij}} \tag{3}$$

$$\rho_{ij} = \frac{r_e - (\kappa D_e/k_e)^{1/2}}{1 - \exp(-r_e/\rho_{ij})}, \quad \kappa = 3.96 \tag{4}$$

where

$$k_e \equiv (d^2U/dr^2)_{r=r_e} \tag{5}$$

is the "force constant" which could be obtained in interpolating the (*monotonous*) electronic energy curve $[U(r) - Z_1Z_2/r]$ corresponding to the RKR potential and adding the point $U(r_e) = -D_e$. In spectroscopy, one usually supposes the validity of the following approximation:

$$k_e = \mu\omega^2 = \mu\omega_e^2 4\pi^2 c^2 \tag{6}$$

where μ is the reduced mass of the molecule; ω is the "harmonic" vibration frequency, determined from the spectra; ω_e is the "harmonic" spectroscopic vibrational constant; c is the velocity of light. This approximation is due to Dunham (1932), a slight correction being omitted, for simplicity. In the cases of interest, this equation may, indeed, be accepted, which is of importance for practical calculations of reduced RKR curves of diatomics. At any rate, it will hold for ground states, where perturbations may be neglected. [To this problem, cf. Jenč (1965b, 1966b).] This definition of the "reduced" potential contains only the main molecular constants that have a direct geometrical meaning with respect to the potential curve.

For $0 \leq \rho_{ij} < r_e$, the reduced quantities fulfil the following conditions:

$$\begin{aligned} \rho &\geq 0 & & \\ \rho &= 0 & \text{for} \quad r &= 0 \\ \rho &= 1 & \text{for} \quad r &= r_e \\ \rho &\to \infty & \text{for} \quad r &\to \infty \\ u &\leq 0 & \text{for} \quad U &\leq 0 \\ u &= 0 & \text{for} \quad U &= 0 \\ u &\to \infty & \text{for} \quad U &\to \infty \\ u &= -1 & \text{for} \quad U &= -D_e \end{aligned} \quad (7)$$

It is clear that these relations — which we naturally expect a reduced potential should fulfil — impose rather restrictive conditions on possible definitions of the reduced potential.

The reduced coordinates of the minimum of a potential curve are $\rho = +1$, $u = -1$; in the figures, we always plot ρ versus $(u + 1)$.

The value of the universal constant, $\kappa = 3.96$, follows from putting $\rho_{ij} = 0$ for the one-electron problem H_2^+; the factor $[1 - \exp(-r/\rho_{ij})]\rho_{ij}$ has been heuristically introduced to take, in a way, account of the effect of electron repulsion. For very heavy molecules, the inequality $\rho_{ij} > r_e$ may occur. Then, the reduced potential curve cannot be employed for *very large* internuclear distances (still up to $u + 1 > 0.9$), which, however, does not play any important role in practical calculation of reduced RKR curves. As has been stressed above, the RKR method (or any similar method using spectral lines) does not permit the construction of the potential curve for large values of internuclear distance, giving in practice only the shape of the *potential well*.

This behavior of the reduced potential curve could be removed through introducing, into ρ, further parameters or through introducing a new term in u (F. Jenč, unpublished)[3]. Such corrections also could still improve

---

[3] For instance, introduction of a constant exponential factor in the function $\rho(r)$ already leads to correct behavior of the RPC for the molecules tested to date. The reduced potential may then oscillate in a *very small* neighborhood of the origin, which, however, is irrelevant for

the coincidence of the reduced curves of various molecules. Nevertheless, for practical applications, the definition of the reduced potential as formulated in Eqs. (2)–(5) proved very useful; its great advantage is its relative *simplicity*.

B. Reduced Ground-State RKRV Potentials

With the definition (2)–(4) of the reduced potential curve (RPC), one obtains the following results for the reduced RKR curves: One does not find a coincidence of the reduced potential curves of all diatomic molecules (which would mean expecting too much, indeed). However, interesting regularities are observed that may be exploited in a number of practical applications. First, we summarize the results obtained for the *ground states*.

(1) The RPCs of different molecules never intersect.

(2) The RPCs of diatomics with only slightly differing values of both atomic numbers of various molecules (i.e., differing by one or two, say) may, in a very good approximation, be identified. The goodness of this approximation improves markedly with growing values of the atomic numbers (heavy molecules).

(3) In general, the picture of the RPC changes with growing atomic numbers as follows: The RPC turns slowly to the right around the minimum and becomes broader, which may be interpreted as a decrease in the "reduced attractive force" for $r > r_e$, and an increase in the "reduced repulsive force" for $r < r_e$, in heavy molecules. From the physical point of view, this overall result seems rather plausible. The differences in the RPCs of molecules the atomic numbers of which differ by more units are more pronounced for light molecules and diminish rapidly with growing weight (increasing values of atomic numbers) of the molecules. This ordering of the reduced potential curves refers to *both* atomic numbers. As yet the results certainly do not permit one to deduce any formulas describing the dependence of the RPC on Z_1 and Z_2.

While keeping one atomic number constant, a considerable change in the value of the other number may have a less pronounced effect than a relatively small change in the values of *both* atomic numbers. This fact is illustrated, e.g., by the RPCs of heavy hydrides in Fig. 8. A safe guide is, at any rate, a comparison of RPCs of molecules with slightly differing atomic numbers.

---

the use of the RPC method. (Even this flaw may be eliminated through adding terms in u.) Direct calculations show, however, that the definition of Eqs. (2)–(5) is in order for $p > 80$, $u + 1 > 0.99$, e.g., for Cl_2 ($Z = 17$) and I_2 ($Z = 53$), where $p_{ij} > r_e$.

These regularities have been verified for a very large number of diatomic molecules, so far, and the results are very convincing. It is certain that they would not be considered fortuitous (cf. also the following paragraph on the sensitivity of the RPC to a change in the values of the molecular constants).

In spite of the large number of molecules tested, the only important deviation from the regularities of the RPC scheme seemed to be found in the ground state of the Cl_2 molecule (Jenč, 1967c). However, Coxon (1971) has shown that this discrepancy was in fact due to an inaccurate analysis of the complicated Cl_2 spectrum. With the new data and the new analysis of the spectrum, the ground state of Cl_2 fits very well into the RPC scheme [see also the new data of Douglas and Hoy (1975)]. (The conjecture that the seeming anomaly of Cl_2 should be due to an unsatisfactory rotational analysis of the spectrum has been communicated to the author by Dr. R. L. Le Roy, private communication, 1968.)

In addition, Coxon has also shown that the ground-state RKRV curve of the Br_2 molecule fits very well into the RPC scheme [Fig. 3 of Coxon (1971)].

To appreciate the merits of the RPC method, which permits a direct comparison of the potentials in a unique scheme, one should compare Figs. 3 and 4, where the RKRV potential curves and the RPCs of some molecules, respectively, are plotted. None of the so-called "empirical potential functions" can permit such a *clear visualization* of the regularities in the properties of the potential functions of diatomics as the RPC method does. One cannot visualize any of such regularities in a simple $r-U$ diagram, where the location of the potential curves of different molecules and their geometry are terribly diversified.

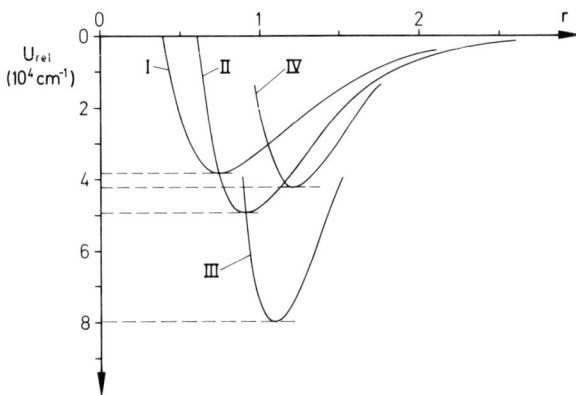

FIG. 3. RKRV potential curves of the molecules: (I) H_2, (II) HF, (III) N_2, (IV) O_2. (The dissociation limits shifted to coincide.)

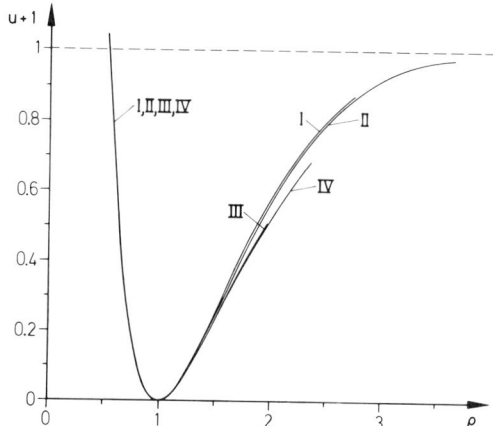

FIG. 4. Reduced RKRV potential curves of the molecules: (I) H_2, (II) HF, (III) N_2, (IV) O_2. III and IV approximately (closely) coincide with the RPCs of other combinations of C, N, and O (C_2, CN, CO, NO).

There, naturally, is an exception to this rule, which also should a priori be expected, namely, the *rare gases* (Jenč, 1964f). The RPCs of rare gases coincide and form the "right-hand side limit" of the "admissible RPC region" in the $p-u$ diagram; the "left-hand side border" of this region is defined by the RPC of the lightest molecule, H_2 (or H_2^+). The following rule applies.

(4) All ground-state RPCs are contained within the *"permissible" RPC region,* the upper and lower bounderies of which are defined by the RPC of H_2 and the coinciding RPCs of the rare gases, respectively.

Since, of course, no RKRV curves are available for rare gases, one employed here empirical potential functions of Lennard-Jones or Buckingham type. For He—He, also *ab initio* calculated potential curves were employed. This result may be summarized by saying that the "reduced attractive force," for $r > r_e$, is least for rare-gas diatomic systems whereas the "reduced repulsive force," for $r < r_e$, is here larger than in any other diatomic system.

Thus, a very plausible physical picture is obtained in the RPC scheme. It is certainly interesting that, *in the reduced form,* potential curves of quite different geometry may coincide (i.e., also the molecular constants r_e, D_e, and k_e have quite different values which, however, are *combined* in such a way that the RPCs of molecules with adjacent atomic numbers coincide — cf. Figs. 3 and 4).

In addition to the rare gases, for more than 30 diatomic molecules the calculations were carried out using the RKRV potentials, which, however, could be constructed in a sufficiently wide range of internuclear distance only for a limited number of molecules because of a lack of reliable experimental data (Jenč, 1964b, 1965c, 1967a, 1968, 1969a; Jenč and Plíva, 1963). For 30 other molecules (Jenč, 1967b) a rough approximative test was performed with the use of the Hulburt–Hirschfelder "empirical" potential function (Hulburt and Hirschfelder, 1941), which is supposed by many authors to yield a certain approximation for not too strongly polar molecules (in particular, for nonmetallic molecules). In fact, there exist discrepancies between reduced RKRV and Hulburt–Hirschfelder potentials that are considerable in some cases. It seems, however, that these discrepancies are of a rather *systematic* character: In the attractive branch, the Hulburt–Hirschfelder potential curve lies below the RKRV curve. This fact makes the Hulburt–Hirschfelder function usable for a *qualitative* test on the laws governing the behavior of the reduced potentials. The advantage is, of course, that the "empirical" potential function may be constructed in a wide range of internuclear distance.

The results for reduced RKRV and Hulburt–Hirschfelder potentials are visualized in Figs. 5–10 and Figs. 11–15, respectively. To make the figures readable, we show in each figure only the RPCs of a certain group of molecules. The RPCs of certain molecules are drawn as solid curves to obtain a standard division of the permissible RPC region in the $p-u$ diagram into various sectors. For this purpose, we usually employ the RPCs of H_2 and of the rare gases as border lines, and the RPCs of the C—N—O group

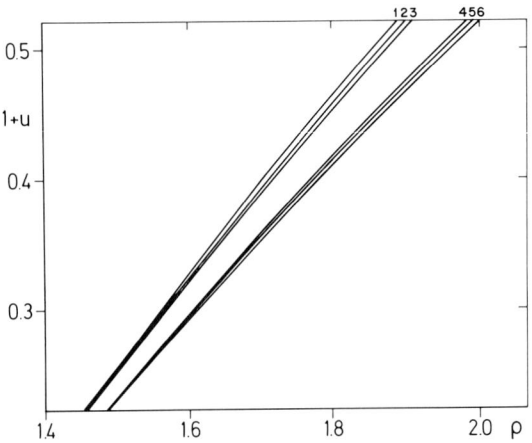

FIG. 5. Reduced ground-state RKRV potential curves for light diatomic molecules: (1) (almost) coinciding RPCs of H_2, LiH, and BeH^+; (2) RPC of OH; (3) RPC of HF; (4) (almost) coinciding RPCs of C_2, N_2, and CN; (5) RPC of O_2; (6) RPC of CO. (From Jenč, 1967b.)

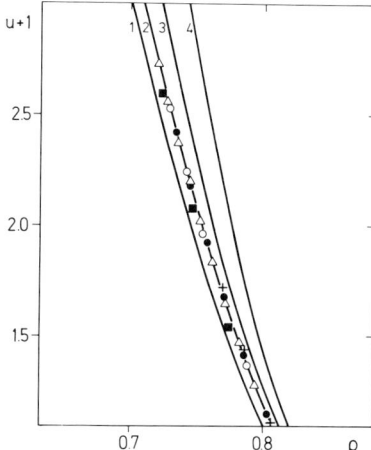

FIG. 6. Reduced ground-state RKRV potential curves of diatomic molecules (repulsive branch): (1) RPC of H_2; (2) RPC of P_2; (3) RPC of I_2; (4) (almost) coinciding RPCs of rare gases. (■) RPC of BeF; (△) RPC of CS; (○) RPC of SiN; (●) RPC of SiS; (+) RPC of GeO. (From Jenč, 1968.)

and of I_2 as sector lines. Since the differences between the RPCs of molecules possessing adjacent values of the atomic numbers are very small (cf. Fig. 4), we intentionally show only parts of the curve, to make the differences visible in this scale. Slight discrepancies, as in the case of CO (Fig. 5), may be due to uncertainties in the experimental values of the molecular constants, in particular D_e (the determination of which is, in the case of CO, rather complicated).

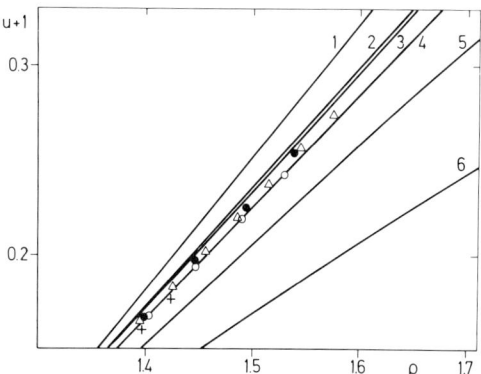

FIG. 7. Reduced ground-state RKRV potential curves of diatomic molecules (attractive branch): (1) RPC of H_2; (2) RPC of BeF; (3) RPC of N_2; (4) RPC of P_2; (5) RPC of I_2; (6) (almost) coinciding RPCs of rare gases. (△) RPC of CS; (●) RPC of SiN; (○) RPC of SiS; (+) RPC of GeO. (From Jenč, 1968.)

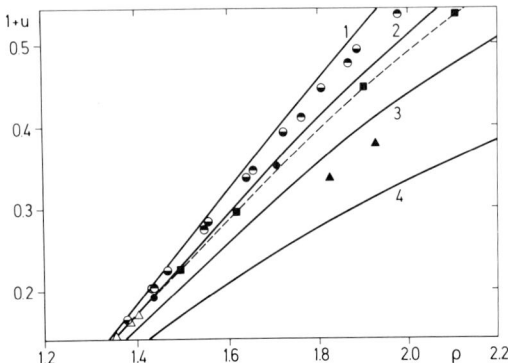

FIG. 8. Reduced ground-state RKRV potential curves for diatomic molecules (attractive branch): (1) RPC of H_2; (2) RPC of N_2; (3) RPC of I_2; (4) (almost) coinciding RPCs of rare gases; (---) RPC of P_2. (◔) RPC of GaH; (◐) RPC of AgH; (△) RPC of SiO; (●) RPC of SO; (■) RPC of S_2; (▲) RPC of Au_2. Observe the coincidence of the RPCs of P_2, S_2, and SiS (Fig. 5). Coxon (1971) has also proved coincidence with the RPC of Cl_2. (From Jenč, 1967a.)

In addition to the RPCs of the rare gases, Figs. 5–10 contain the reduced ground-state RKRV curves of the following systems: H_2, LiH, BeH^+, OH, HF, HCl, HBr, HI, GaH, AgH, BeF, C_2, CN, CO, N_2, O_2, SiN, SiO, CS, SO, SiS, P_2, S_2, GeO, ICl, IBr, I_2, and Au_2.

Figures 11–15 contain the reduced Hulburt–Hirschfelder ground-state potential curves of the following molecules: BH, CH, HF, HBr, HI, KH, GaH, RbH, InH, AuH, BF, CF, PbF, BiF, BO, SiO, PO, SO, TiO, GeO, IO, Li_2, CS, SiS, P_2, S_2, Cl_2, Br_2, and I_2. For the sake of orientation, the reduced

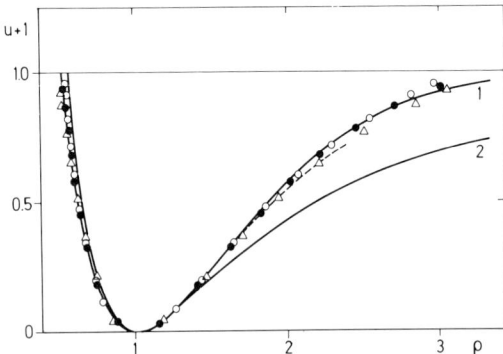

FIG. 9. Reduced ground-state RKRV potential curves of diatomic hydrides: (1) RPC of H_2; (2) RPC of I_2; (---) RPC of AgH. (●) RPC of HCl; (○) RPC of HBr; (△) RPC of HI. The RPCs of halogene hydrides are extrapolated above $(u + 1) = 0.35$. The extrapolation evidently is not accurate enough to reproduce the bent of the curve in the critical zone of large curvature. (From Jenč, 1969a.)

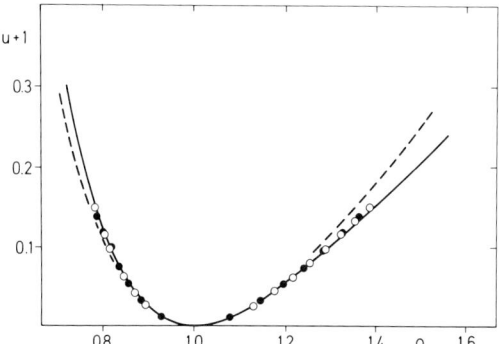

FIG. 10. Reduced ground-state RKRV potential curves for interhalogens: (---) RPC of H_2; (——) RPC of I_2. (○) RPC of IBr; (●) RPC of ICl. (From Jenč, 1965a.)

RKRV curves of H_2 and the rare gases are also drawn as solid curves in these figures. We stress again that the use of the Hulburt–Hirschfelder potential function only serves for a very *rough* test, which, however, qualitatively gives full support to the results obtained for the RKRV potentials. The short vertical lines in Figs. 11 and 12 indicate the discrepancies between the reduced RKRV and the reduced Hulburt–Hirschfelder potentials for some molecules for which reliable RKRV curves may be constructed in a sufficiently wide range of internuclear distance.

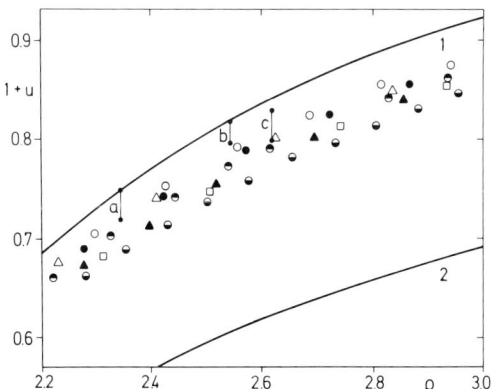

FIG. 11. Reduced ground-state Hulburt–Hirschfelder potential curves of hydrides: (○) BH; (●) CH; (△) KH; (▲) GaH; (◐) RbH; (□) InH; (⊖) AuH. (1) RKRV RPC of H_2; (2) RKRV RPC of I_2. The short vertical lines mark the typical difference between the reduced Hulburt–Hirschfelder and RKRV potential curves for some hydrides: (a) for OH; (b) for H_2; (c) for HF (in all of these cases the RKRV RPC lies above the Hulburt–Hirschfelder RPC). (From Jenč, 1967b.)

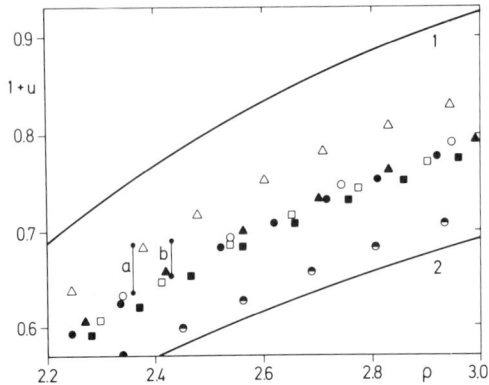

FIG. 12. Reduced ground-state Hulburt–Hirschfelder potential curves of oxides: (△) BO; (○) SiO; (▲) PO; (□) SO; (●) TiO; (■) GeO; (⊖) IO. (1) RKRV RPC of H_2; (2) RKRV RPC of I_2. The short vertical lined show the differences between the reduced Hulburt–Hirschfelder and RKRV potential curves (in these cases, the RKRV RPC lies above the Hulburt–Hirschfelder RPC): (a) for O_2; (b) for NO. (From Jenč, 1967b.)

The regularities reported should, of course, be further verified in calculating the RKR curves and constructing the RPCs for as many diatomic molecules as possible. A lot of new experimental data has appeared which may be considered sufficiently accurate and reliable to permit such a check. In effect, a large-scale program of such computations would seem worthwhile in systematical research on the RPC method. Of course, these calculations [testing also a possible improvement of the definition of the RPC (F. Jenč, unpublished results)] certainly will take some time. It may give a

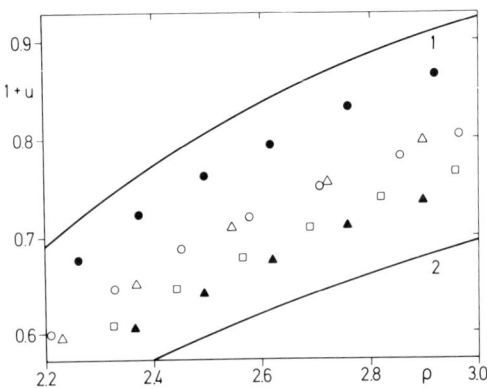

FIG. 13. Reduced ground-state Hulburt–Hirschfelder potential curves of fluorides: (●) HF; (○) BF; (△) CF; (▲) PbF; (□) BiF. (1) RKRV RPC of H_2; (2) RKRV RPC of I_2. (From Jenč, 1967b.)

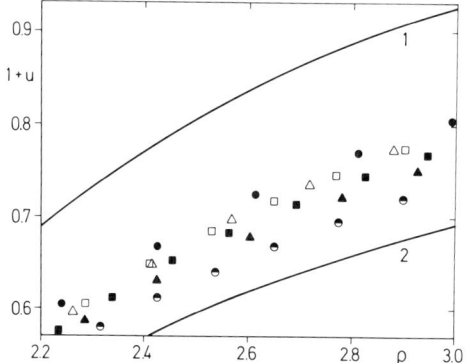

FIG. 14. Reduced ground-state Hulburt–Hirschfelder potential curves: (●) CS; (□) SO; (△) SiS; (■) P_2; (▲) S_2; (◐) Br_2. (1) RKRV RPC of H_2; (2) RKRV RPC of I_2. (From Jenč, 1967b.)

still deeper insight into the significance of the reduced potentials. As yet it is not clear if the bi-alkali molecules constitute a certain exception to the general RPC scheme, coinciding with the H_2 RPC (Jenč, 1965a). This problem may be resolved only when absolutely reliable molecular constants and analysis of the spectra for these molecules are available (one should also analyze if the methods employed for the evaluation of the dissociation energies of these molecules are unchallengeable). It is often the insufficiency of existing rotational data which makes the construction of a reliable RKRV potential difficult, as was also the case for the K_2 molecule (Jenč, 1965a).

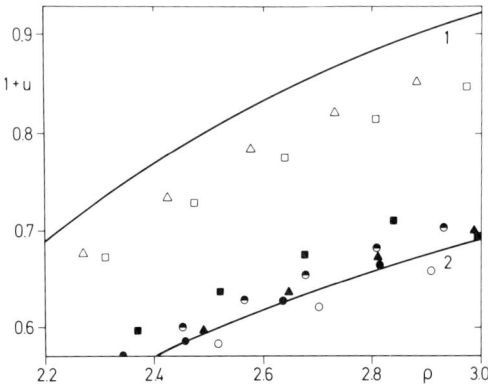

FIG. 15. Reduced ground-state Hulburt–Hirschfelder potential curves of halogen compounds: (△) HBr; (□) HI; (■) Cl_2; (◐) IO; (▲) ClI; (●) BrI; (○) I_2. (1) RKRV RPC of H_2; (2) RKRV RPC of I_2. (From Jenč, 1967b.)

C. Sensitivity of RPC Geometry to Changes in Values of Molecular Constants

The reduces quantities p and u contain the spectroscopic constants r_e, D_e, and k_e. A numerical analysis shows (Jenč, 1964a) that the geometry of the RPC is highly sensitive to a change in the value of r_e and rather sensitive to a change in k_e. Both constants may be, in most cases, determined to a sufficient degree of accuracy from the measured spectra. On the other hand, the sensitivity of the RPC to a change in the value of D_e is much smaller. It is well known that experimental values of this constant often contain small errors (cf. Figs. 16–19). It is worthwhile to emphasize two points:

(1) The sensitivity of the RPC to a change in the values of r_e and k_e makes it improbable that the coincidence of the RPCs could be fortuitous.

(2) A relative insensitivity to small errors (up to 1–2%, say) in the value of D_e is an almost necessary condition for a practically applicable definition of the reduced potential owing to almost unavoidable uncertainties in the experimental value of D_e.

D. Reduced Theoretical Potential Curves

In parallell, one also examined reduced theoretical (*ab initio*-calculated) potential curves of diatomics (Born–Oppenheimer approximation). Calculations of the author and also of others were employed. This work should be

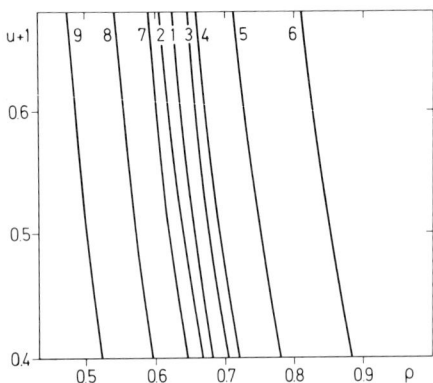

Fig. 16. Effect of a change in the values of the molecular constants (Δx in % of the experimental value). Repulsive branch of the RKRV RPC of O_2 as an example: (1) RKRV RPC of O_2; (2) $\Delta k_e = +15\%$; (3) $\Delta k_e = -15\%$; (4) $\Delta r_e = -2\%$; (5) $\Delta r_e = -5\%$; (6) $\Delta r_e = -10\%$; (7) $\Delta r_e = +2\%$; (8) $\Delta r_e = +5\%$; (9) $\Delta r_e = +10\%$. (From Jenč, 1964a.)

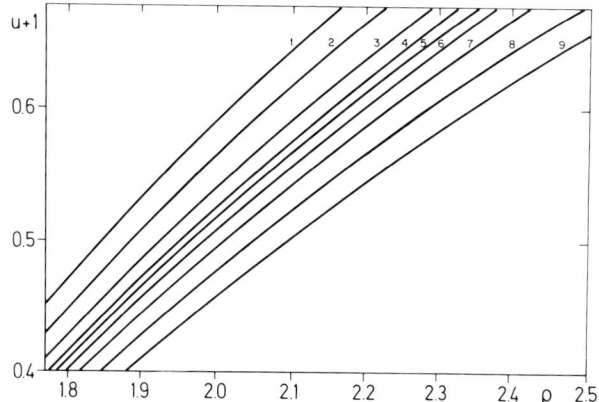

FIG. 17. Effect of a change in the value of k_e (Δk_e in % of the experimental value). Attractive branch (cf. Fig. 16): (1) $\Delta k_e = -15\%$; (2) -10% (3) -5% (4) -2%; (5) 0%; (6) $+2\%$; (7) $+5\%$; (8) $+10\%$; (9) $+15\%$. (From Jenč, 1964a.)

continued on a broader basis when more (usable) *ab initio* calculations become available.

The calculations of the author encompassed light hydride molecules (first-row hydrides). A variational self-consistant field (SCF) and configuration interaction (CI) method was employed using a basis of molecular orbitals calculated with the aid of the molecular orbital, self-consistent field, linear combination of atomic orbitals (MO–SCF–LCAO) method with Slater-type atomic orbitals (Jenč, 1963b). The results obtained for the RPCs of the hydrides were later supported by similar results for the RPCs of other molecules (see Table I).

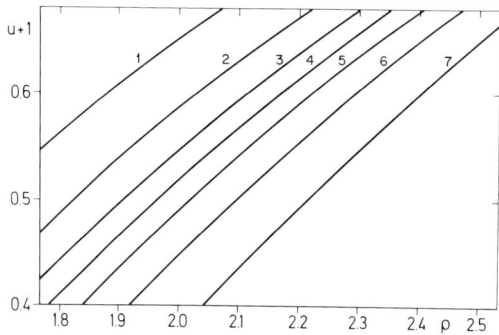

FIG. 18. Effect of change in the value of r_e (Δr_e in % of the experimental value). RKRV RPC of O_2: (1) $\Delta r_e = +10\%$; (2) $+5\%$; (3) $+2\%$; (4) 0%; (5) -2%; (6) -5%; (7) -10%. (From Jenč, 1964a.)

TABLE I

INACCURATE "ab Initio" CALCULATIONS OF POTENTIAL CURVES

Molecule	Method	Reference	D_e(theor.)/D_e(expt.)
H_2	SCF	Fraga and Ransil (1961)	0.75
H_2^+	SCF	Fraga and Ransil (1961)	0.90
H_2	CI	Fraga and Ransil (1961)	0.90
H_2	CI	McLean et al. (1960)	0.95
He_2	SCF	Ransil (1963)	—
HeH^+	CI	Anex (1963)	—
BeH^+	SCF	Jenč (1963b)	0.49
BeH^+	CI	Jenč (1963b)	0.65
CH^+	SCF	Jenč (1963b)	0.47
CH^+	CI	Jenč (1963b)	0.76
HF	SCF	Karo and Allen (1959)	0.33
HF	SCF	Clementi (1962)	0.73
CO	SCF	Lefebvre-Brion et al. (1961)	0.42
CO	2-term CI	Lefebvre-Brion et al. (1961)	0.47

The exact purpose was to show that even *rather inaccurate* calculations may be used in the RPC method for a construction of reliable potentials; i.e., even if a relatively low order of approximation[4] is achieved in calculating the molecular constants, still a coincidence, *in reduced form,* is obtained of the ab initio-calculated and the RKRV potential curves. This result is of great importance for the application of the method to heavy molecules, where a high degree of approximation still *cannot* be expected even in forthcoming calculations.

The following interesting property of the theoretical (*ab initio*-calculated) RPCs then appears if the CI method is employed (Fig. 20): Although rather *inaccurate* approximations to the true values of the molecular constants r_e, k_e, and D_e could be obtained in the calculations—so that the calculated potential curve represents a rather poor approximation of the RKRV curve, too—the *reduced* theoretical curve approximates the reduced RKRV curve to a high degree of accuracy. The important meaning of this fact is the following: The theoretical method may satisfactorily reproduce the essential

---

[4] More accurate calculations could, of course, be performed with methods employed at present. However, the *less accurate* calculations are exactly what one has to test. The reduced ground-state potential curve of H_2 calculated by the CI (MO–SCF) method, where a 94% and a 90% approximation to the experimental value of D_e were obtained, gives, of course, an excellent coincidence with the reduced RKRV curve (cf. Figs. 20 and 21).

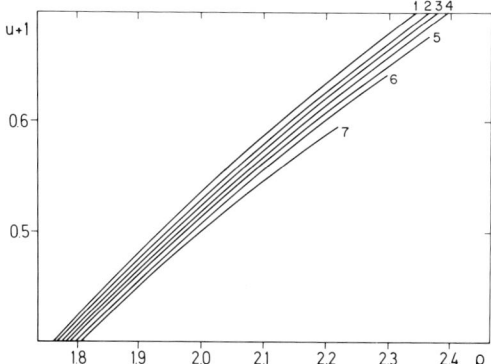

FIG. 19. Effect of a change in the value of D_e (ΔD_e in % of the experimental value). RKRV RPC of O_2: (1) $\Delta D_e = -20\%$; (2) -10%; (3) -5%; (4) 0%; (5) $+5\%$; (6) $+10\%$; (7) $+20\%$. (From Jenč, 1964a.)

structure of the vibrational problem, even though it cannot yield a good approximation to the numerical values of physical quantities (this would, of course, not be possible should the theoretical curve be "parallel" to the RKRV curve). MO–LCAO Slater-type one-electron orbitals (in particular, with optimized exponential parameters) give better results then virtual SCF orbitals or Gauss functions.

On the other hand, the SCF method alone yields a bad approximation also for the reduced potential curve, which is not surprising because of the low degree of approximation in this simple method. It is well known that the SCF method leads to an incorrect asymptotic behavior of the potential curve in most cases,[5] which may be directly seen in the figures. The SCF method cannot correctly describe the *electronic correlation,* therefore it is quite efficient only for rare gases. It is perhaps interesting to note that the reduced SCF curves of first-row hydrides (calculated in a comparable scheme and degree of approximation) practically coincide (Jenč, 1963a, 1964e).

The reduced theoretical potential curve *always* lies on the left-hand side of the reduced RKRV curve and approaches the latter with increasing degree of accuracy of the method of calculation (extension of the basis of atomic orbitals, variation of nonlinear parameters, etc.). This effect may also be observed in the behavior of the reduced SCF curves.

---

[5] In his theoretical study on LiH, Karo (1959) has shown that the SCF-calculated dipole moment asymptotically approaches the value of the dipole moment of ($Li^+ + H^-$), although, in reality, neutral atoms appear as dissociation products of the ground state of LiH.

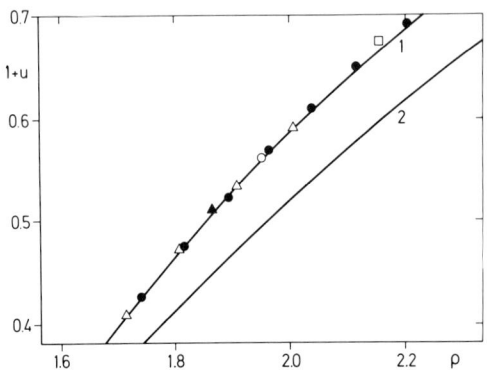

FIG. 20. Comparison of reduced ground-state RKRV potential curves and "ab initio" calculated theoretical potential curves (the number in parentheses means the approximation obtained in the calculation for the value of D_e in % of the experimental value): (1) RKRV RPC of H_2; (2) RKRV RPC of N_2. ($\triangle$) RKRV RPC of BeH^+; ($\bullet$) CI–MO-calculated RPC of H_2 (95%); ($\blacktriangle$) CI–MO-calculated RPC of BeH^+ (64.5%); ($\bigcirc$) CI–MO-calculated RPC of CH^+ (75.5%); ($\square$) MO–SCF (variation of nonlinear parameters) calculated RPC of HF (73%). (From Jenč, 1967a.)

Figures 20–23 illustrate this behavior of reduced *ab initio*-calculated theoretical potential curves on a few examples. For comparison, the corresponding reduced RKRV potential curves are drawn as solid curves in the figures.

Table I contains references to articles the calculations of which were used for the construction of reduced theoretical potential curves and indicates the

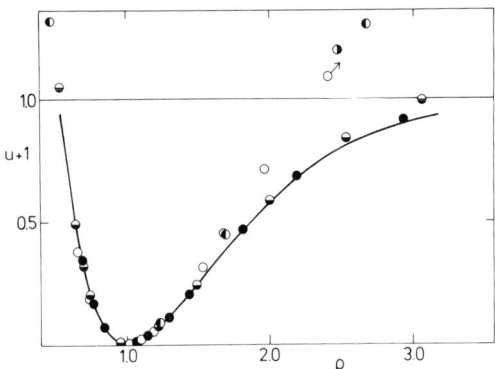

FIG. 21. "*Ab initio*"-calculated ground-state RPCs of light diatomic molecules: (———) RKRV RC of H_2. ($\bigcirc$) MO–SCF-calculated RPC of H_2 (75%); ($\bullet$) CI–MO-calculated RPC of H_2 (90%); ($\ominus$) = MO–SCF-calculated RPC of H_2^+ (90%); ($\obullet$) = MO–SCF-calculated RPC of BeH^+ (49%); ($\circledcirc$) = MO–SCF-calculated RPC of CH^+(47%). (From Jenč, 1964e.)

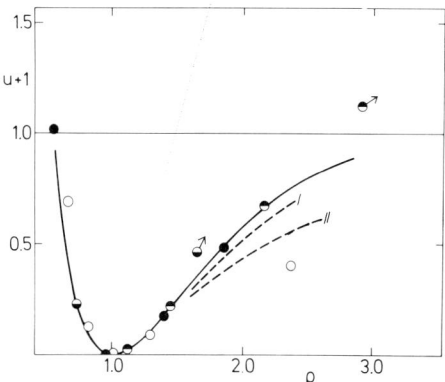

FIG. 22. *"Ab initio"*-calculated ground-state RPCs of light diatomic molecules: (———) RPC of H_2; (I) RPC of N_2; (II) RPC of I_2. (●) CI(MO–SCF)-calculated RPC of HeH^+; (○) MO–SCF-calculated RPC of He—He (variation of exponential parameters); (◐) MO–SCF-calculated RPC of HF (33%); (◑) MO–SCF-calculated (variation of exponential parameters-calculated) RPC of HF (73%). (From Jenč, 1964e.)

degree of approximation obtained for the molecular constant D_e in these calculations (in percent of the experimental value).

At first sight, some doubts on the general validity of this result may arise (as is the case for any method): If, say, the potential curve of a molecule is calculated using CI trial functions of increasing complexity (number of terms in the trial function), only small changes in the calculated values of the molecular constants D_e, r_e, and k_e may be achieved by increasing the number of terms, whereas more pronounced changes in the curve will result

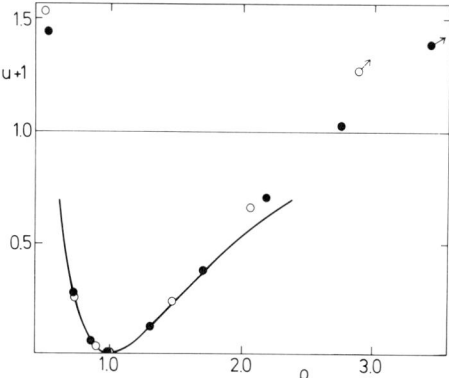

FIG. 23. Effect of the degree of approximation in the *"ab initio"* calculated ground state RPC of CO: (———) RKRV RPC of CO. (○) MO–SCF-calculated RPC of CO ($\approx 41.7\%$ in D_e); (●) MO–SCF-calculated RPC of CO ($\approx 46,7\%$ in D_e)—two-term CI. (From Jenč, 1964e.)

for large values of r (the method having a specific dependence on r). Does this result *in principle* contradict the RPC scheme? The answer is *no*. The analysis of the sensitivity of the shape of the RPC to changes in the values of the molecular constants (see Section III,C) has shown that very small changes in k_e and especially in r_e may lead to significant changes in the geometry of the RPC far from the minimum. In other words, a slight change of the geometry of the potential curve in the neighborhood of the minimum could bring about large changes in the global geometry of the RPC.

The results reported above should, of course, be supported by checking on forthcoming CI calculations for heavy molecules[6] since this property of the variational method could not be deduced from theoretical assumptions and, thus, is a characteristic of the variational method employed. At any rate, the results obtained in several cases cannot be fortuitous (we again remember the sensitivity of the RPC to a change in the values of the molecular constants).

Of course, the calculations must be carefully prepared and analyzed to guarantee reasonable results. Special attention must be payed to the calculation of the constant D_e. Molecular and electronic energies must be scrupulously calculated in the framework of the same approximation. An extrapolation procedure might be rather dubious here (anyway, it should be tested). The experimental value of D_e should be approximated at least to 75%. Preliminary tests seem to support the results obtained to date.

Let us once more summarize this important result: The theoretical *ab initio* calculations (variational method), if carried out to a "sufficient" degree of approximation (CI method), may satisfactorily approximate the potential curve in *reduced* form, although the approximation of the RKR curve (and the molecular constants) may be rather poor. This is also true for rare gases if the Buckingham or Lennard-Jones potential with suitable parameters is taken instead of the RKRV curve, which cannot be constructed in this case (Jenč, 1964f).

E. Mathematical Foundations of RPC Method

The regularities observed in ground-state reduced potential curves of diatomic molecules could easily follow from appropriate mathematical theorems on the lower bounds of the spectra of the corresponding Hamiltonians (Jenč, 1965c). The definition of the mathematical Schrödinger

---

[6] To be sure, for most heavy molecules the spectral data are not sufficient to permit the construction of the RKRV curve in a satisfactorily large range of internuclear distance. The exploitation of even *inaccurate ab initio* calculations is therefore of great interest.

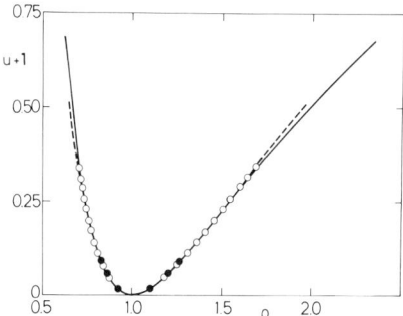

FIG. 24. Comparison of reduced RKRV potential curves for first excited states and ground states: (———) ground-state RKRV RPC of N_2; (---) ground-state RKRV RPC of CN. (●) RKRV RPC for $C_2(X^1\Sigma_g^+)$: (○) RKRV RPC for $C_2(a^3\Pi_u)$. (New ordering of states.) (From Jenč, 1964b.)

parameters Z_1, Z_2, and n. A proof of such theorems seems, however, beyond the scope of the present mathematical methods and would, at any rate, be very complicated since the parameter n determines the representation of the corresponding Hilbert space as a function space of $3n$ coordinates at a time. The relevant mapping between such different function spaces has a very complicated character (Plesner, 1965). Mathematical techniques for the solution of such problems have in fact not yet been developed. Nevertheless, the regularities observed in the RPC scheme indicate at least the possibility of the validity of such theorems and could motivate a search for them. Irrespective of mathematical foundations, the most striking merit of the RPC method is that it seems, indeed, to work — rather generally and accurately. It is expected that the results could still be improved in using new, more accurate experimental data.

F. Excited States

The general validity of the regularities in the behavior of the RPCs observed in ground states of diatomic molecules is, of course, only possible since nonnegligible perturbations are exceptional in a ground state. In excited states, however, perturbations are more probable, hence it is a priori clear that one cannot expect such rules to hold strictly for excited states. Nevertheless, an extended study of a large number of molecules and excited states (Jenč, 1964c,d) leads to the following conclusions visualized in Figs. 24–27:

(1) The effect of excitation is, in principle, similar to the effect of growing weight (growing values of atomic numbers) of the molecules: The "reduced attractive force," for $r > r_e$, decreases whereas the "reduced repulsive force," for $r < r_e$, increases with growing degree of excitation. All the RPCs examined to date lie, however, in the admissible RPC region in the $\rho-u$ diagram (between the RPCs of H_2 and He_2).

(2) The lowest-excited-state RPCs exhibit, in general, only small deviations from the ground-state potential curve in reduced form and, in good approximation, coincide for molecules with adjacent values of the atomic numbers of the molecules.

(3) In those cases where significant perturbations occur, one observes large deviations from the familiar RPC scheme. A salient example is a case where a crossing of potential curves of excited states belonging to the same symmetry could occur; the RKRV method is here *not* applicable.

(4) The excited states may be classified as follows: (a) the RPC of the excited state coincides approximately with the RPC of the ground state (a clear majority of the lower excited states belongs to this group); (b) nonnegligible deviations appear only in the region of high vibrational quantum numbers, which may correspond to a perturbation in this region; (c) the whole system seems perturbed and a global deviation of the RPC of the excited state from the ground-state RPC appears (cf. also Section IV,D on experimental errors, perturbations, and the questionable validity of certain formulas).

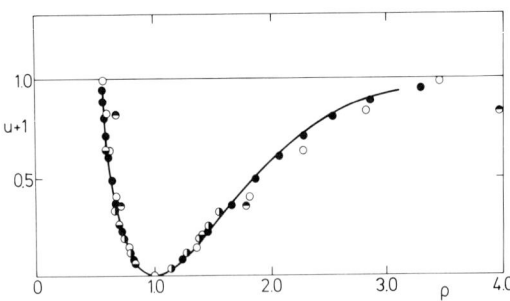

FIG. 25. Low-excited-state reduced RKRV potential curves of hydrides: (———) (almost) coinciding ground-state RKRV RPC of H_2, BeH^+, and CH^+. (●) $OH(A^2\Sigma^+)$; (○) $CH^+(A^1\Pi)$; (◐) $BeH^+(A^1\Sigma^+)$; (⊖) $LiH(B^1\Pi)$; (⊝) *"ab initio"* CI–MO(SCF)-calculated RPC of $BeH^+(A^1\Sigma^+)$. (From Jenč, 1964d.)

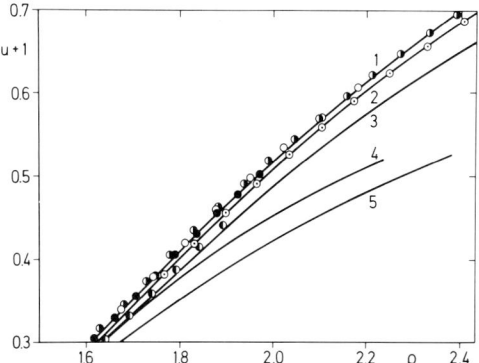

FIG. 26. Excited-state reduced RKRV potential curves of diatomic combinations of C, N, and O atoms: (1) ground-state RPC of CO; (2) CO($a^3\Sigma^+$); (3) CO($A^1\Pi$); (4) $C_2(A^3\Pi_g)$; (5) CN($B^2\Sigma^+$). (●) CN($A^2\Pi_{3/2}$ and $A^2\Pi_{1/2}$); (○) $N_2(A^3\Sigma_u^+$; (◐) NO($X^2\Pi_{3/2}$); (⊙) $N_2(B^3\Pi_g)$; (◐) $N_2(a^1\Pi_g)$. (From Jenč, 1964c.)

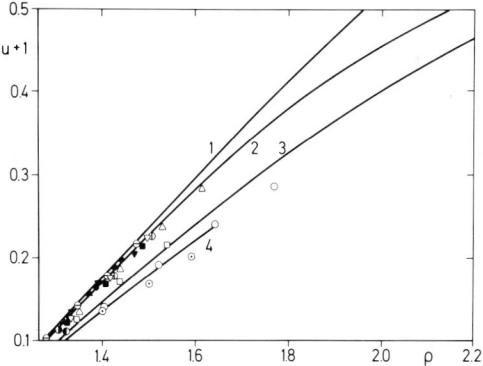

FIG. 27. Excited-state reduced RKRV potential curves for diatomic combinations of C, N, and O atoms: (1) the mean ground-state RKRV RPC for diatomic combinations of C, N, and O atoms; (2) C_2 ($A^3\Pi_g$); (3) C_2 ($e^1\Sigma_g^+$); (4) C_2 ($B^3\Pi_g$). (○) C_2 ($e^1\Sigma_g^+$); (⊙) C_2 ($e^1\Sigma_g^+$); (□) N_2 ($C^3\Pi_u$); (■)O_2 ($b^1\Sigma_g^+$); (▼) NO($A^2\Sigma^+$); (▽) NO($D^2\Sigma^+$); (△⌴) NO($B^2\Delta$); (▲) NO($E^2\Sigma^+$); (●) C_2 ($b^1\Pi_u$); (◐) C_2 ($d^1\Sigma_u^+$); (◐) $C_2(y^3\Sigma_g^-)$; (Φ) CN($J^2\Delta_1$); (⊖) CO($a^3\Pi$). The experimental value of D_e for $C_2(e^1\Sigma_g^+)$ being uncertain, alternatively proposed experimental values were employed: (3) D_e = 2.06 eV; (○) D_e = 3.48 eV; (⊙) D_e = 4.90 eV. [*Old* numbering of states of C_2 as used in the original work (Jenč, 1964c).]

IV. Applications of RPC Method

A. Construction of Internuclear Potentials of Diatomic Molecules in a Sufficiently Wide Range of Internuclear Distance

(1) Because the RPCs of diatomic molecules the atomic numbers of which differ only slightly practically coincide, and the mapping $U \leftrightarrow u$, $r \leftrightarrow \rho$ is one to one in the region of interest, it is possible to calculate the internuclear potential of a molecule A from the RKR potential of another molecule B if the values of the constants r_e, k_e, and D_e of A are known.

The molecule Te_2 is an interesting case where the potential has been calculated using this method (Jenč, 1969b). The internuclear potential for the ground state of Te_2 could be calculated almost up to the dissociation limit from the RKR potential of the ground state of the I_2 molecule, since an exceptionally large number of vibrational levels of I_2 could be determined from the UV resonance spectrum (Verma, 1960). The accuracy of the Te_2 internuclear potential obtained in this way should be better[7] than to 2% in r and $(U + D_e)$. The calculated potential of Te_2 is shown in Table II.

(2) Since the *ab initio*-calculated theoretical potential curves and the RKR curves coincide in reduced form, one could construct the potential curve of a diatomic molecule for which reliable values of the molecular constants r_e, k_e, and D_e are known in calculating *ab initio* the theoretical potential to a "sufficiently" high degree of approximation (the calculated value of D_e should, at any rate, exceed 75% of the experimentally determined value). The CI method seems appropriate.

The internuclear potential may then be calculated from this (rather poor) theoretical approximation using Eqs. (2)–(4) and the experimental values of r_e, k_e, and D_e.

Variation of nonlinear parameters in the atomic orbitals improves the coincidence of the theoretical potential and the RKR curve. In particular, it considerably improves the convergence for large values of the internuclear distance so that the "asymptotic behavior" of the reduced potential fits the measured spectra better. Big computers and some computational tricks make possible sufficiently accurate calculations of theoretical potential curves of (not too) heavy molecules for this use of the RPC method.

The latter method is an alternative to point (1), which seems important since vibrational energy levels can be measured up to the dissociation limit

---

[7] The experimental results of Verma, which served as the basic data for this calculation, have recently been slightly corrected. The corresponding corrections to Table II, however, seem to lie below the limit of error of the RPC method.

TABLE II

GROUND-STATE POTENTIAL CURVE OF Te_2

r (Å)	U (eV)	r (Å)	U (eV)	r (Å)	U (eV)	r (Å)	U (eV)
2.1196	2.7171	2.1918	1.7115	2.4090	0.2068	3.1578	1.1351
2.1207	2.7105	2.2025	1.5765	2.4433	0.1155	3.2265	1.2879
2.1217	2.6961	2.2153	1.4350	2.5049	0.0232	3.2978	1.4350
2.1238	2.6703	2.2303	1.2879	2.6173	0.0232	3.3716	1.5765
2.1281	2.6234	2.2464	1.1351	2.6983	0.1155	3.4511	1.7115
2.1344	2.5667	2.2648	0.9771	2.7534	0.2068	3.5365	1.8400
2.1386	2.5098	2.2864	0.8141	2.7995	0.2970	3.6283	1.9614
2.1429	2.4423	2.2983	0.7309	2.8413	0.3860	3.7316	2.0752
2.1471	2.3647	2.3124	0.6463	2.8799	0.4739	3.8489	2.1807
2.1524	2.2773	2.3266	0.5607	2.9175	0.5607	3.9836	2.2773
2.1588	2.1807	2.3430	0.4739	2.9530	0.6463	4.1414	2.3647
2.1652	2.0752	2.3605	0.3860	2.9875	0.7309	4.3356	2.4423
2.1726	1.9614	2.3825	0.2970	3.0223	0.8141	4.5806	2.5098
2.1811	1.8400			3.0902	0.9771	4.9068	2.5667

only in exceptional cases, for heavy molecules. In most cases, the RKRV curve may be constructed only for $(U + D_e) \leq 0.25 D_e$. It could be shown (Jenč, 1969a) with the help of the RPC method that an extrapolation of the RKRV curve to higher vibrational levels leads to relatively large errors (even if extrapolations using, e.g., a Morse-type function are used). This fact is also illustrated by Fig. 9, where the RKRV curves of halogen hydrides for $(u + 1) > 0.35$ only were extrapolated. One may clearly see that the curves in the upper part unduly follow the course of the lower part and do not bend fast enough in the critical region.

The only alternative permitting the construction of the potential curve in a sufficiently wide range of internuclear distance might be the RPC method as described here.

It is clear that in calculating the potential curve of a molecule to a "sufficient" degree of approximation, one automatically obtains a good approximation to the potentials of all molecules possessing only slightly different atomic numbers, which might save unnecessary work. Thus *ab initio* calculations of potential curves which, so far, have been of rather academic interest (unsatisfactory approximation for heavy molecules) could become of *practical importance.*

This method may be generally applied to ground states. Excited states should be carefully analyzed and the results meticulously checked.

B. Estimation of Molecular Constants

The definition of the RPC [Eqs. (2)–(4)] contains the three molecular constants r_e, k_e, and D_e as parameters. In a case where the RKR curve can be constructed, the values of two of these constants are known to sufficient degree of accuracy, and the value of the third constant is unknown or uncertain, one may obtain an estimate of the value of the third constant by fitting this value so as to make the reduced RKR curve coincide with the reduced RKR curve or the reduced theoretical potential curve of a molecule possessing only slightly different values of atomic numbers. Usually, the values of r_e and k_e are known if the RKR curve may be constructed. Notwithstanding, there might exist exceptional cases where one of these constants could not be determined to a sufficient degree of accuracy. In most cases, the value of D_e (which is not needed in the RKR calculation) is not known to a sufficient degree of accuracy and the results obtained by different experimental methods often exhibit rather large discrepancies. One may then decide between the various values proposed by the experimentalists (assuming the analysis of the spectrum was correct!). Using the RPC method, the error in the estimate of D_e should be, at any rate, smaller than say 5% (the method being rather insensitive to the value of this constant). The RPC method would, however, permit a very accurate estimation of the other two constants (much less than 1% error). Eventually, one could make a simultaneous estimate of two constants taking into account the specific effect of a change in the value of a particular molecular constant on the geometry of the RPC (see Figs. 16–19).

Let us mention that, if (using the RPC method) the potential curve of a molecule may be constructed for sufficiently high values of $(U + D_e)$, one may obtain reliable estimates of the parameter r_0 ($r_0 < r_e$, $U(r_0) = 0$), which is of some interest, e.g., in mass spectroscopy.

C. Classification of "Empirical" Potential Functions

The goodness of various "empirical" potential functions may be tested with the help of the RPC method. Such a test has, for instance, been done by Jenč and Plíva (1963). It appears that, although possibly suitable in an individual case, no "empirical" potential function proposed to date represents a reliable universal approximation to the adiabatic internuclear potential of all diatomic systems.

The RPC method has, in particular, been employed to test various "empirical" potential functions proposed for rare gases. It appears that, with correctly chosen values of the parameters, the reduced Lennard-Jones and Buckingham (1938) potentials of various rare gases coincide in reduced form, to a high degree of accuracy, and coincide also approximately with the *ab initio*-calculated potential curve[8] of He—He (Jenč, 1964f). The *ab initio* calculation for the He—He system (Ransil, 1963) was carried out using a rather sophisticated MO–SCF method. The true RPC of He—He should lie *somewhat lower,* according to the general rule observed in all calculations. However, the effect of the electronic correlation in He—He certainly may be assumed to be very small, hence the error in the MO–SCF approximation may be assumed to be relatively small so that the *ab initio* calculation gives the correct *qualitative* information on the location of the curve in the RPC scheme. At any rate, the RPCs of other rare gases are not allowed to lie *above* the calculated RPC of He—He in the attractive branch. A comparison shows that the Lennard-Jones (9:6)-potential yields worse results than the (12:6)-potential and the Buckingham (exp:6)-potential. For the (exp:6)-potential, the value of the parameter α should probably lie between

---

[8] In all these potential curves for the rare gases a small minimum appears. However, a calculation of the curvature at the minimum shows (in "empirical" as well as theoretical potentials) that the zero energy of the "oscillator" would be larger than the depth of the minimum. Hence, the "oscillator" (molecule) could not be stable.

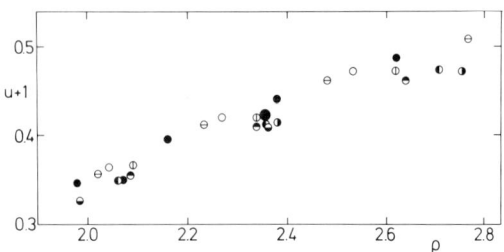

FIG. 28. "Ground-state" reduced Lennard-Jones and Buckingham type potential curves of rare gases compared with the *"ab initio"*-calculated ground-state RPC of He_2: (●) MO–SCF (variation of exponential parameters)-calculated RPC of He_2. (●) Xe_2 (exp:6, $\alpha = 12$); (⊖, ○) Kr_2 (exp:6, $\alpha = 12$); (◐) Xe_2 (exp:6, $\alpha = 13$); (◓) Ne_2 (exp:6, $\alpha = 14.5$); (◒) Xe_2 (exp:6, $\alpha = 14$). (◐) Kr_2 (exp:6, $\alpha = 14$); (◐) Ar_2 (exp:6, $\alpha = 14$). (From Jenč, 1964f.)

14 and 15 (cf. Figs. 28 and 29). The potential function proposed by Guggenheim and McGlashan (1960) on the basis of the work of Rice (1941) appears to be impossible (Fig. 30).

D. Detection of Errors in Experimental Values of Molecular Constants and in Analysis of Spectrum. Detection of Perturbations; Detection of Nonvalidity of Some Commonly Used Formulas

If significant deviations of a reduced RKR curve from the RPC scheme described above are observed, one may guess at several possible reasons:

(1) There may exist errors or inaccuracies in the RKR calculation, which may occur if one does not employ a sufficiently large number of

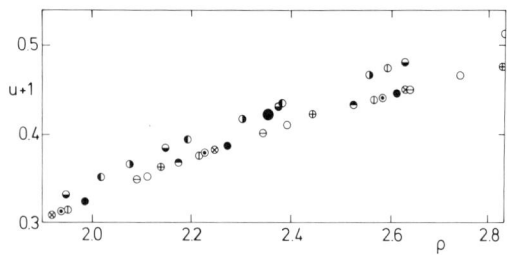

FIG. 29. "Ground-state" reduced Lennard-Jones and Buckingham type potential curves of rare gases as compared with the *"ab initio"*-calculated RPC of He_2: (●) MO–SCF (variation of exponential parameters)-calculated ground-state RPC of He_2. (⊖) Xe_2 (12:6); (●) Ar_2 (12:6); (○, ◐) Kr_2 (12:6); (⊕) Xe_2 (exp:6, $\alpha = 15$); (⊙, ◒) Kr_2 (exp:6, $\alpha = 15$); (⊗) Ar_2 (exp:6, $\alpha = 15$); (◓) Ar_2 (9:6); (◐) Xe_2 (9:6); (◐) Kr_2 (9:6). (From Jenč, 1964f.)

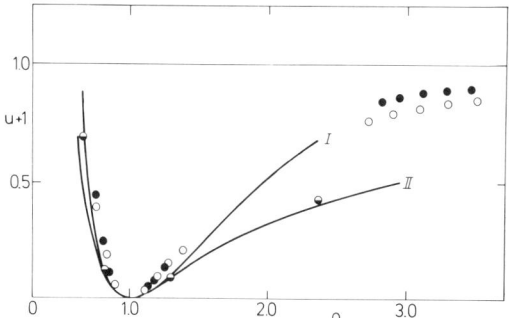

FIG. 30. Comparison of "ground-state" reduced potential curves of rare gases of Lennard-Jones and Buckingham type with those of Rice–Guggenheim–McGlashan and with the *"ab initio"*-calculated ground-state RPC of He_2: (◓) MO–SCF (variation of exponential parameters)-calculated RPC of He_2; (I) mean ground-state RKRV RPC of diatomic combinations of C, N, and O atoms; (II) mean Lennard-Jones (12:6) RPC of rare gases; (●, ○) Rice–Guggenheim–McGlashan RPC for Xe_2 and Ar_2, respectively. (From Jenč, 1964f.)

rotational levels in the RKR calculation or if inadmissible simplifications (approximations) in the energy formulas are made.

(2) There may exist some errors in the values of the molecular constants r_e, k_e, and D_e, or in the analysis of the spectrum.

(3) Perturbations from other states may be present. Hypotheses (1) and must be excluded if hypothesis (3) is being discussed.

The origin of the errors mentioned in hypothesis (3) may be of various types. For instance, such errors may occur if some weak spectral lines have been overlooked [which, e.g., was the cause of a slight error in the D_e value of H_2 obtained through extrapolation of highly accurate data; the existence of this error was indicated by the extremely accurate calculations of Kolos and Wolniewicz (1965)]. A typical example is the SO molecule. In this case, the true transitions to $v'' = 0$ and 1 were not observed in the emission spectrum and were registered only later in flash photolysis of SO_2 by Norrish and Oldershaw (1959; cf. also Powell and Lide, 1964). Therefore, one had to revise the analysis of the spectrum of SO, change the numeration of the vibrational levels, and make a correction in the values of D_e and k_e.

While the old, incorrect analysis of the spectrum of SO entailed some discrepancies in the RPC scheme, the RKRV curve calculated with the use of the corrected experimental data fits very well into the RPC pattern (Jenč, 1967a).

The last possibility mentioned in hypothesis (3) is, of course, more likely to exist in excited states.

A perturbation suspected to exist in the ground state of the Cl_2 molecule

(Jenč, 1967c), as suggested by Mulliken (1930, 1931, 1934, 1940; Schlapp, 1932), seems now improbable: As has already been mentioned, Coxon (1971) has shown that the seeming deviation from the RPC scheme was in fact due to an incorrect analysis of the spectrum and a correspondingly inaccurate RKR curve. Thus, the RPC method helped to *disclose an inadequate analysis of the spectrum* of Cl_2 (and the resulting incorrectness of the RKR potential).

As for the excited states, anomalous behavior was observed for the first excited states (A $^1\Sigma^+$) of LiH and BeH$^+$ (Jenč, 1966a). A similar (though essentially weaker) tendency seems to exist in the first excited states of HF and H_2 (Jenč, 1964d). For BeH$^+$ (A $^1\Sigma^+$), the reduced RKR curve lies outside the RPC region; for LiH (A $^1\Sigma^+$) it cannot be defined ($\rho_{ij} < 0$), which is due to the strikingly anomalous value of some spectroscopic constants, ω_e being extremely small and $\omega_e x_e$ being negative; an extremely low value of k_e then results if Eq. (6) is employed, which makes ρ_{ij} negative.

On the contrary, the reduced *ab initio* calculated potential curves (Born–Oppenheimer approximation) fitted well into the RPC scheme. This result led the present author to the hypothesis of nonnegligible perturbations and the *failure of the Born–Oppenheimer approximation* in the $A^1\Sigma^+$ state of LiH (Jenč, 1966a).

Following this cue, Docken and Hinze (1972a,b) attempted a theoretical verification of this hypothesis in their *ab initio* CI calculation (Born–Oppenheimer approximation) for the LiH molecule, also evaluating the theoretical values of the vibrational term differences $\Delta G_{v+1/2}$ and the corresponding constants ω_e, $\omega_e x_e$; they obtained qualitatively similar anomalous behavior as that obtained in the experiment. However, their conclusion that the Born–Oppenheimer approximation is satisfactorily valid in this $A^1\Sigma^+$ state seems to be discredited by recent calculations of Vidal and Stwaley (1982), which we comment upon in the following discussion. Indeed, perturbations by higher excited states are also surmized by other authors for H_2 and LiH (Kolos and Wolniewicz, 1966; Kolos, 1970, containing other references; Stwaley, 1973).

The hypothesis of the present author seems now to be confirmed by the results of Vidal and Stwaley (1982) who calculated the potential curve of the $A^1\Sigma^+$ state of LiH and the spectroscopic constants from spectral data by using their inverted perturbation approach (IPA) method based on the approach of Kosman and Hinze (1975), thus improving on the RKR construction, which does not seem sufficiently accurate in this anomalous case. They claim to prove a breakdown of the Born–Oppenheimer approximation in the $A^1\Sigma^+$ state of LiH (which also has an anomalous isotope shift). Strikingly enough, the latter authors, although explicitly referring to our paper (Jenč, 1966a) and qualifying a hypothesis of "nonadiabatic" perturbations as false, unfortunately somehow forgot to mention the *correct prediction* of the failure of the Born–Oppenheimer approximation in this state that is unambigously formulated *in the same paper on page 70* and it is absolutely clear from the whole context (it is also correctly quoted by Docken and Hinze, 1972a,b): that certainly is not quite an example of good taste.

The study of the first excited states quoted suggests (does not prove) that the RPC method can work correctly only if the Born–Oppenheimer approximation is valid, which was indeed assumed in the formulation of the method (Jenč 1963a, 1964a). That was also meant in the paper of LiH (Jenč, 1966a) where, of course, the term "adiabatic" approximation (and "nonadiabatic" perturbation) was parallelly employed corresponding, say, to the terminology of the theoretical course of Levich (1973)—rather common terminology in quantum chemistry 20 years ago. It is clear that only in this case a comparison of reduced theoretical potentials calculated in Born–Oppenheimer approximation with reduced RKR curves has a meaning.

If off-diagonal corrections for nuclear motion are large, the concept of the potential curve loses its proper meaning and sufficiently accurate *ab initio* calculations (including diagonal corrections) should show a clear deviation from the formally calculated RKR-type potentials. If only diagonal corrections are significant, the potential curve still has a proper meaning. The perturbation may, however, show up in an anomalous geometry of the potential curve, which may lead to nonvalidity of Dunham formulas. Indeed, spectral lines of the $A^1\Sigma^+$–$X^1\Sigma^+$ transition in LiH can *not* be interpolated by a simple Dunham expansion. The sequence of vibrational term differences $\Delta G_{v+1/2}$ has a pronounced maximum (Crawford and Jörgensen, 1935a,b; Docken and Hinze, 1972b).

It was conjectured by the author (Jenč, 1965b, 1966a,b) that (assuming an accurate analysis of the spectrum) the anomalies discussed could also be due to nonvalidity of the approximation (6) and Dunham formulas in general (Dunham, 1932), hence possibly also an inaccuracy in the RKR curves as far as they rely on Dunham's method (on this point see also Le Roy and Burns, 1968). Thus this problem would lead to a discussion of the fundamentals of theoretical spectroscopy.[9] In a case where the validity of the approximation (6) is doubted, the constant k_e must be determined in interpolating in a suitable way the RKR curve (or the monotonous electronic energy curve), including the point $U(r_e) = -D_e$, if reliable.

Whereas the relation of the force constants of the isoelectronic molecules BeH$^+$ and LiH using Eq. (6) is 2.57 in the ground state $X^1\Sigma^+$, it is as large as

---

[9] A method using the mathematical results of the inverse-scattering-problem solution has been proposed for a discussion of such problems by the author (Jenč, 1965b).

40.80 in the $A^1\Sigma^+$ state; i.e., the value of ω_e(LiH, $A^1\Sigma^+$) is surprisingly small, more than one would expect considering the problems with Dunham interpolation of the spectral lines. Furthermore, the theoretical value of k_e (LiH, $A^1\Sigma^+$) corresponding to Karo's (1959) CI calculation is more than four times larger than the experimental value of Eq. (6) — a result not encountered in any other case (the value of k_e being generally somewhat smaller than $\mu\omega^2$ in similar calculations). Therefore, a really salient deviation from Eq. (6) should be expected if this should be the essential reason of the anomaly. However, preliminary results of Brandt and Jenč (unpublished results) evaluating the potential curves of Vidal and Stwaley (1982) indicate that the striking anomaly of the $A^1\Sigma^+$ state of LiH and the nonexistence of its RPC is *not* due to a substantial failure of the approximation of Eq. (6) (certainly no salient errors of the order mentioned above have been found, though a small deviation seems to exist; in the case of well-behaved molecules or for empirical potential functions, a very good agreement was obtained).

This result is by no means self-evident in this absolutely anomalous case, since it seems clear from mathematical standpoint that Dunham's (1932) approximation method can not have a general validity, neither for well-behaved potentials: A somewhat deformed (continuously deformed and still analytical) spectrally equivalent analytical potential will have a significantly different value of $k_e = (d^2V/dr^2)_{r=r_e}$ than a nominal analytical potential of the equivalence class (Jenč, 1965b). Indeed, a failure of Dunham's method is, in principle, suggested in this case by the difficulties in applying Dunham's expansion for the analysis of the spectrum; this however, does not seem to lead to a substantial invalidation of Eq. (6). It is possible that Dunham's method in general is valid only if the Born–Oppenheimer approximation holds.

V. Comments on Some Misunderstandings about the RPC Method

(1) As has already been stressed, the RPC method permits a *direct visualization* of certain laws governing the internuclear potentials of diatomics. It does *not* aim at the construction of a new universal "empirical" potential function. Nor does the existence of such a function follow from the results reported above (since the RPCs of all diatomics do *not* coincide). The possibility of the existence of such a universal function is not clear.

A misunderstanding of this point is certainly due the remark of Schlier, who mentioned the RPC method in a review article (Schlier, 1969). Schlier —who *unfortunately* reviewed, as he confessed, only a small part of the articles on the RPC method then published—writes: "The present author sees no reason why a universal potential should exist." This author certainly does not intend to make a similar statement.

(2) Since in practice, the spectral lines cannot be measured up to the dissociation limit, the RPC method may be employed for the evaluation of the RKR curves (or any potentials constructed on the basis of a direct use of the spectra) only within the potential well. Therefore, the fact that, for heavy molecules, the present definition of the RPC is not suitable for very large values of r was not too important. As has been emphasized, this definition may easily be improved by introducing, e.g., a constant exponential parameter to make the RPC method usable for all values $r > r_e$, which might be interesting for the analysis of *the theoretical potential curves*. Of course, a direct comparison of reduced RKR potentials can never be made for large values of the internuclear distance, since they cannot be constructed in this region [cf. the criticism of Schlier (1969)].

(3) Another quite surprising misunderstanding shows up in the following rather amusing reasoning of Schlier:

> Mathematically one can, in fact, argue that the three parameters Z_1, Z_2, and n (total number of electrons), together with the prescription to find the lowest eigenvalue of the system, define the potential. But there is no reason that, practically, *continuity* should exist for the dependence of $V(r; Z_1, Z_2, N)$ on its parameters, which are discrete. The very existence of the periodic table should convince the chemist if not the physicist of this fact. (Schlier, 1969, pp. 195–196).

Let us emphasize that in no article of the present author on reduced potentials will such a statement or hypotheses of "continuity" be found. Therefore, it is not clear why the reviewer Schlier made this comment, which, indeed, seems curiously naive. It is clear that, in the atomic region, charges as well as energies or *any other molecular or atomic constants* are, of course, discrete (though not necessarily equidistant, as is the special case for the atomic charges). This fact, however, certainly does not mean that mathematical formulas containing these atomic or molecular parameters should be meaningless—otherwise any methods using, e.g., the atomic numbers as parameters (see, e.g., Hylleraas, 1930) as well as any "empirical" formulas using molecular constants would be meaningless. It appears, however, that certain regularities exist which may, at least approximately, be described by formulas containing such "discrete" parameters. It must be clear to everyone that such formulas are only suitable approximate interpolation formulas for the discrete points (a method commonly used in physics). Similarly, practical spectroscopy of diatomics is based on interpolation of the spectral lines with the aid of a Dunham expansion (in quantum numbers) — an approximate interpolation procedure (not based on *exact* theoretical formula) which may not function impeccably in a particular case. As far as the RPC method is concerned, we only claim to have found some *discrete* regularities in the RPC scheme.

(4) In conclusion, let us say a few critical words on methods of molecular physics in general. It is absolutely clear that the results reported here should be taken with a sound amount of sceptical criticism and that some exceptions to the rules observed may exist. This conclusion holds, of course, for *any other method* as well.

Let us take, by way of example, the "method of effective core potentials." This interesting method has been used with much success for calculations on heavy molecules where the performance of an *"ab initio"* calculation leads to exceedingly large computational complications even with the use of modern computers. In this method the potentials calculated for the atoms are adapted and used for the molecular calculation. However effective this method might prove to be in special examples, there is no *strict theoretical criterion* that could guarantee its validity. Making a calculation for a new molecule simply means having good faith in the validity of the method. There may always exist cases where the approximation is not quite good. Nevertheless, one certainly is willing to employ this method if it proves efficient in a series of examples.

The same holds, in fact, for common variational methods in quantum-chemical calculations. Although "in principle" the method is based on rigorous theory, there is no criterion on how the goodness of the approximation will depend on the number of terms used, say, in a CI method. Anyone involved in such concrete calculations knows that, in practice, the choice of the algorithm of the trial function is, to a high degree, also a matter of experience. Calculating electronic energies and molecular constants for a molecule, for which the experimental data are not known, always presupposes *the faith* that the degree of approximation will be as good as for similar molecules. There may always exist exceptions where the convergence rate may suddenly be different.

Thus, if one is honest, one should recognize this point and handle such methods as useful tools to be used with due reservation. The RPC must be and has been verified on a series of examples, as have the other methods. For none of these methods does absolute certainty exist.

The same holds for the calculations of potential curves using the RKRV of the Vidal method. It is hardly possible to obtain reliable estimates of the deviations of potential curves calculated by these methods from the "true" adiabatic potential. (Reproducing all vibrational *and* rotational energy levels is, of course, an acceptable criterion of correctness.)

V. Conclusions

The knowledge of the adiabatic internuclear potential of diatomic molecules and ions is very important in many physical problems (in a sense, it is also of interest for the research on polyatomic molecules). The RPC method permits a rather accurate *determination of these potentials.* This method also has a number of other interesting applications in molecular physics and spectroscopy. A brief account of the results and prospects of the RPC method have been presented in this article.

A determination of accurate potential curves from *ab initio* calculations for heavy molecules represents, of course, considerably extended computer calculations, which decidedly lies beyond the present scope of the author's capabilities. The concrete examples of all other applications of the RPC method have been given in the references cited in the corresponding sections of this article.[10] It should be clear that the essential assumption for such applications is a supply of sufficiently accurate and reliable experimental data needed for such calculations. As yet, the lack of such data made certain applications feasible only in special cases. Since a much new and accurate experimental (spectroscopic) data have appeared in recent years and computer techniques have evolved, considerably, there seem to exist promising possibilities for applications of the RPC method which as yet, have not be realized. We also hope that other authors might be interested in such problems. Therefore, a publication of this review of the results obtained by the RPC method in a large series of articles, hitherto scattered in various journals, seemed to be of actual interest.

ACKNOWLEDGMENTS

The results depicted in some figures have been essentially published in previous articles of the author scattered in various journals: Figs. 2, 10, 16–19, and 21–30 in the *Collections of Czechoslovak Chemical Communications;* Figs. 5, 9, and 11–15 in the *Journal of Chemical Physics;* Figs. 6 and 7 in *Spectrochimica Acta;* and Figs. 8 and 20 in the *Journal of Molecular Spectroscopy.* Table II appeared in the *Journal of Molecular Structure.* The author gratefully acknowledges the permission of the publishers to use these materials in this article.

---

[10] *Detection of inadequate analysis of the spectrum* and estimation of dissociation energy might be the most current types of such applications.

REFERENCES

Anex, B. G. (1963). *J. Chem. Phys.* **38**, 1651.
Benesch W., Vanderslice, J. T., Tilford, S. G., and Wilkinson, P. G. (1965). *Astrophys. J.* **142**, 1227.
Born, M., and Oppenheimer, R. (1927). *Ann. Phys.* **84**, 457.
Buckingham, R. A. (1938). *Proc. R. Soc. London, Ser. A* **168**, 264.
Clementi, E. (1962). *J. Chem. Phys.* **36**, 33.
Coxon, J. A. (1971). *J. Quant. Spectrosc. Radiat. Transfer* **11**, 443.
Crafword, F. H., and Jörgensen, T., Jr. (1935a). *Phys. Rev.* **47**, 743.
Crawford, F. H., and Jörgensen, T., Jr. (1935b). *Phys. Rev.* **47**, 932.
Docken, K. K., and Hinze, J. (1972a). *J. Chem. Phys.* **57**, 4928.
Docken, K. K., and Hinze, J. (1972b). *J. Chem. Phys.* **57**, 4936.
Douglas, A. E., and Hoy, L. (1975). *Can. J. Phys.* **53**, 1965.
Dunham, J. L. (1932). *Phys. Rev.* **41**, 713.
Fraga, S., and Ransil, B. J. (1961). *J. Chem. Phys.* **35**, 1967.
Frost, A. A., and Musulin, B. (1954). *J. Am. Chem. Soc.* **76**, 2045.
Guggenheim, E. A., and McGlashan, M. L. (1960). *Proc. R. Soc. London, Ser. A* **255**, 456.
Hulburt, H. M., and Hirschfelder, J. O. (1941). *J. Chem. Phys.* **9**, 61.
Hylleraass, E. A. (1930). *Z. Phys.* **65**, 209.
Jenč, F. (1963a). *Collect. Czech. Chem. Commun.* **28**, 2052.
Jenč, F. (1963b). *Collect. Czech. Chem. Commun.* **28**, 2064.
Jenč, F. (1964a). *Collect. Czech. Chem. Commun.* **29**, 1507.
Jenč, F. (1964b). *Collect. Czech. Chem. Commun.* **29**, 1521.
Jenč, F. (1964c). *Collect. Czech. Chem. Commun.* **29**, 1745.
Jenč, F. (1964d). *Collect. Czech. Chem. Commun.* **29**, 2579.
Jenč, F. (1964e). *Collect. Czech. Chem. Commun.* **29**, 2869.
Jenč, F. (1964f). *Collect. Czech. Chem. Commun.* **29**, 2881.
Jenč, F. (1965a). *Collect. Czech. Chem. Commun.* **30**, 3589.
Jenč, F. (1965b). *J. Mol. Spectrosc.* **17**, 188.
Jenč, F. (1965c). *Collect. Czech. Chem. Commun.* **30**, 3772.
Jenč, F. (1966a). *J. Mol. Spectrosc.* **19**, 63.
Jenč, F. (1966b). *J. Mol. Spectrosc.* **21**, 231.
Jenč, F. (1967a). *J. Mol. Spectrosc.* **24**, 284.
Jenč, F. (1967b). *J. Chem. Phys.* **47**, 127.
Jenč, F. (1967c). *J. Chem. Phys.* **47**, 4910.
Jenč, F. (1968). *Spectrochim. Acta, Part A* **24A**, 259.
Jenč, F. (1969a). *J. Chem. Phys.* **50**, 2766.
Jenč, F. (1969b). *J. Mol. Struct.* **4**, 157.
Jenč, F. (1983). To be published.
Jenč, F. and Plíva, J. (1963). *Collect. Czech. Chem. Commun.* **28**, 1449. Ph.D Thesis of F. Jenč.
Karo, A. M. (1959). *J. Chem. Phys.* **30**, 1241.
Karo, A. M., and Allen, L. (1959). *J. Chem. Phys.* **31**, 968.
Klein, O. (1932). *Z. Phys.* **76**, 226.
Kołos, W. (1970). *Adv. Quant. Chem.* **5**, 99.
Kołos, W., and Wolniewicz, L. (1965). *J. Chem. Phys.* **45**, 2429.
Kołos, W., and Wolniewicz, L. (1966). *J. Chem. Phys.* **48**, 3672.
Kołos, W., and Wolniewicz, L. (1975). *Can. J. Phys.* **53**, 2189.
Kosman, W. M., and Hinze, J. (1975). *J. Mol. Spectrosc.* **56**, 93.

Lefebvre-Brion, H., Moser, C., and Nesbet R. K. (1961). *J. Chem. Phys.* **34**, 150.
Le Roy, R. L., and Burns, G. (1968). *J. Mol. Spectrosc.* **25**, 77.
Levich, B. G. (1973). "Theoretical Physics, An Advanced Text," Vol. 3. North-Holland, Amsterdam.
McLean, A. D., Weiss, A., and Yoshimine, M. (1960). *Rev. Mod. Phys.* **32**, 205.
Mulliken, R. S. (1930). *Phys. Rev.* **36**, 699.
Mulliken, R. S. (1931). *Phys. Rev.* **37**, 1412.
Mulliken, R. S. (1934). *Phys. Rev.* **46**, 549.
Mulliken, R. S. (1940). *Phys. Rev.* **57**, 500.
Norrish, R. W. G., and Oldershaw, G. A. (1959). *Proc. R. Soc. London, Ser. A* **249**, 498.
Plessner, A. I. (1965). "Spektralnaja těorija linějnych operatorov." Nauka, Moscow.
Powell, F. X., and Lide, D. R., Jr. (1964). *J. Chem. Phys.* **41**, 1413.
Puppi, G. (1946). *Nuovo Cimento* **3**, 338.
Ransil, B. J. (1963). *J. Chem. Phys.* **34**, 2109.
Rees, A. L. (1947a). *Proc. Phys. Soc., London, Sect. A* **49**, 998.
Rees, A. L. (1947b). *Proc. Phys. Soc., London, Sect. A* **49**, 1008.
Rice, O. K. (1941). *J. Am. Chem. Soc.* **63**, 3.
Rydberg, R. (1931). *Z. Phys.* **73**, 376.
Rydberg, R. (1933). *Z. Phys.* **80**, 514.
Schlapp, R. (1932). *Phys. Rev.* **39**, 806.
Schlier, C. (1969). *Annu. Rev. Phys. Chem.* **20**, 191.
Stwaley, W. X. (1973). *J. Chem. Phys.* **58**, 536.
Vanderslice, J. T., Mason, E. A., Maisch, W. G., and Lippincott, R. E. (1959). *J. Mol. Spectrosc.* **3**, 17.
Varshni, Y. P. (1957). *Rev. Mod. Phys.* **29**, 664.
Verma, A. D. (1960). *J. Chem. Phys.* **32**, 738.
Vidal, C. R., and Scheingraber, H. (1977). *J. Mol. Spectrosc.* **65**, 46.
Vidal, C. R., and Stwaley, W. C. (1982). *J. Chem. Phys.* **77**, 883.
Zare, R. N. (1963). *Univ. Calif. Radiat. Labor. Rep.* **UCRL-10881**.
Zare, R. N. (1964). *J. Chem. Phys.* **40**, 1934.
Zare, R. N., and Cashion, J. H. (1963). *Univ. Calif. Radiat. Lab. Rep.* **UCRL-10925**.
Zare, R. N., Larsson, E. O., and Berg, R. A. (1965). *J. Mol. Spectrosc.* **15**, 117.
Zare, R. N., Schmeltekopf, A. L., Harrop, W. J., and Albritton, A. L. (1973). *J. Mol. Spectrosc.* **46**, 37.

THE VIBRATIONAL EXCITATION OF MOLECULES BY ELECTRON IMPACT

D. G. THOMPSON

Department of Applied Mathematics and Theoretical Physics
The Queen's University of Belfast
Belfast, Northern Ireland

I. Introduction	309
II. Theoretical Considerations	311
A. Molecular Eigenfunctions and Eigenenergies	311
B. The Scattering Formalism	315
III. Applications and Comparison with Experiment	323
A. H_2	324
B. N_2	326
C. CO	330
D. CO_2	332
E. The Hydrogen Halides	335
F. Various Molecules	338
References	340

I. Introduction

In recent years there has been considerable experimental and theoretical interest in the vibrational excitation of molecules by electron impact. Much of the early impetus was due to the work of Schulz (1962) for low-energy scattering by nitrogen molecules. He observed several vibrational transitions and noticed that, as a function of electron energy, each cross section showed a series of sharp peaks separated by a few tenths of an electron volt. It was immediately suggested that this structure was due to resonance effects. When the electron is close to the molecule the system vibrates in a negative ion state; the peak structure of the cross section then corresponds to the positions of the negative ion vibrational states. However, the vibrational excitation process cannot be a straightforward one since it was observed that the peaks in the cross sections did not appear in the same positions for each transition.

The model proposed for the excitation of a neutral molecule M in its

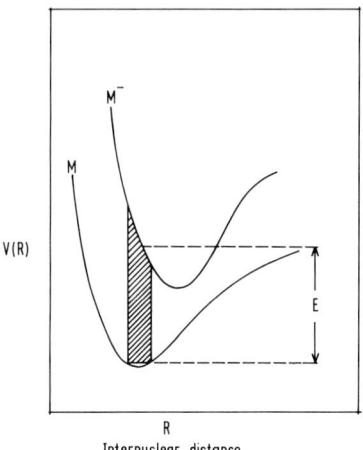

Fig. 1. Vibrational excitation via negative ion compound states.

ground state $v = 0$ was

$$e^- + M(v=0) \to M^- \to M(v>0) + e^-$$

This is illustrated for a diatomic molecule in Fig. 1. The nuclei of the molecule move according to the potential energy curve labeled M. When the projectile electron, initially with energy E, is in the neighborhood of the molecule, it causes an electronic transition to an M^- state. Because the molecule, even its ground state, has a nonzero vibrational energy, transitions can occur over a range of internuclear separations R which, by invoking the Franck–Condon principle (cf., e.g., Herzberg, 1950), we have indicated in an approximate fashion by the shaded area in the figure. Thus, a range of M^- vibrational states can be reached by excitation from the ground state. The system will vibrate according to the M^- potential curve but will eventually decay to one of the excited vibrational levels, usually of the electronic ground state, but perhaps to an electronically excited state.

If it is possible to excite many levels of M^- and if they are long-lived, i.e., the vibrational period of M^-, T, is much greater than its lifetime τ, then the form of the cross section is dominated by the M^- vibrational structure and the cross section for any transition is a series of sharp peaks at the positions of the vibrational levels of the compound state. This is found in the low-energy scattering by O_2 and NO, although the position is complicated by the fact that these two molecules have stable negative ions; i.e., the M^- curve crosses the M curve and has a lower minimum.

If the compound state is not so long-lived, e.g., if $T \approx \tau$, there is still considerable structure in the cross sections but the peaks no longer coincide with the vibrational levels of the compound state. This is just the case for N_2 discussed above, but it also occurs in other molecules, e.g., CO and CO_2.

If the M^- curve has a similar shape to the M curve so that a large range of M^- vibrational states can no longer be excited (the compound state is said to be "weakly coupled" to the molecular state), and if $T \ll \tau$, then the peaked structure disappears. The concept of scattering as a resonant process is now more questionable but it continues to be widely used. Many molecules, including the well-studied H_2, show broad structures, several electron volts wide, which have been interpreted as this kind of resonance scattering.

Large narrow threshold peaks have also been found in a number of molecules. They were first observed in the hydrogen halides, leading to the suggestion that this was a special property of the polar molecules. However, this had to be revised when similar effects were found in the nonpolar molecules CH_4 and SF_6. The resonance model discussed above does not seem to be applicable and the exact mechanism causing these threshold effects is still a matter of active debate.

Figure 1 also shows the connection between vibrational excitation and dissociative attachment for which the resonance model is particularly attractive: $e^- + M \rightarrow M^- \equiv (AB)^- \rightarrow A + B^-$. The M^- state, formed by excitation from the ground state, may have such a large energy that the nuclei do not vibrate in the potential well but can move infinitely far apart; this will always happen if the M^- state is wholly repulsive. We will only discuss dissociative attachment results as they bear on the vibrational excitation process.

Much of the earlier work on vibrational excitation has been comprehensively reviewed by Schulz (1973, 1979) and there are several reviews which cover the whole field of electron–molecule scattering (cf. Brown, 1979; Burke, 1979; Lane, 1980; Resigno et al., 1979; Shimamura and Matsuzawa, 1979). We restrict ourselves in this review to low-energy scattering (< 30 eV) and attempt to show how various theoretical approaches can explain the recent experimental data for a wide variety of molecules.

II. Theoretical Considerations

A. Molecular Eigenfunctions and Eigenenergies

We will make the usual Born–Oppenheimer approximation that because the electron mass is so much smaller than the nuclear mass, the wave function for the molecule, Φ, can be written as the product of an electronic wave function ψ_{el} and a nuclear wave function ψ_n; this also involves the assumption that ψ_{el} varies slowly with changes in the nuclear geometry of the molecule (cf. Born and Oppenheimer, 1927; Wilson et al., 1955; Dennison and Hecht, 1962).

We consider the molecule to have N electrons with coordinates $\mathbf{q}$ ($\equiv \mathbf{q}_1, \ldots, \mathbf{q}_N$) and M nuclei with coordinates $\mathbf{R}$ ($\equiv \mathbf{R}_1, \ldots, \mathbf{R}_M$). Then we make the approximation

$$\Phi(\mathbf{q}, \mathbf{R}) = \psi_{el}(\mathbf{q}; \mathbf{R})\psi_n(\mathbf{R}) \tag{1}$$

The Hamiltonian for the system, $H(\mathbf{q}, \mathbf{R})$, can be written as the sum of electronic and nuclear parts:

$$H(\mathbf{q}, \mathbf{R}) = H_{el}^N(\mathbf{q}; \mathbf{R}) + H_n(\mathbf{R}) \tag{2}$$

where

$$H_{el}^N = -\frac{\hbar^2}{2m}\sum_{i=1}^N \nabla^2_{q_i} - \sum_{i=1}^N \sum_{j=1}^M \frac{Z_j e^2}{|\mathbf{q}_i - \mathbf{R}_j|}$$

$$+ \sum_{\substack{i=1 \\ i<j}}^N \sum_{j=2}^N \frac{e^2}{|\mathbf{q}_i - \mathbf{q}_j|} \tag{3}$$

$$H_n = -\tfrac{1}{2}\hbar^2 \sum_{i=1}^M \nabla^2_{R_i}/M_i + \sum_{\substack{i \\ i<j}} \sum_j Z_i Z_j e^2/|\mathbf{R}_i - \mathbf{R}_j| \tag{4}$$

Nucleus i has mass M_i and charge Z_i; the electron mass is m. In the Born–Oppenheimer approximation we assume that

$$H_n\{\psi_{el}\psi_n\} = \psi_{el}H_n\{\psi_n\} \tag{5}$$

so that the Schrödinger equation for the whole system, $(H - E)\Phi = 0$, separates into two equations, one for the electronic wave function and the other for the nuclear wave function.

$$\{H_{el}^N(\mathbf{q}, \mathbf{R}) - E_n(\mathbf{R})\}\psi_{el}(\mathbf{q}; \mathbf{R}) = 0 \tag{6a}$$

$$\{H_n(\mathbf{R}) + E_n(\mathbf{R}) - E\}\psi_n(\mathbf{R}) = 0 \tag{6b}$$

Equation (6a) is solved for fixed $\mathbf{R}$. The eigenenergies $E_n(\mathbf{R})$ depend parametrically on $\mathbf{R}$ and are separated, typically, by a few electron volts. $E_n(\mathbf{R})$ and the nuclear repulsion term of Eq. (4) become the potential energy surface V on which the nuclei move according to Eq. (6b).

For a polyatomic molecule the motion of the nuclei can be very complicated. For M nuclei there are a total of $3M$ degrees of freedom: three for translation, three for rotation, and $3M$-6 for vibration. (For linear molecules there are only two rotational degrees of freedom and thus $3M$-5 vibrational degrees of freedom.) The problem of the nuclear motion can be simplified by assuming that the vibrational and rotational parts can be separated. This is a good approximation because the vibrational period is much greater than the rotational period and the nuclei make many oscillations before the molecule as a whole rotates appreciably. A full treatment of the separation of rotational and vibrational motions is given in Wilson *et al.* (1955) and we give only a very brief summary here.

Consider two sets of axes, one set (XYZ) centered on the molecule center of mass and fixed in space, the other set (xyz) rotating with the molecule; set (xyz) is taken to be along the three principal axes of the molecule, and the orientation of the (xyz) axes with respect to the (XYZ) axes is given by the three Euler angles $\alpha\beta\gamma\,(\equiv\Omega)$ (cf. Brink and Satchler, 1968).

Let the position vectors of the nuclei, $\mathbf{R}_i$, be with respect to the moving set of axes and introduce the vectors $\mathbf{R}_i^e$, the equilibrium position of the nuclei, and $\boldsymbol{\rho}_i = \mathbf{R}_i - \mathbf{R}_i^e$, the displacement of the nuclei from equilibrium [both with respect to (x, y, z)]. The problem is made more tractable by introducing normal coordinates Q_k ($k = 1, 2, \ldots, 3M$), which are related linearly to the components of $\boldsymbol{\rho}_i$. The relation is particularly simple if we introduce mass-weighted coordinates r_i ($i = 1, 2, \ldots, 3M$), which are obtained by multiplying the components of $\boldsymbol{\rho}_i$ by the square root of the appropriate mass. The transformation between the r_i and Q_k is an orthogonal one which can be chosen so that the classical kinetic energy T becomes

$$T = \sum_{k=1}^{3M} \dot{Q}_k^2 \tag{7}$$

and the potential energy V becomes

$$V = \frac{1}{2} \sum_{k=1}^{3M} \lambda_k Q_k^2 + \text{higher terms in} \quad Q_k \tag{8}$$

Three of the normal coordinates Q_k correspond to translational motion and three correspond to rotational motion.

Wilson *et al.* show how the quantum mechanical Hamiltonian [$H_n + E_n$ of Eq. (6b)] can be separated, approximately, into the sum of a rotational Hamiltonian H_{rot} for a *rigid* rotating body and a vibrational Hamiltonian H_{vib} for a *nonrotating* molecule:

$$H_{\text{rot}} = \frac{1}{2} \left\{ \frac{J_x^2}{I_x} + \frac{J_y^2}{I_y} + \frac{J_z^2}{I_z} \right\} \tag{9}$$

$$H_{\text{vib}} = \frac{1}{2} \sum_{k=1}^{3M-6} P_k^2 + \frac{1}{2} \sum_{k=1}^{3M-6} \lambda_k Q_k^2 + \text{higher terms} \tag{10}$$

where (J_x, J_y, J_z) are the components of the angular momentum operator **J** for the rigid rotor and I_x, I_y, and I_z are the moments of inertia of the molecule about the three principal axes (xyz). (Strictly the moments of inertia depend on the nuclear displacement, but we treat them as constants calculated at the equilibrium configuration.) P_k is the momentum operator conjugate to the vibrational coordinate Q_k.

The ψ_n of Eq. (6b) can be written now as the product of a rotational function $\psi_{rot}(\Omega)$, which is an eigenfunction of H_{rot}, and vibrational function $\psi_{vib}(Q)$ ($Q \equiv Q_1 \cdots Q_{3M-6}$), which is an eigenfunction of H_{vib}.

The rotational wavefunctions depend on the values of the moment of inertias (cf. Wilson et al. or Dennison and Hecht for full details). Most applications have been for diatomics ($I_z = 0$, $I_x = I_y = I$) where the rotational wave function $\psi_{jm}(\Omega)$ is just $Y_j^m(\beta\alpha)$ with energy $\hbar^2 j(j+1)/2I$.

The solution of the vibrational equation depends on the complexity of V. If it is taken to be just $\frac{1}{2}\Sigma \lambda_k Q_k^2$, then the solution is a product of harmonic oscillator functions:

$$\psi_{vib} = \psi_{v_1}(Q_1)\psi_{v_2}(Q_2) \cdots \psi_{v_{3M-6}}(Q_{3m-6}) \tag{11}$$

$$E_{vib} = E_{v_1} + E_{v_2} + \cdots + E_{v_{3M-6}} \tag{12}$$

$$-\frac{\hbar^2}{2}\frac{d^2\psi(Q_k)}{dQ_k^2} + \frac{1}{2}\lambda_k Q_k^2 \psi(Q_k) = E_{v_k}\psi(Q_k) \tag{13}$$

$$E_{v_k} = (v_k + \tfrac{1}{2})h\nu_k \tag{14}$$

The ν_k ($k = 1, 2, \ldots, 3M - 6$) are the frequencies of the classical modes of vibration. The solution of Eq. (13) is

$$\psi(Q_k) = \{\gamma_k^{1/2}/\pi^{1/2} 2^{v_k}(v_k)!\}^{1/2} \exp(-\gamma Q_k^2/2) H_{v_k}(\gamma_k^{1/2} Q_k) \tag{15}$$

where $\gamma_k = 4\pi^2 \nu_k/h$ and $H_n(Z)$ is a Hermite polynomial of degree n.

This harmonic approximation may be satisfactory for low-lying vibrational levels, but for higher levels the equation for ψ_{vib} must be solved with a more accurate V.

The energy separation of the lowest vibrational levels (ΔV) is typically a few tenths of an electron volt, much larger than the spacing of rotational levels (Δr). The ratio $\Delta r/\Delta V$ is largest for H_2, about $\frac{1}{35}$, but is usually much smaller, e.g., the N_2 ratio is $\simeq \frac{1}{500}$ (cf. Herzberg, 1950).

For diatomic molecules there is only one normal coordinate: $Q = \mu^{1/2} R$, where R is the internuclear separation and μ is the reduced mass of the molecule, $M_1 M_2/(M_1 + M_2)$.

B. The Scattering Formalism

In Section I we discussed vibrational excitation as proceeding through a resonant or negative ion compound state. This mechanism can be included explicitly in the scattering formalism by expanding the total wave function for the electron + molecule system in terms of such compound states. This is discussed in Sections II,B,5 and 6, but first we turn to approaches in which the total wave function is expanded in terms of the eigenstates of the molecule. This is needed even for the resonance discussion when the projectile electron is well separated from the molecule, but it can be used in its own right to give cross sections and also, in some cases, information about the compound state responsible for the structure in the cross section.

1. Vibration–Rotation Close Coupling

The Hamiltonian for the complete system of electron + molecule is

$$H(\mathbf{q}, \mathbf{R}, \mathbf{r}) = H_{el}^N(\mathbf{q}, \mathbf{R}) + H_n(\mathbf{R}) + V(\mathbf{q}, \mathbf{R}, \mathbf{r}) - (\hbar^2/2m)\nabla_r^2 \qquad (16)$$

where H_{el}^N and H_n are as defined in Eqs. (3) and (4), and $V(\mathbf{q}, \mathbf{R}, \mathbf{r})$, the interaction between the electron and molecule, is

$$V(\mathbf{q}, \mathbf{R}, \mathbf{r}) = -\sum_{i=1}^{M} e^2 Z_i/|\mathbf{R}_i - \mathbf{r}| + \sum_{i=1}^{N} e^2/|\mathbf{q}_i - \mathbf{r}| \qquad (17)$$

$\mathbf{q}$ and $\mathbf{R}$ are the molecular electron and nuclear coordinates, respectively, as in Section II,A, and $\mathbf{r}$ is the projectile electron coordinate.

The total wave function $\psi(\mathbf{q}, \mathbf{R}, \mathbf{r})$ can be expanded in terms of the eigenfunctions discussed in Section II,A. We retain only the ground electronic state in the expansion; this approximation is commonly corrected by including a polarization potential (cf. Section III and applications). We will also assume for this analysis that the ground state has symmetry $^1A_1$ (or $^1\Sigma$ in the more usual notation for diatomic molecules), corresponding to a closed-shell structure. This covers all the molecules of interest to us except $O_2(^3\Sigma_g^-)$ and $NO(^2\Pi)$.

With these restrictions the angular momentum of the system is carried by the rotational motion of the molecule and the projectile electron. We can introduce functions which are eigenfunctions of J^2 and J_z, $\mathbf{J}$ being the total angular momentum (cf. Arthurs and Dalgarno, 1960:

$$\mathcal{Y}_{ljJM}(\Omega, \hat{\mathbf{r}}) = \sum_{\lambda m} C_{\lambda m M}^{lJM} Y_l^\lambda(\hat{\mathbf{r}}) \psi_j^m(\Omega) \qquad (18)$$

where Y_l^j is the angular part of the projectile electron function and ψ_j^m is the rotational eigenfunction. Defining functions $\Phi_\alpha^{JM}(\Omega, \hat{r}, Q)$ (cf. Henry, 1970),

$$\Phi_\alpha^{JM}(\Omega, \hat{r}, Q) = \mathcal{Y}_{ljJM}(\Omega, \hat{r}) Z_v(Q) \qquad (19)$$

where $Z_v(Q)$ ($Q \equiv Q_1, \ldots, Q_{3M-6}$) is the eigenfunction of the vibrational Hamiltonian H_{vib}, the total wave function can be expanded as

$$\Psi(q, R, r) = \phi_{el}(q, R) r^{-1} \sum_{\alpha JM} \Phi_\alpha^{JM}(\Omega, \hat{r}, Q) F_\alpha^{JM}(r) \qquad (20)$$

Equations for F_α, which are diagonal in J and independent of M, can be obtained from $\langle \psi_{el} \Phi_\alpha | H - E | \Psi \rangle = 0$. If we use the Born–Oppenheimer approximation, i.e., replace $H_n\{\psi_{el} \Phi_\alpha\}$ by $\psi_{el} H_n \Phi_\alpha$, the equations become (using atomic units, $m \equiv 1, \hbar \equiv 1$):

$$\left(\frac{d^2}{dr^2} - \frac{l_\alpha(l_\alpha + 1)}{r^2} + k_\alpha^2 \right) F_\alpha = 2 \sum_{\alpha'} U_{\alpha\alpha'} F_{\alpha'} \qquad (21)$$

where $U_{\alpha\alpha'} = \langle \psi_{el} \Phi_\alpha | V | \Phi_{\alpha'} \psi_{el} \rangle$ and $k_\alpha^2/2$ is the energy of the projectile in channel α. Adding the index $\bar{\alpha}$ to distinguish the independent solutions of the coupled equations, and defining $x_\alpha = k_\alpha r - l_\alpha \pi/2$, the asymptotic form of F can be defined in terms of elements of the S matrix,

$$F_\alpha^{\bar{\alpha}}(r) \xrightarrow{r \to \infty} k_\alpha^{-1/2} \{\delta_{\alpha\bar{\alpha}} \exp(-x_\alpha r) - S_{\alpha\bar{\alpha}} \exp(x_\alpha r)\} \qquad (22)$$

The asymptotic form of the total wave function is

$$\Psi \underset{r \to \infty}{\sim} \psi_{el} \{\psi_{j_1}^{m_1}(\Omega) Z_{v_1}(Q) \exp(i k_1 \cdot r)$$
$$+ r^{-1} \sum_\alpha \psi_j^m(\Omega) Z_v(Q) \exp(i k_\alpha r) f(j_1 m_1 v_1 \to jmv | \hat{r})\} \qquad (23)$$

The scattering amplitude f, for electrons incident along the Z axis and scattered in direction $\hat{r}$ with energy $k_f^2/2$, is

$$\tfrac{1}{2}(k_1 k_f)^{-1/2} \sum (2l + 1)^{1/2} Y_{l'}^{\lambda'}(\hat{r}) (-1)^{l-l-l'}$$
$$\times C_{0m;M}^{lJ,J} C_{\lambda'mM}^{l'jJ} T(j_1 m_1 v_1 \to jmv) \qquad (24)$$

where $T = 1 - S$ and the sum is over $l, l', J, M,$ and λ'. The differential cross section is $k_f |f|^2/k_i$, which is usually averaged over all initial states m_{j_1} and summed over all final states m_j to give a differential cross section for transition $j_1 v_1 \to jv$.

The approximation can be improved by antisymmetrizing the wave function with respect to interchange of electrons. Inclusion of such exchange effects results in the coupled equations (21) becoming integrodifferential equations. This has been carried out in detail for H_2 by Henry (1970) but in many calculations simplifications have been made by introducing a "local" exchange potential. Applications of these potentials will be given later.

2. Vibration Close Coupling

The analysis of Section II,B,1 is straightforward, but the equations are numerically difficult to solve because of the slowness of convergence of the various cross-section expansions with respect of j, l, and v. However, for some applications it is a good approximation to solve for the simpler problem of electron scattering by a nonrotating vibrator. This follows from the fact that the rotational part of the Hamiltonian is negligible compared to other terms when the electron is close to the molecule. When the projectile electron and molecule are well separated this condition no longer applies, but the error in using this approximation for all electron molecule separations is small except for very small energy scattering (≤ 0.1 eV) and for scattering by polar molecules (cf. discussion by Feldt and Morrison, 1982, for H_2).

The wave function for the system of electron and nonrotating vibrator is

$$\Psi(\mathbf{q}, \mathbf{R}, \mathbf{r}) = \phi_{el}(\mathbf{q}, \mathbf{R}) \sum_v Z_v(Q)\chi_v(\mathbf{r}) \tag{25}$$

The projectile electron function χ can be expanded in terms of symmetry-adapted angular functions (cf. Altmann and Cracknell, 1965, and Section II,B,3). For simplicity we refer to diatomic molecules only:

$$\chi_v(r) = r^{-1} \sum_{l\lambda} g_{vl}^\lambda(r) Y_l^\lambda(\hat{\mathbf{r}}) \tag{26}$$

The equations for the g_{vl}^λ are diagonal in λ and are formally the same as Eq. (21) with $\alpha \equiv vl$ and $U_{\alpha\alpha'} = \langle Y_l^\lambda Z_v \psi_{el} | V | \psi_{el} Z_{v'} Y_{l'}^\lambda \rangle$. As in Section II,B,1, we make the Born–Oppenheimer assumption that $H_n\{\phi_{el}Z_v\} = \phi_{el}H_nZ_v$. By arguments analogous to those in the last section, the scattering amplitude for the vibrational transition $v_I \rightarrow v$ is

$$f_{v_I v}(\hat{\mathbf{k}}_{v_I} \cdot \hat{\mathbf{r}}) = 2\pi i(k_{v_I}k_v)^{-1/2} \sum_{ll'\lambda} i^{l-l'} Y_l^{\lambda *}(\hat{\mathbf{k}}_{v_I}) Y_{l'}^\lambda(\hat{\mathbf{r}}) T_{v_I l, v l'} \tag{27}$$

where the electron is incident along the $\hat{\mathbf{k}}_{v_I}$ direction and scattered along the $\hat{\mathbf{r}}$ direction. Note that the spherical harmonics are defined with reference to a set of axes fixed to the molecule. The differential cross section is obtained by referring these spherical harmonics to a set of axes fixed in space and averaging $|f|^2 k_v/k_{v_I}$ over all orientations of the molecule with respect to this set of axes (cf. Salvini and Thompson, 1981a).

Information on rotational transitions is not completely lost using this approach. Because we are assuming the projectile electron energy is much greater than the spacing of the rotational eigenenergies, the nuclear impulse approximation of Chase (1956) should give accurate results. The scattering amplitude of Eq. (27), referred to a set of axes fixed in space, can be written as $f_{vv'}(\hat{\mathbf{k}} \cdot \hat{\mathbf{r}}; \Omega)$, where Ω is the set of Euler angles as in Section II,A. A new scattering amplitude is obtained by averaging f over the initial and final rotational eigenfunctions:

$$f(vjm \to v'j'm'|\hat{\mathbf{k}} \cdot \hat{\mathbf{r}}) = \int \psi_{j'}^{m'*}(\Omega) f_{vv'}(\hat{\mathbf{k}} \cdot \hat{\mathbf{r}}; \Omega) \psi_j^m(\Omega) \, d\Omega \qquad (28)$$

When the molecule has a permanent dipole moment it is found that the whole approach of this section, and subsequent sections where rotation is neglected, is severely restricted. Although momentum transfer cross sections can be calculated, total cross sections are not defined. When H_{rot}, the rotational Hamiltonian, is no longer negligible compared to V, the interaction between electron and molecule, we should go back to the approach described in Section II,B,1. Chang and Fano (1972) in their frame transformation theory have shown how this can be done, but the procedure is extremely cumbersome. Alternative methods are available which, while not being exact, are easy to apply (cf. e.g., Norcross and Padial, 1982).

3. Adiabatic Fixed-Nuclei Approximation

Even with neglect of rotation many coupled equations must be solved to obtain full convergence with respect to v and l. In some instances, however, the problem can be simplified still further by first solving for the scattering of an electron by a *rigid* fixed molecule and then applying the impulse approximation to the vibrational as well as the rotational motion. This approximation is only valid when the time spent by the projectile electron in the vicinity of the molecule is small compared to the vibrational period of the molecule. It can be applied to scattering via a compound state only if the lifetime of this state is very short.

For the scattering of an electron by a rigid fixed molecule the total wave function is the product of ψ_{el} and χ. For diatomic and linear molecules $\chi(\mathbf{r})$ is expanded in terms of spherical harmonics Y_λ^μ and the equations for the radial functions are diagonal in λ, as in Section II,B,2. For polyatomic molecules the generalization is that χ is expanded in terms of symmetry-adapted functions $X_{lh}^{p\mu}(\hat{\mathbf{r}})$. The $X(\hat{\mathbf{r}})$ belong to the μth component of the pth irreducible representation of the molecular point group; h is required to distinguish between functions with the same $(p\mu l)$. The radial equations are now diagonal in p and independent of μ (cf. Gianturco and Thompson, 1980). When the electron is incident along $\hat{\mathbf{k}}$ and scattered along the $\hat{\mathbf{r}}$ direction, the scattering amplitude for closed-shell molecules is

$$f(\hat{\mathbf{k}} \cdot \hat{\mathbf{r}}) = 2\pi i k^{-1} \sum X_{lh}^{p\mu}(\hat{\mathbf{k}}) X_{l'h'}^{p\mu}(\hat{\mathbf{r}}) i^{l-l'} \{\delta_{ll'}\delta_{hh'} - S_{lh,l'h'}^{p\mu}\} \qquad (29)$$

where the sum is over l, l', h, h', p, and μ

The behavior of the **S** matrix may lead to information about the negative ion states. For given scattering state p (or λ for diatomics) **S** can be diagonalized by a unitary matrix (cf. Hazi, 1979) to give matrix Λ. Writing the diagonal elements of Λ as $\exp(2i\eta_i)$, we can define an "eigenphase sum" $\Delta = \Sigma \eta_i$. Hazi shows that if a compound state exists with symmetry p, then Δ can be separated into the sum of "resonant" and "nonresonant" parts:

$$\Delta = \Delta_0 + \arctan\{\Gamma/2(E_0 - E)\} \qquad (30)$$

where Δ_0 is the nonresonant contribution, a slowly varying function of the electron energy E, Γ is the "width" of the resonance, inversely proportional to the lifetime of the state, and E_0 is the position of the resonance. The "resonant" part of Δ increases by π as E goes through the resonance position. This is discussed further in Section II,B,6.

Vibrational cross sections are obtained by calculating the scattering amplitude in Eq. (29) at several nuclear geometries Q and then determining $f_{vv'}$ corresponding to Eq. (27) from

$$f_{vv'}(\hat{\mathbf{k}} \cdot \hat{\mathbf{r}}) = \int Z_v(Q) f(\hat{\mathbf{k}} \cdot \hat{\mathbf{r}}; Q) Z_{v'}(Q) \, dQ \qquad (31)$$

Rotational transitions can also be included through Eq. (28).

4. Hybrid Approximation

The approach of Section II,B,3 is straightforward and easy to apply, but it is not successful for molecules like N_2 where there is a compound state whose lifetime is comparable to the vibrational period. Chandra and Temkin (1976a) suggested that the approach of Section II,B,3 be used for all scattering states except the resonance state, which should be treated by the method of Section II,B,2. Thus in the "hybrid" approximation we calculate $f_{vv'}$ from Eq. (31) using all elements of the **S** matrix, $S_{ll'}^\lambda$ for diatomic molecules, *except* those corresponding to the compound state, for which we use the vibration close coupling matrix elements $S_{vl,v'l'}^\lambda$. Rotation is included using the impulse approximation.

5. R-Matrix Method

Expansion of the total wave function in terms of molecular eigenstates certainly makes sense physically for the region of space where electron and molecule are well separated. However, this may not be so when electron and molecule are close together; particularly if the electron spends a lot of time near the molecule, it would seem more appropriate to expand in terms of negative ion states.

In the R-matrix method space is divided into "internal" and "external" regions; negative ion states are used in the internal region and molecular states in the external region.

In the approach introduced by Schneider et al. (1979a,b) and Le Dourneuf et al. (1979) for diatomic molecules, the electronic part of the total wave function is calculated for the internal region, $r < a$ (say), by an equation which is similar to that for the electronic states of the molecule given in Eq. (6a):

$$\{H_{el}^{N+1}(\mathbf{q}, \mathbf{r}; \mathbf{R}) - E_m(\mathbf{R}) + \mathcal{L}\}\psi_m(\mathbf{q}, \mathbf{r}; \mathbf{R}) = 0 \tag{32}$$

The Hamiltonian H_{el}^{N+1} is defined by Eq. (2). The eigenfunctions ψ_m the eigenenergies E_m are calculated for fixed nuclear geometry $\mathbf{R}$, as in a bound-state calculation, except for the boundary conditions imposed at $r = a$. This can be illustrated using the approximation for ψ used by Schneider et al. Let ψ_m be expressed as the product of an N-electron wave function $\tilde{\phi}_{el}(\mathbf{q}, \mathbf{R})$ and extra electron function. If $\tilde{\phi}_{el}$ has $^1\Sigma$ symmetry, Eq. (32) will give solutions ψ_m^λ, where

$$\psi_m^\lambda(\mathbf{q}, Q, \mathbf{r}) = \tilde{\phi}_{el}(\mathbf{q}, Q) \sum_l Y_l^\lambda(\hat{\mathbf{r}}) r^{-1} f_l^{m\lambda}(r, Q) \tag{33}$$

The R-matrix boundary conditions are

$$\left(\frac{d}{dr} - \frac{b}{r}\right) f_l^{m\lambda}(r, Q) = 0 \quad \text{at} \quad r = a \tag{34}$$

Schneider et al. enforce these conditions by using the Block operator $\mathcal{L}$:

$$\mathcal{L} \equiv \sum_l |\tilde{\phi}_{el} Y_l^\lambda\rangle \delta(r - a) \left(\frac{d}{dr} - \frac{b-1}{r}\right) \langle \tilde{\phi}_{el} Y_l^\lambda | \tag{35}$$

The total wave function for the inside region can be expanded in terms of the eigenfunctions ψ_m^λ:

$$\Psi^\lambda(\mathbf{q}, Q, \mathbf{r}) = \sum_m \psi_m^\lambda(\mathbf{q}, Q, \mathbf{r}) \xi_m^\lambda(Q) \tag{36}$$

and if we make the Born–Oppenheimer assumption that $H_n\{\psi, \xi\} = \psi H_n\{\xi\}$, the equation for the nuclear function ξ can be easily obtained:

$$\{H_n + E_n^\lambda(Q) - E\}\xi_m^\lambda(Q) = \langle \psi_m | \mathcal{L} | \Psi \rangle \tag{37}$$

In the external region $r > a$, the Ψ^λ can be expanded as in Section II,B,2:

$$\Psi^\lambda(\mathbf{q}, Q, \mathbf{r}) = \phi_{el}(\mathbf{q}; Q) \sum_v Z_v(Q) Y_l^\lambda(\hat{\mathbf{r}}) F_{vl}^\lambda(r) \tag{38}$$

Using the expansion in Eq. (36) and the solution of Eq. (37), we obtain

$$F_{vl}^\lambda(a) = \sum_{v'l'} R_{vl,v'l'}^\lambda \left(\frac{d}{dr} - \frac{b}{a}\right) F_{v'l'}(a) \tag{39}$$

where the elements of the R-matrix are

$$R_{vl,v'l'}^\lambda = \sum_m \int f_l^{m\lambda}(a, Q) Z_v(Q) G_m^\lambda(Q, Q') Z_v(Q')$$

$$\times f_{l'}^{m\lambda}(a, Q') \, dQ \, dQ' \tag{40}$$

G_m^λ is the Green's function for Eq. (37). Using this R-matrix, we can solve the vibration close coupling equations for F_{vl}^λ in the external region to give the S matrix and then the cross section.

6. Resonance Models—Local Complex Potential

The method of expanding the wave function in the internal region in terms of negative ion states was first applied to electron–molecule collisions by Herzenberg and Mandl (1962). The main developments of the "local complex potential" approximation have been by Herzenberg and collaborators (cf. Bardsley *et al.*, 1966a,b; Herzenberg, 1968; Birtwistle and Herzenberg, 1971; Dubé and Herzenberg, 1979).

The approach is formally similar to that described for the R-matrix method. Considering diatomic molecules only, we look for compound-state eigenfunctions ψ_m^λ and eigenenergies E_m^λ which are solutions of Eq. (32) in the internal region $r < a$. However, we drop the Block operator $\mathcal{L}$ and impose the boundary condition at $r = a$ that ψ_m^λ should contain outgoing waves only. This resonance prescription has usually been effected by the methods of Kapur and Peierls (1938) and of Siegert (1939). The principle of the methods can be seen clearly by considering a simple one-dimensional example. We look for solutions $\psi(r)$ of the equations

$$[(d^2/dr^2) + V(r) + k^2]\psi(r) = 0; \quad V(r) = 0 \quad \text{for} \quad r > a \tag{41}$$

in terms of the eigenfunctions $\psi_i(r)$ of

$$[(d^2/dr^2) + V(r) + k_i^2]\psi_i(r) = 0 \quad \text{for} \quad r \leq a \tag{42}$$

The Kapur–Peierls and Siegert boundary conditions are that $\psi_i(a)$ should have the values $\exp(ika)$ and $\exp(ik_ia)$, respectively. Writing the solution in the region $r > a$ as $\exp(-ikr) + S\exp(ikr)$, we see that the Siegert energies correspond to the poles of the S matrix.

It is important to note that solutions of Eq. (32) with such resonance boundary conditions can only be obtained if the eigenenergy E_m^λ is complex. If the real part of E_m^λ is E_0, and the imaginary part is $-\Gamma/2$, then E_0 and Γ are the position and width as defined in Eq. (30). We note that state Ψ with energy E_m has time dependence $\exp(-iE_m t/\hbar)$; i.e., $|\Psi|^2$ will be proportional to the decaying exponential $\exp(-\Gamma_m t/\hbar)$. $\tau = \hbar/\Gamma_m$ is known as the lifetime of the state.

Neglecting rotation, the total wave function Ψ in the internal region can be expressed in terms of these compound-state functions as in Eq. (36). If only one compound state ψ_m^λ is physically significant for the vibrational process, we can approximate Ψ by the expression

$$\Psi = \exp(i\mathbf{k}_i \cdot \mathbf{r})\phi_{el}Z_v + \psi_m^\lambda \xi_m^\lambda(Q) \tag{43}$$

where the molecule is initially in electronic state ϕ_{el} and vibrational state Z_v, and the projectile electron has initial energy $k_i^2/2$. The equation for the nuclear function ξ can be easily derived:

$$[H_n + E_m^\lambda(Q) - E]\xi_m^\lambda(Q) = \zeta_m^\lambda(Q)Z_v(Q) \tag{44}$$

where $\zeta_m^\lambda = \langle \psi_m | V | \exp(i\mathbf{k}_i \cdot \mathbf{r}) \phi_{el} \rangle$ and the integrals are over the internal region only; we have made the usual Born–Oppenheimer assumptions that $H_n\{\phi_{el}Z\} = \phi_{el}H_n\{Z\}$ and $H_n\{\psi_m\xi\} = \psi_m H_n\{\xi\}$.

Assuming the electronic state does not change, the differential cross section for transition $v \rightarrow v'$ can be calculated from

$$(k_f/4\pi^2 k_i)|\langle \exp(i\mathbf{k}_f \cdot \mathbf{r})\phi_{el}Z_{v'}|V|\Psi\rangle|^2 \tag{45}$$

where $k_f^2/2$ is the final energy of the electron. The resonant contribution to the cross section, i.e., that part of Eq. (45) which involves the compound state, comes from the integral $\langle Z_{v'}|\bar{\zeta}_m^\lambda|\xi_m^\lambda\rangle$, where $\bar{\zeta}$ is the ζ defined following Eq. (44) with k_i replaced by k_f. Provided the electron energy is not close to the vibrational threshold, we can put $\bar{\zeta}$ equal to ζ.

It can be shown that ζ is related to Γ. Thus, if we know E_m^λ as a function of Q, we can calculate first ξ_m^λ and then the cross section. However, in most applications E_m^λ has been treated as a parameter which has been chosen by appeal to the cross section data. This "fitting" procedure can be very good.

7. Resonance Models—Further Discussion

The main feature of the resonance treatment in Section II,B,6 is that when the electron is close to the molecule, the nuclei move in a local complex potential which depends only on the nuclear geometry. A more complete resonance discussion shows that this is only an approximation and the question immediately arises of how applicable the model really is.

There have been several formulations (e.g., Chen, 1964a,b; Bardsley, 1968; Fiquet-Fayard, 1974) based on the method of Fano (1961) in which one considers the interaction between a bound state embedded in a continuum of states. The bound state becomes a "resonance" state able to "decay" into the continuum. There is a shift and a width attached to this state, their values depending on the strength of the coupling between the bound and continuum states. Domcke and Cederbaum have made a full study of vibrational excitation, in an extension of this formalism (cf. Domcke *et al.*, 1979; Domcke and Cederbaum 1977, 1980, 1981; Cederbaum and Domcke, 1981; Domcke, 1981). They have shown that the theory reduces under certain conditions to the local complex potential treatment of Section II,B,6 and have successfully applied this restricted analysis to N_2 and CO_2. However, in the main they have not tried to make a direct comparison with experiment but have considered a number of simple models with exact solutions. They have constructed a "nonadiabatic" theory in which the complex potential depends on the energies of the projectile electron and the nuclei, and have illustrated how important such a theory may be for a proper understanding of scattering close to thresholds.

The concept of a local complex potential has been questioned by Nesbet (1981), who, while mentioning the work of Cederbaum and Domcke for vibrational excitation, widened the discussion to include the processes of dissociative attachment and electron detachment in heavy-particle collisions. He suggested that there is need for a fuller theory to cover all the processes. Berman *et al.* (1983) have discussed the use of local complex potentials for the description of broad resonances and conclude that there are too many ambiguities to make the procedure worthwhile. However, despite these qualifications the model described in Section II,B,6 has proved most satisfactory in explaining vibrational excitation at electron energies not too close to threshold and for transitions $0 \to v$, where v is not too large.

Fabrikant (1976, 1977a,b, 1978, 1980) has considered the threshold behavior of cross sections for the scattering of electrons by polar molecules. He has looked at the solution of the vibration rotation and vibration close coupling equations (cf. Sections II,B,1 and 2) in the asymptotic region where the $1/r^2$ potential is dominant. The short-range problem is not considered, its effect being considered given by some boundary condition such as an R-matrix. This approach has been able to explain some of the threshold effects but does not have sufficient applicability to explain all the details.

III. Applications and Comparison with Experiment

We now describe some of the recent work on vibrational excitation by electron impact. Our aim is to show when the various approaches discussed in Section II are valid, and to illustrate how successful they are by comparing their results with the most recent experimental data. A complete theoretical treatment presents severe computational problems for large molecules and there have not been many *ab initio* calculations. Several groups are now in a position to do these within the fixed nuclei + impulse approximation, and more complete treatments needed to reproduce fine structure should soon follow the initial work of Schneider *et al.* (1979b). However, there have been quite a number of "model" calculations and much work has also been done using the local complex potential treatment, usually with a single compound state whose width depends on the position of the nuclei.

A. H_2

Bardsley and Wadehra (1979) have investigated the cross sections in the energy range up to 10 eV from the point of view of the compound state analysis of Section II,B,6. This is an extension of earlier work by Bardsley *et al.* (1966b), who analyzed the experiment of Schulz (1964) with a single compound state which was short-lived and weakly coupled to the ground state.

Bardsley and Wadehra also wished to explain the dissociative attachment process which required the inclusion of a second compound state. Thus, they introduced $^2\Sigma_u^+$ $(1\sigma_g^2)(1\sigma_u)$ and $^2\Sigma_g^+$ $(1\sigma_g)(1\sigma_u^2)$ H_2^- states which are associated with the X $^1\Sigma_g^+$ ground state and the repulsive b $^3\Sigma_u^+$ state, respectively, of H_2. Both H_2^- states dissociate at large internuclear separations into H(1s) and H^- $^1S(1s^2)$. Whereas the $^2\Sigma_u^+$ state can decay only to the ground state, the $^2\Sigma_g^+$ state can also decay to the excited state; partial widths for each decay must be introduced. Bardsley and Wadehra chose values for the real and imaginary parts of the local complex potentials by a mixture of calculation and appeal to experiment (mainly the dissociative attachment process).

The result obtained by Bardsley and Wadehra for the $v = 0 \to 1$ transition agrees to within about 20% with the experimental values of Linder and Schmidt (1971a) shown in Fig. 2 (the crosses). They found that the contribution of the $^2\Sigma_g^+$ state is only significant for the transitions $0 \to v$, $v \geqslant 3$. A similar enhancement in the cross sections at about 10 eV has been seen in the experiments of Hall (1973) on electron scattering by D_2; they also give considerable fine structure for energies above 11 eV, unfortunately not reproducible by the inclusion of just two compound states.

This resonance analysis shows that the lifetime of the $^2\Sigma_u$ compound state is short compared to its vibrational period and, thus, the fixed-nuclei plus impulse approximation of Section II,B,3 should give an adequate description of the scattering process. This was partially confirmed by Faisal and

Temkin (1972), but the first real test of the method was by Henry and Chang (1972), whose results for the $v = 0 \to 1$ transition are given in Fig. 2 (dotted curve). Exchange effects were included by properly antisymmetrizing the total function, but uncertainties in the polarization potential may have contributed to the disagreement with experiment.

We saw in Section II,B,3 that the eigenphase sum in the fixed-nuclei approximation can give information on the compound state. The calculations of Henry and Chang and also of Buckley (1978) have shown that the $^2\Sigma_u^+$ scattering state is a resonance state, the $l = 1$ partial wave being predominant. Although the nonresonant scattering is so great that it is difficult to extract a position and a width, the results are consistent with those of Bardsley and Wadehra.

Klonover and Kaldor (1979a,b) have carried out the most accurate calculations to date within the fixed-nuclei plus impulse approximation. They have used the T-matrix expansion method of Resigno et al. (1974a,b, 1975) and have properly included exchange and polarization effects. Their results for the $v = 0 \to 1$ transition are shown in Fig. 2 (full line). They have also shown that the proper inclusion of polarization is vital, and that exchange effects alone do not reproduce integrated or differential cross sections very accurately.

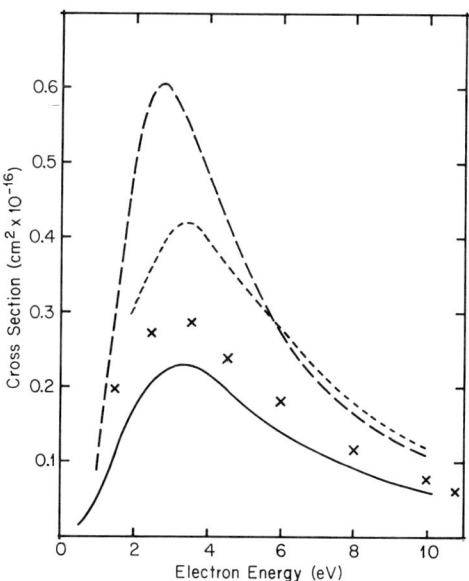

FIG. 2. Integrated cross section for transition $v = 0 \to 1$ of H_2: ———, theory, Klonover and Kaldor (1979b); ----, theory, Henry and Chang (1972); – – – –, theory, Henry (1970); ××××, experiment, Linder and Schmidt (1971a).

We have also included in Fig. 2 the older results of Henry (1970) (dashed curve) which were calculated using the vibration–rotation close coupling equations of Section II,B,1. The large disagreement between theory and experiment is probably due in part to errors in the approximate polarization potential but is also caused by the truncation in the sums over rotation and vibration states.

B. N_2

1. The 1.5–3.5-eV Region

In this low-energy region the cross sections are dominated by a series of narrow peaks separated by just a few tenths of an electron volt. This is illustrated for the $v = 0 \to 1$ and 2 transitions in Figs. 3 and 4. The compound state responsible for this structure has been identified as having $^2\Pi_g$ symmetry. This follows from the structure of the ground state of N_2, which is $X\ ^1\Sigma_g^+\ (1\sigma_g^2,\ 1\sigma_u^2,\ 2\sigma_g^2,\ 2\sigma_u^2,\ 3\sigma_g^2,\ 1\pi_u^4)$, if we assume the additional electron goes into the first vacant orbital, which happens to be $1\pi_g$. It is also confirmed by the differential cross section measurements (Erhardt and Willman, 1967), which have a very marked $l = 2$ behavior.

The original resonance analysis (Section II,B,6) of Herzenberg (1968) and of Birtwistle and Herzenberg (1971) showed that the experimental data could be analyzed in terms of a $^2\Pi_g$ compound state having a lifetime comparable to the vibrational period. The calculation has been refined by Dubé and Herzenberg (1979) to such an extent that the N_2^- parameters

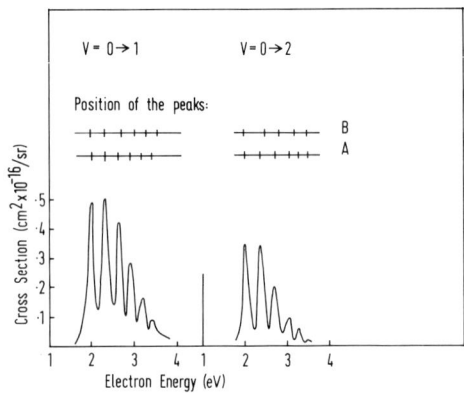

FIG. 3. Differential cross section at 90° in neighborhood of Π_g resonance for N_2: transitions $v = 0 \to 1$ and $0 \to 2$. Position of the peaks: (A) experiment, Wong; Dube and Herzenberg (1979); (B) Hazi et al. (1981).

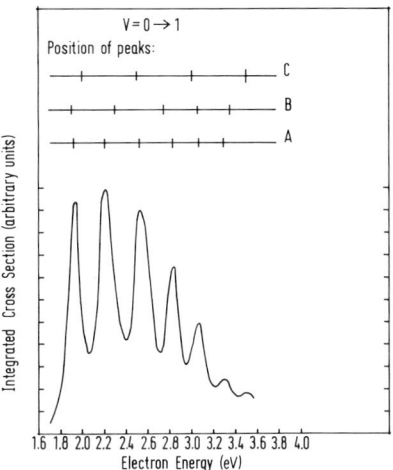

FIG. 4. Integrated cross section in neighborhood of Π_g resonance transition $v = 0 \to 1$. Position of peaks: (A) experiment, Erhardt et al. (1968); (B) Schneider et al. (1979b); (C) Chandra and Temkin (1976b.)

obtained by fitting to some experiments of Wong (cf. Dubé and Herzenberg, 1979) enabled a large number of cross sections to be reproduced with little error.

Hazi et al. (1981) have carried out an *ab initio* calculation within the formalism of this local complex potential theory. The N_2^- parameters were obtained from an R-matrix fixed-nuclei calculation (cf. Schneider et al., 1979b) performed at several internuclear separations. The agreement with experiment is not as good as the Dubé and Herzenberg fitting exercise, but the error is small and reflects the uncertainties in the R-matrix scattering results. In Fig. 3 we have indicated the position of the peaks determined by these two calculations for the $v = 0 \to 1$ and 2 transitions. The results of Dubé and Herzenberg agree exactly with experiment; those of Hazi et al. are slightly shifted.

Calculations have also been carried out within the R-matrix approach of Section II,B,5 and the hybrid method of Section II,B,4. In their calculation Schneider et al. (1979b) obtained $^2\Pi_g$ R-matrix eigenstates for the $(N_2 + e^-)$ system with the single configuration $(1\sigma_g^2, 1\sigma_u^2, 2\sigma_g^2, 2\sigma_u^2, 3\sigma_g^2, 1\pi_u^4, 1\pi_g)$. Exchange effects were included by antisymmetrizing the wave function. The total wave function for the inner region was matched to the outer region solution, where the interaction was assumed to be zero, to enable the resonance parameters to be determined. The eigenstates were also used to calculate the nuclear wave function and the vibrational cross section.

The hybrid method as applied by Chandra and Temkin (1976a,b) contained simplifications to the exchange and polarization interactions. Instead of antisymmetrizing the wave function, exchange effects were approximated by an orthogonalization technique (cf. Burke and Chandra, 1972). The polarization potential, correcting for the truncation in the electronic state expansion, had the correct $1/r^4$ asymptotic form but also had a short-range cutoff, arbitrarily chosen to give the correct position of the $^2\Pi_g$ resonance.

Results for these two methods are illustrated in Fig. 4, where we have indicated the positions of the cross section peaks for the $v = 0 \rightarrow 1$ transition. The error in the R-matrix calculation is small and may be due to the quality of the N_2 bound wave functions. The error in the hybrid calculation is more severe and becomes even greater for other transitions. As discussed previously in Section II,B, the calculations of Chandra and Temkin show the great difficulty of obtaining properly convergent results with respect to vibration and projectile angular states.

2. The 8-12-eV Region

In most of the processes discussed in this review the molecular electronic state does not change. However, recent measurements by Heutz et al. (1980; cf. also earlier work of Mazeau et al., 1973) are of interest in that they have shown considerable vibrational structure in the cross sections for excitation of the A $^3\Sigma_u^+$ and B $^3\Pi_g$ states, which are 6.22 and 7.39 eV, respectively, above the ground state. Transitions up to $v = 24$ have been observed for both states, each cross section being characterized by a large number of closely spaced oscillations. Heutz et al. have extended the work of Čadež and Fiquet-Fayard (1973) by analyzing the A $^3\Sigma_u^+$ data in terms of the complex potential resonance model. The configuration of this state is ($1\sigma_g^2$, $1\sigma_u^2$, $2\sigma_g^2$, $2\sigma_u^2$) $1\pi_u^3$, $3\sigma_g^2$, $1\pi_g$ and the negative ion state formed by filling the first empty orbital will have the configuration (. . .) $1\pi_u^3 3\sigma_g^2 1\pi_g^2$, $^2\pi_u$. This compound state can decay into the A $^3\Sigma_u^+$ parent and also the X $^1\Sigma_g^+$ ground state. However, the fitting analysis was simplified by assuming that the partial width for the decay to the ground state was negligible compared to that for the excited state.

Heutz et al. were very successful in fitting the positions of the peaks in the cross sections, but had less success with the absolute magnitude of the oscillations.

The $^2\Pi_u$ compound state also has an effect on transitions where the electronic state does not change. Again, there is considerable vibrational structure which can be qualitatively reproduced by a resonance analysis, although the procedure is not as straightforward as for the electronically excited state.

3. Intermediate Energies

The cross sections for the intermediate energies (15–30 eV) are smaller than at the lower energies, but it was noticed by Pavlovic et al. (1972) that they exhibit a broad peak between 20 and 30 eV. This has been confirmed by the experiments of Tronc et al. (1980) and of Tanaka et al. (1981). Pavlovic et al. suggested that the peak is due to many overlapping resonance states, but the more recent work, experimental and theoretical, points to just one broad resonance with $^2\Sigma_u$ symmetry.

On the theoretical side there are two calculations which use the fixed-nuclei plus impulse approximation of Section II,B,3. The continuum multiple scattering model (CMSP) of Dill and Dehmer (1974) has been applied to several molecules including nitrogen. The scattering problem is simplified by constructing spherical regions around each nucleus, and then a larger sphere around these two, or more, regions. If the nuclear separation for N_2 is R, then there are spheres of radii $R/2$ centered in each nucleus and a sphere R centered on the center of mass. In the initial work the potentials inside the inner spheres and outside the outer sphere were assumed to be spherically symmetric, and the potential between the spheres was assumed to be constant. The Schrödinger equation was solved in each region and then the wave function for all space was constructed by enforcing continuity at the spherical boundaries. In later work (Siegel and Dill, 1976) the restriction on the use of spherical potentials was dropped. For N_2 (Siegel et al., 1980; Dehmer et al., 1980) direct potentials were constructed in the inner regions from N_2 wave functions and exchange was introduced through the Hara local potential (Hara, 1967). [For a recent discussion of the application of local exchange potentials in electron–molecule scattering, see Morrison and Collins (1981) and Salvini and Thompson (1981b).] Resonance effects were seen in the Δ_g state at 13 eV and the Σ_u state of 26 eV, as well as the Π_g at 2–4 eV. In view of the fact that the breadth of the Σ_u resonance was found to be several electron volts, the impulse approximation should be quite satisfactory for the vibrational excitation process. Dehmer et al. (1980) calculated cross sections in this approximation and we give their results for the $v = 0 \rightarrow 1$ transition in Fig. 5. The peak is almost wholly due to the Σ_u scattering state; the effect of the Δ_g resonance at 13 eV is not apparent in the cross section.

A second calculation in the fixed-nuclei plus impulse approximation has been carried out by Rumble et al. (1981) for several energies between 5 and 50 eV. The exchange interaction was again simplified by the local potential approximation; a polarization potential with short-range cutoff, adjusted to give the position of the Π_g resonance, was also employed. Rumble et al. also

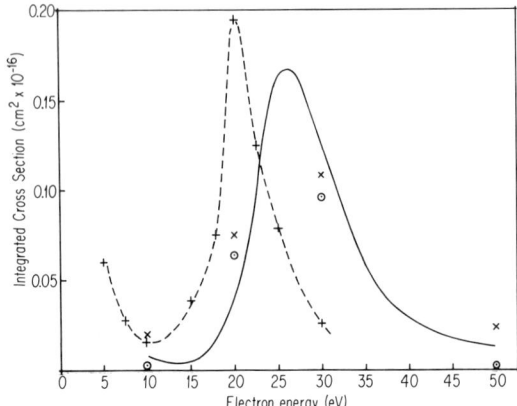

FIG. 5. Integrated cross section at intermediate energies for N_2: transition $v = 0 \rightarrow 1$. ——, theory, Dehmer et al. (1980); ×, theory, Rumble et al. (1981), all states included; ⊙, theory, Rumble et al. (1981), only state Σ_u included; +--+, experiment, Tanaka et al. (1981).

found the contribution to the vibration cross sections from the Σ_u state to be much larger than all the other states combined in the 20–30 eV range. Their results for the $v = 0 \rightarrow 1$ transition are shown in Fig. 5.

These calculations are compared in Fig. 5 with the experiments of Tanaka et al. (1981). The dotted line is a rough interpolation between the experimental points, which are quoted as having a 20% error in the 15–30 eV range. Theory and experiment agree as to peak size, but there is quite a large difference for the peak position. Tanaka et al. also give differential cross sections; these show an f-wave character at 20 eV which is consistent with the Σ_u designation of the compound state.

There is another calculation which we have not given in Fig. 5. Onda and Truhlar (1980) have solved the rotational close-coupling equations (neglect vibration in Section II,B,1) and then applied the vibration impulse approximation. The peak in the $v = 0 \rightarrow 1$ cross section is at 30 eV and has a value about half that of the data in the figure.

The above conclusion on the Σ_u resonance is confirmed by some fixed-nuclei calculations using the R-matrix method (Burke et al., 1983).

C. CO

The molecule CO is isoelectronic with N_2 and the cross sections of the two molecules are very similar. Between 1 and 3 eV, the cross sections are enhanced and show considerable structure; the $v = 0 \rightarrow 1$ transition resembles the experimental curves in Figs. 3 and 4, except that the size of the peaks is somewhat reduced. As we would expect, the local complex potential resonance analysis has been successful in analyzing the data. The ground state of CO is $^1\Sigma(1\sigma^2, 2\sigma^2, 3\sigma^2, 4\sigma^2, 5\sigma^2, 1\pi^4)$ and the compound state formed by the additional electron occupying the first vacant orbital has symmetry $^2\Pi$.

Zubek and Szmytkowski (1979) have obtained a good fit to the experiments of Erhardt et al. (1968). The CO^- potential which they determined shows quite a small width, certainly too small for the impulse approximation to be applicable.

Poe (1979) has carried out a calculation within the hybrid approximation, with the additional simplifications of local exchange potential and polarization potential with short-range cutoff. He obtained some vibrational structure, but agreement with experiment was not good.

As for N_2, there is evidence for a further resonance in the intermediate energy range. The experiments of Tronc et al. (1980) and of Chutjian and Tanaka (1980) have obtained enhancement in the cross sections at about 20 eV. Their differential cross section results, for the transition $v = 0 \rightarrow 1$, at this energy and the energy of the Π resonance are given in Fig. 6. The 1.83 eV curves do have a p-wave character, but the 19.5 eV curves are more f-wave. By analogy with N_2, we would expect the resonance at this higher

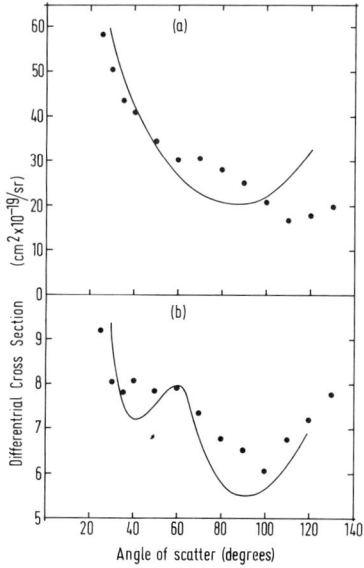

FIG. 6. Differential cross section for CO at energies in neighborhoods of (a) Π resonance, 1.83 eV and (b) Σ resonance 19.5 eV: ——, experiment, Tronc et al. (1980); ●, experiment, Chutjian and Tanaka (1980).

D. CO_2

In its electronic ground state, CO_2 is a linear molecule and therefore has four vibrational degrees of freedom. The three modes of vibration, symmetric and asymmetric stretch, and bending, which is twofold degenerate, are shown in Fig. 7. On the right of the figure we indicate how each normal coordinate transforms under symmetry operations of the molecule. States will be referred to by (V_s, V_b, V_a), the three numbers being respectively the eigenvalues for symmetric stretch, bending, and asymmetric stretch.

1. The 3–5-eV Region

Experimental results (cf. e.g., Čadež et al., 1974, 1977; Boness and Schulz, 1974) for this energy region show considerable structure, like N_2 and CO. Many transitions have been clearly identified, mainly for the symmetric stretch mode, but also for the bending mode. As for N_2 and CO, we expect the transitions to be governed by a compound state, some 3–4 eV above the CO_2 ground state, and with a width comparable to the vibrational transition energy.

Such a state was confirmed by the calculations of Claydon et al. (1970). The ground state of CO_2 is $^1\Sigma_g^+$ and the first compound state, with the linear geometry of the ground state, is $^2\Pi_u$. The potential energy curve for this state as a function of symmetric stretch coordinate, keeping the bending and asymmetric stretch coordinate fixed at zero, showed a minimum, as for the compound state in Fig. 1. Thus, the CO_2^- state formed by excitation of the ground state in the Franck–Condon region can vibrate along the symmetric stretch coordinate. However, the position is complicated by the extra vibrational degrees of freedom. Keeping the symmetric and asymmetric stretch coordinates zero, calculations of the CO_2^- potential curve as a function of bending coordinate showed that the state which is Π_u at 180° bending angle splits into two states with symmetries $^2A_1$ and ^2B. The ^2B has a minimum at

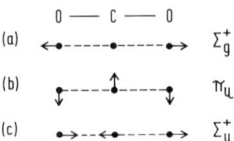

FIG. 7. Modes of vibration for CO_2: (a) symmetric stretch; (b) bending, (c) asymmetric stretch.

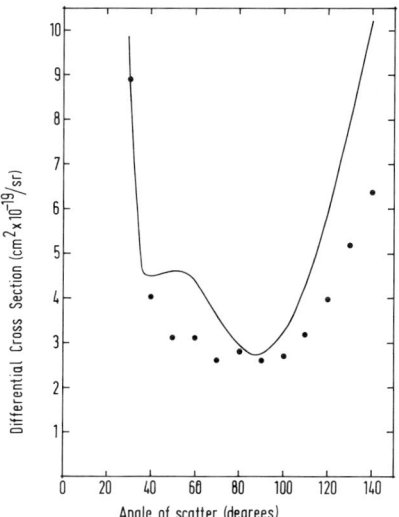

FIG. 8. Differential cross section for CO_2 at 10 eV for asymmetric stretch mode; transition $(000) \rightarrow (001)$: ———, theory, Thirumalai et al. (1980); ●, experiment, Register et al. (1980).

180°, but the $^2A_1$ has a minimum at approximately 140°; i.e., once the CO_2^- state is formed in the Franck–Condon region, the nuclei can begin to move away from the 180° configuration and begin to vibrate along the $^2A_1$ bending coordinate, as well as the stretching coordinate.

We can see that any resonance analysis along the lines of the Herzenberg method should be generalized. However, simplifications can be made in the case of CO_2 because the bending motion is much slower than the stretching, and the CO_2^- decays back to the ground state before much bending can occur. [It is interesting to note that the peaks in the cross sections for the $(n, 0, 0)$ and $(n, 1, 0)$ transitions are at the same positions (cf. Čadež et al., 1977), suggesting they have been formed by the same compound-state process.]

The structure in the symmetric stretch cross sections has been analyzed by two groups of authors, assuming for the reasons given above that the motion in the compound state was only along the symmetric stretch coordinate. The theory is just that given in Section II,B,6 with the internuclear separation replaced by the symmetric stretch coordinate $S = (r_1 + r_2)/2^{1/2}$, where r_1 and r_2 are the extensions of the two CO bonds. The more recent calculation was by Čadež et al. (1977) using their own experiments. The fitting procedure was very good when they assumed a $^2\Pi_u$ compound state which was predominantly p-wave, $l = 1$. Another calculation by Szmytkowski and Zubec (1977) used earlier results of Čadež et al. (1974); although their fit also was good, they obtained a larger width and found that the Π_u state was predominantly $l = 3$. Domcke and Cederbaum (1977) have also shown that structure can be obtained using a resonance model which is equivalent to the local complex potential model (cf. Section II,B,7).

2. The <2-eV Region

At energies for which there is no long-lived resonance state, we have seen that the vibrational excitation cross sections can be calculated using the fixed-nuclei and impulse approximations. This has been done by Morrison and Lane (1979) for the first symmetric stretch transition. They used the model of Hara local exchange and an empirical polarization potential chosen to give a resonance in the Π_u scattering state at the correct energy of 4 eV (cf. previous section). The impulse approximation breaks down at threshold, but the calculations do give evidence for a sharp increase in the cross section as the electron energy decreases; Morrison and Lane quoted tentative experimental observations by Wong that confirm a strong enhancement in the cross section at threshold.

These results are relevant to the general interpretation of threshold peaks. The CO_2 calculations are consistent with a "virtual state" interpretation. For low energies the main scattering state is Σ_g. The eigenphase sum is predominantly s-wave in character and exhibits a very strong energy dependence (with negative slope) as the energy tends to zero. Perhaps a similar interpretation applies to molecules both with and without dipole moments. We consider this question again in later sections.

3. Intermediate Energies

Dill et al. (1979) have applied their continuum multiple-scattering model (cf. Section III,B,3) to CO_2 in the energy range 0–35 eV. From consideration of the eigenphase sums they obtained the following resonance states (at equilibrium geometry): Π_u at 3.4 eV, Δ_g at 7 eV, Σ_g at 13.5 eV, and Σ_u at 29.5 eV. [The earlier *ab initio* calculations of Claydon et al. (1970) gave the Π_u state, as described above, and also gave a number of other states up to 15 eV. However, the symmetries are not in agreement with those of Dill *et al.*]

Dill et al. calculated vibrational cross sections for the first symmetric stretch transition in the fixed-nuclei and impulse approximations, obtaining enhancements at the positions of the Π_u, Σ_g, and Σ_u resonances. The approximation is not valid for the Π_u state, as we have seen, but is probably satisfactory for the other two states since the width is much larger. Agreement with the experiment of Tronc et al. (1979), who obtained enhancement of the 90° differential cross section at 10.8 and 30 eV, is quite good.

An extensive calculation has been carried out by Thirumalai et al. (1980, 1981; Thirumalai and Truhlar, 1981) for the excitation of the first asymmetric mode, but at only one energy, 10 eV. This energy was chosen because it is well away from resonance regions and the vibration impulse approximation should be valid. However, this was not applied as in Section II,B,3. We have already noted that problems arise in the fixed-nuclei approach for molecules with a permanent dipole moment (cf. Section II,B,2), and this is the case for CO_2 once we consider the asymmetric stretch coordinate. Thirumalai et al. overcame this problem by solving the rotational close-coupling equations, i.e., by using the theory of Section II,B,1 but with the omission of the expansion over the vibrational states. The rotational close-coupling equations gave a scattering amplitude for the $0-j'$ rotational transition which enabled $(vj) = (00) \rightarrow (1j')$ amplitudes to be calculated with the impulse approximation. The values of j' are restricted by the symmetry requirements on the total wave function. Since the oxygen nuclei have zero spin, the total wave function must be symmetric for their interchange. For the molecule in its ground electronic Σ_g^+ state and first vibrationally excited asymmetric state, which has the same symmetry as the normal coordinate Q_3, i.e., Σ_u^+, j' must be restricted to odd values only. Thirumalai et al. calculated cross sections for $j' = 1, 3, 5, 7$ and thus were able to obtain a rotationally summed $v = 0 \rightarrow 1$ cross section. Their result is shown in Fig. 8 together with the experiment of Register et al. (1980). Agreement is very good despite the approximation of the model, which included a local exchange potential and adiabatic polarization potential.

E. The Hydrogen Halides

The cross sections for HF, HCl, and HBr are characterized by large narrow threshold peaks where the scattering is almost isotropic [cf. Rohr and Linder (1975, 1976) and Rohr (1977a), but note that, according to Azria et al. (1980a), some of the HBr peaks are due to negative ion contributions from the associative attachment process]. At a few electron volts above threshold there is a broader, much less pronounced peak (not present in HF). Some of the cross sections show structure at energies where new vibrational channels become open.

If the broad feature is due to the effect of a compound state, it will be very short-lived. Gianturco and Thompson (1977) calculated eigenphase sums and differential cross sections for HCl held fixed at its equilibrium geometry.

Their model was a simple one with exchange approximated by an orthogonalization technique and with a polarization potential which had an adjustable parameter chosen to give a $^2\Sigma$ resonance state at 3.6 eV and a $^2\Pi$ resonance state at about 10 eV. The width of the $^2\Sigma$ state was found to be several electron volts. Because of the difficulty of separating resonant and nonresonant effects, the authors had no success in obtaining similar resonance parameters for HF, although Chang (1977), using work based on the frame transformation theory of Chang and Fano (1972), was able to use their results to fit the experimental differential cross section.

The stabilization method of Taylor *et al.* (1966) has been applied to HCl by Taylor *et al.* (1977) and Goldstein *et al.* (1978) and to HF by Segal and Wolf (1981). [There is some incompatibility between the HCl results and the dissociative attachment experiments of Azria *et al.* (1980b) which is explained by Segal and Wolf as being due to computational restrictions.] Taylor *et al.* obtained a $^2\Sigma$ HCl$^-$ state at about 4 eV above the ground state, in the Franck–Condon region, which could be responsible for the broad feature in the cross sections. Segal and Wolf obtained a similar state for HF at about 6 eV, where there are no experimental results.

Model calculations have been carried out for HCl and HF in an attempt to explain the threshold peaks. Dubé and Herzenberg (1977) considered the scattering by a nonrotating vibrator and solved the close-coupling equations for the asymptotic region, $r > r_0$, with just the long-range dipole interaction. No calculations were done for the internal region $r < r_0$; boundary conditions at $r = r_0$ were chosen by fitting to the experimental results of Rohr and Linder (1975). As well as obtaining a reasonably good fit, as can be seen in Fig. 9, for the $v = 0 \rightarrow 2$ transition, Dubé and Herzenberg showed that the boundary condition required at $r = r_0$ implied that there was considerable enhancement of the s-wave part of the electron function inside the molecule. This is consistent with our understanding of a "virtual state" discussed previously for CO_2.

Rudge (1980) has calculated cross sections for HF within the close-coupling formalism of Section II,B,1. The scattering potential $V_{\alpha\alpha'}$, which is used in the scattering equations (2), was calculated by replacing the integral $\langle \Psi_{el} V \Psi_{el} \rangle$ by a simple model potential which contained an adjustable parameter. Agreement with experimental integrated and differential cross sections is quite good; comparison with the results of Rohr and Linder (1976) for the transition $v = 0 \rightarrow 2$ is shown in Fig. 9.

Resigno *et al.* (1982) have made an *ab initio* determination of the HF cross sections within the fixed-nuclei plus impulse approximation of Section II,B,3. Exchange effects were included exactly but polarization was neglected. Their results showed a threshold peak which was due to the rapid variation of the $^2\Sigma^+$ state K matrices with respect to internuclear distance; the $^2\Sigma^+$ eigenphases also varied rapidly with energy as the energy tended to 0, which is again consistent with the virtual state interpretation. There was no evidence for a broad resonance state a few electron volts above threshold.

Their results for the $v = 0 \to 2$ transition are shown in Fig. 9 and are considerably lower than the experimental values. Resigno et al. have suggested that the disagreement may be due to uncertainties in the experimental data but neglect of polarization is also probably important. We recall that Klonover and Kaldor (1979b) showed the importance of polarization for H_2, results with polarization included being about 50% higher than the static exchange values.

The stabilization calculations have also given information on the threshold region. Negative ion states have been obtained very close to the ground state and the obvious suggestion has been made that they are responsible for the narrow peaks. Nesbet (1977) has suggested that these results are consistent with a virtual state interpretation. He argued that the stabilization technique tries to reproduce the real part of the position of the S-matrix pole in the complex momentum plane; but this is zero for a virtual state since the pole lies on the negative imaginary axis.

Domcke and Cederbaum (1981) have suggested that threshold behavior can be analyzed in terms of a discrete s-state, several electron volts above threshold, interacting with the (electron + molecule) continuum. As we have seen from the above discussion, it is still debatable whether this model is applicable to HF, but Domcke and Cederbaum have constructed with some success an exactly solvable model to mimic the HCl case. They did not attempt to make a complete fit to the experimental data but we have included their results for the $v = 0 \to 2$ transition in Fig. 9 to show the considerable structure obtainable by this model.

Domcke and Cederbaum also considered the fixed-nuclei problem within their model and calculated the energies at which the resonant phase shift would be equal to $\pi/2$. They showed that because of the long-range dipole potential there are three such energies, one corresponding to the usual resonance state resulting from the interaction of a discrete state with a continuum and leading to the broad peak in the cross sections, a second corresponding to a bound state, and a third which, although leading to an enhancement of the cross section near threshold, is not a resonance state since the phase shift has negative slope. The authors also suggested that these results are qualitatively consistent with the stabilization calculations. Taylor et al. obtained two HCl^- states just above the ground state. Domcke and Cederbaum suggested that maybe the ground-state energy was slightly in error and should be shifted so that one HCl^- state becomes bound; the other

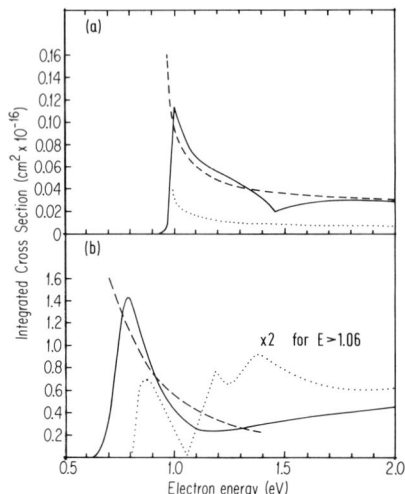

FIG. 9. Integrated cross section for HF (a) and HCl (b) near threshold: transition $v = 0 \rightarrow 2$. (a) ——, Experiment, Rohr and Linder (1976); ----, theory, Rudge (1980); ·····, theory, Resigno et al. (1982). (b) ——, Experiment, Rohr and Linder (1975); ----, theory, Dubé and Herzenberg (1977); ·····, theory, Domcke and Cederbaum (1981).

state now corresponds to the third case above and the threshold behavior of the two calculations looks similar.

Fabrikant (1980) mentioned that his effective range theory can reproduce the angular distributions in the threshold peaks but no numbers were given.

F. Various Molecules

1. O_2 and NO

An extensive resonance analysis has been carried out by Parlant and Fiquet-Fayard (1976) (cf. also Fiquet-Fayard, 1974) for O_2 and by Teillet-Billy and Fiquet-Fayard (1978) for NO. O_2 has a Σ_g^- ground state and a stable $^2\Pi_g$ negative ion; NO has a $^2\Pi$ ground state and a stable $^3\Sigma^-$ negative ion. The cross sections for excitation of both molecules at low projectile energies are series of sharp peaks at the positions of the vibrational levels of the negative ions (cf. Linder and Schmidt, 1971b; Tronc and Azria, 1979). This can only occur if the lifetime of the compound states is long compared to the vibrational period of these states.

Recent experiments by Tronc and Azria (1979) have shown a broad enhancement of cross sections at 10 eV, attributed to a short-lived X $^4\Sigma_u^-$ compound state, with fine structure attributed to Feshbach resonances.

2. N_2O

N_2O is a linear asymmetric molecule, N-N-O, with ground state $^1\Sigma^+$, and therefore we might expect the vibrational structure to be like CO_2. However, although the experiments of Azria *et al.* (1974) show an enhancement in the cross section around 2–3 eV, there is no fine structure. This suggests that the resonance state responsible is short-lived.

Dubé and Herzenberg (1975) analyzed these results using the local complex potential model on the supposition that the compound state has symmetry $^2\Sigma$. The analysis of Section II,B,6 was complicated by the extra degrees of freedom but simplified by neglecting the kinetic energy term from the equation for the nuclear motion [cf. Eq. (44)]. The fitting procedure by this "impulse" approximation was satisfactory.

3. H_2O and H_2S

H_2O (H—O—H) and H_2S (H—S—H) are nonlinear molecules and therefore have three vibrational degrees of freedom. The three modes of vibration, symmetric stretch (A_1 symmetry), bending (A_1 symmetry), and asymmetric stretch (B symmetry) are generalizations of those shown in Fig. 7 for CO_2. Scattering experiments cannot distinguish between the two stretch modes because their frequencies are almost equal.

Cross sections for excitation of the stretch modes and the bending mode have been measured by Seng and Linder (1974, 1976) and by Rohr (1977b, 1978). They have similar characteristics to those described for HCl and HBr, a large narrow peak at threshold followed by a much broader peak. At energies away from the threshold peak excitation of the H_2O bending mode was interpreted as a nonresonant scattering process, but the other transitions were interpreted in terms of short-lived A_1 resonance states.

Gianturco and Thompson (1980) carried out fixed-nuclei calculations for both H_2O and H_2S. For H_2O the dominance of the nonresonant contribution prevented resonance parameters being obtained for the A_1 state, but a very broad resonance was found in the B_2 state (NB: we are employing the usual literature notation for B states, not that of Gianturco and Thompson). Jain and Thompson (1983a) have extended the calculations to obtain vibrational excitation cross sections for the symmetric stretch and bending modes within the fixed-nuclei plus impulse approximations. They have considered the variation of the fixed-nuclei scattering amplitude with respect to just one normal coordinate, the other two being put equal to zero. The result for the excitation of the first symmetric stretch mode has the same shape as, but only about one quarter the magnitude of, the experimental result, which is for the sum of symmetric and asymmetric stretch excitations. The bending mode results agreed closely with the experimental differential cross sections away from threshold, and were about 50% higher than the corresponding Born approximation results (cf. Itikawa, 1974). Only two states, A_1 and B_2, contributed significantly to the cross sections in the energy range, threshold to 10 eV, A_1 being predominant near the threshold peak but B_2 being the main contribution at other energies.

For H_2S Gianturco and Thompson (1980) were able to obtain resonance parameters for two A_1 states and one B_1 state (note there is an error in the designation of this state in Gianturco and Thompson). However, more recent improved calculations by Jain and Thompson (1983b) suggested there is only one A_1 resonance, rather broad at about 7 eV, and a narrower B_2 resonance at about 2 eV. Again, these two states dominated the vibrational excitation cross sections for symmetric stretch and bending modes. Agreement with experiment was good, but the physics of of the process may still be incomplete since the calculations do not take into account the B_1 Feshbach resonance discussed by Trajmar and Hall (1974) for the dissociative attachment process.

Fabrikant (1980) used his effective range theory for H_2O to obtain a fairly satisfactory fit to the differential cross sections at 0.61 eV, the energy of the threshold peak.

4. SF_6 and CH_4

Rohr (1977c, 1980) has observed that SF_6 and CH_4, both nonpolar molecules, have threshold structure like the hydrogen halides. Domcke and Cederbaum (1980) have shown that for a simple model a nonadiabatic resonance treatment of a discrete state embedded in a continuum can lead to narrow threshold peaks. Their analysis may be applicable to SF_6 and CH_4.

5. H_2^+

Robb and Collins (1980) have calculated cross sections at low energies in the fixed-nuclei plus impulse approximation. Their model included exchange and gave results which increased monotonically as the energy decreased, agreeing qualitatively with some Coulomb–Born calculations of Boĭkova and Ob″edkov (1968).

REFERENCES

Altmann, S. L., and Cracknell, A. P. (1965). *Rev. Mod. Phys.* **37**, 19.
Arthurs, A. M., and Dalgarno, A. (1960). *Proc. R. Soc. London, Ser. A* **256**, 540.

Azria, R., Wong, S. F., and Schulz, E. J. (1974). *Phys. Rev. A* [3] **11**, 1309.
Azria, R., Le Coat, Y., and Guillotin, J. P. (1980a). *J. Phys. B* [1] **13**, L505.
Azria, R., Le Coat, Y., Simon, D., and Tronc, M. (1980b). *J. Phys. B* [1] **13**, 1909.
Bardsley, J. N. (1968). *J. Phys. B* [1] **1**, 349.
Bardsley, J. N., and Wadehra, J. M. (1979). *Phys. Rev. A* [3] **20**, 1398.
Bardsley, J. N., Herzenberg, A., and Mandl, F. (1966a). *Proc. Phys. Soc., London* **89**, 305.
Bardsley, J. N., Herzenberg, A., and Mandl, F. (1966b). *Proc. Phys. Soc., London* **89**, 321.
Berman, M., Cederbaum, L. S., and Domcke, W. (1983). *J. Phys. B.* [1] **16**, 875.
Birtwistle, D. T., and Herzenberg, A. (1971). *J. Phys. B* **4**, 53.
Boĭkova, R. F., and Ob″edkov, V. D. (1968). *Sov. Phys.—JETP (Engl. Transl.)* **27**, 772.
Boness, M. J. W., and Schulz, G. T. (1974). *Phys. Rev. A* [3] **9**, 1969.
Born, M., and Oppenheimer, J. R. (1927). *Ann. Phys. (Leipzig)* [4] **84**, 457.
Brink, D. M., and Satchler, G. R. (1968). "Angular Momentum." Oxford Univ. Press (Clarendon), London and New York.
Brown, S. C., ed. (1979). "Electron-Molecule Scattering." Wiley, New York.
Buckley, B. D. (1978). *Electron. At. Collisions, Proc. Int. Cont., 10th, 1977* Abstracts, p. 140.
Burke, P. G. (1979). *Adv. At. Mol. Phys.* **15**, 471.
Burke, P. G., and Chandra, N., (1972). *J. Phys. B* [1] **5**, 1696.
Burke, P. G., Noble, C. J., and Salvini, S. (1983). *J. Phys. B.* [1] **16**, L113.
Čadež, I., and Fiquet-Fayard, F. (1973). *Phys. Electron. At. Collisions, Invit. Pap., Prog. Rep. Int. Conf., 8th, 1973* Abstracts, p. 454.
Čadež, I., Tronc, M., and Hall, R. I. (1974). *J. Phys. B* [1] **7**, L132.
Čadež, I., Gresteau, F., Tronc, M., and Hall, R. I. (1977). *J. Phys. B* [1] **10**, 3821.
Cederbaum, L. S., and Domcke, W. (1981). *J. Phys. B* **14**, 4665.
Chandra, N., and Temkin, A. (1976a). *Phys. Rev. A* [3] **13**, 188.
Chandra, N., and Temkin, A. (1976b). *Phys. Rev. A* [3] **14**, 507.
Chang, E. S. (1977). *J. Phys. B* **10**, L395.
Chang, E. S., and Fano, U. (1972). *Phys. Rev. A* [3] **6**, 173.
Chase, D. M. (1956). *Phys. Rev.* **104**, 838.
Chen, J. C. Y. (1964a). *J. Chem. Phys.* **40**, 3507.
Chen, J. C. Y. (1964b). *J. Chem. Phys.* **40**, 3513.
Chutjian, A., and Tanaka, H. (1980). *J. Phys. B* [1] **13**, 1901.
Claydon, C. R., Segal, G. A., and Taylor, H. S. (1970). *J. Chem. Phys.* **52**, 3387.
Dehmer, J. L., Siegel, J., Welch, J., and Dill, D. (1980). *Phys. Rev. A* [3] **21**, 101.
Dennison, D. M., and Hecht, K. T. (1962). *In* "Quantum Theory" (D. R. Bates, ed.), Vol. 2, pp. 246–322. Academic Press.
Dill, D., and Dehmer, J. L. (1974). *J. Chem. Phys.* **61**, 692.
Dill, D., Welch, J., Dehmer, J. L., and Siegel, J. (1979). *Phys. Rev. Lett.* **43**, 1236.
Domcke, W. (1981). *J. Phys. B* [1] **14**, 4889.
Domcke, W., and Cederbaum, L. S. (1977). *Phys. Rev. A* [3] **16**, 1465.
Domcke, W., and Cederbaum, L. S. (1980). *J. Phys. B* [1] **13**, 2829.
Domcke, W., and Cederbaum, L. S. (1981). *J. Phys. B* [1] **14**, 149.
Domcke, W., Cederbaum, L. S., and Kapur, F. (1979). *J. Phys. B* **12**, L359.
Dubé, L., and Herzenberg, A. (1975). *Phys. Rev. A* [3] **11**, 1314.
Dubé, L., and Herzenberg, A. (1977). *Phys. Rev. A* [3] **15**, 820.
Dubé, L., and Herzenberg, A. (1979). *Phys. Rev. A* [3] **20**, 194.
Erhardt, H., and Willman, K. (1967). *Z. Phys.* **204**, 462.
Erhardt, H., Langhans, L., Linder, F., and Taylor, H. S. (1968). *Phys. Rev.* [2] **173**, 222.
Fabrikant, I. I. (1976). *Sov. Phys.—JETP (Engl. Transl.)* **44**, 77.
Fabrikant, I. I. (1977a). *Sov. Phys.—JETP (Engl. Transl.)* **46**, 693.

Fabrikant, I. I. (1977b). *J. Phys. B* [1] **10**, 1761.
Fabrikant, I. I. (1978). *J. Phys. B* [1] **11**, 3621.
Fabrikant, I. I. (1980). *Phys. Lett. A* **77A**, 421.
Faisal, F. H. M., and Temkin, A. (1972). *Phys. Rev. Lett.* **28**, 203.
Fano, U. (1961). *Phys. Rev.* **124**, 1866.
Feldt, A. N., and Morrison, M. A. (1982). *J. Phys. B* [1] **15**, 301.
Fiquet-Fayard, F. (1974). *Vacuum* **24**, 533.
Gianturco, F. A., and Thompson, D. G. (1977). *J. Phys. B* [1] **10**, L21.
Gianturco, F. A., and Thompson, D. G. (1980). *J. Phys. B* [1] **13**, 613.
Goldstein, E., Segal, G. A., and Wetmore, R. W. (1978). *J. Chem. Phys.* **68**, 271.
Hall, R. (1978). *Phys. Electron. At. Collisions, Invit. Pap. Prog. Rep. Int. Conf., 10th, 1977* pp. 25–41.
Hara, S. (1967). *J. Phys. Soc. Jpn.* **22**, 710.
Hazi, A. U. (1979). *Phys. Rev. A* [3] **19**, 920.
Hazi, A. U., Resigno, T. N., and Kurilla, M. (1981). *Phys. Rev. A* [3] **23**, 1089.
Henry, R. J. W. (1970). *Phys. Rev. A* [3] **2**, 1349.
Henry, R. J. W., and Chang, E. S. (1972). *Phys. Rev. A* [3] **5**, 276.
Herzberg, G. (1950). "Molecular Spectra and Molecular Structure." Van Nostrand-Reinhold, Princeton, New Jersey.
Herzenberg, A. (1968). *J. Phys. B* [1] **1**, 548.
Herzenberg, A., and Mandl, F. (1962). *Proc. R. Soc. London, Ser. A* **270**, 48.
Heutz, A., Čadež, I., Gresteau, F., Hall, R. I., Vichon, D., and Mazeau, J. (1980). *Phys. Rev. A* [3] **21**, 622.
Itikawa, Y. (1974). *J. Phys. Soc. Jpn.* **36**, 1127.
Jain, A. K., and Thompson, D. G. (1982). *J. Phys. B* [*1*] **15**, L631.
Jain, A. K., and Thompson, D. G. (1983a). *J. Phys. B* [*1*] **16**, L347.
Jain, A. K., and Thompson, D. G. (1983b). *J. Phys. B*. In press.
Kapur, P. L., and Peierls, R. (1938) *Proc. R. Soc. London, Ser. A* **166**, 277.
Klonover, A., and Kaldor, U. (1979a). *J. Phys. B* [1] **12**, 323.
Klonover, A., and Kaldor, U. (1979b). *J. Phys. B* [1] **12**, 3797.
Lane, N. F. (1980). *Rev. Mod. Phys.* **52**, 29.
Le Dourneuf, M., Vo Ky Lan, and Schneider, B. I. (1979). *In* "Symposium on Electron-Molecule Collisions—Invited Papers" (I. Shimamura and M. Matsuzawa, eds.), Appendix. University of Tokyo, Tokyo.
Linder, F., and Schmidt, H. (1971a). *Z. Naturforsch., A* **26A**, 1603.
Linder, F., and Schmidt, H. (1971b). *Z. Naturforsch., A* **26A**, 1617.
Mazeau, J., Gresteau, F., Hall, R. I., Joyez, G., and Reinhardt, J. (1973). *J. Phys. B* [1] **6**, 862.
Morrison, M. A., and Collins, L. A. (1981). *Phys. Rev. A* [3] **23**, 127.
Morrison, M. A., and Lane, N. F. (1979). *Chem. Phys. Lett.* **66**, 527.
Nesbet, R. K. (1977). *J. Phys. B* [1] **10**, L739.
Nesbet, R. K. (1981). *Comments At. Mol. Phys.* **11**, 25.
Norcross, D. W., and Padial, N. T. (1982). *Phys. Rev. A* [3] **25**, 226.
Onda, K., and Truhlar, D. G. (1980). *J. Chem. Phys.* **72**, 5249.
Parlant, G., and Fiquet-Fayard, F. (1976). *J. Phys. B* [1] **9**, 1617.
Pavlovic, Z., Boness, M. J. W., Herzenberg, A., and Schulz, G. J. (1972). *Phys. Rev. A* [3] **6**, 676.
Poe, R. T. (1979). *In* "Symposium on Electron-Molecule Collisions—Invited Papers" (I. Shimamura and M. Matsuzawa, eds.), pp. 49–54. University of Tokyo, Tokyo.
Register, D. F., Nishimura, H., and Trajmar, S. (1980). *J. Phys. B* [1] **13**, 1651.
Resigno, T. N., McCurdy, C. W., and McKoy, V. (1974a). *Chem. Phys. Lett.* **27**, 401.

Resigno, T. N., McCurdy, C. W., and McKoy, V. (1974b). *Phys. Rev. A* [3] **10,** 2240.
Resigno, T. N., McCurdy, C. W., and McKoy, V. (1975). *Phys. Rev. A* [3] **11,** 825.
Resigno, T. N., McKoy, V., and Schneider, B. (1979). "Electron-Molecule and Photon-Molecule Collisions." Plenum, New York.
Resigno, T. N., Orel, A. E., Hazi, A. U., and McKoy, B. V. (1982). *Phys. Rev. A* [3] **26,** 690.
Robb, W. D., and Collins, L. A. (1980). *Phys. Rev. A* [3] **22,** 2474.
Rohr, K. (1977a). *J. Phys. B* [1] **10,** L399.
Rohr, K. (1977b). *J. Phys. B* [1] **10,** L735.
Rohr, K. (1977c). *J. Phys. B* [1] **10,** 1175.
Rohr, K. (1978). *J. Phys. B* [1] **11,** 4109.
Rohr, K. (1980). *J. Phys. B* [1] **13,** 4897.
Rohr, K., and Linder, F. (1975). *J. Phys. B* [1] **8,** L200.
Rohr, K., and Linder, F. (1976). *J. Phys. B* [1] **9,** 2521.
Rudge, M. R. H. (1980). *J. Phys. B* [1] **13,** 1269.
Rumble, J. R., Truhlar, D. G., and Morrison, M. A. (1981). *J. Phys. B* [1] **14,** L301.
Salvini, S. A., and Thompson, D. G. (1981a). *Comput. Phys. Commun.* **22,** 49.
Salvini, S. A., and Thompson, D. G. (1981b). *J. Phys. B* [1] **14,** 3797.
Schneider, B. I., Le Dourneuf, M., and Burke, P. G. (1979a). *J. Phys. B* [1] **12,** L365.
Schneider, B. I., Le Dourneuf, M., and Vo Ky Lan (1979b). *Phys. Rev. Lett.* **43,** 1926.
Schulz, G. J. (1962). *Phys. Rev.* [2] **125,** 229.
Schulz, G. J. (1964). *Phys. Rev. A* [2] **135,** 988.
Schulz, G. J. (1973). *Rev. Mod. Phys.* **45,** 423.
Schulz, G. J. (1979). *In* "Electron-Molecule Scattering" (S. C. Brown, ed.), pp. 1–56. Wiley, New York.
Segal, G. A., and Wolf, K. (1981). *J. Phys. B* [1] **14,** 2291.
Seng, G., and Linder, F. (1974). *J. Phys. B* [1] **7,** L509.
Seng, G., and Linder, F. (1976). *J. Phys. B* [1] **9,** 2539.
Shimamura, I., and Matsuzawa, M., eds. (1979). "Symposium on Electron-Molecule Collisions—Invited Papers." University of Tokyo, Tokyo.
Siegel, J., and Dill, D. (1976). *J. Chem. Phys.* **64,** 3204.
Siegel, J., Dehmer, J. L., and Dill, D., (1980). *Phys. Rev. A* [3] **21,** 85.
Siegert, A. J. F. (1939). *Phys. Rev.* **56,** 750.
Szmytkowski, C., and Zubek, M. (1977). *J. Phys. B* [1] **10,** L31.
Tanaka, H., Yamamoto, T., and Okada, T. (1981). *J. Phys. B* [1] **14,** 2081.
Taylor, H. S., Nazaroff, G. V., and Golebiewski, A. (1966). *J. Chem. Phys.* **45,** 2872.
Taylor, H. S., Goldstein, E., and Segal, G. A. (1977). *J. Phys. B* [1] **10,** 2253.
Teillet-Billy, D., and Fiquet-Fayard, F. (1978). *Electron. At. Collisions, Proc. Int. Conf., 10th, 1977* Abstracts, p. 150.
Thirumalai, D., and Truhlar, D. G. (1981). *J. Chem. Phys.* **75,** 5207.
Thirumalai, D., Onda, K., and Truhlar, D. G. (1980). *J. Phys. B* [1] **13,** L619.
Thirumalai, D., Onda, K., and Truhlar, D. G. (1981) *J. Chem. Phys.* **74,** 6792.
Trajmar, S., and Hall, R. I. (1974). *J. Phys. B* [1] **7,** L458.
Tronc, M., and Azria, R. (1979). *In* "Symposium on Electron-Molecule Collisions—Invited Papers" (I. Shimamura and M. Matsuzawa, eds.), pp. 105–110. University of Tokyo, Tokyo.
Tronc, M., Azria, R., and Paineau, R. (1979). *J. Phys. Lett.* **40,** L323.
Tronc, M., Azria, R., and Le Coat, Y. (1980). *J. Phys. B* [1] **13,** 2327.
Wilson, E. B., Decius, J. C., and Cross, P. C. (1955). "Molecular Vibrations." McGraw-Hill, New York.
Zubec, M., and Szmytkowski, C. (1979). *Phys. Lett. A* **74,** 60.

VIBRATIONAL AND ROTATIONAL EXCITATION IN MOLECULAR COLLISIONS

MANFRED FAUBEL

Max-Planck-Institut für Strömungsforschung
Göttingen, Federal Republic of Germany

I. Introduction 345
II. Theoretical Methods 347
    A. *Ab Initio* and Model Potentials 347
    B. Computation of Cross Sections 351
    C. The Sudden Collision "Factorization Relations" 353
III. Experimental Techniques 354
    A. Survey of Methods 354
    B. Molecular Beams and Detectors 357
IV. Studies of Rotational Scattering Cross Sections 362
    A. Na_2–Ar 362
    B. Rotational Rainbows in Na_2–Ar and Na_2–Ne 364
    C. He–N_2, CO, O_2, and CH_4 367
    D. Exact versus Approximate He–N_2 Cross Sections 374
    E. D_2–CO Diatom–Diatom Scattering 378
V. Studies of Vibrational Excitation 380
    A. Vibrational Excitation in Diatomics 380
    B. Mode-Selective Excitation of CO_2 by H^+ and D^+ 381
    C. High-Overtone Mode-Selective and Excitation
       of CF_4 by Ions 382
VI. Summary of Detailed Scattering Experiments and
    Concluding Remarks 385
    References 389

I. Introduction

When two molecules collide, they usually change their direction of flight and, often, also their internal state of rotation and vibration. Depending on the collision energy and on the collision partners involved, other inelastic processes might be electronic transitions, possibly accompanied by electron or photon emission, a chemical reaction, or the dissociation of a molecule. However, these latter processes will practically always involve some rotational or vibrational motion of the nuclei, whereas isolated rotational and vibrational excitation or energy exchange can be observed in a large class of molecules. Thus, the isolated investigation of vibrational and rotational inelastic energy transfer can be justified as a process interesting in itself as well as a precursory study for the whole complex of inelastic and reactive molecular collisions.

A theoretical approach to inelastic molecular collisions requires a knowledge of the molecular interaction potential. In a second step, scattering theory leads through usually very involved computations to state-to-state differential scattering cross sections. These are generally considered as the most detailed observable quantities of the molecular collision process. During the last 10 years sensitive experimental tests on the quality of theoretical predictions for potentials and cross sections became available with vibrational and rotational resolved differential cross section measurements for molecular collisions at thermal velocities. Surveys on the early development of these scattering experiments and on their theoretical interpretation are given in reviews by Toennies (1976), by Faubel and Toennies (1978), and by Bernstein (1979). The literature up to the end of 1979 is covered in the recent reviews by Loesch (1980) and by DePristo and Rabitz (1980).

The first successful vibrational-state-resolved low-energy ion–molecule scattering experiments were performed in the early 1970s. Subsequently, the angular dependence of individual rotational state transitions was observed in ion–hydrogen molecule scattering experiments and about five years ago also in neutral–neutral crossed-molecular-beam experiments with the H_2 isotopic species. In 1978 the first example for fully resolved rotational state-to-state transitions of a nonhydrogen molecule became available with the measurement of differential inelastic cross sections for the scattering of Na_2 by rare gases. Details on these experiments are covered in the above reviews and will not be repeated here except for a short general discussion in Section III and for the literature survey given in Table III at the end of the present review.

Meanwhile — as illustrated by Table III — the number of detailed angular-resolved scattering experiments with at least partial internal vibrational and rotational state resolution has increased to more than 40 different combinations of collision partners. These include the scattering of ions and atoms from diatomic and small polyatomic molecules and rotationally inelastic diatomic–diatomic molecule collisions.

The present review will discuss a few crossed-molecular-beam experiments which in the authors view are typical examples for the experimental developments and for the interaction of theory and experiment in the time period from 1979 to about June 1982. Section II gives a short introduction to the theoretical methods employed for the interpretation of experimental inelastic scattering data. General experimental methods and some technically important details of state-resolving crossed-molecular-beam experiments are discussed in Section III. Recent high-resolution rotationally and vibrationally inelastic scattering experiments are described in Sections IV and V. The experimental results are compared with *ab initio* theory when available. Additionally, an attempt is made to interpret characteristic cross section phenomena in terms of simplified physical models for the collision dynamics.

II. Theoretical Methods

A. *Ab Initio* AND MODEL POTENTIALS

The dynamics of a molecular collision is completely governed by the interaction forces between and within the colliding molecules. For given molecular electronic states these can be summarized in potential energy functions of the respective internuclear distances and, when excluding electronic transitions, only one potential energy surface is relevant for the molecular collision. Theoretically justified by the Born–Oppenheimer approximation for separating nuclear motion and electronic wave functions, a potential surface can be derived and computed from first principles by solving the electronic quantum eigenvalue equation with the essentially electrostatic interaction of electrons and nuclei for a sufficiently large number of sets of fixed nuclear coordinates.

Reviews on this approach to potential surfaces have been given by Kutzelnigg (1977) and Schaefer (1979). But, because of the high numerical accuracy required for deducing the interaction potential as a rather small change of the total binding energy of the two molecules, this quantum chemical approach rapidly becomes a numerically formidable task as the number of electrons or, even worse, the number of nuclei in the two molecules increases. Thus, for larger molecules and for potentials with extremely shallow van der Waals attractive wells, resort must be had to more approximate, but practicable, predictive potential models.

One such model, appealing because its computational speed is orders of magnitude higher, is the modernized version of the Drude electron gas model introduced by Gordon and Kim (1972) and recently revised by Waldman and Gordon (1979a). Although this model predicts potential wells that are notoriously too deep by 15–20% and up to 100% for light molecules, some of its problems could be remedied by additional correction terms with only a moderate sacrifice in computational speed (Rae, 1973; Waldman and Gordon, 1979a,b).

An intermediate position is taken by potential models starting from the two asymptotic regions of small and very large intermolecular distances where *ab initio* calculations work best. In these regions, respectively, the large repulsive interaction energy can be obtained with high relative accuracy from Hartree–Fock self-consistent field (SCF) calculations, while the long-range attractive part is developed into an inverse power series of induction and dispersion interaction energies as, e.g., in Buckingham (1967) or Kihara (1978) model potentials. However, these latter terms are here derived from a perturbation treatment of the separate molecule's SCF *ab initio* eigenfunctions. The remaining intermediate region near the potential well is then interpolated by appropriate semitheoretical or semiempirical cutoff and correction functions which are characteristic for the respective model potentials proposed by Tang and Toennies (1977, 1978, 1982) and by Hepburn *et al.* (1975), Douketis *et al.* (1982), and Rodwell and Scoles (1981).

Well-known potential surface parts relevant for the rotational excitation of a closed-shell diatomic molecule by a structureless atom are shown in Fig. 1 for the $Li^+ - H_2$ ion–molecule interaction and for the rare gas–hydrogen $Ne - H_2$ potential. The $Li^+ - H_2$ potential was obtained from full-scale *ab initio* calculations by Lester (1970) and Kutzelnigg *et al.* (1973) and the

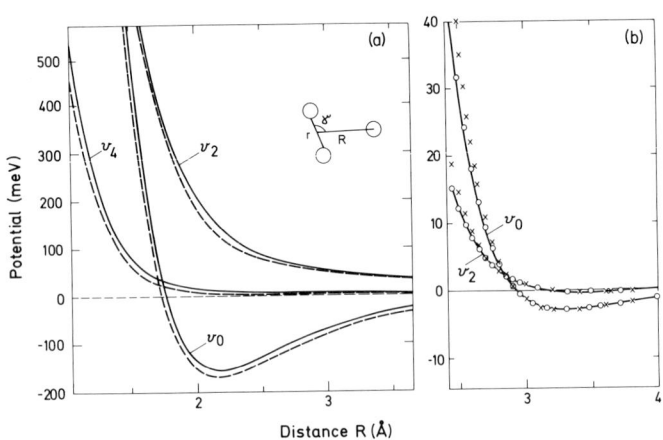

FIG. 1. Legendre polynomial expansion terms $v_\lambda(R)$ of the $H_2 - Li^+$ (a) and the $H_2 - Ne$ (b) intermolecular potentials. For $H_2 - Li^+$ SCF (solid lines, Lester, 1970) and CI (dashed lines, Kutzelnigg *et al.*, 1973) *ab initio* results are shown. For $H_2 - Ne$ the continuous line represents the best-fit experimental potential (Buck, 1982). The predictions of two potential models are shown: (×) Tang and Toennies (1978) and (○) Rodwell and Scoles (1981).

Ne–H_2 potential is an example of model potential predictions (Tang and Toennies, 1978; Rodwell and Scoles, 1981). Both theoretical potentials are confirmed by detailed experimental and theoretical rotationally inelastic scattering investigations. Remaining descrepancies are smaller than 10–15% for Li^+–H_2 (Faubel and Toennies, 1979; Schaefer and Lester, 1975) and, similarly, the experiments for Ne–H_2 by Zandee and Reuss (1977) and by Buck et al. (1980) agree. For Ne–H_2 even a direct inversion of elastic and rotationally inelastic experimental cross sections was possible within an exponential distorted wave approximation (Gerber et al., 1980).

As usual, for closed-shell atom–diatom interaction the potential is represented as a function of the distance R between the atom and the center of mass of the molecule, the intramolecular bond length r, and the orientation angle γ, illustrated by the insert in Fig. 1a. The angular dependence of the potential $V(R, r, \gamma)$ is then conventionally expanded into a series of Legendre polynomials:

$$V(R, r, \gamma) = \sum_\lambda v_\lambda(R, r) P_\lambda(\cos \gamma) \qquad (1)$$

although expansions in terms of an angular-dependent well depth $\epsilon(\gamma)$ and well position $r_m(\gamma)$ (Pack, 1978; Keil et al., 1978) were found to be more convenient for fitting interaction potentials.

For the remaining bond length dependence of the expansion terms $v_\lambda(R, r)$ often a lowest-order Taylor expansion from the intramolecular vibrational equilibrium distance r_e is sufficiently accurate to describe changes in this essentially harmonic part of the potential. For pure rotational excitation of a rigid rotor, r is kept constant at $r = r_e$ and these are the potential expansion terms $v_\lambda(R)_{r=r_e}$ previously shown in Fig. 1. Because of the symmetry of the H_2 molecule, only even-numbered terms $v_2(R)$ appear in the potential expansion. For both potentials the contributions of the spherical part $v_0(R)$ and the $v_2(R)$ term are by far dominant over higher-order terms in the Legendre expansion and thus the spatial form of the potential can easily be reconstructed for given orientation angles γ as, e.g., $V(R, \gamma = 0) = v_0(R) + v_2(R)$ and $V(R, \gamma = 90°) = v_0(R) - \frac{1}{2}v_2(R)$ for the collinear and for the perpendicular configuration, respectively.

The most dramatic difference between the ionic and the rare gas interaction with H_2 is in the average well depth ϵ_0. Caused by the ion-charge-induced polarization of the H_2, Li^+–H_2 has an $\epsilon_0 \approx 150$ meV compared to an $\epsilon_0 \approx 2.9$ meV for the van der Waals interaction in Ne–H_2, making the Li^+–H_2 a strongly attractive and the Ne–H_2 a shallow, mainly repulsive potential for thermal collision energies in the range of some ten to several hundred millielectron volts. Differences are also noted in the mean radii of the potentials ($R_m \approx 2.1$ Å for Li^+–H_2 and $R_m \approx 3.2$ Å for the Ne–H_2). Additionally, the v_2 term in the Li^+–H_2 interaction is at all distances rather large in comparison to the v_0 term, whereas in Ne–H_2 $v_2(R)$ is everywhere small compared to the isotropic part of the potential. Therefore, one might distinguish between the strong unlocalized anisotropic interaction for the case of Li^+–H_2 and a weak unlocalized interaction for Ne–H_2. A third case of a strong interaction localized at the repulsive barrier will be discussed in Section IV,D for the He–N_2, O_2 potentials and is also observed to some extent in the H_2–CO potential shown in Fig. 2.

A formal representation of the interaction of an atom with a polyatomic molecule and of two diatomic or polyatomic molecules is easily obtained as a straightforward extension of Eq. (1) by replacing the Legendre polynomials $P_\lambda(\cos \gamma)$ with spherical harmonics or an orthogonal set of products of spherical harmonics for describing the multidimensional orientation dependence of the respective molecular interactions. As inelastic scattering experiments were recently undertaken with H_2–H_2 and CO–H_2 (see Section IV,E and Table III), the first five radial expansion terms for an *ab initio* CO–H_2 potential (Flower *et al.*, 1979) are shown in Fig. 2 as an example for the interaction potential of two rigid diatomic molecules. With the unit vectors $\hat{r}_1$ and $\hat{r}_2$ for the orientation of the internuclear axes of the H_2 and the CO and the vector **R** connecting the centers of mass of the two molecules, the potential is expanded in body-fixed coordinates where the intermolecular axis R is one of the axes of the coordinate system:

$$V(\hat{r}_1, \hat{r}_2, R) = \sum_{\lambda_1 \lambda_2 \mu \geq 0} v_{\lambda_1 \lambda_2 \mu}(R) \mathcal{Y}_{\lambda_1 \lambda_2 \mu}(\hat{r}_1, \hat{r}_2) \qquad (2a)$$

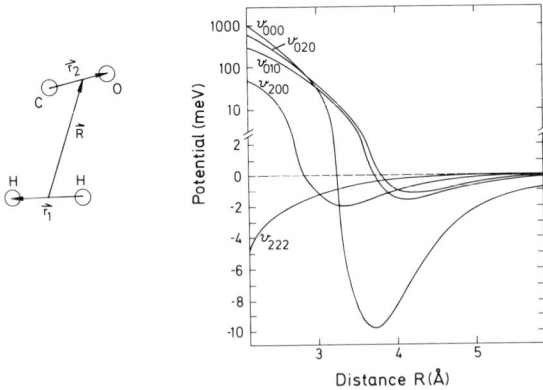

FIG. 2. Expansion terms $v_{\lambda_1 \lambda_2 \mu}(R)$ for the H_2–CO rigid rotor, SCF long-range dispersion interaction potential of Flower *et al.* (1979) in body-fixed coordinates [Eq. (2a)]. The diagram on the left illustrates the definition of the vector coordinates $\mathbf{r}_1$, $\mathbf{r}_2$, and **R** used for the potential expansion into products of spherical harmonics. The energy scale is logarithmic for $v_{\lambda_1 \lambda_2 \mu} > 2.7$ meV.

with the orthogonalized products of spherical harmonics:

$$\mathcal{Y}_{\lambda_1\lambda_2\mu}(\hat{r}_1, \hat{r}_2) = 4\pi[2(1 + \delta_{\mu 0})]^{-1/2}$$
$$\times [Y_{\lambda_1\mu}(\hat{r}_1)Y_{\lambda_2-\mu}(\hat{r}_2) + Y_{\lambda_1-\mu}(\hat{r}_1)Y_{\lambda_2\mu}(\hat{r}_2)]$$

Slightly different body-fixed coordinates have been used by Alexander and DePristo (1976) for the HF–HF potential. Currently in wider use are space-fixed coordinate systems with triple products of spherical harmonics for taking account of the orientation of the intermolecular axis R [see Green (1977); and, e.g., for CO–H_2, Green and Thaddeus (1976); for H_2–H_2 Schaefer and Meyer (1979); and for N_2–N_2, Berns and van der Avoird (1980), van Hemert and Berns (1982)]:

$$V(\hat{r}_1, \hat{r}_2, \mathbf{R}) = \sum_{\lambda_1\lambda_2\lambda} A_{\lambda_1\lambda_2\lambda}(R)$$
$$\times \left[\sum_{m_1 m_2 m} \langle \lambda_1 m_1 \lambda_2 m_2 | \lambda m \rangle Y_{\lambda_1 m_1}(\hat{r}_1) Y_{\lambda_2 m_2}(\hat{r}_2) Y^*_{\lambda m}(\hat{R}) \right] \quad (2b)$$

with $\langle \lambda_1 m_1 \lambda_2 m_2 | \lambda m \rangle$ being a Clebsch–Gordon coefficient.

The two representations are interrelated by

$$A_{\lambda_1\lambda_2\lambda}(R) = 4\pi[4\pi/(2\lambda + 1)]^{1/2} \sum_{\mu \geq 0} v_{\lambda_1\lambda_2\mu}(R)$$
$$\times \langle \lambda_1 \mu \lambda_2 - \mu | \lambda 0 \rangle [2/(1 + \delta_{\mu 0})]^{1/2} \quad (2c)$$

for $\lambda_1 + \lambda_2 + \lambda$ even and $A_{\lambda_1\lambda_2\lambda} = 0$ otherwise (Flower et al., 1979).

In the body-fixed expansion for H_2–CO, shown in Fig. 2, $v_{000}(R)$ is the isotropic part of the potential; v_{010} and v_{020} contribute to rotational excitation of the CO and v_{200} to rotational excitation of the H_2 molecule, only. Odd values of λ_1 are forbidden by the H_2 symmetry and v_{222} causes simultaneous excitation of the H_2 and CO molecules. The isotropic part of the potential has a well depth of 9.8 meV at $R_m \approx 3.7$ Å. The asymmetry of the potential with respect to rotations of the CO molecules is quite strong at the repulsive barrier, where up to $v_{000}(R) \approx 50$ meV the CO rotation terms $v_{010}(R)$ and $v_{020}(R)$ are larger than the isotropic potential.

B. Computation of Cross Sections

In order to treat the collision dynamics, the Schrödinger equation for the motion of the molecular nuclei on the given potential energy surface has to be solved and from its asymptotic solutions the experimentally observable differential state-to-state scattering cross sections are obtained. A complete recent survey on the presently available theoretical methods is given by a number of extensive review articles collected in a book on atom molecule collision theory, edited by Bernstein (1979). Therefore the discussion of details of scattering theory can here be restricted to a few points of immediate relevance for the experiments to be discussed in the following sections.

As with the computation of potentials from the electronic Schrödinger equation, an exact, but computationally problematic procedure—the "coupled channels" or "close coupling" formalism—is also known for the scattering theory (see, e.g., Light, 1979; Secrest, 1979). Numerous approximate schemes range from quantum approximations (see Kouri, 1979) over semiclassical to completely classical Monte Carlo trajectory treatments (see Gentry, 1979; Pattengill, 1979).

Within the exact coupled-channels treatment a complete orthogonal basis for the scattering function is constructed from products of the separate system molecular eigenfunctions, spherical harmonics, and unknown radial eigenfunctions. This expansion leads to an infinite set of coupled radial equations which has to be truncated to approximately 100 or less coupled channels to be solvable with presently available algorithms and computers. An estimate of the number of coupled channels contributing to an inelastic collision can be obtained by counting the number of separate system molecular eigenstates with energy levels smaller than the total collision energy. In addition to these energetically open channels, closed channels also may have to be included if deep attractive wells are present in the interaction potential. The actual number of channels required for convergence is usually checked out numerically. Because of the large magnetic quantum number degeneracy of higher rotational states, only a maximum of about 10–15 rotational levels of one molecule can be included. Thus, the coupled-channels treatment of the pure rotational excitation in atom–linear molecule collisions could be extended to diatom–diatom scattering, as for H_2–H_2 (Green, 1975), and to collisions of an atom with symmetrical and spherical top molecules, as for He–H_2CO (Garrison et al., 1976), He–NH_3 (Green, 1980), and Ar–CH_4 (Smith and Secrest, 1981), with a much higher computational complexity. But until now, combined rotational and vibrational excitation cross sections could be computed by exact coupled-channels calculations only for H_2, with its extremely wide rotational energy levels (e.g., see Schaefer and Lester, 1975).

Therefore, a heavy demand exists for either drastically speeding up the coupled-channels calculation, for which the computational effort increases with the third power of the number of included channels, or for finding acceptable approximate methods.

The numerically most successful of all quantum or classical approximations developed is the "infinite order sudden" (IOS) approach, which consists of two approximations. The orbital angular momentum l in the exact coupled-channels equations is kept fixed at a constant value (the "coupled states" approximation) and, additionally, the wave number k_{jn} of each rotation/vibration channel is also kept at a fixed average j value for a given vibrational state n, corresponding to a rotational "energy sudden" approximation. The resulting, only vibrationally coupled, coupled-channels equations can be solved with a much lower computational effort for a number of fixed potential orientation angles. These asymptotic solutions are finally superimposed to give magnetic quantum number-averaged scattering cross sections. For pure rotational excitation only one equation remains from the original set of coupled equations and the problem reduces to the straightforward solution of the well-known elastic partial waves equations (see, e.g., Parker and Pack, 1978; Kouri, 1979). The validity of the IOS approach is limited by these two approximations. The fixed-l or "coupled states" approximation is expected to be always quite good with predominantly repulsive potentials when the change of the turning point position with a change in l is not large. It is poor if there are three turning points, as associated with a deep potential well. The energy sudden approximation works best for small relative energy changes in the collision and negligible contributions of closed channels.

C. The Sudden Collision "Factorization Relations"

Beyond the benefit of great computational speed, very interesting analytical relations among different rotational cross sections could be derived within the IOS approximation (Goldflam et al., 1977), and also within the "l_z-conserving" energy sudden approximation which is less restrictive on deep potential wells (Khare, 1978). For different initial rotational states in atom–linear, rigid rotor scattering cross sections a "factorization relation" could be established which reads as follows:

$$\frac{d\sigma}{d\omega}(j_i \to j_f) = \frac{k_0^2}{k_{j_i}^2}(2j_f + 1) \sum_{j_L} \begin{pmatrix} j_i & j_L & j_f \\ 0 & 0 & 0 \end{pmatrix}^2 \frac{d\sigma}{d\omega}(0 \to j_L) \qquad (3)$$

This equation reduces differential cross sections from arbitrary initial states j_i to linear, $3j$ symbol $\begin{pmatrix} abc \\ 000 \end{pmatrix}$ weighted superpositions of rotationally inelastic cross sections $d\sigma/d\omega$ ($0 \to j_L$) for the rotational ground state. The restrictive energy sudden requirement of small relative rotational energy transfer in Eq. (3) can be somewhat relaxed by additional "adiabaticity" factors in the "energy corrected sudden" approximation by DePristo et al. (1979). An expression quite similar to Eq. (3) holds for diatom–diatom collisions (Goldflam and Kouri, 1979). In a generalization to atom–symmetric top molecules, corresponding relations were found to hold between T-matrix elements only and not for the cross sections themselves [Green (1979); see also Beard *et al.* (1982) for a recent discussion].

A second analytical relation within the energy sudden approximation states the conservation of the sum of rotational cross sections over all final states j_f, which is found to be independent of the initial j_i state. Within the IOS approximation this total rotational cross section is further shown to be (Parker and Pack, 1978)

$$\sum_{j_f} \frac{d\sigma}{d\omega}(j_i \to j_f, n_i \to n_f) = \frac{d\sigma}{d\omega}(n_i \to n_f)$$

$$= \frac{1}{2}\int_{-1}^{1}\frac{d\sigma}{d\omega}(n_i \to n_f; \gamma)\,d(\cos\gamma) \quad (4)$$

for any j_i. The integrand on the right-hand side is the differential cross section calculated for a vibrational transition $n_i \to n_f$ only at a fixed molecular orientation angle γ. Thus, Eq. (4) shows explicitly how to treat rotationally averaged vibration independently of the rotation, and also shows that for a given elastic or vibrationally inelastic transition the cross section is preferentially weighted for the perpendicular ($\gamma = 90°$) approach to the molecule.

Even if such factorization relations are only valid in an approximate fashion and for a probably limited range of collision conditions, they have the great practical value of reducing the number of cross sections which are actually to be considered as sources of independent information in a study of rotational collision phenomena. Thus, they were successfully used by Smith and Pritchard (1981) for reducing and fitting observed rotational state-to-state integral cross sections and rate constants by simple scaling laws. The factorization can also be expected to be very helpful in the inversion of scattering data to interaction potentials, where, for example, in a "sensitivity analysis" of scattering cross sections versus potential features by Eno and Rabitz (1981) attention was limited to the angular dependence of a few rotational transitions.

III. Experimental Techniques

A. Survey of Methods

Experimental potential energy surfaces, or at least parts thereof, can be determined by the inversion of spectroscopic vibration–rotation data, thus by-passing completely the collision dynamics. This has long been done for stable molecules, although problems still exist, for example, for weakly bound triatomic molecules as discussed by Mills (1977). Recent methods of determining intermolecular potentials by the spectroscopy of van der Waals dimer molecules were reviewed by LeRoy and Carley (1980).

Less detailed but direct average information on the interaction potentials of gases is also available through the temperature dependence of second virial coefficients (see, e.g., Maitland *et al.,* 1981).

Experiments on vibrational and rotational transitions by molecular collisions always exhibit some averaging over the collision energy, scattering angles, and internal initial and final states of the participating molecules. Thus, a whole hierarchy of cross sections of various degrees of averaging has been defined and experimentally investigated.

The most extensive source of data on a very wide variety of collision phenomena comes from bulk gas phase experiments where rotational and vibrational energy transfers are observed as relaxation processes, in spectral line shifts and line broadening, and through influences on the transport properties of gases.

Detailed discussions of energy transfer phenomena are found in monographs by Lambert (1977) and by Yardley (1980), and of transport properties, in the books by Hirschfelder *et al.* (1954) and by Maitland *et al.* (1981) and in a review by Beenakker and McCourt (1970). A compact survey on various other collision-induced bulk-phase phenomena was given some time ago by Gordon *et al.* (1968). Particularly by using laser techniques for observing state-to-state molecular energy transfer, many individual level relaxation times were investigated. Some have very practical consequences like the near-resonant energy transfers for the pumping of lasers or for the investigation of state-selective chemical reactions (Kneba and Wolfrum, 1980). An example of a detailed forward comparison of *ab initio* theory with time-resolved, infrared double-resonance gas phase experiments for the rotational level-to-level relaxation of CO by H_2 is found in a study by Brechignac *et al.* (1980). But the thermal averaging over translational and, often, also internal states in bulk cross sections limits the extraction of finer collision details as illustrated in the work by Eno and Rabitz (1980) or by Smith and Tindell (1982) on the determination of interaction potentials from macroscopic data.

Most of the thermal averaging present in bulk-phase collision data can be removed in crossed-molecular-beam experiments where a defined collision energy and true single-collision conditions can be insured.

In integral total-cross-section measurements the averaging extends over final states and scattering angles. One molecular beam is usually rotational and magnetic quantum number state selected by electric or magnetic deflection and its attenuation by a second molecular beam is measured as a function of the collision energy and the molecular orientation. Typical molecules suited for Electric focusing are CsF and TlF in electric quadrupole and NO in electric six-pole fields. H_2 molecules have been magnetically state selected. The orientational dependence of the glory structures in total cross sections yields information on the anisotropic interaction potential. These are investigated for collisions with many common gas molecules and were reviewed by Thuis et al. (1979) and by Loesch (1980).

When the final internal state of molecules leaving from the scattering center is also identified, state-to-state integral cross sections are obtained. The first examples of state-to-state cross sections of alkali halides gave usually a "partial" integral cross section only integrated over the forward scattering angular region transmitted by the state-analyzing device (see Borkenhagen et al., 1979). Integral rotational state-to-state cross sections for LiH on He, HCl, DCl, and HCN were investigated, using electric state focusing for the selection of the initial j-state and laser-induced fluorescence (LIF) for the state-selective detection of the final rotational states of LiH (see Dagdigian, 1980). Selective excitation of one initial vibration–rotation state of HF with an infrared chemical laser followed by probing of the infrared fluorescence spectrum of the scattered hydrogen halide allowed the measurement of rotational state-to-state integral cross sections of HF on Ne, Ar, and Kr in the vibrational states $n = 1$ and 2 of the HF molecule (Barnes et al., 1982). Also recently reported were Ar–NO integral rotational state-to-state cross section measurements by LIF in the ultraviolet combined with pulsed nozzles for cooling the NO to the rotational ground state. Here, interestingly, a strong alternation between odd and even level transition cross sections was observed and is attributed to the asymmetry of the slightly heteronuclear NO molecule (Andresen et al., 1982).

A third class of averaged crossed-molecular-beam experiments is total differential scattering cross sections where the angular distribution of scattered molecules is measured, but their final internal state is left undetermined. Here, the quenching of the angular diffraction and rainbow structures can be exploited for information on the aspherical part of the interaction potential. These experiments are available for a very large class of molecular collisions. Up to about 1979 they are summarized in the review of Loesch (1980). For He–N_2, NO, CO, O_2, CO_2, CH_4, N_2O, and C_2N_2 they are presented and evaluated for potentials in a recent series of articles by the Kuppermann group, the latest article being by Parker et al. (1983).

Differential vibrationally and rotationally inelastic scattering experiments measure the angular dependence of fully resolved state-to-state cross sections or, when internal state resolution cannot be completely achieved, inelastic excitation cross sections for energetically close groups of states. Early vibrational-state-resolved scattering experiments at collision energies of a few electron volts were reported by David et al. (1973) for Li^+-H_2 and by Udseth et al. (1973) for H^+-H_2. Resolution of rotational states was achieved first for Li^+-H_2 (van den Bergh et al., 1973) and for hydrogen–rare gas neutral scattering (Buck et al., 1977; Gentry and Giese, 1977). Meanwhile (see also Table III in Section VI), rotational and vibrational inelastic differential scattering measurements are available for some 50 different combinations of collision partners. Current studies investigate the rotational excitation of heavier diatomic as well as of polyatomic molecules by atoms and by diatomic molecules (Section IV,E). Interesting features of vibrational mode selectivity emerged also in the scattering of ions from polyatomic molecules (see Section V).

B. Molecular Beams and Detectors

Crossed-molecular-beam experiments for high translational and internal state resolution require dedicated and complicated equipment, typically representing investments equivalent to 10,000 working hours, for producing collimated, velocity- and state-selected molecular beams and for detecting the translational and internal state distributions. Because of the widely varying physical properties of molecules and the still serious signal-to-noise problems in inelastic scattering cross-section measurements, no true "universal" molecular beam machine has yet been built. Instead, matching the particular needs of a particular scattering experiment, a wide variety of devices have been developed and combined, ranging from continuously cooled and heated molecular, atom, and ion beam sources, to pulsed nozzle beams, velocity selectors, single and random chopping time-of-flight techniques, state focusing and state labeling devices, to species-sensitive and state-sensitive detectors. Many of these techniques have been described in detail in a textbook by Fluendy and Lawley (1973) or are covered in the reviews by Faubel and Toennies (1978) and by Loesch (1980).

One component common to practically all present experiments is the nozzle beam source for intense molecular beams [see Anderson (1974) for a review]. In these nozzles a gas is expanded from a high-pressure region through a small hole into a vacuum. In addition to giving a high beam brightness of typically $10^{18}-10^{21}$ molecules/sec/sr, nozzle expansions act by isentropic hydrodynamic cooling in a very efficient manner as a velocity-defining device and, because of internal state cooling, they also can produce molecular beams which are almost completely enriched in their lowest rotational state. The terminal translational energy E_{trans} or velocity u_∞ of a nozzle beam is determined by the nozzle temperature T_0, or more precisely by the stagnation gas enthalpy $H_0(T_0)$, and approaches the value $E_{trans} \propto$

$H_0(T_0) = 5/2kT_0$ and $7/2kT_0$, respectively, for ideal monatomic and diatomic gases. For gas mixtures the heavy component when present in small concentrations can be accelerated up to $E_{trans} = (m_H/m_L)H_0(T_0)$, where m_H/m_L is the mass ratio of the heavy and light components. Thus, by varying nozzle temperatures and gas mixtures the molecular beam energy can be tuned from a few millielectron volts up to several electron volts.

The degree of internal cooling and the velocity resolution of a nozzle beam increases with the product of the nozzle pressure p_0 and the nozzle diameter D_N, which determines the absolute number of collisions of a molecule in the hydrodynamic expansion before the free molecular flow region is entered. By making $p_0 D_N$ large enough, a nozzle expansion can always be driven to the point where condensation starts to heat up the beam. Examples of such high velocity resolutions, observed near the condensation limit, are given in Table 1 for a few, continuous molecular nozzle beams. For He, the most favorable case, a full width at half maximum velocity (FWHM) selection value $\Delta v_\parallel/u_\infty \leqslant 0.007$ can be obtained with nozzle openings of 20 μm and pressures of 200 atm. The example of H_2 beams for different nozzle temperatures between 300 and 26 K shows that the ultimate translational cooling characterized by the temperature $T_\parallel$ derived from the velocity spread of the beam is fairly independent of the source temperature T_0. For most other common gases $\Delta v_\parallel/u_\infty$ values are near 10%. Also shown in Table I are rotational temperatures T_{rot} and rotational state populations for the p_0 and D_N values listed in the last two columns.

The condensation is known (see, e.g., Knuth, 1977; Hagena, 1974) to scale with $p_0^2 D_N$, whereas the relaxation goes with $p_0 D_n$. Therefore, a somewhat higher translational and rotational cooling—down to values of 1 K for most gases—can be achieved with larger nozzles at the expense of a higher gas flux proportional to $p_0 D_N^2$ into the vacuum system, creating pumping capacity and residual gas background problems. Considerably higher nozzle gas fluxes up to 10^{23} molecules/sec can be handled by pulsed nozzle beam sources without overloading the vacuum system. These were surveyed by Gentry (1980b). Theoretical limits on the shortest pulse length were derived by Saenger (1981), giving values near 100 μsec for a full expansion of common gases. Pulse repetition rates as high as 750 Hz were recently reported by Cross and Valentini (1982).

Further useful discussions, references, and engineering formulas for the nozzle beam translational and internal state cooling are given by Poulsen and Miller (1977) and for beam intensities and skimmer design by Beijerinck and Verster (1981).

The density of molecules in a well-collimated and velocity-selected free molecular beam cannot be much higher than some 10^{13} molecules/cm^3, the equivalent of a room temperature gas at a pressure of 10^{-3} Torr, before the

TABLE I

PROPERTIES OF SOME TYPICAL NOZZLE BEAMS

Molecule	$\Delta v_\parallel / U_\infty$	$T_\parallel$(K)	T_{rot}(K)	Rotational state population [experimental method]	T_0(K)	p_0(Torr)	D_N(cm)
He[a,b]	0.007	—	—	—	300	1.5×10^5	0.002
H_2	0.02^c / 0.04^c / $\leq 0.15^d$	0.12 / 0.12 / ≤ 0.5	120	92% in $j=0$ for p-H_2^c [energy balance]	293 / 80 / 26	1.3×10^5	0.006
N_2^e	0.07	3	7	$j=0$: 52%; $j=1$: 33%; $j=2$: 15%^{f} [electron beam fluorescence]	300	5×10^3	0.01
O_2	—	—	13	$j=1$: 70%; $J=3$: 30%^{g} [molecular beam magnetic resonance]	300	9×10^2	0.01
CH_4	—	—	13	All ground state (see Fig. 7) except 10–20% left in $j=2$ level of F modification^h [inverse Raman]	300	2.8×10^3	0.01
CF_4^e	0.1^e	4–5	—	—	300	5×10^3	0.01
Na_2^j (10% Na_2 in Na)	0.15 / 0.16 / to 0.22	10	14 / 40 K for $j \leq 17$ and 75 K for $j > 17$ [$T_{vib}(Na_2) = 150$ K for $j \leq 28$]	—	300 / 300 / 800	3×10^3 / 19 for Na	0.01 / 0.05

^{a} Campargue et al. (1977). ^{b} Brusdeylins et al. (1977). ^{c} Winkelmann (1979). ^{d} van Deursen and Reuss (1973). ^{e} Brusdeylins and Meyer (1979). ^{f} Faubel and Weiner (1981). ^{g} Amirav and Even (1980). ^{h} Valentini et al. (1980). ^{i} Bergmann et al. (1980).

beam starts to suffer from self-collisions, even for the translationally cooled nozzle beams. Using a conservative value of 10^{12} molecules/cm^3 and a length of 1 cm for the interaction zone of two beams in a scattering experiment, a primary beam attenuation of 1% can be expected for a typical molecular integral cross section of 100 Å^2. The differentially scattered intensity flowing into a solid angle element of 10^{-2} sr, extended by a detector with 6° angular resolution, is then a fraction of 10^{-6} of the primary beam flux for a typical differential cross section of 1 Å^2/sr. The beam flux for a primary beam with a molecular velocity of 10^5 cm/sec and density of 10^{12} molecules/cm^3 through an area of 1 cm^2 is 10^{17} molecules/sec. Thus, the scattered intensity for the above example becomes 10^{11} molecules/sec and the scattered particle's density is $10^6 - 10^4$ molecules/cm^3 in distances of 10 cm to 1 m from the scattering center. Whereas this estimate is at most too pessimistic by more than a factor of 100, these numbers can easily decrease by many orders of magnitude, e.g., for energy change experiments [see the discussion in Faubel and Toennies (1978)] when an angular resolution narrower than the above modest 6° is required, or when only a small fraction of beam molecules is in the state of interest and state selection or labeling must be applied.

Molecular beam detectors for use in scattering experiments should, therefore, be sensitive to $< 10^4$ molecules/cm^3 and beam fluxes $\leqslant 10^9$ molecules/sec. The properties of a number of currently used detectors are listed and compared in Table II. Detection of particles is easiest in the case of ions or metastable molecules because these can trigger directly a secondary electron multiplier. Thus, they can be detected with very low spurious background, short detector response times and with detection efficiencies close to unity. The detector length, of interest for time-of-flight experiments, is shorter than 1 mm, the detector area can be matched to any large or small angular aperture, and using channel plates, it can even be made position sensitive, allowing measurement of very many scattering angles at one time (Wijnaendts van Resandt and Los, 1980). Because of these advantages, many neutral molecule detection schemes convert the neutral molecules into ions. Examples are the long-known hot ribbon surface ionizers for alkalis and for halides which were recently used, for example, for the efficient detection of UF_6 molecules (Dittner and Datz, 1978). More "universal" are electron bombardment ionizers with ionization probabilities on the order of 10^{-4} for an ionizer length of 1 cm and an open area of 1 cm^2 (Lantzsch, 1974). As intense and tunable laser light sources became available, two-photon (e.g., Herrmann et al., 1977) and resonance-enhanced multiphoton ionization (Zandee and Bernstein, 1979) were also explored. Both methods can be made state and species selective and the two-photon ionization has very promising efficiencies of 20%, although it has not yet been used in a

TABLE II

PROPERTIES OF MOLECULAR BEAM DETECTORS

Detector	Detected molecules	Detection efficiency (counts/molecule)	Detector length	Response time
Open secondary-electron multiplier	Ions, metastables	$\leqslant 1$	1 mm	nsec
Surface ionization	Alkalis, halides	$\leqslant 0.1\text{–}1$	1 mm	μsec to msec
Electron bombardment ionization	All	$\sim 10^{-4}$	1 cm	μsec
Laser-induced fluorescence	Na_2, LiH, NO, ...	0.02 $10^{5\,a}$	1 cm–1 μm ·/·	nsec[b] sec[b]
Bolometer and tunable infrared laser	HF, CO, NH_3, ...			
Cw two-photon and pulsed multiphoton ionization	Na_2, K_2, large organic molecules	0.2 (TPI) 10^{-4} (MPI)	—	nsec[b]

[a] Noise equivalent estimated for 2 sec averaging time and 0.2 eV/molecule.
[b] Species and state selective.

scattering experiment. Laser-induced fluorescence requires only one laser and is a state- and species-selective flux-sensitive detector. The overall efficiency of 2% (Bergmann *et al.*, 1980) results from photon losses in the fluorescence light collection system and on the photocathode of the photomultiplier. A further, new, detection scheme combines a bolometer which is sensitive to the total energy of a molecular beam with a tunable infrared laser for exciting beam molecules before they hit the bolometer from their original rotational state to a vibrationally excited state (Gough *et al.*, 1981). Because vibrationally excited molecules live a long time before they radiate, almost all molecules deposit this additional vibrational energy on the bolometer. By scanning the laser frequency through the rotation vibration lines of the molecule, the flux of beam molecules in each rotation vibration state is read from the bolometer. The first state-to-state scattering experiments with this detection method were reported by Boughton *et al.* (1982) for the rotation–vibration dependence of HF–He integral cross section.

IV. Studies of Rotational Scattering Cross Sections

A. Na_2–Ar

The Na_2 molecule with a ground-state binding energy of $D_e^0 = 0.72$ eV is readily produced in fractions of 10–20% Na_2 dimers in supersonic beams of sodium. The molecule has a rotational constant of $B_e = 0.15$ cm^{-1} = 0.019 meV and an also quite narrow spacing of vibrational energy levels with $\omega_e = 159$ cm^{-1} = 19.7 meV (Huber and Herzberg, 1979). An absorption band system near 600 nm makes the molecule ideally suited for LIF studies with presently available lasers.

Because of a natural linewidth of about 20 MHz, the fluorescence intensities are very susceptible to small frequency shifts of either the laser or of the resonant line by the Doppler shift of molecules moving in different directions. The latter, commonly as undesirable as the former, was recently used by Kinsey and Pritchard in a very elegant and efficient way for measuring angular distributions from the doppler shift (ADDS) (Phillips *et al.*, 1978).

In order to illustrate the ADDS technique, Fig. 3a shows a Newton diagram of the velocities of Na_2 scattered from an argon beam with the respective initial velocities $v_{Na_2} = 1.55 \times 10^5$ cm/sec and $v_{Ar} = 5.6 \times 10^4$ cm/sec used in the actual experiment (Serri *et al.*, 1980, 1981). All molecules scattered with a fixed energy loss ΔE for a given internal transition end up with velocities $u_{cm}(\Delta E)$ on a sphere centered at the constant center of mass

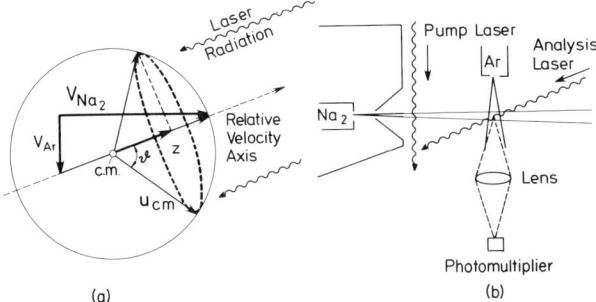

FIG. 3. Velocity diagram for Na$_2$–Ar illustrating the principle (a) and the actual design (b) of the ADDS experiment.

(c.m. in Fig. 3a) motion of the colliding molecules. Furthermore, molecules with a fixed center of mass scattering angle ϑ are all found on a Newton circle—the dashed out-of-plane circle in Fig. 3a—which is characterized by a constant velocity component $u_z(\vartheta) = u_{cm}(\Delta E) \cos \vartheta$ along the axis of relative motion. When the scattering region is illuminated with laser light parallel to the direction of the relative velocity vector, only molecules with selected scattering angle ϑ given by the Doppler frequency shift $\Delta \nu = u_z/\lambda = u_{cm}(\cos \vartheta)/\lambda$ will be excited and produce a fluorescence signal. The scattering angle is then changed, without any moving parts in the actual scattering experiment, by scanning the laser through a frequency range of $2u_{cm}\lambda \approx 3$ GHz for a typical velocity $u_{cm} = 10^5$ cm/sec and for a laser wavelength near $\lambda = 600$ nm.

The actual design of this experiment is shown in Fig. 3b. The only major problem of the method is the requirement of good velocity resolution in both colliding beams for obtaining an acceptable angular resolution. This was solved by using a seeded beam of 30 Torr Na/Na$_2$ in 5 atm of a Ne/He carrier gas mixture to give the Na$_2$ a velocity spread of slightly better than 10% half-width. After two stages of collimation the Na$_2$ primary beam intersects an argon nozzle beam. The fluorescence light produced in the intersection region by the analysis laser is collected onto a photomultiplier for single photon counting. Before the Na$_2$ beam enters the scattering region, an individual initial rotational state of Na$_2$ is selectively depleted by a second "pump" laser. The state-to-state cross section is measured as the signal difference with the pump laser on and the pump laser off.

State-to-state transitions from the initial rotational state $j_i = 7$ and vibrational state $n_i = 0$ were observed at a collision energy $E = 0.3$ eV for a large number of final states up to $\Delta j = 80$. An example is given in Fig. 4 showing the differential cross sections for $\Delta j = 8, 28, 42, 62,$ and 72 for $\Delta n = 1$ transitions (Serri et al., 1981). The vertical scale gives relative cross-section

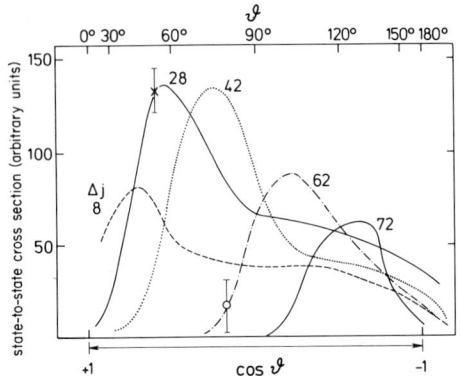

FIG. 4. Na$_2$–Ar rotational state-to-state differential cross sections (Serri *et al.*, 1981) for the Na$_2$ transitions $(n_i = 0, j_i = 7) \rightarrow (n_f = 1, j_f = 7 + \Delta j)$. The collision energy is $E = 0.3$ eV. The bottom horizontal axis shows the cosine of the (c.m.) scattering angle ϑ, which is directly proportional to the ADDS Doppler frequency shift.

values in counts per second normalized to one transition, i.e., corrected for the laser intensity and the excitation probabilities of different levels. Scattering angles are plotted on the horizontal axis in terms of cos ϑ. The actual angles are indicated in the top scale of Fig. 4. The measured cross-section points were interpolated by smooth lines. Two typical error bars for best and worst conditions are shown. The angular distributions range beyond cos $\vartheta = \pm 1$ because of finite angular resolution with an estimated best value $\Delta\vartheta_{FWHM} \approx 10°$ at $\vartheta = 120°$. Similar angular distributions were observed as well for the $\Delta n = 0$ transition (Serri *et al.*, 1980).

Characteristic for the cross section of a given Δj transition is a maximum at a specific scattering angle. This maximum shifts toward larger scattering angles for increasing Δj. A given transition, e.g., $\Delta j = 62$, shows practically no rotational excitation at smaller scattering angles. The $\Delta j = 62$ excitation cross section suddenly rises to the maximum at $\vartheta = 100°$ and then slowly levels off toward larger angles. These features are characteristic for "rotational rainbows" (Thomas, 1977; Schinke and McGuire, 1979; Bowman, 1979).

B. Rotational Rainbows in Na$_2$–Ar and Na$_2$–Ne

The origin of rotational rainbows can be explained in a straightforward manner within the classical trajectory picture shown in Fig. 5 (Schepper *et al.*, 1979). Just as a hard sphere model of a molecule is an acceptable first approximation for the description of atom–atom scattering, a deformed

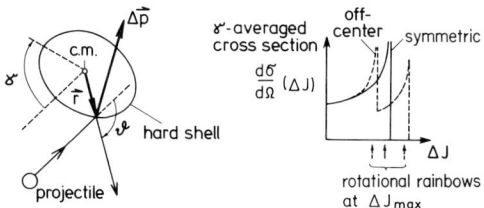

FIG. 5. Classical rotational rainbows: In a sudden collision of a projectile with a nonspherical hard shell molecule the recoil momentum $\Delta \mathbf{p}$ results in an angular momentum transfer $\Delta \mathbf{J} = \mathbf{r} \times \Delta \mathbf{p}$. For a given scattering angle ϑ, $|\Delta \mathbf{J}|$ has at least one maximum value with respect to the molecular orientation angle γ. These maxima $\Delta J_{max}(\gamma, \vartheta = \text{const})$ give rise to rainbow-like singularities in the γ-averaged classical cross sections for rotational excitation. One rotational rainbow occurs for an ellipsoidal symmetric molecule and two singularities are present when the center of mass is different from the center of symmetry. These types are shown as the solid and dashed line cross sections, respectively.

hard sphere like the ellipsoid shown on the left side of Fig. 5 can account for rotationally inelastic scattering in atom–molecule collisions. In the energy sudden limit, valid for collision times short with respect to the rotational period of the molecule and for small energy transfers, a projectile hitting the hard shell surface will be elastically scattered and transfer a recoil momentum $\Delta \mathbf{p}$ to the molecule. This recoil momentum usually does not point toward the center of mass of the ellipsoid and thus an angular momentum $\Delta \mathbf{J} = \Delta \mathbf{p} \times \mathbf{r}$ is transferred to the molecule. $\Delta \mathbf{J}$ is both a function of the scattering angle ϑ and the molecular orientation angle γ. For a fixed scattering angle ϑ the values of $|\Delta \mathbf{J}(\gamma, \vartheta)|$ as a function of the molecular orientation γ have at least one maximum $\Delta J_{max}(\vartheta = \text{const})$. Therefore, as shown in Fig. 5, the classical cross section for rotational excitation $d\sigma/d\Omega$ ($\Delta J; \vartheta = \text{const}$) with averaging over all molecular orientations γ will exhibit one singularity at $\Delta J = \Delta J_{max}$ ($\vartheta = \text{const}$) for symmetrical molecules and two "rotational rainbow" singularities for molecules with the center of symmetry different from the center of mass. This is in perfect qualitative agreement with the observation in the (symmetric molecule) Na_2–Ar cross sections of Fig. 4, where for a fixed angle e.g., at $\vartheta \approx 75°$ the rotational excitation cross section is steadily increasing from the $\Delta j = 8$ to the maximum value at the $\Delta j = 42$ transition, whereas the larger $\Delta j = 62$ and 72 transitions have vanishingly small cross sections.

Additionally, as first pointed out by Schepper et al. (1979), the classical ellipsoidal hard shell picture also allows one to derive analytical relations between the occurrence of the rotational rainbow singularites and the major and minor semiaxes A and B of the ellipsoid. For the symmetric molecule case (and when the ratio μ/I, where μ is the reduced mass of the collision partners and I is the moment of inertia of the ellipsoid, is small) the maximum excited rotational state $j_R(\vartheta)$ is given by the following simple expression:

$$J_R(\vartheta) = 2k(A - B)\sin(\vartheta/2) \tag{5}$$

i.e., is a function of the scattering angle ϑ and the collision wave number k(Bosanac, 1980; Bowman and Schinke, 1983). In the Na_2-Ar experiment at $E = 0.3$ eV the wave number is $k \approx 55$ Å^{-1} and Eq. (5) applied to the cross-section data of Fig. 4 allows one to estimate a deformation $(A - B) \approx 0.65$ Å for the repulsive barrier of the Na_2-Ar interaction potential. This gives j_R values of 18, 30, 50, and 62 at $\vartheta = 30, 50, 90,$ and $120°$, respectively, and roughly reproduces the increase of the rotational rainbow with the scattering angle of the Fig. 4 cross sections. Unfortunately, an accurate Na_2-Ar potential energy surface is not yet available for comparison. However, the Na_2-Ne ab initio potential which was reported by Schinke et al. (1982) shows a negligibly shallow potential well with a greatest depth of only 0.3 meV at $R_m \approx 7$ Å and thus justifies a scattering treatment with a purely repulsive potential. The equipotential lines of this (rigid rotor) Na_2-Ne potential are fairly ellipsoidal with deformations $(A - B) \approx 0.7$ and 1.1 Å at the potential energies $V(R, \gamma) = 50$ and 300 meV, respectively. Slightly smaller deformations could well be expected for the larger Ar atom.

For the Na_2-Ne rotational excitation also IOS approximation, quantum scattering cross sections have been calculated and are shown in Fig. 6a. These theoretical cross sections, at a slightly smaller energy of $E = 0.175$ eV,

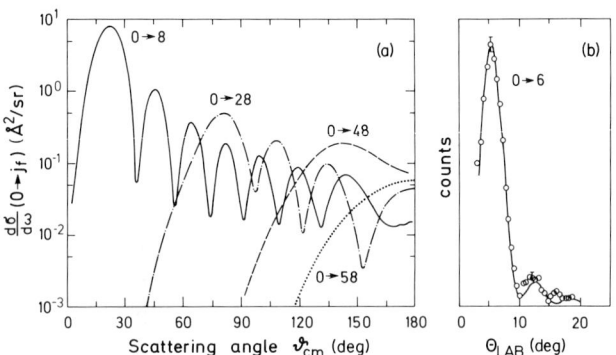

FIG. 6. Na_2-Ne rotational state-to-state cross sections at $E = 0.175$ eV (Hefter et al., 1981). (a) The theoretical (IOS) cross sections for the $j_i = 0 \rightarrow j_f = 8, 28, 48, 58$ rotational transitions show the rotational rainbow structure. Additional oscillatory structures—the "supernumerary rotational rainbows"—result from quantum interferences. (b) Experimentally observed (○) supernumerary rotational rainbow oscillations in the $0 \rightarrow 6$ differential cross section are compared with the ab initio prediction (———). Note that the vertical scale is logarithmic in (a), but linear in (b).

show the gross rotational rainbows structure that was observed with the Na_2–Ar measurements. Additionally, however, the lower rotational excitation cross sections $0 \rightarrow 8$ and $0 \rightarrow 28$ in Fig. 6a have superimposed an oscillatory angular structure. These oscillations result from the quantum interference of different trajectories leading to the same scattering angle and rotational state. They are thus closely related to the rotational rainbow phenomenon and can be called "supernumerary rotational rainbows," in analogy to the spherical potential scattering.

These supernumerary rotational rainbow oscillations could not be observed in the Ar–Na_2 cross-section measurements because the angular resolution was too low. In addition to this experimental lack of resolution, the Na_2–Ar cross sections were measured for the initial state $j_i = 7$. For rotationally excited initial states, however, the factorization formula for rotational cross sections [Eq. (3), discussed in Section II,C) shows that angular structure present for ground state $j_i = 0$ scattering will be considerably averaged and washed out because in this equation cross sections for initially rotating states are a weighted sum of many ground-state rotational excitation cross sections. The supernumerary rotational rainbow oscillations were therefore searched for and found in the $j_i = 0$ ground-state scattering of Na_2 on Ne. Figure 6b shows the $0 \rightarrow 6$ rotational transition measured with $\Delta\theta_{LAB} \approx 2°$ in the high-angular-resolution LIF apparatus of the Bergmann group (Hefter et al., 1981). The experimental $0 \rightarrow 6$ state-to-state cross-section points are shown here as open circles for the laboratory scattering angles $\theta_{LAB} = 3-18°$. The first oscillation maximum at $\theta_{LAB} = 5°$ and the much smaller second maximum near $\theta_{LAB} = 12°$ are well reproduced by the smooth line theoretical, infinite order sudden approximation cross sections calculated from the *ab initio* potential for Na_2–Ne. The slight remaining discrepancies can be largely attributed to the IOS approximation as was recently discussed by Schinke et al. (1982).

C. He–N_2, CO, O_2, and CH_4

Na_2, as just discussed, is a typical representative of a still small number of molecules where laser state-selective methods are applicable. But in the more conventional velocity and energy change experiments, the energy level resolution could also be increased and is now close to 1 meV (= 8 cm^{-1}). This allowed detailed studies of the rotational excitation for a number of common gas molecules such as N_2, O_2, CO, and CH_4.

As is shown by the energy level diagram in Fig. 7, these molecules have rotational energy level spacings wider by a factor of ≈ 10 than those of Na_2 and narrower by an order of magnitude than those of the hydrogen isotopic

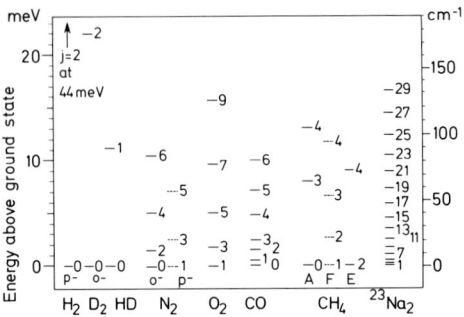

FIG. 7. Rotational energy levels of some common molecules.

species investigated in the earliest rotationally resolved scattering experiments (see Table III, Section VI). N_2, O_2, and CO have fairly similar rotational constants and the differences in their rotational energy level schemes result largely from nuclear spin alignment and exchange symmetry (see, e.g., Herzberg, 1950). The homonuclear N_2, like H_2, D_2, and Na_2, exists in two modifications which have either even- or odd-numbered rotational levels only and cannot interconvert by collisions. The higher populated modification is always called the ortho and the other the para state. In N_2 gas these cannot be separated and, therefore, in energy loss scattering experiments a resolution of 1 meV is required in order to distinguish the $0 \rightarrow 2$ rotational transition with an energy gap of $\Delta E_{0 \rightarrow 2} = 1.5$ meV from the $1 \rightarrow 3$ transition of $p - N_2$ with $\Delta E_{1 \rightarrow 3} = 2.5$ meV.

The O_2 molecule has a slightly smaller rotational constant than N_2 but the ^{16}O nucleus has a spin of zero. Thus, O_2 has only one modification with odd-numbered rotational levels $j = 1, 3, 5, \ldots$ because, additionally, the O_2 electronic shell has an electronic spin of unity. The energy spacing of the lowest $1 \rightarrow 3$ rotational transition for O_2 is $\Delta E_{1 \rightarrow 3} = 1.7$ meV.

The rotational constant of the CO molecule is almost identical to that of N_2. In this heteronuclear molecule, however, all rotational states are accessible with a smallest energy gap of only 0.5 meV for the $0 \rightarrow 1$ transition.

Finally, methane is an example of a more complicated, five-atom molecule. This tetrahedral spherical top has three different rotational state ladders, called A, F, and E modifications according to their symmetries (see Table VII of Herzberg, II, 1950). In the most abundant F modification, populated by $\frac{9}{16}$ of the molecules, all rotational states with $j \geq 1$ are allowed. In the A species with a $\frac{5}{16}$ fractions of the CH_4 molecules, the lowest rotational state is $j = 0$. The states $j = 1$ and 2 are forbidden by their symmetry and $j = 3$ and 4 are the first two excited rotational states. The E modification starts with $j = 2$ and the first allowed excited level is $j = 4$. The transition with the lowest energy gap is the $j = 1 \rightarrow 2$ excitation in the F modification of CH_4. But, for separating the $j = 1 \rightarrow 3$ transition of the F from the $j = 0 \rightarrow 3$ transition of the A modification, again an energy loss resolution of 1 meV is required.

The well-known principle and an actual design of a time-of-flight experiment for high-resolution inelastic-scattering cross-section measurements are shown in Fig. 8a and b, respectively. One of the colliding beams is chopped by slits in a fast spinning wheel into bursts of about 5-μsec duration and the different flight times of elastically and inelastically scattered molecules are measured at selected scattering angles. The mass spectrometer time-of-flight detector at the right-hand side of Fig. 8b is 165 cm from the scattering center, giving a typical velocity resolution of $\Delta v/v \approx 0.7\%$. Because the intrinsic, kinematical energy resolution of a scattering experi-

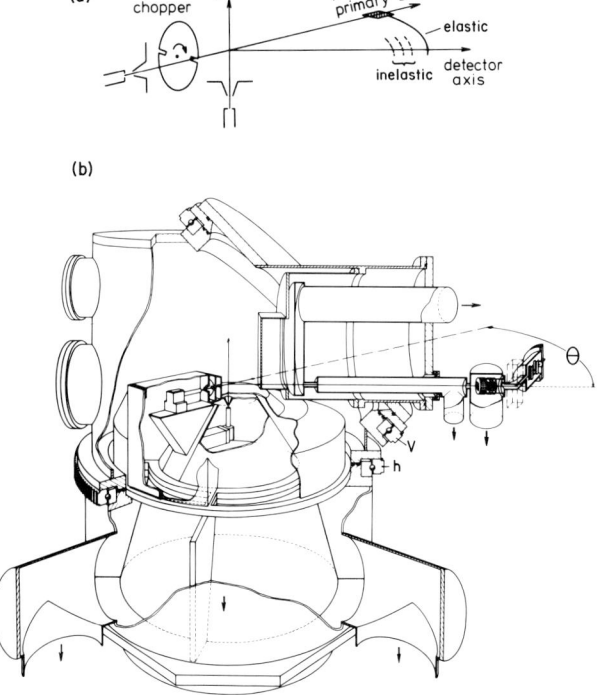

FIG. 8. The principle (a) and an actual design (b) of a high-resolution crossed-molecular-beams time-of-flight scattering apparatus (Faubel et al., 1982). The mass spectrometer detector is at the right-hand side of (b) at a distance of 165 cm from the scattering center. Principal components and beam axes are as shown in (a). Visible pumping ports are indicated by arrows (↓).

ment depends not only on the velocity resolutions of both beams but also on all six angular apertures of the two beams and of the detector, the scattering intensity decreases with the sixth power of the energy resolution (see Faubel and Toennies, 1978). In order to cope with this anticipated intensity loss in high-resolution experiments, two large diffusion pumps with maximum gas throughputs of 5 and 2 mol/hr, respectively, are used for producing the vertical target nozzle beam and the horizontal primary helium beam. Eight additional differential pumping stages are used for collimating the colliding beams and for buffering the detector ultrahigh vacuum (a partial pressure of 10^{-16} Torr at mass helium) from the normal high vacuum (10^{-6} Torr) in the scattering region. For changing the scattering angle, the detector can be moved by the two gears and rotatable vacuum seals designated "v" and "h" in Fig. 8b. For reasons of resolution, in the subsequent scattering measurements only perpendicular out-of-plane scattering angles θ_{LAB} were used. The primary He beam was collimated to 0.75° in-plane and 2° out-of-plane and has a velocity of 950 m/sec with a spread of ±3 m/sec obtained by cooling the nozzle to 88 K and operating with 20-μm diameter and 100-atm pressure. The target beam divergence was 2° in both angles and the 100-μm target nozzle was kept at 300 K in order to prevent the condensation of the N_2, O_2, or CH_4 beams. With target nozzle pressures of 6 atm the target molecules were essentially in their respective ground states and the actual populations were estimated from Table I of Section III,B.

Four examples of time-of-flight spectra of He scattered from N_2, CO, O_2, and CH_4 are shown in Figure 9 (Faubel *et al.*, 1980; Kohl, 1982). The flight time axis in the Figure 9 is divided into 140 channels of 4-μsec width and the vertical axis shows the number of helium atoms detected in each time-of-flight channel during a total accumulation time of 10–30 hr. The He–N_2 spectrum in Fig. 9a for a collision energy $E = 27.7$ meV and at a scattering angle $\theta_{LAB} = 39.5°$ ($\vartheta_{cm} = 35°$) shows a large peak of elastically scattered helium at a flight time of 1.82 msec followed by two smaller peaks resulting from the $0 \rightarrow 2$ and $1 \rightarrow 3$ rotational excitations of the N_2 molecule with energy losses of 1.5 and 2.5 meV, respectively. The half-width of the isolated elastic peak is only about half the spacing between the $0 \rightarrow 2$ transition and the elastic peak and corresponds to an energy level resolution of 0.7–0.8 meV (≈ 6 cm^{-1}). In Fig. 9b, showing a He–CO time-of-flight spectrum at the same energy and scattering angle, the valley between the elastic and the $0 \rightarrow 2$ transitions is filled up by the additional $0 \rightarrow 1$ and $1 \rightarrow 2$ rotational transitions of this heteronuclear molecule. These have energy spacings of only 0.5 meV (see Fig. 7) and, thus, cannot be separated with the present resolution. But, with the peak shape of the isolated transitons being known from the He–N_2 scattering in Fig. 9a, the He–CO spectrum can be deconvoluted as shown by the smooth Gaussian-shaped

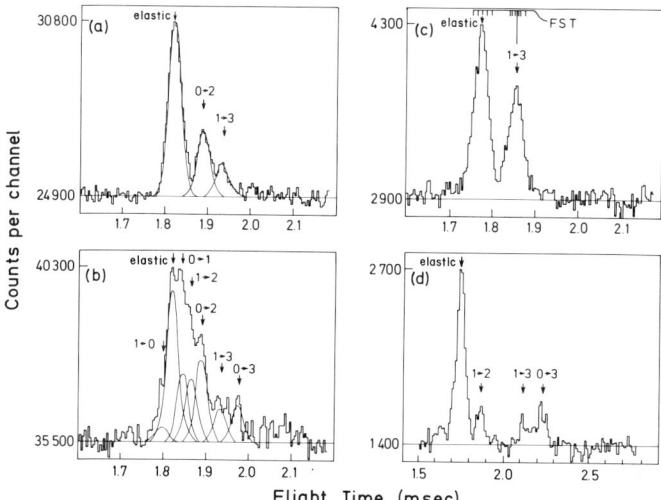

FIG. 9. Rotation-resolved time-of-flight spectra of He scattered from (a) N_2 ($\theta = 39.5°$, $E = 27.7$ meV), (b) CO ($\theta = 39.5°$, $E = 27.7$ meV), (c) O_2 ($\theta = 21.6°$, $E = 27.1$ meV), and (d) CH_4 ($\theta = 19.6°$, $E = 34.8$ meV). Expected flight times of individual rotational transitions are indicated by vertical arrows. For the He–N_2 (a) and He–CO (b) spectra a deconvolution of individual transitions is also shown (smooth Gaussian-shaped curves). In the top of the He–O_2 spectrum (c) a time-of-flight scale for the (not yet resolved) O_2 fine-structure transitions (FST) is shown.

lines. However, instabilities in the least-squares-fit deconvolution procedure limit the accuracy of individual transition amplitudes of the He–CO spectrum to about 30%, whereas from the well-separated He–N_2 spectra rotational state-to-state transition probabilities can be easily determined with better than 5% accuracy.

In Fig. 9c the scattering of He from O_2 is shown for an angle $\theta_{LAB} = 21.6°$ and for the collision energy $E = 27.1$ meV. This time-of-flight spectrum is particularly simple because here more than 90% of all O_2 molecules are cooled into the rotational ground state $j = 1$ and, essentially, only the $1 \to 1$ elastic and $1 \to 3$ rotational transitions are present and fully separated. The O_2 molecule however, exhibits, another complication in the fine-structure splitting of the rotational states, caused by the electronic spin 1 of the molecule. This "Hund's case b" triplet splitting has particularly wide energy spacings of 2 and 4 cm^{-1} in the rotational ground state $j = 1$ (see, e.g., Fig. 103 of Herzberg I, 1950; Krupenie, 1972). The locations where such fine-structure transitions would occur are therefore also indicated in the He–O_2 time-of-flight spectrum by short vertical bars at the top of the $1 \to 1$ and $1 \to 3$ rotational transition peaks. Although for a separation of fine-structure transitions a resolution at least a three times higher than the present 6 cm^{-1} would be required, a strong occurrence of fine-structure transitions should lead to a broadening or a noticeable shift of the rotational transition peaks which are actually not observed in the present experiment. But, this realignment process of molecular nuclear rotation with respect to the electronic spin is certainly worth additional theoretical (Smith and Giraud, 1979) and future experimental investigations.

The last time-of-flight spectrum in Fig. 9d shows the scattering of He from CH_4 at $\theta_{LAB} = 20°$ and for $E = 34.1$ meV. Here the channel width was increased to 10 μsec and a flight time range wider by a factor of 2.5 is displayed. The CH_4 is almost completely cooled into its respective rotational ground states $j_i = 0, 1,$ and 2 (cf. Table I). Except for the large elastic peak, the $1 \rightarrow 2$ and $1 \rightarrow 3$ transitions of the F modification and the $0 \rightarrow 3$ transition of the A modification of CH_4 are excited with transition amplitudes of 10–20% of the elastic peak. This appears to be remarkably large when comparing the energy gap of 7.8 meV for the $0 \rightarrow 3$ transition with the collision energy of 34.8 meV.

The rotational excitation probabilities can change dramatically with the scattering angle as shown in Fig. 10 by He–N_2 time-of-flight spectra at six different scattering angles from $\theta_{LAB} = 13°$ to 92° for the collision energy $E = 27.7$ meV. In the range from 13° to 19.6° the rotational excitation is quite small. With only 1° change of the scattering angle from 19.6° to 20.6°,

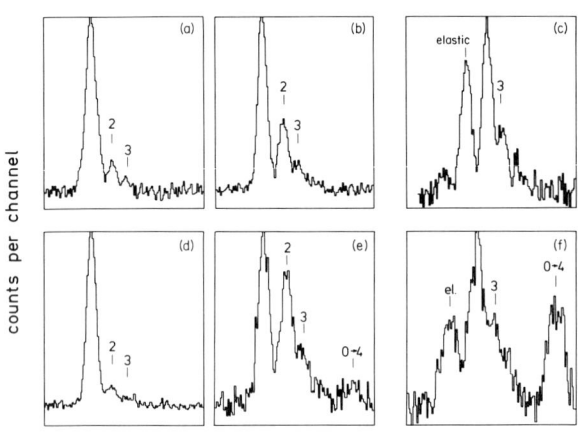

FIG. 10. He–N_2 time-of-flight spectra at a number of laboratory scattering angles θ_{LAB}: (a) 13.0°, (b) 19.6°, (c) 20.6°, (d) 22.6°, (e) 56.1°, (f) 92.0° (Faubel et al., 1982). The collision energy is $E = 27.7$ meV. Note the sudden increase of the $0 \rightarrow 2$ and $1 \rightarrow 3$ rotational transitions at 20.6°.

however, the 0 → 2 and 1 → 3 excitations increase to the dominant structure of the time-of-flight spectrum, and as suddenly as they increased, these transitions fall back to very small excitation amplitudes at 22.6°. A second 0 → 2 and 1 → 3 rotational excitation maximum was observed near θ_{LAB} = 29° (not shown here). With still larger scattering angles the inelastic transitions begin to increase steadily with respect to the elastic peak and at 56.1° and 92° also the 0 → 4 rotational transition is measurably excited.

The center-of-mass differential cross sections derived from these and many more time-of-flight spectra are shown in Fig. 11 for the He–N_2, He–O_2, and He–CH_4 rotationally inelastic and the total scattering (Faubel et al., 1981; Kohl, 1982). Whereas the rotationally inelastic cross sections

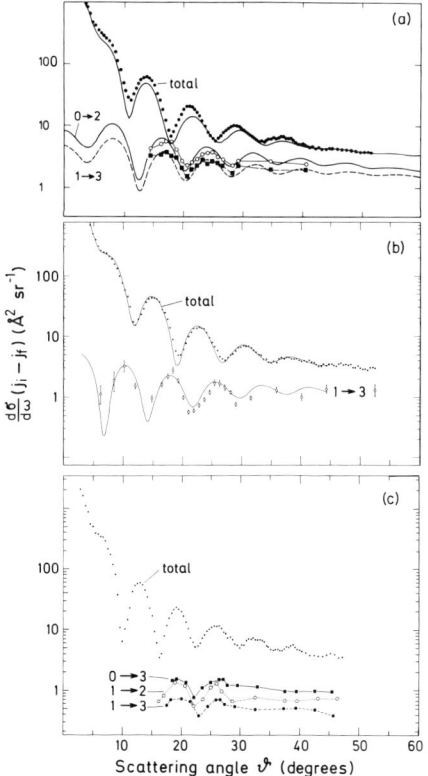

FIG. 11. Experimental rotationally inelastic and total differential cross sections for (a) He–N_2 at E = 27.7 meV, (b) He–O_2 at E = 27.1 meV, and (c) He–CH_4 at E = 34.8 meV in the center-of-mass system. The smooth line cross sections are (a) the coupled-channels prediction from the "HTT" He–N_2 model potential and (b) an IOS best fit for the He–O_2 experimental data.

are true state-to-state cross sections, the total cross sections given in Fig. 11 are weighted sums over the experimental initial-state populations because the elastic scattering on different initial target molecule states could not be distinguished in the time-of-flight spectra. But because of Eq. (4) (Section II,C), the total cross section can be expected to be at least approximately independent of the initial state. All cross sections in Fig. 11 show quite similar angular oscillation structures which are damped with increasing scattering angles. These are Fraunhofer diffraction patterns and not supernumerary rotational rainbows as in the previously shown Na_2– rare gas cross sections. The shoulder in the total cross sections at angles between $\vartheta = 5$ and $10°$ is a last remainder of the conventional, angular rainbow scattering structure which is expected there for the shallow, ≈ 2-meV-deep He-molecule potentials, but is almost completely washed out by the widely spaced diffractions.

The diffraction oscillations in the rotationally inelastic cross sections are in phase with the elastic- and the total-cross-section oscillations for the He–CH_4 scattering. For He–N_2 and He–O_2, however, the oscillations in the $\Delta j = 2$ rotational transition cross sections are phase-shifted $180°$ with respect to the total cross section and, thus, give rise to the very sudden change of the ratio between elastic and inelastic cross sections near the total-cross-section diffraction minima as, e.g., at $\theta_{LAB} = 20.6°$ ($\vartheta = 18°$) in the previous Fig. 10. Finally, in the He–N_2 scattering the $1 \rightarrow 3$ rotational cross section in Fig. 11a is observed to be almost proportional to the $0 \rightarrow 2$ cross section. This phenomenon is just an experimental confirmation of the sudden approximation factorization formula [Eq. (3) of Section II,C]. Because at scattering angles of up to $40°$ the $0 \rightarrow 4$ rotational excitation is negligibly small, the only term remaining in the sum on the right-hand side of Eq. (3) is the term with the $0 \rightarrow 2$ differential cross section and the $1 \rightarrow 3$ differential cross section becomes $d\sigma/d\omega$ $(1 \rightarrow 3) = 0.6$ $d\sigma/d\omega$ $(0 \rightarrow 2)$.

D. Exact versus Approximate He–N_2 Cross Sections

In the above scattering experiments of He on N_2, O_2, and CH_4 the collision energies are low enough with respect to the rotational energy level spacings to allow exact, converged coupled-channels cross-section calculations when an interaction potential is available. For He–N_2 a "Tang–Toennies," model potential (see Section II,A) has recently been derived by Habitz *et al.* (1982) and its Legendre expansion terms $v_0(R)$, $v_2(R)$, and $v_4(R)$ are shown in Fig. 12a. In the repulsive barrier region of this "HTT" potential, the v_2 term is considerably larger then the v_0 term and the potential is in this respect quite different from the H_2–Li^+ and H_2–Ne

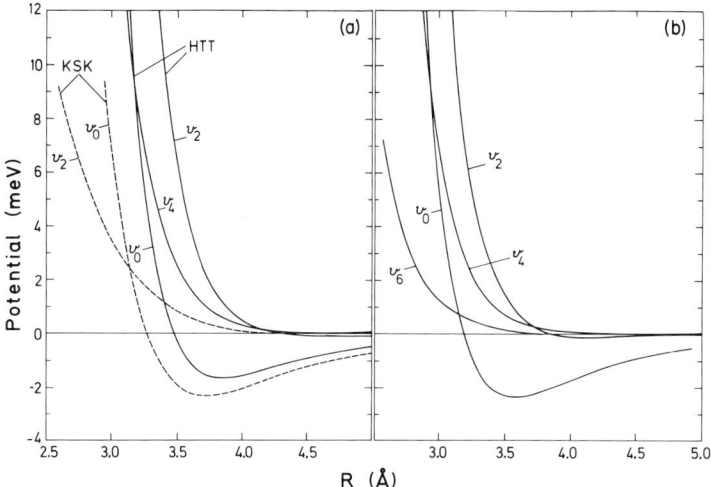

FIG. 12. Legendre expansion terms for the He–N$_2$ and He–O$_2$ rigid rotor interaction potentials. (a) The He–N$_2$ v_0, v_2, and v_4 terms predicted by the Tang–Toennies potential model (Habitz *et al.*, 1982) are compared with a previous experimental potential ("KSK") determined from total cross-section measurements only (Keil *et al.*, 1979). (b) The He–O$_2$ potential was obtained by fitting the rotationally inelastic cross-section measurements of Fig. 11b (Gianturco, 1982).

potentials discussed previously in Section II,A. Near the experimental collision energy, at $E = 27.3$ meV, coupled-channels calculations with 25 coupled channels were carried out with this potential by Tang and Yung. These theoretical cross sections are shown as the smooth lines in Fig. 11a and are in a quite acceptable overall agreement with the respective experimental cross sections in Fig. 11a. In more detail this study (Faubel *et al.*, 1982) showed that the potential well of the HTT potential has to be lowered by approximately 20% or 0.3 meV and the repulsive parts have to be slightly shifted inward by about 0.1 Å. But the aspherical deformation of the equipotential lines of the repulsive core cannot be changed by more than 0.02 or 0.03 Å before noticeable discrepancies are encountered in the relative amplitudes of the total and the rotationally inelastic cross sections and also in the relative phases of the diffraction oscillations.

With the He–N$_2$ scattering at $E = 27.7$ meV being dominated by the repulsive part of the potential, the numerically much faster IOS approximation can be expected to work well and IOS cross sections for the 0 → 0 and the 0 → 2 transition are compared in Fig. 13a with the respective (dashed lines) coupled-channel results. In the angular range from $\vartheta = 10°$ to $30°$ the IOS cross sections are seen to be almost identical with the exact ones, with typical deviations of less than 10% in the amplitudes of the oscillation

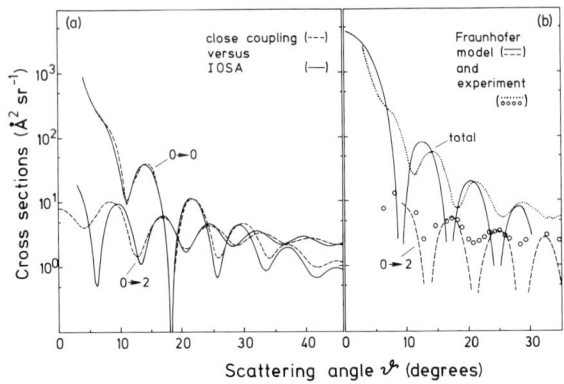

FIG. 13. He–N_2 rotational state-to-state cross sections at $E = 27.7$ meV. (a) Comparison of exact close-coupling cross sections (----) with the IOS approximation (———). The "HTT" potential of Fig. 12a was used. (b) Comparison of experimental cross sections (····, ○) with the inelastic Fraunhofer model (———, ----), Eq. (6), for the total and Eq. (8a) for the $0 \rightarrow 2$ rotational transition, respectively. The model parameters used are $R_0 = 3.5$ Å, $\delta_2 = 0.65$ Å, and $k = 6.7$ Å^{-1}.

maxima. For larger scattering angles the IOS approximation increasingly underestimates the elastic cross section. At small scattering angles below $\vartheta = 10°$, where the long-range parts of the interaction potential gain influence on the cross section, a major discrepancy in the $0 \rightarrow 2$ rotational inelastic cross sections is present. Thus, when respecting these limitations on the angular range of validity, the IOS approximation can be well used for fitting improved potentials to the He–N_2 and to the qualitatively very similar He–O_2 scattering data. Such an iterative fitting procedure was used by Gianturco (1982) to determine a He–O_2 potential shown in Fig. 12b from the experimental $1 \rightarrow 1$ and $1 \rightarrow 3$ cross sections shown in Fig. 11b. The smooth lines represent the final best-fit IOS cross sections. A preliminary comparison of this experimental He–O_2 potential with fully *ab initio* (CEPA) potential points calculated by Jaquet and Staemmler (1982) showed qualitative agreement. The attractive well of the experimental potential, however, is almost twice as deep as the *ab initio* value of $\epsilon_0 \approx 1.5$ meV.

Quite instructive also is a comparison of the approximately correct He–N_2 model potential in Fig. 12a with an "effective" potential derived from total differential cross section measurements only (Keil *et al.*, 1979). This "KSK" potential, for which the Legendre expansion v_0 and v_2 terms are shown by the dashed lines in Fig. 12a, has a considerably smaller asymmetry, giving too small inelastic cross sections. The KSK spherical term $v_0(R)$ also differs appreciably from the correct spherical part of the He–N_2 potential, but reproduces as well the total cross section (Habitz *et al.*, 1982).

Thus, this example illustrates that the effective spherical potential derived from total cross section measurements is not always identical to the true spherical part $v_0(R)$ of the potential. The observation that the repulsive barrier of the effective potential appears at smaller intermolecular distances than the actual $v_0(R)$ rather suggests that the perpendicular configuration of the actual potential, $V(R, \gamma = 90°) \approx v_0(R) - \frac{1}{2}v_2(R)$ is preferentially weighted in the scattering cross sections. This had been supposed by Monchick and Mason (1961) from phase space arguments and was recently discussed by Parker and Pack (1978) within the IOS approximation of Eq. (4).

An only qualitative but rather direct analytical relationship between the He–N$_2$ scattering cross-section features and specific details of the interaction potential can be established by using an inelastic Fraunhofer scattering model (Drozdov, 1955; Blair, 1966). The He–N$_2$ scattering cross sections are dominated in the investigated angular range by quantum interference structures. As shown in Fig. 13b the oscillations in the total scattering cross section are approximately correctly reproduced by the simple Fraunhofer cross section formula for the scattering of a plane wave on a "black" disk or a sphere:

$$\frac{d\sigma}{d\omega}(\vartheta)\bigg|_{\text{Fraunhofer}} \approx (kR_0^2)^2 \frac{2}{\pi x^3} \cos^2\left(x + \frac{\pi}{4}\right) \quad (6)$$

Here R_0 is the radius of the sphere, k ($=6.7$ Å^{-1} for the experimental collision energy $E = 27.7$ meV) is the wave number, and $x = kR_0\vartheta$ is the reduced angular variable. A sphere radius $R_0 = 3.5$ Å was estimated from the experimental spacings of the diffraction oscillations and is in rough agreement with the location of the repulsive barrier in the actual He–N$_2$ potential of Fig. 12a.

When the scattering object is an axially symmetric deformed sphere with the surface given in a spherical harmonics expansion:

$$R(\gamma) = R_0 + \sum_L \delta_L Y_L^0(\gamma, 0) \quad (7)$$

the Fraunhofer scattering yields in the limit of small deformations $\delta_L \ll R_0$ and within the energy sudden approximation the m-quantum averaged $0 \to I$ rotational excitation cross sections:

$$\frac{d\sigma}{d\omega}(0 \to I) \approx (kR_0)^2 \frac{\delta_I^2}{2\pi^2 x} \begin{cases} \sin^2(x + \pi/4) & \text{for } I \text{ even} \quad (8a) \\ \cos^2(x + \pi/4) & \text{for } I \text{ odd} \quad (8b) \end{cases}$$

For a transition to the rotational level I these first-order excitation cross sections are proportional to the square of only one — the I th — multipole deformation value δ_I of the expansion equation (7). From the ratio of the total (approximately elastic) cross section to the $0 \rightarrow 2$ rotationally inelastic cross sections of the $He-N_2$ measurements, a value of $\delta_2 \approx 0.65$ Å was estimated. With this value the inelastic Fraunhofer model cross section from Eq. (8a) reproduces well in Fig. 13b both the phase shifts and the angular dependence of the oscillation maxima of the experimental $0 \rightarrow 2$ differential cross sections. Interestingly, the odd–even diffraction alternation of the cross sections as given by Eq. (8) reverses the position of maxima and minima in a second-order treatment of the inelastic Fraunhofer scattering which has to be used when δ_I is no longer small compared to R_0 [see Blair (1966) for this and also for a rationalization of the phase shift effect for the diffraction on a narrow rim]. This was also observed in exact cross-section test calculation with only slightly larger dislocations between the $v_0(R)$ and the $v_2(R)$ repulsive barriers of the $He-N_2$ potential (Faubel et al., 1982) and, thus, the $He-N_2$ scattering conditions seem just to hit the upper end of the validity range of Eq. (8).

The $He-N_2$ Fraunhofer model value $\delta_2 = 0.65$ Å is approximately equal to the differences of the radii $R(\gamma)$ of the deformed sphere expansion [Eq. (7)] for the orientation angles $\gamma = 0$ and $\gamma = \pi/2$ and is within 20% in agreement of the deformation of an equipotential line with $V(R, \gamma) \approx 27$ meV of the actual $He-N_2$ potential of Fig. 12a. The information extracted by Eqs. (6) and (8a) from the $He-N_2$ cross sections is thus very similar to the information obtained by Eq. (5) in Section IV,B from the location of rotational rainbows, when present. The deviations of the Fraunhofer model cross sections from the experimental ones in Fig. 13b show also that additional information on the interaction potential is contained in the cross sections, e.g., at scattering angles $< 10°$ and $> 30°$.

E. D_2-CO Diatom–Diatom Scattering

Scattering studies with resolved rotational structures in more complicated than triatomic collision systems are now feasible as was shown in Fig. 11c with the cross sections for $He-CH_4$. Here also the theoretical scattering formalism is fully developed, but cross-section calculations thus far are only available for $Ar-CH_4$ model potentials (Smith and Secrest, 1981). Besides atom–polyatomic molecule experiments, rotationally inelastic scattering experiments were also extended to collisions of diatomic with diatomic molecules. The most extensive and detailed studies are for hydrogen–hydrogen scattering, where a quite satisfactory agreement from *ab initio* potentials through exact coupled-channels calculations to rotationally resolved scattering experiments has been established (see Buck, 1982).

Another case of diatom–diatom scattering was recently investigated with D_2–CO. Here the rotational energy levels of the CO molecule are considerably narrower than those of the D_2. The collision energy in this time-of-flight experiment by Andres *et al.* (1982) was $E = 87.2$ meV and measured center-of-mass energy loss spectra at six different scattering angles from $\vartheta = 36°$ to $105°$ are displayed in Fig. 14 as continuous lines. In each energy loss spectrum a higher peak centered at energy losses on the order of 2–5 meV is observed and is accompanied by a second shoulder near $\Delta E \approx 20$ meV, the latter increasing in relative amplitude with the scattering angle. But, with the energy resolution of 8–10 meV in this experiment, individual rotational transitions could not be resolved. Because unresolved energy loss spectra cannot be decomposed into state-to-state differential cross sections, a comparison with theory was made here in the energy loss spectra after convoluting the individual theoretical cross sections with experimental resolution and weighting functions. The theoretical cross sections were calculated from the SCF long-range dispersion potential shown already in Fig. 2. The large number of CO states were treated in the IOS approximation and the $0 \rightarrow 2$ rotational transition of D_2 in the coupled-states approximation. From the calculations the $0 \rightarrow 2$ transition of the D_2 (shown as dotted lines in Fig. 14) is found to not contribute noticeably to the excitation probabilities of the second peak or to the shoulder at energy losses near 20 meV. The dashed line in Fig. 14 is the composed energy loss spectrum for both D_2 and CO

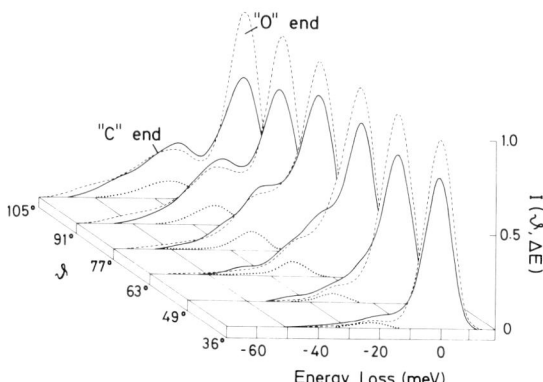

FIG. 14. Center-of-mass system energy loss spectra for D_2–CO scattering with $E = 87.2$ meV (Andres *et al.*, 1982). The experiment (———) is compared with theory (---). Dotted lines (···) represent the (theoretical) isolated $0 \rightarrow 2$ rotational transitions of the D_2 molecule. The two observed rotational rainbow maxima correspond to collisions with the "C" end and with "O" end of the CO molecule (see text).

excitation and reproduces, roughly, the measured (continuous line) structures.

Qualitatively, the double-peak structure in the CO excitation spectrum could be interpreted as rotational rainbows. From a closer inspection of the excitation function for different molecular orientation angles of the CO molecule the weak peak at larger energy losses could be attributed to collisions of the D_2 with the "C" end and the larger rotational rainbow peak at small energy losses to collisions with the "O" end of the CO molecule. Therefore, the quite large discrepancies between theory and experiment at the low-energy-loss peak allowed the conclusion that the H_2-CO potential needs most improvement at the "O" end of the CO molecule.

V. Studies of Vibrational Excitation

A. Vibrational Excitation in Diatomics

With increasing collision energies vibrational channels open and vibrational cross sections reach significant, measurable levels.

For the weakly bound Na_2 molecules (as discussed before in Section IV,A) the threshold for vibrational excitation is only 20 meV. Thus, in the Na_2-Ar scattering experiments at $E = 300$ meV by Pritchard and Kinsey and already shown in Fig. 4, both rotational and vibrational $\Delta n = 1$ transitions were investigated, the cross sections showing a rather wide, rotational rainbow, distribution of rotational transitions superimposed on the vibrational transition. Although here detailed scattering calculations are not yet available and an exact coupled-channel treatment in the near future is prohibited by the large number of coupled channels, the findings seem to support theoretical pictures which decouple the vibrational and rotational motions.

In contrast to Na_2, most common diatomic molecules have wider vibrational energy gaps of several hundred millielectron volts (up to 0.516 eV for the H_2 molecule), and still higher collision energies of several electron volts are required to excite noticeable vibrational transitions. These higher collision energies are more easily achieved and controlled in ion beams and vibrational excitation is thus still a domain of ion energy change or time-of-flight scattering experiments, as previously reviewed, e.g., by Faubel and Toennies (1978).

In the scattering of light ions such as H^+, D^+, or Li^+ from diatomic molecules such as H_2, N_2, CO, or O_2 (see Table III in Section VI) often an almost pure vibrational excitation is observed, although there are exceptions like the scattering of Li$^+$ on CO and H$_2$ where for $E = 3-5$ eV at small scattering angles rotational transitions with energy gaps comparable to the vibrational transitions are strongly excited. But, at the same energy, the Li$^+$–H$_2$ scattering, for example, shows almost pure vibrationl excitation for scattering angles larger than 150° (David et al., 1973; Faubel et al., 1975). Thus, experiments were undertaken to explore whether under somewhat favorable conditions the feasibility of observing rather isolated vibrational excitation is persistent for polyatomic molecules where the situation is further complicated by the existence of several vibrational normal modes.

B. Mode-Selective Excitation of CO$_2$ by H$^+$ and D$^+$

Already, a rather simple triatomic molecule like CO$_2$ has three vibrational normal modes: the symmetric stretch (v_1, 0, 0), the double degenerated bending mode (0, v_2, 0), and the asymmetric stretch (0, 0, v_3) with the respective vibrational energy quanta of 0.172, 0.083, and 0.291 eV.

In the scattering of Li$^+$ ions from CO$_2$ and, similarly, from N$_2$O at $E = 10$ eV (Eastes et al., 1977) only very little rotational excitation was present and only the bending vibration was found to be collisionally excited. When scattering H$^+$ and D$^+$ from CO$_2$ at somewhat higher collision energies near 50 eV, Linder (1980) also found negligible rotational excitation. However, as seen in the H$^+$–CO$_2$ energy loss spectrum of Fig. 15 at the scattering angle $\vartheta = 0$, in addition to the strongest — bending mode — excitation, a number of asymmetric stretch vibrational levels are also populated. Furthermore, a combined excitation of bending and stretch modes is observed,

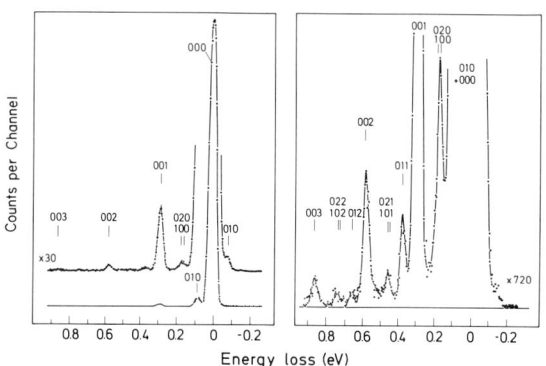

Fig. 15. Mode-selective vibrational excitation of CO$_2$ by protons for $E = 54$ eV, $\vartheta = 0°$ (Linder, 1980). The energy loss spectrum is shown in three different vertical scales ($\times 1$, $\times 30$, and $\times 720$). The most strongly excited modes are the bending mode (010) and the asymmetric stretch (0, 0, v_3). Virtually no rotational excitation is present.

some 200–400 times weaker than the (010) bending transition probability, but nevertheless well-resolved and above noise. In contrast, at larger angles in the rainbow region near $\vartheta = 7°$ at $E = 50$ eV, the asymmetric stretch mode $(0, 0, v_3)$ up to $v_3 = 3$ was found to be by far dominant over all other modes, including the bending excitation.

The predominantly excited bending and asymmetric stretch modes are both infrared active and their selective excitation was therefore interpreted as resulting from the ion-charge-induced forces on the CO_2 molecule. A first qualitative model interpretation of these data was given by Krutein and Linder (1977a) within the Born approximation and considering the dipole potential term as the only interaction. Qualitatively, collision time effects seem to control the ratio $R = I_{010}/I_{001}$ of excitation of the 010 and 001 transitions, a maximum of energy being transferred to one mode if the collision time duration t_{coll} equals $\frac{1}{2}t_{vib}$ where t_{vib} is the vibrational period.

In a refined treatment with time-dependent perturbation theory, for the scattering angle $\vartheta = 0$ the ratios $R = I_{010}/I_{001}$ were found to be only dependent on the relative velocity g of the collision partners with the excitation ratio:

$$R(g) = \exp(a/g^{3/2}) \qquad (9)$$

where a is a constant parameter. This relation was found to be in excellent agreement both with H^+- and D^+–CO_2 scattering data over a velocity range from 5×10^4 to 12×10^4 m/sec, where this ratio R changed by a factor of 20 (Bischof et al., 1982; Richards, 1982).

C. High-Overtone Mode-Selective Vibrational Excitation of CF_4 by Ions

Larger polyatomic molecules—the spherical tops CH_4, SF_6, and CF_4—were investigated by Gentry et al. (1975). Eastes et al. (1979a), Ellenbroek et al. (1980, 1982), and Gierz et al. (1982). In the experiments these molecules showed mode selectivity when vibrationally excited by collisions with H^+, D^+, and Li^+ ions. For example, the tetrahedral molecule CF_4, presently the most striking example of mode-selective excitation, has four different vibrational modes. These are illustrated qualitatively in Fig. 16, with the arrows on the respective atoms indicating the direction and the relative amplitudes of the nuclei in the respective normal vibration (Herzberg II, 1950). On the right-hand side of Fig. 16 also a vibrational energy level diagram for CF_4 is given with the first vibrationally excited state of each mode only labeled explicitly. Above 0.159 eV, the energy gap for the excitation of the 0010 level of the widest-spaced v_3 mode, the energy level diagram shows a quite regularly distributed and dense ladder of higher vibrational states.

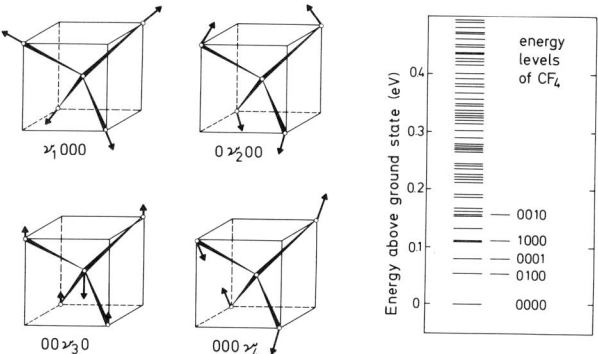

FIG. 16. Normal mode vibrations and vibrational energy levels of the CF_4 tetrahedral molecule.

Nevertheless, surprisingly well-separated peak structures in the time-of-flight spectrum Fig. 17 of H^+ scattered from CF_4 with $E = 9.7$ eV and at $\vartheta = 10°$ show very clearly that only one, the v_3 mode, is here strongly excited up to the vibrational level $v_3 = 5$. The deconvolution of this time-of-flight spectrum, with the known Gaussian-shaped experimental resolution function represented by smooth lines in Fig. 17, shows, additionally, the presence of a much weaker second series of peaks which have been assigned to the excitation of the v_3–v_4 mode intercombination vibration ($00v_31$), with energy levels indicated by the second row of vertical bars at the top of the energy loss spectrum (Gierz et al., 1982). The relative contribution of this latter excitation to the overall inelastic scattering increases rapidly with the

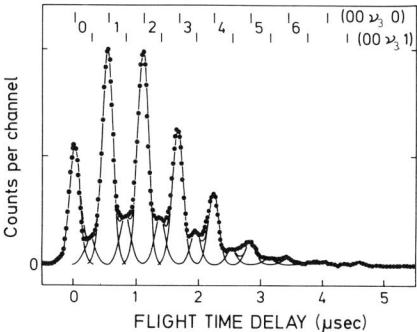

FIG. 17. High overtone mode-selective excitation in a H^+–CF_4 time-of-flight spectrum for $E = 9.7$ eV, $\vartheta = 10°$ (Gierz et al., 1982). The two scales at the top of the spectrum give the flight time delays corresponding to excitation of vibrational quanta of the $(00v_30)$ mode and of the v_3–v_4 intercombination mode $(00v_31)$. By far the strongest is the excitation of the pure v_3 mode. The deconvolution shows that additional structure in the valleys between the large v_3-mode transitions is readily explained by the assumption of $(00v_31)$ transitions only.

scattering angle in the region from 5° to 20°, and at $\vartheta = 25°$ only an essentially structureless broad energy loss distribution could be observed. Scattering of D^+ led to qualitatively similar results. But, less-resolved structure was present in Li^+-CF_4 spectra, showing at lower collision energies of 4 eV preferential of the Raman-active v_2 mode, whereas at higher energies the infrared active v_3 mode is dominant as with H^+ and D^+ (Ellenbroek et al., 1982).

A qualitative explanation of the strong vibrational mode selectivity in these scattering experiments could be given within a simple forced harmonic oscillator model (Ellenbroek et al., 1980). As with many diatomic molecules (see, e.g., Gentry, 1979), the relative intensities or populations $P(0 \to v_i)$ of different vibrational levels of one selectively excited mode v_i such as the v_3 mode of CF_4 in Fig. 17 were found to follow closely a Poisson distribution:

$$P(0 \to v_i) = (1/v_i!)(\epsilon^{v_i})(\exp \epsilon) \tag{10}$$

with

$$\epsilon = \overline{\Delta E}/\hbar \omega_i = \sum_{v_i=1}^{\infty} v_i P(0 \to v_i)$$

being the mean energy transfer $\overline{\Delta E}$ into the vibrational levels of the mode v_i divided by the vibrational energy level spacing $\hbar \omega_i$. This result is typical for the vibrational level population in a forced oscillator excitation model. In this model the mean energy transfer to one component k of a degenerate normal mode k is obtained by classical mechanics as

$$\overline{\Delta E_k} = \frac{1}{2m} \left| \int_{-\infty}^{\infty} F_k(t) \exp(-i\omega_k t) \, dt \right|^2 \tag{11}$$

The driving force $F_k(t)$ acting along a time-dependent classical trajectory can be derived to first order from a Taylor expansion of the interaction potential in the normal vibration coordinates Q_k of the molecule. Since little is known about the exact interaction potentials for larger molecules, the long-range part of the potential surface, probed by trajectories with small scattering angles, was roughly estimated from the interaction of the ion charge with the static and induced molecular multipole moments. An R^{-8} repulsive barrier and the well depth were fitted with an $R_m = 1.4$ Å and $\epsilon_0 = 3.6$ eV for H^+-CF_4 (but, $R_m = 2.8$ Å and $\epsilon_0 = 0.6$ eV for Li^+-CF_4) to match the scattering data. For the infrared-active v_3 and v_4 modes of the CF_4 molecule the dominant part in the driving force $F_k(t)$ is proportional to the derivatives of the CF_4 dipole moment μ with respect to the normal coordinates. These can be derived from infrared intensity data. Their values are $\partial \mu / \partial Q_3 = 2.31 \times 10^{-6}$ A·s·kg$^{-1/2}$ and $\partial \mu / \partial Q_4 = 0.26 \times 10^{-6}$ A·s·kg$^{-1/2}$ for the v_3 and the v_4 mode, respectively, and can be shown to be independent of the molecular orientation (Ellenbroek and Toennies, 1982). According to Eq. (11) the selection of predominantly excited modes v_k now depends on two factors: the maximum amplitude or strength of the driving force and the duration of a collision — given by the half-width of $F_k(t)$ — compared to the vibrational period $2\pi/\omega_k$ of the mode to be excited. Thus, by the strength of the driving force, the ω_3 mode of CF_4, with an order-of-magnitude-larger dipole moment derivative $\partial\mu/\partial Q_3$, is favored over the v_4 mode. Because the strength of the driving force is also proportional to the electric field exerted on the CF_4 molecule, the average energy transfer also must increase with increasing scattering angles, which correspond to smaller impact parameters and a closer approach of the ion to the molecule.

A closer examination of the collision dynamics, of course, reveals a more complicated actual situation. Higher multipole derivatives, too, have to be included. For example, the Raman-active v_1 and v_2 modes are also slightly excited, and the anisotropy of the repulsive barrier leads to some rotational excitation for large scattering angles. This and more detailed comparisons with the experimental results are discussed by Ellenbroek and Toennies (1982) and by Gierz et al. (1982).

VI. Summary of Detailed Scattering Experiments and Concluding Remarks

A short summary of recent molecular beam scattering experiments on vibrationally and rotationally inelastic collisions is given in Table III. The symbols "V" and "R" in the third column of the table denote the observation of resolved vibrational or rotational transitions for a pair of collision partners. The "R" has been set in parentheses when unresolved rotational structure was observed. Literature references are given for the first and for the most recent publication as of this writing. Also indicated in Table III (footnotes c to g) is the occurrence of the particular cross-section features discussed in the preceding sections.

A variety of experimental techniques were used in these scattering experiments. The ion experiments in the first two groups of Table III employ electrostatic energy selectors or a time-of-flight method for measuring the energy change in an inelastic collision. Time-of-flight measurement with mechanical chopper wheels and mass spectrometer-type detectors in combination with velocity- and state-defining nozzle beam sources (described in Section IV,C) is presently a widespread technique for inelastic neutral–neutral scattering. Pulsed nozzle beams and time-of-flight state change detec-

TABLE III

HIGH-RESOLUTION DIFFERENTIAL SCATTERING EXPERIMENTS ON VIBRATIONALLY AND ROTATIONALLY INELASTIC COLLISIONS

Collision partner[a]	Collision partner	Transition[b]	Reference
D^+, H^+	H_2	V	Udseth et al. (1973), Linder (1980)
		R	Rudolph and Toennies (1976), Krutein and Linder (1979)
	D_2	V	Udseth et al. (1973)
	HD	V	Udseth et al. (1973)
	N_2	V	Krutein and Linder (1979), Gianturco et al. (1981)
	O_2	V	Gianturco et al. (1981)
	CO	V	Krutein and Linder (1979), Gianturco et al. (1981)
	NO	V	Krutein and Linder (1979), Gianturco et al. (1981)
	Na_2	—	
	CO_2	V[c]	Krutein and Linder (1977b), Bischof et al. (1982)
	N_2O	V	Gierz and Toennies (1982)
	NH_3	—	
	CH_4	V[c]	Ellenbroek et al. (1982)
	CF_4	V[c]	Ellenbroek et al. (1982)
	SF_6	V[c]	Ellenbroek et al. (1980, 1982)
Li^+	H_2	V	David et al. (1973)
		R	van den Bergh et al. (1973), Faubel and Toennies (1979)
	D_2	—	
	HD	—	
	N_2	(R)V	Böttner et al. (1976)
	O_2	—	
	CO	(R)V	Böttner et al. (1976)
		R	Eastes et al. (1979a)
	NO	—	
	Na_2	—	
	CO_2	(R)V	Eastes et al. (1977)
	N_2O	(R)V	Eastes et al. (1977)
	NH_3	—	
	CH_4	(R)V	Eastes et al. (1979a), Ellenbroek et al. (1982)
	CF_4	V[c]	Ellenbroek et al. (1982)
	SF_4	V[c]	Ellenbroek, et al. (1982)
He	H_2	—	
	D_2	R	Gentry (1980a)
	HD	R	Gentry and Giese (1977)
	N_2	R[d]	Faubel et al. (1980, 1982)
	O_2	R[d]	Faubel et al. (1981)
	CO	(R)	Faubel et al. (1980)
	NO	—	

TABLE III (continued)

Collision partner[a]	Collision partner	Transition[b]	Reference
Ne	Na_2^e	R^f	Bergmann et al. (1978, 1980)
	CO_2	—	
	N_2O	—	
	NH_3	—	
	CH_4	R	Faubel et al. (1980, 1981)
	CF_4	—	
	SF_6	—	
	H_2	—	
	D_2	R	Andres et al. (1980a), Buck (1982)
	HD	R	Buck et al. (1977), Buck (1982)
	N_2	—	
	O_2	—	
	CO		
	NO	—	
Ar	Na_2^e	R^g	Bergmann et al. (1981), Hefter et al. (1982)
	CO_2	—	
	N_2O	—	
	NH_3	—	
	CH_4	(R),R	Buck et al. (1982)
	CF_4	—	
	SF_6	(V)	Eccles et al. (1981)
	H_2	—	
	D_2	R	Andres et al. (1980b), Buck (1982)
	HD	—	
	N_2	—	
	O_2	—	
	CO		
	NO	—	
K	Na_2^e	R^f	Serri et al. (1980), Bergmann et al. (1981)
		R,V^f	Serri et al. (1981)
	CO_2	(R)	Loesch (1976)
	N_2O	—	
	NH_3	—	
	CH_4	(R),R	Buck et al. (1982)
	CF_4	—	
	SF_6	(V)	Eccles et al. (1981)
	H_2	—	
	D_2	—	
	HD	—	
	N_2	$(R)^f$	Schepper et al. (1979)
	O_2	$(R)V^f$	Beck et al. (1979)
	CO	$(R)^f$	Schepper et al. (1979)
	NO	—	
	Na_2	—	
	CO_2	$(R)V^f$	Ross et al. (1981)
	N_2O	—	
	NH_3	(R)	Kusunoki (1974)

(continued)

TABLE III (continued)

Collision partner[a]	Collision partner	Transition[b]	Reference
	CH_4	—	
	CF_4	—	
	SF_6	—	
D_2	H_2	R	Buck (1982)
	D_2	—	
	HD	(R,R)	Gentry (1980a), Buck (1982)
	N_2	—	
	O_2	—	
	CO	(R,R)[f]	Andres et al. (1982)
	NO	—	
	Na_2	—	
	CO_2	—	
	N_2O	—	
	NH_3	—	
	CH_4	—	
	CF_4	—	
	SF_6	—	

[a] Collision energies are in the range 25 meV $\leq E \leq$ 100 meV except for D^+, H^+ ($E \geq 4$ eV); Li^+ ($E > 0.5$ eV); and K (0.15 eV $< E < 2$ eV).

[b] V, resolved vibrational transition; R, resolved rotational transition; parentheses indicate unresolved rotational structures.

[c] Mode-selective excitation.

[d] Characteristic shifts of diffraction pattern.

[e] Collision energy, 0.15 eV $\leq E \leq$ 0.3 eV.

[f] Rotational rainbow structure.

[g] Supernumerary rotational rainbow structure.

tion were employed in the experiments of Gentry (1980a) and of Gentry and Giese (1977). Mechanical Fizzeau-type velocity selectors and surface ionization detectors are characteristic for the potassium scattering experiments in Table III. Finally, laser state labeling and LIF state-sensitive detection (with an example discussed in Section IV,A) were applied in all investigations with Na_2 as a collision partner.

Qualitative models or approximate quantitative theoretical descriptions of observed scattering features could be provided with almost all experiments listed in Table III and reveal interesting collision dynamics phenomena like rotational rainbows, factorization relations, or mode-selective excitation. Precision evaluations using converged close-coupling cross sections are presently only practical for the H_2-isotope experiments in the second column of the table and for the low-energy helium, neon and argon scattering experiments. As shown in the preceding sections, such detailed cross-section data provide a very sensitive test on theoretical *ab initio* or model potential predictions. When used for an improved fitting of the scattering relevant parts of the interaction potential, predictions of at least rotationally inelastic cross sections become feasible on a level of accuracy of a few percent. Thus, the joint experimental and theoretical studies of microscopic vibration–rotation inelastic cross sections seem also well suited to bring the understanding and quantitative prediction of macroscopic phenomena such as gas-phase transport properties and internal-state relaxation processes of nonspherical molecules onto firmer microscopic grounds.

Future inelastic scattering experiments should (and, after some additional work, probably will) be undertaken to complement already available theoretical studies on low-energy collisions of atoms with symmetrical top molecules as, e.g., NH_3. Also desirable would be experimental rotational state-resolved scattering data on atom–molecule and diatom–diatom collisions in deeper attractive potential wells because, incidentally, all neutral scattering studies of Table III probe primarily the shape of the repulsive barrier part of the respective interaction potentials, although the average well depth of these rather shallow potentials wells could be determined to within a few tenths of a millielectron volt. Certainly, neutral scattering experiments will soon also cover a wider range of vibrational excitation phenomena, including possibly the study of V–V and V–R near-resonance effects with vibrationally excited colliding beams. What is to be expected from R–R rotational energy transfer can already be answered to some extent from the diatom–diatom factorization relations for energy sudden collisions. However, from the theoretical point of view the larger number of energetically accessible internal states present for higher excited molecules will remain problematic for collisions which cannot be described in a sudden approximation and exceed the limit of a few hundred coupled channels.

ACKNOWLEDGMENT

The author is very grateful to Dr. R. B. Doak and to Professor J. P. Toennies for comments and suggestions on the manuscript.

REFERENCES

Alexander, M. H., and DePristo, A. E. (1976). *J. Chem. Phys.* **65**, 5009.
Amirav, A., and Even, U. (1980). *J. Appl. Phys.* **51**, 1.

Anderson, J. B. (1974). *In* "Molecular Beams and Low Density Gas Dynamics" (P. P. Wegener, ed.), p. 1. Dekker, New York.
Andres, J., Buck, U., Huisken, F., Schleusener, J., and Torello, F. (1980a). *J. Chem. Phys.* **73**, 5620.
Andres, J., Buck, U., Huisken, F., Schleusener, J., and Torello, F. (1980b). *In* "Electronic and Atomic Collisions" (N. Oda and K. Takayanagi, eds.), p. 531. North-Holland Publ., Amsterdam.
Andres, J., Buck, U., Meyer, H., and Meyer, J. (1982). *J. Chem. Phys.* **76**, 1417.
Andresen, P., Joswig, H., Pauly, H., and Schinke, R. (1982). *J. Chem. Phys.* **77**, 2204.
Barnes, J. A., Keil, M., Kutina, R. E., and Polanyi, J. C. (1982). *J. Chem. Phys.* **76**, 913.
Beard, L. H., Kouri, D. J., and Hoffman, D. K. (1982). *J. Chem. Phys.* **76**, 3623.
Beck, D., Ross, U., and Schepper, W. (1979). *In* "Electronic and Atomic Collisions" (K. Takayanagi and N. Oda, eds.), p. 808. North-Holland Publ., Amsterdam.
Beenakker, J. J. M., and McCourt, F. R. (1970). *Annu. Rev. Phys. Chem.* **21**, 47.
Beijerinck, H. C. W., and Verster, N. F. (1981). *Physica C (Amsterdam)* **111C**, 327.
Bergmann, K., Engelhardt, R., Hefter, U., Hering, P., and Witt, J. (1978). *Phys. Rev. Lett.* **40**, 1446.
Bergmann, K., Hefter, U., and Witt, J. (1980). *J. Chem. Phys.* **72**, 4777.
Bergmann, K., Hefter, U., Mattheus, A., and Witt, J. (1981). *Chem. Phys. Lett.* **78**, 61.
Berns, R. M., and van der Avoird, A. (1980). *J. Chem. Phys.* **72**, 6107.
Bernstein, R. B. (1979). *In* "Atom—Molecule Collision Theory" (R. B. Bernstein, ed.), p. 1. Plenum, New York.
Bischof, G., Hermann, V., Krutein, J., and Linder, F. (1982). *J. Phys. B* [1] **15**, 249.
Blair, J. S. (1966). *Lect. Theor. Phys.* **8, C**, 343.
Borkenhagen, U., Malthan, H., and Toennies, J. P. (1979). *J. Chem. Phys.* **71**, 1722.
Bosanac, S. (1980). *Phys. Rev. A* [3] **22**, 2617.
Böttner, R., Ross, U., and Toennies, J. P. (1976). *J. Chem. Phys.* **65**, 733.
Boughton, C. V., Miller, R. E., and Watts, R. O. (1982). To be published.
Bowman, J. M. (1979). *Chem. Phys. Lett.* **62**, 309.
Bowman, J. M., and Schinke, R. (1982). *Top. Curr. Phys.* **33** (J. M. Bowman, ed.) p. 61. Springer Verl., Berlin.
Brechignac, P., Picard-Bersellini, A., Charneau, R., and Launay, J. M. (1980). *Chem. Phys.* **53**, 165.
Brusdeylins, G., and Meyer, H. D. (1979). *In* "Rarefield Gas Dynamics" (R. Campargue, ed.), Vol. II, p. 919. Comissariat Energie Atomique, Paris.
Brusdeylins, G., Meyer, H. D., Toennies, J. P., and Winkelmann, K. (1977). *In* "Rarefied Gas Dynamics" (J. L. Potter, ed.), Vol. 2, p. 1047. AIAA Publ., New York.
Buck, U. (1982). *Faraday Discuss. Chem. Soc.* **73**, 187.
Buck, U., Huisken, F., Schleusner, J., and Pauly, H. (1977). *Phys. Rev. Lett.* **38**, 680.
Buck, U., Huisken, F., Schleusener, J., and Schäfer, J. (1980). *J. Chem. Phys.* **72**, 1512.
Buck, U., Kohlhase, A., and Meyer, H. (1982). To be published.
Buckingham, A. D. (1967). *Adv. Chem. Phys.* **12**, 107.
Campargue, R., Lebehot, A., and Lemonnier, J. C. (1977). *In* "Rarefied Gas Dynamics" (J. L. Potter, ed.), Vol. 2, p. 1033. AIAA Publ., New York.
Cross, J. B., and Valentini, J. (1982). *Rev. Sci. Instrum.* **53**, 38.
Dagdigian, P. J. (1980). *In* "Electronic and Atomic Collisions" (K. Oda and K. Takayanagi, eds.), p. 513. North-Holland Publ., Amsterdam.
David, R., Faubel, M., and Toennies, J. P. (1973). *Chem. Phys. Lett.* **18**, 87.
DePristo, A. E., and Rabitz, H. (1980). *Adv. Chem. Phys.* **52**, 271.

DePristo, A. E., Augustin, S. D., Ramaswamy, R., and Rabitz, H. (1979). *J. Chem. Phys.* **71**, 850.
Dittner, P. F., and Datz, S. (1978). *J. Chem. Phys.* **68**, 2451.
Douketis, C., Scoles, G., Marchetti, S., Zen, M., and Thakkar, A. J. (1982). *J. Chem. Phys.* **76**, 3057.
Drozdov, S. I. (1955). *Zh. Eksp. Teor. Fiz.* **28**, 734.
Eastes, W., Ross, U., and Toennies, J. P. (1977). *J. Chem. Phys.* **66**, 1919.
Eastes, W., Ross, U., and Toennies, J. P. (1979a). *J. Chem. Phys.* **70**, 1652.
Eastes, W., Ross, U., and Toennies, J. P. (1979b). *Chem. Phys.* **39**, 407.
Eccles, J., Pfeffer, G., Piper, E., Ringer, G., and Toennies, J. P. (1981). *Europhys. Conf. Abstr.* **5A**, 668.
Ellenbroek, T., and Toennies, J. P. (1982). *Chem. Phys.* **71**, 309.
Ellenbroek, T., Gierz, U., and Toennies, J. P. (1980). *Chem. Phys. Lett.* **70**, 459.
Ellenbroek, T., Gierz, U., Noll, M., and Toennies, J. P. (1982). *J. Phys. Chem.* **86**, 1153.
Eno, L., and Rabitz, H. (1980). *J. Chem. Phys.* **72**, 2314.
Eno, L., and Rabitz, H. (1981). *J. Chem. Phys.* **74**, 3859.
Faubel, M., and Toennies, J. P. (1978). *Adv. At. Mol. Phys.* **13**, 229.
Faubel, M., and Toennies, J. P. (1979). *J. Chem. Phys.* **71**, 3770.
Faubel, M., and Weiner, E. R. (1981). *J. Chem. Phys.* **75**, 641.
Faubel, M., Rudolph, K., and Toennies, J. P. (1975). *In* "Electronic and Atomic Collisions" (J. S. Risley and R. Geballe, eds.), p. 49. Univ. of Washington Press, Seattle.
Faubel, M., Kohl, K. H., and Toennies, J. P. (1980). *J. Chem. Phys.* **73**, 2506.
Faubel, M., Kohl, K. H., and Toennies, J. P. (1981). *In* "Electronic and Atomic Collisions" (S. Datz, ed.), p. 935. Oak Ridge Natl. Lab., Oak Ridge, Tennessee.
Faubel, M., Kohl, K. H., Toennies, J. P., Tang, K. T., and Yung, Y. Y. (1982). *Faraday Discuss. Chem. Soc.* **73**, 205.
Flower, D. R., Launay, J. M. Kochanski, E., and Prisette, J. (1979). *Chem. Phys.* **37**, 355.
Fluendy, M. A. D., and Lawley, K. P. (1973). "Chemical Applications of Molecular Beam Scattering." Chapman & Hall, London.
Garrison, B. J., Lester, W. A., and Miller, W. H. (1976). *J. Chem. Phys.* **65**, 2193.
Gentry, W. R. (1979). *In* "Atom Molecule Collision Theory" (R. B. Bernstein, ed.), p. 391. Plenum, New York.
Gentry, W. R. (1980a). *In* "Electronic and Atomic Collisions" (N. Oda and K. Takayanagi, eds.), p. 807. North-Holland Publ., Amsterdam.
Gentry, W. R. (1980b). *Comments At. Mol. Phys.* **9**, 113.
Gentry, W. R., and Giese, C. F. (1977). *J. Chem. Phys.* **67**, 5389.
Gentry, W. R., Udseth, H., and Giese, C. F. (1975). *Chem. Phys. Lett.* **36**, 671.
Gerber, R. B., Buch, V., and Buck, U. (1980). *J. Chem. Phys.* **72**, 3596.
Gianturco, F. A. (1982). *Faraday Discuss. Chem. Soc.* **73**, 282.
Gianturco, F. A., Gierz, U., and Toennies, J. P. (1981). *J. Phys. B* [1] **14**, 667.
Gierz, U., and Toennies, J. P. (1982). To be published.
Gierz, U., Noll, M., and Toennies, J. P. (1982). To be published.
Goldflam, R., and Kouri, D. J. (1979). *J. Chem. Phys.* **70**, 5076.
Goldflam, R., Green, S., and Kouri, D. J. (1977). *J. Chem. Phys.* **67**, 4149, 5661.
Gordon, R. G., and Kim, Y. S. (1972). *J. Chem. Phys.* **56**, 3122.
Gordon, R. G., Klemperer, W., and Steinfeld, J. I. (1968). *Annu. Rev. Phys. Chem.* **19**, 215.
Gough, T. E., Miller, R. E., and Scoles, G. (1981). *Faraday Discuss. Chem. Soc.* **71**, 77.
Green, S. (1975). *J. Chem. Phys.* **62**, 2271.
Green, S. (1977). *J. Chem. Phys.* **67**, 715.

Green, S. (1979). *J. Chem. Phys.* **70**, 816.
Green, S. (1980). *J. Chem. Phys.* **73**, 2740.
Green, S., and Thaddeus, P. (1976). *Astrophys. J.* **205**, 766.
Habitz, P., Tang, K. T., and Toennies, J. P. (1982). *Chem. Phys. Lett.* **85**, 461.
Hagena, O. F. (1974). *In* "Molecular Beams and Low Density Gas Dynamics" (P. P. Wegener, ed.), p. 93. Dekker, New York.
Hefter, U., Jones, P. L., Mattheus, A., Witt, J., Bergmann, K., and Schinke, R. (1981). *Phys. Rev. Lett.* **46**, 915.
Hepburn, J., Scoles, G., and Penco, R. (1975). *Chem. Phys. Lett.* **36**, 451.
Herrmann, A., Leutwyler, S., Schumacher, E., and Wöste, L. (1977). *Chem. Phys. Lett.* **52**, 418.
Herzberg, G. (1950). "Molecular Spectra and Molecular Structure." Van Nostrand-Reinhold, Princeton, New Jersey.
Hirschfelder, J. O., Curtiss, C. F., and Bird, R. B. (1954). "Molecular Theory of Gases and Liquids." Wiley, New York.
Huber, K. P., and Herzberg, G. (1979). "Constants of Diatomic Molecules." Van Nostrand-Reinhold, Princeton, New Jersey.
Jaquet, R., and Staemmler, V. (1982). To be published.
Keil, M., Parker, G. A., and Kuppermann, A. (1978). *Chem. Phys. Lett.* **59**, 443.
Keil, M., Slankas, J. T., and Kuppermann, A. (1979). *J. Chem. Phys.* **70**, 541.
Khare, V. (1978). *J. Chem. Phys.* **68**, 4631.
Kihara, T. (1978). "Intermolecular Forces." Wiley, New York.
Kneba, M., and Wolfrum, J. (1980). *Annu. Rev. Phys. Chem.* **31**, 47.
Knuth, E. L. (1977). *J. Chem. Phys.* **66**, 3515.
Kohl, K. H. (1982). Ph.D. Thesis, University of Göttingen.
Kouri, D. J. (1979). *In* "Atom Molecular Collision Theory" (R. B. Bernstein, ed.), p. 301. Plenum, New York.
Krupenie, P. H. (1972). *J. Phys. Chem. Ref. Data* **1**, 423.
Krutein, J., and Linder, F. (1977a). *Chem. Phys. Lett.* **51**, 597.
Krutein, J., and Linder, F. (1977b). *J. Phys. B* [1] **10**, 1363.
Krutein, J., and Linder, F. (1979). *J. Chem. Phys.* **71**, 599.
Kusunoki, I. (1977). *J. Chem. Phys.* **67**, 2224.
Kutzelnigg, W. (1977). *Faraday Discuss. Chem. Soc.* **62**, 185.
Kutzelnigg, W., Staemmler, V., and Hoheisel, K. (1973). *Chem. Phys.* **1**, 27.
Lambert, J. D. (1977). "Vibrational and Rotational Relaxation in Gases." Oxford Univ. Press (Clarendon), London and New York.
Lantzsch, B. (1974). *Ber.* **120**. *Max-Planck-Inst. Stroemungsforsch.*
LeRoy, R. J., and Carley, J. S. (1980). *Adv. Chem. Phys.* **42**, 353.
Lester, W. A. (1970). *J. Chem. Phys.* **53**, 1511.
Light, J. C. (1979). *In* "Atom–Molecule Collision Theory" (R. B. Bernstein, ed.), p. 239. Plenum, New York.
Linder, F. (1980). *In* "Electronic and Atomic Collisions" (N. Oda and K. Takayonaki, eds.), p. 535. North-Holland, Publ., Amsterdam.
Loesch, H. J. (1976). *Chem. Phys.* **18**, 431.
Loesch, H. J. (1980). *Adv. Chem. Phys.* **52**, 421.
Maitland, G. C., Rigby, M., Smith, E. B., and Wakeham, W. A. (1981). "Intermolecular Forces." Oxford Univ. Press (Clarendon), London and New York.
Mills, I. M. (1977). *Faraday Discuss. Chem. Soc.* **62**, 7.
Monchick, L., and Mason, E. A. (1961). *J. Chem. Phys.* **35**, 1676.

Pack, R. T. (1978). *Chem. Phys. Lett.* **55**, 197.
Parker, G. A., and Pack, R. T. (1978). *J. Chem. Phys.* **68**, 1585.
Parker, G. A., Keil, M., and Kuppermann, A. (1983). *J. Chem. Phys.* **78**, 1145.
Patengill, M. D. (1979). *In* "Atom–Molecule Collision Theory" (R. B. Bernstein, ed.), p. 359. Plenum, New York.
Phillips. W. D., Serri, J. A., Ely, D. J., Pritchard, D. E., Way, K. R., and Kinsey, J. L. (1978). *Phys. Rev. Lett.* **41**, 937.
Poulsen, P., and Miller, D. R. (1977). *In* "Rarefied Gas Dynamics" (J. L. Potter, ed.), Vol. 2, p. 899. AIAA Publ., New York.
Rae, A.I.M. (1973). *Chem. Phys. Lett.* **18**, 574.
Richards, D. (1982). *J. Phys. B* [1] **15**, 1499.
Rodwell, W. R., and Scoles, G. (1981). Chem. Phys. Res. Rep. No. CP-165. University of Waterloo, Waterloo, Ontario.
Ross, U., Schepper, W., Schulze, T., and Beck, D. (1981). *In* "Electronic and Atomic Collisions" (S. Datz, ed.), p. 939. Oak Ridge Natl. Lab., Oak Ridge, Tennessee.
Rudolph, K., and Toennies, J. P. (1976). *J. Chem. Phys.* **65**, 4483.
Saenger, K. L. (1981). *J. Chem. Phys.* **75**, 2467.
Schaefer, H. F., III (1979) *In* "Atom–Molecule Collision Theory" (R. B. Bernstein, ed.), p. 45. Plenum, New York.
Schaefer, J., and Lester, W. A., Jr. (1975). *J. Chem. Phys.* **62**, 1913.
Schaefer, J., and Meyer, W. (1979). *J. Chem. Phys.* **70**, 344.
Schepper, W., Ross, U., and Beck, D. (1979). *Z. Phys. A* **290**, 131.
Schinke, R., and McGuire, P. (1979). *J. Chem. Phys.* **71**, 4201.
Schinke, R., Müller, W., and Meyer, W. (1982). *J. Chem. Phys.* **76**, 895.
Secrest, D. (1979). *In* "Atom–Molecule Collision Theory" (R. B. Bernstein, ed.), p. 265. Plenum, New York.
Serri, J. A., Morales, A., Moskowitz, W., Pritchard, D. E., Becker, C. H., and Kinsey, J. L. (1980). *J. Chem. Phys.* **72**, 6304.
Serri, J. A., Becker, C. H., Elbel, M. B., Kinsey, J. L., Moskowitz, W. P., and Pritchard, D. E. (1981). *J. Chem. Phys.* **74**, 5116.
Smith, E. B., and Tindell, A. R. (1982). *Faraday Discuss. Chem. Soc.* **73**, 221.
Smith, E. W., and Giraud, M. (1979). *J. Chem. Phys.* **71**, 4209.
Smith, L. N., and Secrest, D. (1981). *J. Chem. Phys.* **74**, 3882.
Smith, N., and Pritchard, D. E. (1981). *J. Chem. Phys.* **74**, 3939.
Tang, K. T., and Toennies, J. P. (1977). *J. Chem. Phys.* **66**, 1496; for errata, see **67**, 375 (1977); **68**, 786 (1978).
Tang, K. T., and Toennies, J. P. (1978). *J. Chem. Phys.* **68**, 5501.
Tang, K. T., and Toennies, J. P. (1982). *J. Chem. Phys.,* **76**, 2524.
Toennies, J. P. (1976). *Ann. Rev. Phys. Chem.* **27**, 225.
Thomas, L. D. (1977). *J. Chem. Phys.* **67**, 5224.
Thuis, H., Stolte, S., and Reuss, J. (1979). *Comments At. Mol. Phys.* **8**, 123.
Udseth, H., Giese, C. F., and Gentry, W. R. (1973). *Phys. Rev. A* [1] **8**, 2483.
Valentini, J. J., Esherick, P., and Owyoung, A. (1980). *Chem. Phys. Lett.* **75**, 590.
van den Bergh, H. E., Faubel, M., and Toennies, J. P. (1973). *Faraday Discuss. Chem. Soc.* **55**, 203.
van Deursen, A., and Reuss, J. (1973). *Int. J. Mass. Spectrom. Ion Phys.* **11**, 483.
van Hemert, M. C., and Berns, R. M. (1982). *J. Chem. Phys.* **76**, 354.
Waldman, M., and Gordon, R. G. (1979a). *J. Chem. Phys.* **71**, 1325.
Waldman, M., and Gordon, R. G. (1979b). *J. Chem. Phys.* **71**, 1340.

Wijnaendts van Resandt, R. W., and Los, J. (1980). *In* "Electronic and Atomic Collisions" (N. Oda and K. Takayanaki, eds.), p. 831. North-Holland Publ., Amsterdam.

Winkelmann, K. (1979). *In* "Rarefied Gas Dynamics" (R. Campargue, ed.), Vol. II, p. 899. Comissariat Energie Atomique, Paris.

Yardley, J. T. (1980). "Introduction to Molecular Energy Transfer." Academic Press, New York.

Zandee, L., and Bernstein, R. B. (1979). *J. Chem. Phys.* **71,** 1359.

Zandee, L., and Reuss, J. (1977). *Chem. Phys.* **26,** 327, 345.

SPIN POLARIZATION OF ATOMIC AND MOLECULAR PHOTOELECTRONS

N. A. CHEREPKOV

A. F. Ioffe Physico-Technical Institute
Academy of Sciences of the USSR
Leningrad, USSR

I. Introduction	395
II. Theory of Spin Polarization Phenomena in Atoms	397
A. Polarization Caused by the Fine-Structure Splitting of Atomic Levels	397
B. Analysis of Expressions for Degree of Polarization	401
C. Fano Effect	404
D. Simultaneous Inclusion of the Spin–Orbit Interaction in Both Discrete and Continuous Spectra	407
E. On the Complete Quantum-Mechanical Experiment	410
F. Some Generalizations	413
III. Comparison with Experiment and Applications	416
A. Rare Gas Atoms	416
B. Fano Effect in ns and ns^2 Subshells	421
C. Atoms with One Outer p electron	425
D. Method for Identification of Autoionization Resonances	429
E. Sources of Polarized Electrons	433
IV. Spin Polarization of Molecular Photoelectrons	434
A. General Derivation	434
B. Molecules Having a Plane of Symmetry	438
C. Optically Active Molecules	440
V. Conclusion	442
References	443

I. Introduction

Photoionization is one of the simplest processes of atomic collisions which allows a detailed investigation of atomic structure. Therefore, much attention was given to it by both theorists and experimentalists. Regularly published reviews (see, for example, Bethe, 1933; Hall, 1936; Seaton, 1951; Fano and Cooper, 1968; Hudson and Kieffer, 1971; Pratt *et al.*, 1973;

Amusia and Cherepkov, 1975; Berkowitz, 1979; Kelly and Carter, 1980; Amusia, 1981; Samson, 1982; Starace, 1982) give a complete account of these investigations.

Although Sauter (1931) pointed out that atomic photoelectrons can be spin-polarized, nobody studied this question until the late 1950s. The first systematical investigation of polarization of atomic photoelectrons ejected from unpolarized atoms was carried out by Nagel (1960) and by Nagel and Olsson (1960). They considered the K-shell photoionization cross section in the Coulomb approximation with relativistic corrections taken into accuont up to second-order terms in αZ, where $\alpha = e^2/\hbar c$ is the fine-structure constant and Z is the nuclear charge. It was shown that in the low-energy limit when $v/c \ll 1$, v being the photoelectron velocity, the spin polarization of electrons is described by terms proportional to $(\alpha Z)^2$. This result is quite natural since the operator of interaction between electron and electromagnetic field does not have a direct influence upon the electron spin, and the polarization of photoelectrons appears only through a spin–orbit interaction. The latter, as is well known (Landau and Lifshitz, 1976), is proportional to $(\alpha Z)^2$. Therefore, the polarization of atomic photoelectrons was considered to be a relativistic effect which is important for high photon energies when either photoelectron velocity is high ($v/c \sim 1$) or the charge of nucleus is large ($\alpha Z \sim 1$). Up to now the only experimental measurement of the polarization of K-shell photoelectrons produced by high-energy photons (662 keV) in gold has been performed by Gomes and Byrne (1980). Their result coincides with theoretical prediction (Hultberg et al., 1968).

It was at first pointed out by Fano (1969a) that in the nonrelativistic energy region photoelectrons can also be highly polarized. Previously, Kessler observed that in low-energy elastic electron scattering on atoms the polarization maxima usually occur at cross section minima (Kessler, 1969, 1982). In the photoionization cross section of alkali atoms near thresholds there are also minima caused by the change of sign of dipole matrix elements (Seaton, 1951), which are called the Cooper minima (Cooper, 1962). Due to the spin–orbit interaction in the continuous spectrum, the dipole matrix elements corresponding to the $ns \rightarrow \epsilon p_{1/2}$ and $ns \rightarrow \epsilon p_{3/2}$ transitions change sign at different points in energy. In the vicinity of these points the difference between matrix elements is not small, and at some point the degree of polarization reaches 100% (the Fano effect). The smallness of order $(\alpha Z)^2$ disappears from the degree of polarization, but appears instead in the cross section magnitude.

The most striking case proved to be subshells with $l > 0$. Photoelectrons corresponding to different fine-structure components of an initial atomic state and/or a final ionic state appear to be polarized irrespective of photon energy. The degree of polarization can be as high as 100% even in cross section maxima (Cherepkov, 1972, 1973, 1980b). Smallness of order $(\alpha Z)^2$ defines in this case the fine-structure splitting and does not enter the expressions for the degree of polarization.

Photoelectron polarization measurements give qualitatively new information on atomic structure and, being combined with the usually measured partial photoionization cross section and angular distribution of photoelectrons, allow a complete quantum-mechanical experiment to be performed. This fact shows the fundamental importance of photoelectron polarization measurements for atomic physics and justifies construction of the quite sophisticated experimental apparatus required for this kind of investigations.

In this review the theory of polarization phenomena in the photoionization of atoms and molecules is presented. Theoretical predictions are compared with existing experimental results. Some future prospects of polarization investigations are also discussed. The consideration is restricted by nonrelativistic photon energies where the dipole approximation is applicable. Only one-photon processes with the ejection of one electron are discussed.

II. Theory of Spin Polarization Phenomena in Atoms

A. Polarization Caused by the Fine-Structure Splitting of Atomic Levels

Let us consider absorption of light of a given polarization by unpolarized atoms. For the complete quantum-mechanical description of a photoionization process it is necessary to find the probability of ejection of electrons in a given direction κ with the spin oriented in some other direction $\mathbf{s}$ (κ and $\mathbf{s}$ are the unit vectors). The interaction operator between electron and electromagnetic field in the dipole approximation is proportional to the dipole operator

$$\hat{d}_\lambda = \left(\frac{4\pi}{3}\right)^{1/2} \sum_{i=1}^{N} r_i Y_{1\lambda}(\theta_i, \varphi_i) \tag{1}$$

where N is the number of electrons in atom, $\lambda = \pm 1$ is the photon helicity for circularly polarized light, and $\lambda = 0$ for linearly polarized light. Since the dipole operator given by Eq. (1) does not act on the electron spin variables, the photoelectron polarization can appear only through the spin–orbit interaction.

For simplicity, we shall consider at first the photoionization of a one-electron atomic subshell with $l \neq 0$, taking into account the spin–orbit interaction in the initial state only, i.e., the fine-structure splitting of atomic levels. Suppose also that one fine-structure component is populated. The initial state of the atomic electron is described in this case by quantum numbers $nljm_j$, where n is the principal quantum number, l is the orbital momentum, $j = l \pm \frac{1}{2}$ and m_j are the total angular momentum and its projection, respectively. The initial state wave function is

$$\psi_{nljm_j}(\mathbf{r}) = \sum_{m,\mu} [j]^{1/2} (-1)^{1/2-l-m_j} \begin{pmatrix} l & \frac{1}{2} & j \\ m & \mu & -m_j \end{pmatrix} \varphi^j_{nlm\mu}(\mathbf{r}) \quad (2)$$

where $[j] \equiv (2j + 1)$,

$$\varphi^j_{nlm\mu}(\mathbf{r}) = R^j_{nl}(r)\, Y_{lm}(\theta, \varphi)\, \chi_\mu \quad (3)$$

$R^j_{nl}(r)$ is the radial part of a one-electron wave function, which depends also on the total momentum j, and χ_μ is the spin wave function, μ being the spin projection on the Z axis. It should be noted that for linearly and circularly polarized light the coordinate systems are defined differently. For linearly polarized light the Z axis is directed along the photon polarization vector $\mathbf{e}$, whereas for circularly polarized and unpolarized light it coincides with the direction of a photon beam.

The final-state electron wave function contains in the asymptotic region the superposition of a plane wave propagating in the direction of the electron momentum $\mathbf{p}$ and a converging spherical wave, and can be expanded in partial waves as usual (Sobel'man, 1972):

$$\psi^-_{\mathbf{p}\mu_1}(\mathbf{r}) = \frac{2\pi}{\sqrt{p}} \sum_{l_1, m_1} (i)^{l_1}$$

$$\times \exp\{-i\delta_{l_1}\} R_{\epsilon l_1}(r)\, Y_{l_1 m_1}(\theta, \varphi)\, Y^*_{l_1 m_1}(\hat{\kappa})\, \chi_{\mu_1} \quad (4)$$

where $\kappa = \mathbf{p}/|p|$, $\epsilon = p^2/2$ (atomic units $\hbar = m = e = 1$ are used in this article), and the radial wave functions $R_{\epsilon l_1}(r)$ are normalized to an energy δ-function. Using Eqs. (2)–(4), it is easy to calculate the matrix element of the dipole operator given by Eq. (1):

$$\langle \psi^-_{\mathbf{p}\mu_1} | \hat{d}_\lambda | \psi_{nljm_j} \rangle$$

$$= \frac{2\pi}{\sqrt{p}} \sum_{l_1, m_1} \sum_m [j]^{1/2} (i)^{-l_1} (-1)^{1/2-l-m_j-m_1} \exp\{i\delta_{l_1}\} Y_{l_1 m_1}(\hat{\kappa})$$

$$\times \begin{pmatrix} l & \frac{1}{2} & j \\ m & \mu_1 & -m_j \end{pmatrix} \begin{pmatrix} l_1 & 1 & l \\ -m_1 & \lambda & m \end{pmatrix} \langle \epsilon l_1 \| d \| nlj \rangle \quad (5)$$

Here the reduced dipole matrix element is defined as follows:

$$\langle \epsilon l_1 \| d \| nlj \rangle = [l, l_1]^{1/2} \begin{pmatrix} l & 1 & l_1 \\ 0 & 0 & 0 \end{pmatrix} \int_0^\infty R_{nl}^j(r) R_{\epsilon l_1}(r) r^3 dr \quad (6)$$

The squared dipole matrix element of Eq. (5) gives the probability of ejection of an electron in the direction κ with a spin oriented along the Z axis. The differential photoionization cross section for the ejection of photoelectrons in some direction κ with the spin oriented along another direction s is given by the expression

$$I_j^\lambda(\kappa, s) = \frac{\alpha \omega p}{[j]} \sum_{\mu_1, \mu_2} \sum_{m_j} \langle \psi_{nljm_j} | d_\lambda^* | \psi_{\mathbf{p}\mu_1}^- \rangle$$
$$\times \tfrac{1}{2}(1 + s\sigma)_{\mu_1\mu_2} \langle \psi_{\mathbf{p}\mu_2}^- | d_\lambda | \psi_{nljm_j} \rangle \quad (7)$$

where ω is the photon energy and $\tfrac{1}{2}(1 + s\sigma)$ is the spin projection operator with σ being the Pauli matrix vector. This operator can be written explicitly as

$$(1 + s\sigma)_{\mu_1\mu_2} = \delta_{\mu_1\mu_2}[1 + (-1)^{\mu_1 - 1/2} \cos \alpha]$$
$$+ (1 - \delta_{\mu_1\mu_2}) \sin \alpha \exp\{-2i\varphi\mu_1\} \quad (8)$$

where α and φ characterize the direction of s in our coordinate system (see Fig. 1). For umambiguous definition of the coordinate system we suppose that κ always lies in the XOZ plane. Summation over all projections in Eq. (7) gives (Cherepkov, 1972, 1979)

$$I_j^\lambda(\kappa, s) = \frac{\sigma_{nlj}(\omega)}{8\pi} [1 + \tfrac{1}{2}(2 - 3\lambda^2)\beta^j P_2(\cos \theta) + \lambda A^j \cos \alpha$$
$$- \lambda \gamma^j P_2(\cos \theta) \cos \alpha - \tfrac{1}{2}\lambda \gamma^j P_2^1(\cos \theta) \sin \alpha \cos \varphi$$
$$- \tfrac{1}{3}(2 - 3\lambda^2)\eta^j P_2^1(\cos \theta) \sin \alpha \sin \varphi] \quad (9)$$

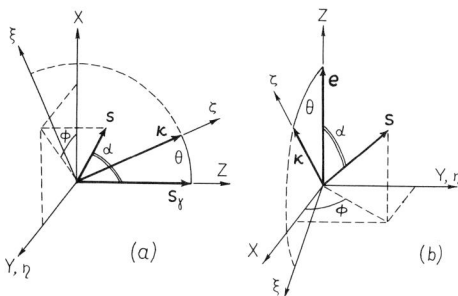

FIG. 1. Coordinate systems in the cases of circularly polarized (a) and linearly polarized (b) light.

Here P_2 and P_2^1 are the Legendre and associate Legendre polynomials, respectively,

$$\sigma_{nlj}(\omega) = \frac{4\pi^2 \alpha \omega N_j}{3(2l+1)} \sum_{l_1} |\langle \epsilon l_1 \| d \| nlj \rangle|^2 \tag{10}$$

N_j is the number of electrons in the subshell with given nlj, β^j is the usual angular asymmetry parameter (Cooper and Zare, 1968):

$$\beta^j(\omega) = \frac{(l-1)(d^j_{l-1})^2 + (l+2)(d^j_{l+1})^2 + 6[l(l+1)]^{1/2} d^j_{l-1} d^j_{l+1} \cos(\delta_{l+1} - \delta_{l-1})}{(2l+1)[(d^j_{l-1})^2 + (d^j_{l+1})^2]} \tag{11}$$

with

$$d^j_{l\pm 1} \equiv \langle \epsilon, l \pm 1 \| d \| nlj \rangle \tag{12}$$

and the parameters A^j, γ^j, and η^j define the polarization of photoelectrons,

$$A^j(\omega) = \frac{(-1)^{j-l-1/2}}{[j]} \frac{l(d^j_{l+1})^2 - (l+1)(d^j_{l-1})^2}{(d^j_{l-1})^2 + (d^j_{l+1})^2} \tag{13}$$

$$\gamma^j(\omega) = \frac{2(-1)^{j-l-1/2}}{[j,l]}$$

$$\times \frac{l(l+2)(d^j_{l+1})^2 - (l^2-1)(d^j_{l-1})^2 - 3[l(l+1)]^{1/2} d^j_{l-1} d^j_{l+1} \cos(\delta_{l+1} - \delta_{l-1})}{(d^j_{l-1})^2 + (d^j_{l+1})^2} \tag{14}$$

$$\eta^j(\omega) = \frac{(-1)^{j-l-1/2}}{[j]} \frac{3[l(l+1)]^{1/2} d^j_{l-1} d^j_{l+1} \sin(\delta_{l+1} - \delta_{l-1})}{(d^j_{l-1})^2 + (d^j_{l+1})^2} \tag{15}$$

For general analysis it is more convenient to deal with expressions which are independent of a particular choice of coordinate system. To this end, we introduce in Eq. (9) the unit vectors κ and s defined above, the polarization vector e for linearly polarized light, and the vector of photon spin s_γ for circularly polarized light, which is directed along the vector of photon momentum for $\lambda = +1$ and opposite to it for $\lambda = -1$. From Fig. 1 one can conclude that

$$(\kappa s) = \cos\theta \cos\alpha + \sin\theta \sin\alpha \cos\varphi$$
$$(es) = (ss_\gamma) = \cos\alpha, \quad (\kappa e) = (\kappa s_\gamma) = \cos\theta \tag{16}$$
$$(s[e\kappa]) = (s[s_\gamma \kappa]) = \sin\theta \sin\alpha \sin\varphi$$

Using these definitions one obtains from Eq. (9) that

$$I_j^{\pm 1}(\boldsymbol{\kappa}, \mathbf{s}) = \frac{\sigma_{nlj}(\omega)}{8\pi} \left\{ 1 - \frac{\beta^j}{2}[\tfrac{3}{2}(\boldsymbol{\kappa}\mathbf{s}_\gamma)^2 - \tfrac{1}{2}] + A^j(\mathbf{s}\mathbf{s}_\gamma) \right.$$
$$\left. - \gamma^j[\tfrac{3}{2}(\boldsymbol{\kappa}\mathbf{s}_\gamma)(\boldsymbol{\kappa}\mathbf{s}) - \tfrac{1}{2}(\mathbf{s}\mathbf{s}_\gamma)] - \eta^j(\mathbf{s}[\boldsymbol{\kappa}\mathbf{s}_\gamma])(\boldsymbol{\kappa}\mathbf{s}_\gamma) \right\} \quad (17)$$

for circularly polarized light and

$$I_j^0(\boldsymbol{\kappa}, \mathbf{s}) = \frac{\sigma_{nlj}(\omega)}{8\pi} \{1 + \beta^j[\tfrac{3}{2}(\boldsymbol{\kappa}\mathbf{e})^2 - \tfrac{1}{2}] + 2\eta^j(\mathbf{s}[\boldsymbol{\kappa}\mathbf{e}])(\boldsymbol{\kappa}\mathbf{e})\} \quad (18)$$

for linearly polarized light. Presenting unpolarized light as a superposition of left and right circularly polarized light with equal weights, one obtains from Eq. (17) (Lee, 1974) that

$$I_j^{\mathrm{un}}(\boldsymbol{\kappa}, \mathbf{s}) = \frac{\sigma_{nlj}(\omega)}{8\pi} \left\{ 1 - \frac{\beta^j}{2}[\tfrac{3}{2}(\boldsymbol{\kappa}\mathbf{k})^2 - \tfrac{1}{2}] - \eta^j(\mathbf{s}[\boldsymbol{\kappa}\mathbf{k}])(\boldsymbol{\kappa}\mathbf{k}) \right\} \quad (19)$$

where **k** is the unit vector in the direction of a photon momentum.

In relativistic approximation, when higher multipoles are taken into account (Pratt *et al.*, 1964; Hultberg *et al.*, 1968), Eqs. (17)–(19) still remain valid if the parameters σ_{nlj}, β^j, A^j, γ^j, η^j are considered functions of the photoelectron ejection angle. Furthermore, in Eq. (18) an additional term appears which is proportional to the vector combination $(\mathbf{s}[\mathbf{e}\mathbf{k}])(\boldsymbol{\kappa}\mathbf{e})$ with a new coefficient. This coefficient (C_{21} and C_{23} in the notation of Pratt *et al.*, 1964) is of order $h\nu/mc^2$ and tends to zero in the nonrelativistic limit, whereas all other coefficients in this limit remain of order $(\alpha Z)^2$ or even of order 1.

B. Analysis of Expressions for Degree of Polarization

Consider at first the spin polarization of the total photoelectron flux. Integration over electron ejection angles in Eqs. (17)–(19) gives

$$I_j^{\pm 1}(\mathbf{s}) = \frac{\sigma_{nlj}(\omega)}{2} \{1 + A^j(\mathbf{s}\mathbf{s}_\gamma)\}, \qquad I_j^0(\mathbf{s}) = I_j^{\mathrm{un}}(\mathbf{s}) = \frac{\sigma_{nlj}(\omega)}{2} \quad (20)$$

From the definition of the degree of polarization of photoelectrons

$$P_j^\lambda(\mathbf{s}) = \frac{I_j^\lambda(\mathbf{s}) - I_j^\lambda(-\mathbf{s})}{I_j^\lambda(\mathbf{s}) + I_j^\lambda(-\mathbf{s})} \quad (21)$$

it follows that

$$P_j^{\pm 1}(\mathbf{s}) = A^j(\mathbf{s}\mathbf{s}_\gamma), \qquad P_j^0(\mathbf{s}) = P_j^{\mathrm{un}}(\mathbf{s}) = 0 \quad (22)$$

In other words, the parameter A^j is equal to the degree of polarization of the total photoelectron flux in the direction of the photon spin $\mathbf{s}_\gamma$ under absorption of circularly polarized light. For absorption of linearly polarized or unpolarized light the total photoelectron flux is unpolarized.

The degree of polarization of photoelectrons ejected at a definite angle can easily be found using Eqs. (17)–(19) from the expressions analogous to Eq. (21), and is equal to

$$P_j^{\pm 1}(\boldsymbol{\kappa}, \mathbf{s}) = \frac{A^j(\mathbf{s}\mathbf{s}_\gamma) - \gamma^j[\frac{3}{2}(\boldsymbol{\kappa}\mathbf{s}_\gamma)(\boldsymbol{\kappa}\mathbf{s}) - \frac{1}{2}(\mathbf{s}\mathbf{s}_\gamma)] - \eta^j(\mathbf{s}[\boldsymbol{\kappa}\mathbf{s}_\gamma])(\boldsymbol{\kappa}\mathbf{s}_\gamma)}{1 - (\beta^j/2)[\frac{3}{2}(\boldsymbol{\kappa}\mathbf{s}_\gamma)^2 - \frac{1}{2}]} \quad (23)$$

for circularly polarized light,

$$P_j^0(\boldsymbol{\kappa}, \mathbf{s}) = \frac{2\eta^j(\mathbf{s}[\boldsymbol{\kappa}\mathbf{e}])(\boldsymbol{\kappa}\mathbf{e})}{1 + \beta^j[\frac{3}{2}(\boldsymbol{\kappa}\mathbf{e})^2 - \frac{1}{2}]} \quad (24)$$

for linearly polarized light, and

$$P_j^{un}(\boldsymbol{\kappa}, \mathbf{s}) = - \frac{\eta^j(\mathbf{s}[\boldsymbol{\kappa}\mathbf{k}])(\boldsymbol{\kappa}\mathbf{k})}{1 - (\beta^j/2)[\frac{3}{2}(\boldsymbol{\kappa}\mathbf{k})^2 - \frac{1}{2}]} \quad (25)$$

for unpolarized light.

The expression in square brackets after γ^j in Eq. (23) coincides with the Legendre polynomial $P_2(\boldsymbol{\kappa}\mathbf{s}_\gamma)$ for $\mathbf{s} \parallel \mathbf{s}_\gamma$ and is equal to zero when $\mathbf{s}$ is perpendicular to the reaction plane, defined by $\mathbf{s}_\gamma$ and $\boldsymbol{\kappa}$. Thus, the γ^j parameter gives the polarization of photoelectrons along the X axis (see Fig. 1):

$$P_j^{\pm 1}(\boldsymbol{\kappa}, \mathbf{s} \parallel X \text{ axis}) = -\tfrac{3}{2}\gamma^j \frac{\cos\theta \sin\theta}{1 - (\beta^j/2)P_2(\cos\theta)} \quad (26)$$

Terms proportional to η^j in Eqs. (23)–(25) give the component of the electron spin polarization perpendicular to the reaction plane (parallel to the Y axis in Fig. 1), which for linearly polarized and unpolarized light is defined by the vectors $\boldsymbol{\kappa}$ and $\mathbf{e}$ or $\boldsymbol{\kappa}$ and $\mathbf{k}$, respectively. For absorption of linearly polarized and unpolarized light this is the only component of spin polarization of photoelectrons (in the dipole approximation), like in differential elastic scattering of unpolarized electrons from atoms [Mott scattering; see Mott and Massey (1965)].

The degree of spin polarization along the photon spin $\mathbf{s}_\gamma$ can be readily obtained from Eq. (23):

$$P_j^{\pm 1}(\boldsymbol{\kappa}, \mathbf{s} \parallel \mathbf{s}_\gamma) = \frac{A^j - \gamma^j P_2(\cos\theta)}{1 - (\beta^j/2)P_2(\cos\theta)} \quad (27)$$

For $\theta = 0$ and $\theta = \pi$ it gives

$$P_j^{\pm 1}(0, \mathbf{s} \| \mathbf{s}_\gamma) = P_j^{\pm 1}(\pi, \mathbf{s} \| \mathbf{s}_\gamma) = \frac{A^j - \gamma^j}{1 - \frac{1}{2}\beta^j} = (-1)^{j-l+1/2} \frac{2}{[j]} \quad (28)$$

For $l = 1$, $j = \frac{1}{2}$ photoelectrons are totally polarized in the direction $\mathbf{s}_\gamma$ irrespective of photon energy (Brehm, 1971; Cherepkov, 1972, 1973):

$$P_{1/2}^{\pm 1}(0, \mathbf{s} \| \mathbf{s}_\gamma) = P_{1/2}^{\pm 1}(\pi, \mathbf{s} \| \mathbf{s}_\gamma) = 1 \quad (29)$$

This condition directly follows from the dipole selection rules. Indeed, from properties of the spherical functions $Y_{l,m_l}(\theta, \varphi)$ it follows that the contribution to the outgoing wave in Eq. (4) for $\theta = 0$ and $\theta = \pi$ is given by the states with $m_l = 0$ only. From the dipole selection rules the final state with $m_l = 0$ can be reached from the initial state with $m = -\lambda$. For $j = \frac{1}{2}$ we have $|m_j| = |m + \mu| = \frac{1}{2}$; therefore $\mu_1 = \mu = \frac{1}{2}\lambda$, that is, the electron spin is always directed along the photon spin.

From Eq. (13) the following conditions can be derived (Cherepkov, 1973):

$$\begin{aligned} -\tfrac{1}{2} \leq A^j \leq (l+1)/2l, & \quad j = l - \tfrac{1}{2} \\ -\tfrac{1}{2} \leq A^j \leq l/[2(l+1)], & \quad j = l + \tfrac{1}{2} \end{aligned} \quad (30)$$

and A^j can reach the largest value of 1 for $l = 1$, $j = \frac{1}{2}$ only. For the η^j parameter one obtains from Eq. (15) (Cherepkov, 1979) that

$$|\eta^j| \leq 3[l(l+1)]^{1/2}/2(2j+1) \quad (31)$$

and the largest value $|\eta^{1/2}| = \tfrac{3}{2}\sqrt{2}$ is again reached for $l = 1$, $j = \tfrac{1}{2}$.

If we neglect the dependence of dipole matrix elements on j [which is defined by terms proportional to $(\alpha Z)^2$], the polarization parameters A^j, γ^j, and η^j appear to be inversely proportional to the statistical weights of states with given l and different j and have the opposite signs:

$$\frac{A^{l+1/2}}{A^{l-1/2}} = \frac{\gamma^{l+1/2}}{\gamma^{l-1/2}} = \frac{\eta^{l+1/2}}{\eta^{l-1/2}} = -\frac{l}{l+1} \quad (32)$$

From Eqs. (23)–(25) analogous conditions follow for the degree of polarization:

$$\frac{P_{l+1/2}^\lambda(\kappa, \mathbf{s})}{P_{l-1/2}^\lambda(\kappa, \mathbf{s})} = \frac{P_{l+1/2}^{\mu n}(\kappa, \mathbf{s})}{P_{l-1/2}^{\mu n}(\kappa, \mathbf{s})} = -\frac{l}{l+1} \quad (33)$$

Corresponding photoionization cross sections in the same approximation are proportional to the statistical weights:

$$\sigma_{nl,l+1/2}(\omega)/\sigma_{nl,l-1/2}(\omega) = (l+1)/l \quad (34)$$

Therefore if the fine-structure levels are populated statistically and the photoelectrons corresponding to different j are not separated, the spin polarization of photoelectrons in this approximation will be equal to zero as if the spin-orbit interaction were neglected from the very beginning:

$$\bar{P}^\lambda(\kappa, s) = \frac{\sigma_{nl,l-1/2}P^\lambda_{l-1/2}(\kappa, s) + \sigma_{nl,l+1/2}P^\lambda_{l+1/2}(\kappa, s)}{\sigma_{nl,l-1/2} + \sigma_{nl,l+1/2}} = 0 \qquad (35)$$

Actually the radial wave functions and the dipole matrix elements depend on j, the conditions Eqs. (32)–(34) are not fulfilled, and the degree of electron polarization even after summation over j will be different from zero, but is usually small, of the order $(\alpha Z)^2$.

In the case of a closed subshell with $l > 0$, photoelectrons will be also spin-polarized. All equations starting from Eq. (9) remain fully applicable for closed shells except that the total angular momentum $j = l \pm \frac{1}{2}$ refers now to the final state of the ion (Cherepkov, 1973). Photoelectrons corresponding to the different ionic fine-structure levels have different energies and must be separated for the electron spin polarization being observed.

C. Fano Effect

For subshells with $l = 0$ the fine-structure splitting is absent, and the photoelectron spin polarization can appear to be due to only the spin-orbit interaction in the continuous spectrum. For K-shells and high photon energies this question has been studied in detail by Nagel (1960) and by Nagel and Olsson (1960; see also Pratt *et al.*, 1973) using the Coulomb wave functions with relativistic corrections. They have shown that photoelectrons will be spin-polarized in the direction perpendicular to the scattering plane for absorption of unpolarized light and will have all three components of polarization for absorption of circularly polarized and linearly polarized light. For low electron velocities the polarization parameters are proportional to $(\alpha Z)^2$, that is, they are usually small.

This result is valid for any ns or ns^2 subshell irrespective of photon energy, but with the exception of the Cooper minimum region. Here, as was first shown by Fano (1969a), the degree of electron polarization under absorption of circularly polarized light is not small and can reach 100%. This is possible since in the Cooper minimum a dipole matrix element itself can be smaller than the difference between matrix elements corresponding to the $ns \to \epsilon p_{1/2}$ and $ns \to \epsilon p_{3/2}$ transitions, caused by the spin-orbit interaction in the continuous spectrum. Because the Cooper minimum does not appear in K-shells (Cooper, 1962) or in any other shell with the hydrogen-like wave functions (Bethe, 1933; Hall, 1936), more realistic approximations should be used for studying the Fano effect.

Let us consider the angular distribution of photoelectrons with defined spin polarization ejected from an ns subshell. The initial-state wave function in this case is given by Eq. (3) with $l = m = 0, j = \frac{1}{2}$ (j will be omitted below). In the final state wave function of Eq. (4) it is necessary to add the orbital angular momentum l_1 to the photoelectron spin into the total angular momentum j_1, for the spin–orbit interaction in the continuous spectrum to be taken into account:

$$\psi_{\mathbf{p}\mu_1}^-(\mathbf{r}) = \frac{2\pi}{\sqrt{p}} \sum_{l_1,m_1} \sum_{j_1,m_{j1}} (i)^{l_1} \exp\{-i\delta_{l_1}\} Y^*_{l_1 m_1}(\hat{\boldsymbol{\kappa}}) [j_1]^{1/2}$$

$$\times (-1)^{1/2-l_1-m_{j1}} \begin{pmatrix} l_1 & \frac{1}{2} & j_1 \\ m_1 & \mu_1 & -m_{j1} \end{pmatrix} \varphi_{\epsilon l_1 j_1 m_{j1}}(\mathbf{r}) \qquad (36)$$

Due to the spin–orbit interaction the radial part of a photoelectron wave function, $R_{\epsilon l_1}(r)$ and the phase shift δ_{l_1} in Eq. (36) depend on the total angular momentum j_1, which will be noted below by the corresponding superscript. Finally, we get from Eq. (36) that

$$\psi_{\mathbf{p}\mu_1}^-(\mathbf{r}) = \frac{2\pi}{\sqrt{p}} \sum (i)^{l_1} [j_1] \exp\{-i\delta_{l_1}^{j_1}\} Y^*_{l_1 m_1}(\hat{\boldsymbol{\kappa}}) R_{\epsilon l_1}^{j_1}(r)$$

$$\times Y_{l_1 m'}(\theta, \varphi) \chi_{\mu'} \begin{pmatrix} l_1 & \frac{1}{2} & j_1 \\ m_1 & \mu_1 & -m_{j1} \end{pmatrix} \begin{pmatrix} l_1 & \frac{1}{2} & j_1 \\ m' & \mu' & -m_{j1} \end{pmatrix} \qquad (37)$$

where the summation is over $l_1, m_1, j_1, m_{j1}, m', \mu'$. Calculating the matrix element of the dipole operator of Eq. (1) between the wave functions of Eqs. (3) and (37) and inserting it into Eq. (7), one arrives again at expressions (17)–(19) for the angular distribution of photoelectrons with defined spin polarization, where

$$\sigma_{ns}(\omega) = \frac{4\pi^2 \alpha \omega}{3} N_{ns} (\tfrac{1}{3}|d_1|^2 + \tfrac{2}{3}|d_3|^2) \qquad (38)$$

$$\beta = \frac{2|d_3|^2 + 4 \operatorname{Re}\{d_1 d_3^* \exp[-i(\delta_3 - \delta_1)]\}}{|d_1|^2 + 2|d_3|^2} \qquad (39)$$

$$A = \frac{5|d_3|^2 - |d_1|^2 - 4 \operatorname{Re}\{d_1 d_3^* \exp[-i(\delta_3 - \delta_1)]\}}{3(|d_1|^2 + 2|d_3|^2)} \qquad (40)$$

$$\gamma = \frac{2|d_3|^2 - 4|d_1|^2 + 2 \operatorname{Re}\{d_1 d_3^* \exp[-i(\delta_3 - \delta_1)]\}}{3(|d_1|^2 + 2|d_3|^2)} \qquad (41)$$

$$\eta = \frac{3 \, \mathrm{Im}\{d_1 d_3^* \exp[-i(\delta_3 - \delta_1)]\}}{|d_1|^2 + 2|d_3|^2} \tag{42}$$

$$d_{2j_1} = \langle \epsilon p j_1 \| d \| ns \rangle = \int_0^\infty R_{\epsilon p}^{j_1}(r) \, R_{ns}(r) \, r^3 \, dr \tag{43}$$

and $\delta_{2j_1} \equiv \delta_{l_1}^{j_1}$. If the spin–orbit interaction is absent, then $d_3 = d_1$, $\delta_3 = \delta_1$, $\beta = 2$, and $A = \gamma = \eta = 0$.

In his original paper Fano (1969a) considered only the polarization of the total electron flux ejected by circularly polarized light from alkali atoms, which is defined by the A parameter. The angular asymmetry parameter β near the Cooper minimum was first discussed by Heinzmann et al. (1970) and was calculated for the alkalis by Walker and Waber (1973) and by Marr (1974). The γ parameter was introduced by Brehm (1971), whereas the η parameter has not been treated to date.

The general properties of the angular distributions of Eqs. (17)–(19), up to Eq. (29), are valid for the Fano effect also. As before, photoelectrons ejected by circularly polarized light at the $\theta = 0$ and $\theta = \pi$ angles are completely polarized in the direction of the photon spin (Nagel and Olsson, 1960). However, the physical reason for this property is now quite different. Indeed, for $\theta = 0$ and $\theta = \pi$ the contribution is given by the partial waves with $m_1 = 0$ only, but in the initial state now $m = 0$ too. Therefore, from the conservation law for the angular momentum projection it follows that $\mu + \lambda = \mu_1$, $\mu = -\lambda/2 = -\mu_1$, and the whole contribution is given by the spin-flip process owing to the spin–orbit interaction in the continuous spectrum.

The greatest value of A ($A = 1$) is reached near the Cooper minimum at a point where $d_3 = -2d_1$ (supposing that $\delta_3 = \delta_1$) (Fano, 1969a). Smallness of order $(\alpha Z)^2$ appears here not in the polarization parameters but in the photoionization cross section, which in K, Rb, and Cs is of order 10^{-20} cm^2, or two orders of magnitude lower than in Li, where the Cooper minimum lies in the discrete energy region (Hudson and Kieffer, 1971).

Far from the Cooper minimum the polarization will be of order $(\alpha Z_{\mathrm{eff}})^2$, where Z_{eff} is the effective charge of the subshell under consideration. But even for the outermost electron in alkalis this effective charge is close to the charge of nucleus. For example, estimation of the phase shift difference $\delta_3 - \delta_1$ for the 6s electron of Cs from the difference of the corresponding quantum defects $\tau_3 - \tau_1$ gives (Sobel'man, 1972; Fano, 1969b)

$$\delta_3 - \delta_1 \cong \pi(\tau_3 - \tau_1) \cong -0.1 \tag{44}$$

This value is proportional to $(\alpha Z_{\mathrm{eff}})^2$ with $Z_{\mathrm{eff}} \sim 44$, which is close to the charge of the nucleus rather than to unity.

In the photoionization cross section of the outer ns^2 subshells of alkali earth and rare gas atoms, and probably of a majority of other atoms, there is also the Cooper minimum above the ionization threshold (Amusia and Cherepkov, 1975). The Fano effect appears there also and is described by the same Eqs. (39)–(43) (Cherepkov, 1973).

D. Simultaneous Inclusion of the Spin–Orbit Interaction in Both Discrete and Continuous Spectra

In the previous sections the spin–orbit splitting of atomic levels and the spin–orbit interaction in the continuous spectrum were treated separately. As a matter of fact, they both are present simultaneously except for subshells with $l = 0$. Therefore, it is the purpose of this section to take them both into account for an arbitrary open subshell $(nl)^N$. Supposing that LS-coupling is applicable, the initial-state wave function can be presented as

$$\Psi^N_{LSJM} = \sum_{M_L,M_S} [J]^{1/2}(-1)^{S-L-M} \begin{pmatrix} L & S & J \\ M_L & M_S & -M \end{pmatrix} \Psi^N_{LM_LSM_S} \quad (45)$$

where J is the total angular momentum of an atom and M is its projection (closed subshells do not take part in the process and are not considered here).

The final state of the ion is characterized by the quantum numbers $L'S'J'M'$, whereas the ejected electron wave function is given by Eq. (37). Therefore, the final state of the total system is described by the following wave function:

$$\Psi_f = \frac{1}{\sqrt{N}} \sum_{i=1}^N (-1)^{N-i} \frac{2\pi}{\sqrt{p}} \sum_{l_1,m_1,j_1,m_{j1},M'_L \atop m',\mu',M'_S} (i)^{l_1} \exp\{-i\delta^{j_1}_{l_1}\}(-1)^{S'-L'-M'}$$

$$\times [j_1][J']^{1/2} \begin{pmatrix} L' & S' & J' \\ M'_L & M'_S & -M'_J \end{pmatrix} \begin{pmatrix} l_1 & \tfrac{1}{2} & j_1 \\ m_1 & \mu_1 & -m_{j1} \end{pmatrix} \begin{pmatrix} l_1 & \tfrac{1}{2} & j_1 \\ m' & \mu' & -m_{j1} \end{pmatrix}$$

$$\times \varphi^{j_1}_{\epsilon l_1,m'\mu'}(\mathbf{r}_i)\Psi^{N-1}_{L'M'_LS'M'_S}(\mathbf{r}_1,\ldots,\mathbf{r}_{i-1},\mathbf{r}_{i+1},\ldots,\mathbf{r}_N) \quad (46)$$

The matrix element of the dipole operator of Eq. (1) between these wave functions is calculated using the standard technique (Sobel'man, 1972):

$$\langle \Psi_f | \hat{d}_\lambda | \Psi^N_{LSJM} \rangle$$

$$= \frac{2\pi}{\sqrt{p}} \sum_{l_1,m_1} \sum_{j,m_j} \sum_{j_1,m_{j1}} [j,j_1]^{1/2}(i)^{-l_1}(-1)^{j-j_1-J-1+M'+m_j}$$

$$\times Y_{l_1 m_1}(\hat{\kappa}) \begin{pmatrix} l_1 & \tfrac{1}{2} & j_1 \\ m_1 & \mu_1 & -m_{j1} \end{pmatrix} \begin{pmatrix} j_1 & j & 1 \\ m_{j1} & -m_j & -\lambda \end{pmatrix} \begin{pmatrix} J & j & J' \\ -M & m_j & M' \end{pmatrix}$$

$$\times \langle J'l_1 j_1 \| d \| Jlj \rangle \quad (47)$$

where

$$\langle J'l_1 j_1 \| d \| Jlj \rangle = \sqrt{N} \, [L, SJ, J', j, j_1]^{1/2} \exp\{i\delta_{l_1}^{j_1}\}$$

$$\times (l^{N-1}L'S', l|\}l^N LS) \begin{Bmatrix} j_1 & j & 1 \\ l & l_1 & \frac{1}{2} \end{Bmatrix} \begin{Bmatrix} J & J' & j \\ S & S' & \frac{1}{2} \\ L & L' & l \end{Bmatrix}$$

$$\times \langle \epsilon l_1 j_1 \| d \| nlj \rangle \tag{48}$$

and the reduced dipole matrix element $\langle \epsilon l_1 j_1 \| d \| nlj \rangle$ differs from that of Eq. (6) only by a dependence of the radial wave function on j_1. Inserting Eq. (47) into Eq. (7) and performing summations over projections, one obtains instead of Eq. (17) (Cherepkov, 1980b; see also Lee, 1974; Huang, 1980; Klar, 1980)

$$I_j^{\pm 1}(\kappa, s) = \frac{\sigma_{nl}^J(\omega)}{8\pi} \left\{ 1 - \frac{\beta^J}{2} \left[\tfrac{3}{2}(\kappa s_\gamma)^2 - \tfrac{1}{2}\right] + A^J(ss_\gamma) \right.$$

$$\left. - \gamma^J[\tfrac{3}{2}(\kappa s)(\kappa s_\gamma) - \tfrac{1}{2}(ss_\gamma)] - \eta^J(s[\kappa s_\gamma])(\kappa s_\gamma) \right\} \tag{49}$$

and analogously for linearly polarized and unpolarized light, where

$$\sigma_{nl}^J(\omega) = \sum_j \sigma_{nlj}^J(\omega) \tag{50}$$

$$\beta^J = (\sigma_{nl}^J)^{-1} \sum_j \beta^j \sigma_{nlj}^J(\omega), \qquad A^J = (\sigma_{nl}^J)^{-1} \sum_j A^j \sigma_{nlj}^J(\omega) \tag{51}$$

and the same for γ^J and η^J. Here

$$\sigma_{nlj}^J(\omega) = \frac{4\pi^2 \alpha \omega}{3} \sum_{l_1, j_1} |\langle J'l_1 j_1 \| d \| Jlj \rangle|^2 \tag{52}$$

$$\beta^j = \frac{\sqrt{30}}{B} \sum_{l_1, l_2} \sum_{j_1, j_2} [l_1, l_2, j_1, j_2]^{1/2} (-1)^{j+j_1+j_2+(1/2)(l_1-l_2+1)}$$

$$\times \begin{pmatrix} l_1 & l_2 & 2 \\ 0 & 0 & 0 \end{pmatrix} \begin{Bmatrix} 2 & 1 & 1 \\ j & j_1 & j_2 \end{Bmatrix} \begin{Bmatrix} j_1 & j_2 & 2 \\ l_2 & l_1 & \frac{1}{2} \end{Bmatrix} \langle Jlj \| d^* \| J'l_1 j_1 \rangle$$

$$\times \langle J'l_2 j_2 \| d \| Jlj \rangle \tag{53}$$

$$A^j = \frac{3}{B} (-1)^{j-l-1/2} \sum_{l_1, j_1, j_2} [j_1, j_2]^{1/2} \begin{Bmatrix} 1 & 1 & 1 \\ j & j_1 & j_2 \end{Bmatrix}$$

$$\times \begin{Bmatrix} \frac{1}{2} & 1 & \frac{1}{2} \\ j_1 & l_1 & j_2 \end{Bmatrix} \langle Jlj \| d^* \| J'l_1 j_1 \rangle \langle J'l_1 j_2 \| d \| Jlj \rangle \tag{54}$$

$$\gamma^j = \frac{\sqrt{5}}{8B} (-1)^{j-1/2} \sum_{l_1,l_2} \sum_{j_1,j_2} [l_1, l_2, j_1, j_2]^{1/2} \begin{pmatrix} l_1 & l_2 & 2 \\ 0 & 0 & 0 \end{pmatrix}$$

$$\times (-1)^{j_1+j_2+1+(1/2)(l_1-l_2)} \langle Jlj \| d^* \| J'l_1 j_1 \rangle \langle J'l_2 j_2 \| d \| Jlj \rangle$$

$$\times \sum_{x,y} [x,y]^{3/2} \begin{Bmatrix} \tfrac{1}{2} & 1 & x \\ j & l_1 & j_1 \end{Bmatrix} \begin{Bmatrix} x & 2 & y \\ l_2 & j & l_1 \end{Bmatrix} \begin{Bmatrix} y & 1 & \tfrac{1}{2} \\ j_2 & l_2 & j \end{Bmatrix} \quad (55)$$

$$\eta^j = \frac{3\sqrt{10}}{8B} (-1)^{j-l-1/2} \sum_{l_1,l_2} \sum_{j_1,j_2} [l_1, l_2, j_1, j_2]^{1/2}$$

$$\times (i)^{l_1+l_2+1}(-1)^{j_1+j_2} \langle Jlj \| d^* \| J'l_1 j_1 \rangle \langle J'l_2 j_2 \| d \| Jlj \rangle \quad (56)$$

$$\times \begin{pmatrix} l_1 & l_2 & 2 \\ 0 & 0 & 0 \end{pmatrix} \sum_{x,y} (-1)^{x-1/2} [x,y](1-\delta_{xy}) \begin{Bmatrix} \tfrac{1}{2} & 1 & x \\ j & l_1 & j_1 \end{Bmatrix}$$

$$\times \begin{Bmatrix} x & 2 & y \\ l_2 & j & l_1 \end{Bmatrix} \begin{Bmatrix} y & 1 & \tfrac{1}{2} \\ j_2 & l_2 & j \end{Bmatrix} \quad (56)$$

$$B = \sum_{l_1,j_1} |\langle J'l_1 j_1 \| d \| Jlj \rangle|^2 \quad (57)$$

In the particular case of $l = 0$ these equations coincide with Eqs. (38)–(42), whereas by neglecting the spin–orbit interaction in the continuous spectrum one obtains Eqs. (10)–(15).

Equations (30) and (31) are not fulfilled now, and it is possible only to say that $|A^j| \leq 1$, $|\eta^j| \leq \tfrac{3}{2}\sqrt{2}$.

TABLE I

Terms of the Final Ion States of Atoms with $(np)^q$ Outer Subshells, for Which the Degree of Polarization Can Be High[a] and Cannot Be High[b]

Configuration	Term of initial state	Terms of final ion states	
		With no sum over j	With a sum over j
np^1	$^2P_{1/2}$	$^1S_0$	—
np^2	$^3P_0$	$^2P_{1/2}, {}^2P_{3/2}$	—
np^3	$^4S_{3/2}$	$^3P_0$	$^3P_1, {}^3P_2$
np^4	$^3P_2$	—	$^4S_{3/2}, {}^2P_{1/2}, {}^2P_{3/2}, {}^2D_{3/2}$
np^5	$^2P_{3/2}$	$^1S_0, {}^3P_0$	$^3P_1, {}^3P_2, {}^1D_2$
np^6	$^1S_0$	$^2P_{1/2}, {}^2P_{3/2}$	—

[a] There is no sum over j.
[b] There is a sum over j.

If J or J' is equal to zero, then in Eqs. (50) and (51) only one term remains in the sum over j, otherwise the contribution is given by both terms with $j = l \pm \frac{1}{2}$. Since the conditions (32) and (34) are approximately fulfilled in all atoms (Cherepkov, 1980a; Johnson and Cheng, 1979), the A^J, γ^J, and η^J parameters are expected to be small in all cases when they are defined by the sum of two terms with $j = l \pm \frac{1}{2}$, except probably for the region of autoionization resonances and the Cooper minima. Therefore, the most interesting are transitions between the states where J or J' is equal to zero. All these transitions for the $(np)^N$ subshells are listed in Table I. In all subshells except for the $(np)^4$ one, there is at least one transition of this type, and it is necessary to analyze the photoelectron energies in order to separate this transition from others. Obviously, the same consideration is valid for negative and positive ions (Rau and Fano, 1971).

E. On the Complete Quantum-Mechanical Experiment

In the dipole approximation there are at most five independent values which completely define the atomic photoionization process, namely: (*i*) three dipole matrix elements, corresponding to the transitions

$$
\begin{array}{l}
 \nearrow \epsilon,\, l-1,\, j-1 \\
nlj \to \epsilon,\, l-1,\, j \\
 \searrow \epsilon,\, l+1,\, j+1
\end{array}
$$

when $j = l - \frac{1}{2}$, or

$$
\begin{array}{l}
 \nearrow \epsilon,\, l-1,\, j-1 \\
nlj \to \epsilon,\, l+1,\, j \\
 \searrow \epsilon,\, l+1,\, j+1
\end{array}
$$

when $j = l + \frac{1}{2}$, and (*ii*) two differences of the phase shifts corresponding to three outgoing partial waves (the absolute values of phase shifts have no physical meaning and could not influence the process). For the initial states with $j = \frac{1}{2}$ the number of allowed transitions is reduced to two, whereas the number of independent theoretical values is reduced to three.

The expressions (17) and (49) for the angular distribution of photoelectrons with defined spin polarization ejected by circularly polarized light contain just five independent vector combinations: a constant, $(\mathbf{s}\mathbf{s}_y)$, $(\kappa \mathbf{s}_y)^2$, $(\kappa \mathbf{s})(\kappa \mathbf{s}_y)$, and $(\mathbf{s}[\kappa \mathbf{s}_y])(\kappa \mathbf{s}_y)$. Therefore, five measurements enable, in principle, determination of all theoretical values and thus constitute the complete quantum-mechanical experiment in the dipole approximation (Lee, 1974; Cherepkov, 1979; Heinzmann, 1980b; Kessler, 1981). Strictly speaking, since the phase shift difference enters only the argument of the sine or cosine [see Eqs. (11), (14), and (15)], one measurement does not suffice for its unambiguous determination and some additional information is needed. This information can be gained from experiments with polarized or aligned atoms (Kollath, 1980; Klar and Kleinpoppen, 1982), or from the condition for the phase shift difference to be a smooth function of energy and to coincide with the difference of the corresponding quantum defects, multiplied by π, at the threshold (Heinzmann, 1980b).

To perform the complete quantum-mechanical experiment, it is necessary to measure, for example, the following:

(1) the partial photoionization cross section of a subshell under consideration, σ_{nlj};
(2) the angular distribution of photoelectrons, which gives β^j;
(3) the degree of polarization of the total electron flux ejected by circularly polarized light, which gives A^j;
(4) the degree of transverse polarization of photoelectrons ejected, for instance, by unpolarized light under some angle $\theta \neq 0$, $\pi/2$, or π, which gives η^j;
(5) the degree of polarization in the direction of the X axis (see Fig. 1) of photoelectrons ejected by circularly polarized light under some angle $\theta \neq 0$, $\pi/2$, or π, which gives γ^j.

As follows from Eq. (27), the parameter A^j can be found also as the degree of polarization of photoelectrons ejected by circularly polarized light under the magic angle $\theta = 54.7°$ [for which $P_2(\cos \theta) = 0$].

Surely, it is possible to use another set of five independent parameters. For example, Lee (1974) have chosen the three polarization parameters to be proportional to the degree of spin polarization of photoelectrons along the axes of the rotated coordinate system $\xi\eta\zeta$ of Fig. 1. The relations between polarization parameters introduced by different authors are presented in Table II. Then, in the complete experiment it is possible, instead of (3) and (5), to measure the degree of polarization along the ξ and ζ axes of the rotated system. In any case, for the complete experiment to be achieved a highly sophisticated measurement should be performed: One must observe the spin polarization of photoelectrons with given energy ejected under a definite angle by circularly polarized light (angle- and spin-resolved photoelectron spectroscopy). Up to now this kind of experiment has been achieved with linearly polarized and unpolarized light only (Heinzmann, 1980c; Schönhense, 1980).

TABLE II

Relations between Polarization Parameters Introduced by Several Authors

This work	Lee (1974)	Huang (1980)	This work	Lee (1974)	Huang (1980)
A^j	$\frac{1}{3}(\gamma_k + 2\delta_k)$	$\delta = \frac{1}{3}(\zeta - 2\xi)$	$-(A^j + \frac{1}{2}\gamma^j)$	$-\delta_k$	ξ
γ^j	$-\frac{2}{3}(\gamma_k - \delta_k)$	$-\frac{2}{3}(\zeta + \xi)$	η^j	$2\xi_k$	η
η^j	$2\xi_k$	η	$A^j - \gamma^j$	γ_k	ζ

For subshells with $j = \frac{1}{2}$ only three of five parameters $\sigma_{nl1/2}$, $\beta^{1/2}$, $A^{1/2}$, $\gamma^{1/2}$, $\eta^{1/2}$ are independent. Therefore, there are two relations between them, the first of which immediately follows from Eqs. (28) and (29):

$$\gamma^{1/2} = A^{1/2} + \tfrac{1}{2}\beta^{1/2} - 1 \tag{58}$$

and the second can be proved by substitution (Huang, 1980):

$$(A^{1/2} + \tfrac{1}{2}\gamma^{1/2})^2 = (1 - \tfrac{1}{2}\beta^{1/2})(1 + \beta^{1/2}) - (\eta^{1/2})^2 \tag{59}$$

If the spin–orbit interaction in the continuous spectrum is neglected, as was done in Sections II,A and B, the number of independent parameters is equal to three for all values of j. Two relations between the parameters in this case are a direct generalization of Eqs. (58) and (59):

$$\gamma^j = A^j + (-1)^{j-l-1/2}(1 - \tfrac{1}{2}\beta^j)\frac{2}{[j]} \quad (l \neq 0) \tag{60}$$

$$(A^j + \tfrac{1}{2}\gamma^j)^2 + (\eta^j)^2 = \frac{2l(l+1)}{[j]^2}(1 + \beta^j)(1 - \tfrac{1}{2}\beta^j) \tag{61}$$

In the exact treatment of subshells with $l > 0$ all five aforementioned parameters are independent, but they can vary in the limits defined by a value of β^j (Huang, 1980). To see it, let us consider the transverse polarization of photoelectrons ejected by linearly polarized light [Eq. (24)]:

$$P_j^0(\theta) = -\frac{2\eta^j \sin\theta \cos\theta}{1 + \beta^j P_2(\cos\theta)} \tag{62}$$

The maximal value of $P_j^0(\theta)$ is reached at angles

$$\theta = \arccos\{\pm[(2 - \beta^j)/(4 + \beta^j)]^{1/2}\} \tag{63}$$

and should be less than or equal to 1, from which it follows that (Huang, 1980)

$$|\eta^j| \leq [\tfrac{1}{2}(1 + \beta^j)(2 - \beta^j)]^{1/2} \tag{64}$$

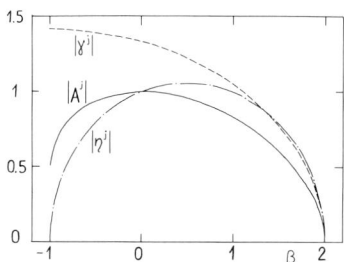

Fig. 2. Maximum ranges of spin polarization parameters plotted against the angular asymmetry parameter β.

For absorption of circularly polarized light the degree of polarization along the X axis [Eq. (26)] reaches the maximal value at angles

$$\theta = \arccos\{\pm[(4 + \beta^j)/(8 - \beta^j)]^{1/2}\} \tag{65}$$

and also cannot be greater than 1, which gives

$$|\gamma^j| \leq \tfrac{1}{3}[(4 + \beta^j)(4 - 2\beta^j)]^{1/2} \tag{66}$$

Finally by using Eqs. (58) and (59) it is possible to show that

$$|A^j| \leq |A^{1/2}| \leq \frac{2 - \beta^j}{6} + \tfrac{2}{3}[\tfrac{1}{2}(2 - \beta^j)(1 + \beta^j)]^{1/2} \tag{67}$$

These relations are demonstrated in Fig. 2. They can serve as checks on the self-consistency in measurements of the angular asymmetry and spin polarization parameters. As is seen from Fig. 2, when β reaches its maximum value 2 the spin polarization of photoelectrons disappears, and when β attains its minimum value -1 the transverse spin polarization perpendicular to the reaction plane disappears.

F. Some Generalizations

Up to now we have considered the ejection of photoelectrons from unpolarized atoms by circularly or linearly polarized light. In a real experiment the light can be polarized elliptically or only partially, and atoms also can be (partially) polarized or aligned. For the most general treatment of the problem it is necessary to use density matrices rather than wave functions for the initial states of an atom and a photon beam. Using the Stokes parameters ξ_1, ξ_2, ξ_3, the density matrix of a photon beam in the helicity representation can be written as (Blum, 1981)

$$\rho_{\lambda\lambda'} = \tfrac{1}{2} \begin{pmatrix} 1 + \xi_2 & -\xi_3 + i\xi_1 \\ -\xi_3 - i\xi_1 & 1 - \xi_2 \end{pmatrix} \tag{68}$$

Here $\tfrac{1}{2}(1 + \xi_3)$ is the probability of linear polarization along the X axis, $\tfrac{1}{2}(1 + \xi_1)$ is the probability of linear polarization along an X' axis making an angle of 45° with the X axis, and $\tfrac{1}{2}(1 + \xi_2)$ is the probability of right circular polarization. Making the substitution $d_\lambda^* d_\lambda \to \Sigma_{\lambda\lambda'} d_{\lambda'}^* d_\lambda \rho_{\lambda\lambda'}$ in Eq. (7), one obtains, instead of Eqs. (17)–(19) (Cherepkov, 1979),

$$\begin{aligned} I_j(\mathbf{\kappa}, \mathbf{s}) = \frac{\sigma_{nlj}(\omega)}{8\pi} \Bigg\{ & 1 - \frac{\beta^j}{2} P_2(\cos\theta) + \tfrac{3}{4}\beta^j \sin^2\theta \\ & \times (\xi_1 \sin 2\varphi + \xi_3 \cos 2\varphi) + \xi_2 A^j(\mathbf{s}\mathbf{e}_z) \\ & - \gamma^j \xi_2 [\tfrac{3}{2}(\mathbf{\kappa}\mathbf{e}_z)(\mathbf{\kappa}\mathbf{s}) - \tfrac{1}{2}(\mathbf{s}\mathbf{e}_z)] \\ & - \xi_2 \eta^j (\mathbf{e}_z[\mathbf{s}\mathbf{\kappa}])(\mathbf{\kappa}\mathbf{e}_z) + 2\eta^j [\xi_1 (\mathbf{e}_x'[\mathbf{s}\mathbf{\kappa}])(\mathbf{\kappa}\mathbf{e}_x') \\ & + \tfrac{1}{2}(1 - \xi_1 - \xi_2 + \xi_3) \times (\mathbf{e}_x[\mathbf{s}\mathbf{\kappa}])(\mathbf{\kappa}\mathbf{e}_x) \\ & + \tfrac{1}{2}(1 - \xi_1 - \xi_2 - \xi_3) \times (\mathbf{e}_y[\mathbf{s}\mathbf{\kappa}])(\mathbf{\kappa}\mathbf{e}_y)] \Bigg\} \end{aligned} \tag{69}$$

where $\mathbf{e}_x, \mathbf{e}_y, \mathbf{e}_z$, and $\mathbf{e}_x'$ are unit vectors in the directions of the X, Y, Z, and X' axes, respectively, and the angles θ and φ are specified in Fig. 3. The degree of electron polarization in the directions of the X, Y, and Z axes can be easily found from Eq. (69):

$$P_{jx}(\theta, \varphi) = -\frac{\sin\theta \cos\theta}{F(\theta, \varphi)} [(\tfrac{3}{2}\gamma^j\xi_2 + \eta^j\xi_1) \cos\varphi \\ + (1 - \xi_3)\eta^j \sin\varphi] \tag{70}$$

$$P_{jyY}(\theta, \varphi) = \frac{\sin\theta \cos\theta}{F(\theta, \varphi)} [(\eta^j\xi_1 - \tfrac{3}{2}\gamma^j\xi_2) \sin\varphi \\ + (1 + \xi_3)\eta^j \cos\varphi] \tag{71}$$

$$P_{jz}(\theta, \varphi) = \frac{1}{F(\theta, \varphi)} \{\xi_2[A^j - \gamma^j P_2(\cos\theta)] \\ + \eta^j \sin^2\theta(\xi_1 \cos 2\varphi - \xi_3 \sin 2\varphi)\} \tag{72}$$

where

$$F(\theta, \varphi) = 1 - \frac{\beta^j}{2} P_2(\cos\theta) \\ + \tfrac{3}{4}\beta^j \sin^2\theta(\xi_1 \sin 2\varphi + \xi_3 \cos 2\varphi) \tag{73}$$

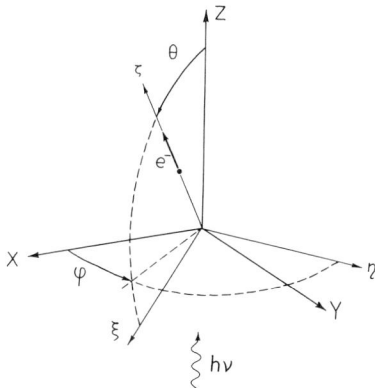

FIG. 3. Coordinate system in the case of partially polarized light.

Expressions for the degree of polarization along the ξ, η, ζ axes of the rotated coordinate system (see Fig. 3), which can be also obtained from Eq. (69), were derived by Huang (1980) using another parametrization of the photon density matrix.

Since the electrons ejected by linearly polarized light can be only transversely polarized, it follows immediately from Eqs. (28) and (72) that for partially polarized light the degree of longitudinal polarization of the electrons ejected along the Z axis is proportional to the ξ_2 parameter, giving the degree of circular polarization of the light (Cherepkov, 1979):

$$P_{jz}(0) = P_{jz}(\pi) = \xi_2(-1)^{j-l+1/2}2/[j] \qquad (74)$$

This fact can be used to measure the degree of circular polarization of the light in a broad energy region from $\sim$ 10 eV up to several kiloelectron-volts where the dipole approximation is applicable.

If atoms in the initial state are polarized or aligned, one must introduce in Eq. (7) the corresponding atomic density matrix as was done above with the photon density matrix (Jacobs, 1972; Kabachnik and Sazhina, 1976; Parzyński, 1980; Klar and Kleinpoppen, 1982; Huang, 1982). For the initially polarized atoms the complete experiment can be performed by measuring the angular distribution of photoelectrons only (Klar and Kleinpoppen, 1982). It is possible also to combine different kinds of experiments. For example, Kollath (1980) and Kaminski et al. (1980), in their complete experiment for the excited 7 $^2P_{1/2}$ state of Cs atoms, have used the measurement of the total photoionization cross section of the aligned atoms instead of one spin polarization measurement.

III. Comparison with Experiment and Applications

All fundamental polarization phenomena in atoms were first predicted theoretically and then verified experimentally. Up to now the spin polarization measurements have been performed for alkali atoms (Heinzmann *et al.*, 1970; Baum *et al.*, 1970; Drachenfels *et al.*, 1977), rare gases (Heinzmann, 1980a,b; Heinzmann *et al.*, 1979a,b, 1980b; Heinzmann and Schäfers, 1980; Schönhense, 1980), Tl (Heinzmann *et al.*, 1975, 1976), Pb (Heinzmann, 1978), Ag (Heinzmann *et al.*, 1980c), Hg (Schönhense *et al.*, 1982a,b; Schäfers *et al.*, 1982), Cd (Schäfers *et al.*, 1982), and for the 6 $^2D_{3/2,5/2}$ (Kaminski *et al.*, 1979) and 7 $^2P_{1/2,3/2}$ (Kollath, 1980; Kaminski *et al.*, 1980) excited states of Cs atoms.

A. Rare Gas Atoms

Rare gas atoms are especially suitable for theoretical investigations since the random phase approximation with exchange (RPAE) — a very efficient method of calculation of atomic collision processes taking into account many-electron correlations — works well in these atoms (Amusia *et al.*, 1971; Amusia and Cherepkov, 1975). The first systematical calculation of polarization parameters for subshells with $l \neq 0$ were performed in RPAE for Ar and Xe (Cherepkov, 1979). As an example, Fig. 4 shows the parameters β, $A^{1/2}$, $\eta^{1/2}$, $\gamma^{1/2}$ for the $5p^6$ subshell of Xe calculated in RPAE. Experimental points for β and $\eta^{1/2}$ are also shown. There is good agreement

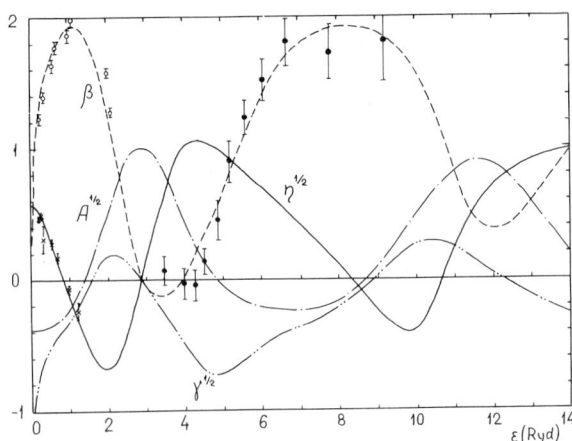

FIG. 4. The parameters β, $A^{1/2}$, $\gamma^{1/2}$, $\eta^{1/2}$ versus the photoelectron energy ϵ for the $5p^6(^2P_{1/2})$ subshell of Xe, calculated in the RPAE using the matrix elements from Amusia and Ivanov (1977). Experimental results for β: (○) Dehmer *et al.* (1975); (●) Torop *et al.* (1976); for $\eta^{1/2}$: (×) Heinzmann (1980a).

between theory and experiment. Results of recent calculations of the polarization parameters for Ar, Kr, and Xe (Huang et al., 1981) in the relativistic random phase approximation (RRPA), which differs from the RPAE in its use of Dirac-Fock wave functions instead of Hartree-Fock ones (Johnson and Lin, 1979), are quite close to the RPAE results and will be discussed later.

From Eq. (15) it immediately follows that η^j turns out to be equal to zero at each point where either d_{l+1} or d_{l-1} is equal to zero, and also where the phase shift difference $(\delta_{l+1} - \delta_{l-1})$ is divisible by π. In Fig. 4 the first zero of $\eta^{1/2}$ at $\epsilon = 0.84$ Ryd is due to the change of sign of $\sin(\delta_{l+1} - \delta_{l-1})$, whereas the second zero at $\epsilon = 2.88$ Ryd corresponds to the Cooper minimum of the cross section, where d_{l+1} changes sign. The sign changes of $\eta^{1/2}$ at $\epsilon = 8.3$ and 10.7 Ryd are caused by changes of sign of d_{l+1} which appear to be due to the strong influence of the $4d^{10}$ subshell near its threshold on the $5p^6$ one. In the Hartree-Fock approximation the $\eta^{1/2}$ parameter in this energy region is everywhere positive. Therefore, the experimental measurement of $\eta^{1/2}$ for energies $\epsilon > 8$ Ryd can give direct information on the change of sign of the dipole matrix element and can be used to prove the existence of the strong intershell interaction in Xe.

Since η^j in Eq. (15) is proportional to the sine of the phase shift difference whereas β^j in Eq. (11) contains the same term but with the cosine, there is a correlation in their behavior: Near each zero of η^j the β^j parameter has an extremum, and consecutive zeroes of η^j correspond in turn to a maximum and a minimum of β^j. This correlation is clearly demonstrated by Fig. 4. It is seen also from this figure that the polarization parameters $A^{1/2}$, $\gamma^{1/2}$, $\eta^{1/2}$ tend to zero at energies where β approaches 2, in accord with the relations presented in Fig. 2.

The measured values of $\sigma_{1/2}$ (West and Morton, 1978), $\beta^{1/2}$ (Dehmer et al., 1975; Torop et al., 1976) and $\eta^{1/2}$ (Heinzmann et al., 1979b; Heinzmann, 1980a) for the $^2P_{1/2}$ ionic state of Xe constitute the complete quantum-mechanical experiment and therefore enable the determination of the matrix elements and phase shift difference directly from the experiment. For unambiguous determination of the phase shift difference its threshold value was set approximately equal to the difference of the corresponding quantum defects $(\tau_2 - \tau_0)$ calculated by several authors (Dill, 1973; Lee, 1974; Geiger, 1977), multiplied by π. Figure 5 compares the dipole matrix elements and the quantum defect difference, determined in this way by Heinzmann (1980b), with theoretical values calculated in the RPAE (Amusia and Ivanov, 1977). Since in the treatment of Heinzmann it was supposed that the dipole matrix elements are real, the complex matrix elements $D_l(\omega)$ in the RPAE were presented in the form

$$D_l(\omega) = |D_l(\omega)|e^{i\Delta_l} \qquad (75)$$

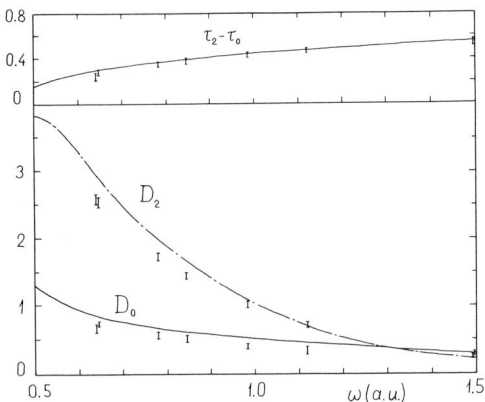

FIG. 5. Dipole matrix elements and quantum defect difference ($\tau_2 - \tau_0$) for the $5p^6(^2P_{1/2})$ subshell of Xe deduced from experimental results by Heinzmann (1980b) (error bars) compared with the RPAE results of Amusia and Ivanov (1977) (curves).

and an additional phase shift Δ_l was included in δ_l. Figure 5 demonstrates the effectiveness of the complete experiment in determining the theoretical values from the experiment. Some discrepancy in magnitude between the RPAE matrix elements and those deduced from the experiment is connected with deviation of the RPAE cross section for the $5p^6$ subshell of Xe from the experimental one near the threshold (Amusia and Cherepkov, 1975).

In the case of $j = \frac{1}{2}$ and in all cases when the spin-orbit interaction in the continuous spectrum is neglected, three parameters—σ, β^j, and η^j—exhaust the complete set of independent parameters; therefore, A^j could not be independent. The corresponding relation between A^j, β^j, and η^j can be derived from Eqs. (60) and (61):

$$(\eta^j)^2 = \frac{l(l+1)}{[j]^2}(2-\beta^j)(1+\beta^j)$$
$$-\left[\tfrac{3}{2}A^j + \frac{(-1)^{j-l-1/2}}{[j]}(1-\tfrac{1}{2}\beta^j)\right]^2 \qquad (76)$$

The parameters presented in Fig. 4 fulfill this relation.

The $A^{3/2}$ and $\eta^{3/2}$ parameters for np subshells in the RPAE are directly connected with $A^{1/2}$ and $\eta^{1/2}$ by Eq. (32), whereas both in the RRPA and in the experiment Eq. (32) is not fulfilled. However, the deviation from Eq. (32) is expected to be small, of order $(\alpha Z)^2$. Figure 6 shows the parameter η^j for the $3p^6$ subshell of Ar calculated both in the RPAE and in the RRPA together with experimental points. Results for the $\eta^{3/2}$ parameter are multiplied by -2 to visualize the deviation from Eq. (32). It is seen that the

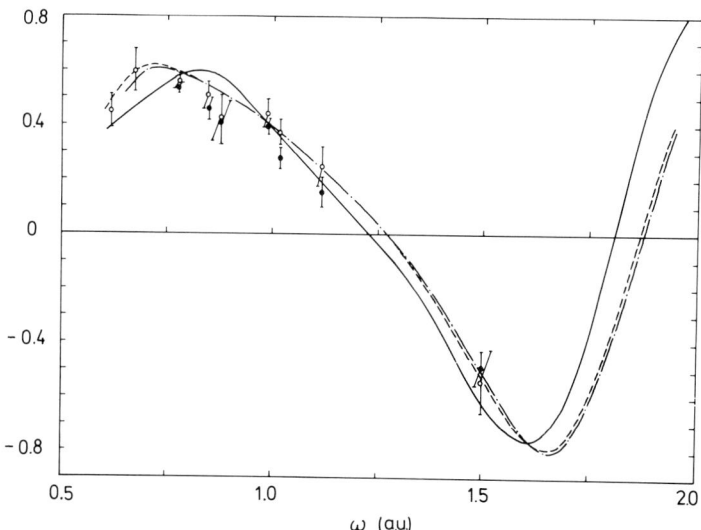

FIG. 6. Spin polarization parameter η^j for the $3p^6$ subshell of Ar: the RPAE result of Cherepkov (1979) (———); the RRPA results of Huang et al. (1981) for $-2\eta^{3/2}$ (----) and $\eta^{1/2}$ (— · — ·); the experimental results of Heinzmann et al. (1980b) for $-2\eta^{3/2}$ (○) and $\eta^{1/2}$ (●).

difference between $\eta^{1/2}$ and $-2\eta^{3/2}$ in the RRPA, which appears to be due to only the spin–orbit interaction in the continuous spectrum, is quite small. Some discrepancy between the RRPA and RPAE curves is connected apparently with other relativistic effects. In Xe the difference between $\eta^{1/2}$ and $-2\eta^{3/2}$ in the RRPA is more pronounced (Huang et al., 1981; Cherepkov, 1980a), but still it is smaller than the experimental error bars (Heinzmann et al., 1979b). Corresponding results for the $A^{1/2}$ and $-2A^{3/2}$ parameters in the $5p^6$ subshell of Xe in the RRPA and RPAE are presented in Fig. 7. Figures 6 and 7 demonstrate the applicability of the nonrelativistic RPAE theory for both prediction and quantitative description of polarization phenomena, the very existence of which is caused by the spin–orbit interaction neglected in the RPAE.

Figure 8 shows the angular dependence of the degree of polarization of photoelectrons ejected from the $3p^6$ subshell of Ar by the light of different polarization with wavelength $\lambda = 58.43$ nm. The experimental points for linearly polarized light (Schönhense, 1980) are in good agreement with the RPAE result.

There are polarization phenomena in subshells with $l > 0$ which could not be described by nonrelativistic theories. Unfortunately, among them are all phenomena investigated experimentally with circularly polarized light in rare gases: polarization of photoelectrons at photon energies between two thresholds, $^2P_{3/2}$ and $^2P_{1/2}$, and polarization of the total electron flux from

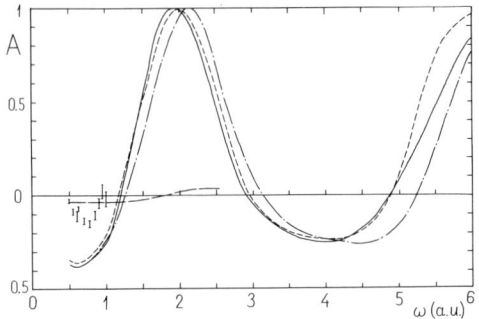

FIG. 7. Spin polarization parameter A^j for the 5p^6 subshell of Xe: the RPAE result of Cherepkov (1979) (———); the RRPA results of Huang et al. (1981) for $-2A^{3/2}$ (----) and $A^{1/2}$ (— · — ·). Experimental points: the $\bar{A}$ parameter measured by Heinzmann (1980a). The long-dashed curve is the RRPA result for $\bar{A}$ calculated using Eq. (77) with the branching ratio from Johnson and Cheng (1979).

both $^2P_{1/2}$ and $^2P_{3/2}$ levels without energy separation. Calculations of polarization parameters in the region of autoionization resonances between two thresholds were performed by Lee (1974) for Ar and Xe using the multichannel quantum defect method, and by Johnson et al. (1980) for Ar, Kr, and Xe by the same method but using the RRPA matrix elements and phase shifts. Results of these calculations, and especially of the last one, are in good agreement with experimental points for $A^{3/2}$ in Kr (Heinzmann and Schäfers, 1980) and Xe (Heinzmann et al, 1979a), whereas in Ar the experimental curve (Heinzmann and Schäfers, 1980) is too smooth as compared with theoretical predictions.

The degree of polarization of the total electron flux from both fine-struc-

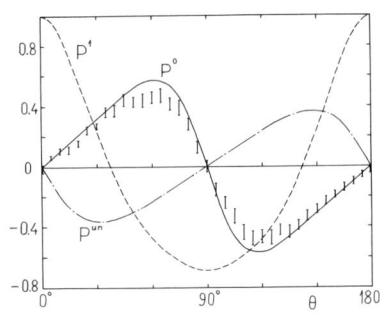

FIG. 8. The angular dependence of the polarization of photoelectrons ejected from the 3p^6($^2P_{1/2}$) subshell of Ar by linearly polarized light (solid curve, $s \perp e$, $s \perp \kappa$), unpolarized light (dot–dash curve, $s \perp \kappa$, $s \perp k$), and circularly polarized light (dashed curve, $s \| s_\gamma$) of wavelength $\lambda = 58.43$ nm ($\omega = 21.2$ eV). Experimental points for linearly polarized light are from Schönhense (1980).

ture levels $\bar{A}$ is given by equation

$$\bar{A} = \frac{A^{l+1/2}\sigma_{l+1/2} + A^{l-1/2}\sigma_{l-1/2}}{\sigma_{l+1/2} + \sigma_{l-1/2}} = \frac{fA^{l+1/2} + A^{l-1/2}}{1+f} \quad (77)$$

where f is the branching ratio, $f = \sigma_{l+1/2}/\sigma_{l-1/2}$. In the nonrelativistic approximation it follows from Eqs. (32) and (34) that $\bar{A} = 0$. Using the values of A^j and f calculated in the RRPA for the $5p^6$ subshell of Xe (Johnson and Cheng, 1979; Huang et al., 1981), one obtains $\bar{A}$ as presented in Fig. 7, which is much lower than the experimental values (Heinzmann, 1980a), also shown in Fig. 7. By an order of magnitude, $\bar{A}$ should be proportional to $A^{l-1/2}(\alpha Z)^2$. In Ar and Kr the experimental values of $\bar{A}$ (Heinzmann and Schäfers, 1980) are quite close to zero and do not contradict theoretical estimations.

B. Fano Effect in ns and ns² Subshells

The Fano effect in alkali atoms (Fano, 1969a) has been studied by many authors both experimentally (Heinzmann et al., 1970; Baum et al., 1970, 1972; Drachenfels et al., 1977) and theoretically (Weisheit, 1972; Chang and Kelly, 1972; Norcross, 1973) and has been reviewed by Kessler (1976) and by Delone and Fedorov (1979). As an example, Fig. 9 shows the dipole matrix elements d_1 and d_3 [see Eq. (43)], the photoionization cross section σ and the A parameter calculated by Weisheit (1972) in a model polarization potential, and β and η parameters calculated using his data. For σ and A there is good agreement with the experiment. The β parameter has not been measured, although it has been calculated by different authors (Walker and Waber, 1973; Marr, 1974; Huang and Starace, 1979; Ong and Manson, 1979). The η parameter previously has been neither calculated nor measured. Its numerical value is essentially defined by the phase shift difference, which according to Eq. (44) is small, and is constant in a broad region near the threshold (Ong and Manson, 1979). Therefore, this parameter in the Cooper minimum remains as small [of order $(\sigma Z)^2$] as is its value far from the Cooper minimum.

Results for the Fano effect in the $5s^2$ subshell of Xe are shown in Fig. 10. The Dirac–Slater approximation fails to reproduce the correct position of the Cooper minimum (Walker and Waber, 1974), which is defined by the many-electron correlations with the $5p^6$ and $4d^{10}$ subshells. In the RPAE approximation the position of the Cooper minimum is correct (Amusia and Cherepkov, 1975), but the spin–orbit interaction is not taken into account. To incorporate the spin–orbit interaction into the RPAE results, the corresponding dipole matrix elements were split using the relations (Cherepkov, 1978)

$$d_1(\omega) = D(\omega + \tfrac{1}{2}\Delta\omega), \quad d_3(\omega) = \tfrac{3}{4}D(\omega - \tfrac{1}{2}\Delta\omega)$$

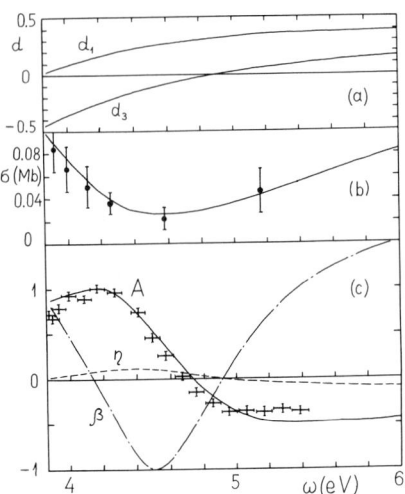

FIG. 9. Fano effect in the Cs atom. (a) The dipole matrix elements d_1 and d_3 deduced from the results of model calculations of Weisheit (1972). (b) The photoionization cross section calculated by Weisheit (1972) and measured by Cook et al. (1977). (c) The polarization parameter A, calculated by Weisheit (———), compared with experimental results of Heinzmann et al. (1970). The parameters β (— · — ·) and η (----), calculated using the matrix elements from the upper part of this figure and with the constant phase shift $(\delta_1 - \delta_3) = 0.1$.

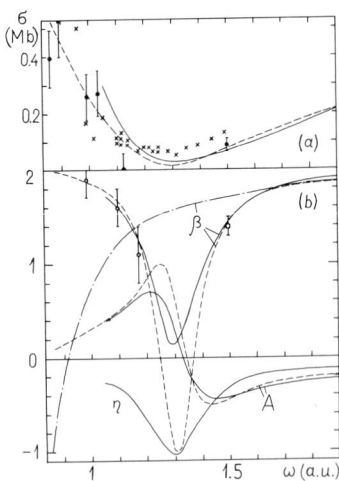

FIG. 10. Fano effect in the $5s^2$ subshell of Xe. The photoionization cross section σ (a), and the parameters β, A, and η (b), calculated in the RRPA by Johnson and Cheng (1978) and by Cheng et al. (1980) (———), and estimated in the RPAE (see text for explanation) by Cherepkov (1978) (----); (— · — ·) the result of Walker and Waber (1974) for β in the Dirac–Slater approximation. Experimental points for σ: (●) Samson and Gardner (1974); (✕) Gustafsson (1977); for β: (□) Dehmer and Dill (1976); (○) White et al. (1979).

where $\Delta\omega = 0.35$ Ryd, while it was supposed that $\delta_3 = \delta_1$. The significant difference between the results of this estimation and calculations in the RRPA (Johnson and Cheng, 1978; Cheng et al., 1980), both shown in Fig. 10, appears to be due to nonzero values of the phase shift difference $(\delta_3 - \delta_1)$ in the RRPA. In particular, a nonzero result for η is possible only if $(\delta_3 - \delta_1) \neq 0$.

The situation near the Cooper minimum in both ns and ns^2 subshells is quite simple and can be analyzed in terms of two independent parameters: (*i*) the ratio of moduli of the dipole matrix elements defined by Eq. (75), $a = |d_1|/|d_3|$, and (*ii*) the phase shift difference Δ, which includes the additional

$$\Delta = \delta_3 + \Delta_3 - \delta_1 - \Delta_1 \tag{78}$$

Using them, one obtains from Eqs. (39)–(42)

$$\beta = \frac{2 + 4a \cos \Delta}{2 + a^2}, \quad A = \frac{5 - a^2 - 4a \cos \Delta}{2 + a^2}$$

$$\eta = \frac{3a \sin \Delta}{2 + a^2} \tag{79}$$

and the cross section σ of Eq. (38) alone depends on the absolute values of the dipole matrix elements. Typical behavior of a and Δ near the Cooper minimum is presented in Fig. 11. The spin–orbit interaction corresponds to attraction for the $\epsilon p_{1/2}$ state and to repulsion for the $\epsilon p_{3/2}$ state (Sobel'man, 1972); therefore, $(\delta_3 - \delta_1) < 0$, and d_1 always changes its sign at a lower energy than does d_3. It means that $a < 1$ below the Cooper minimum and $a > 1$ above it. In alkali atoms the Cooper minimum of the valence shell appears at an energy below the next threshold; therefore, the dipole matrix elements are real and the phase shifts Δ_3 and Δ_1 have the jumps of $\pm \pi$ at the points where d_3 or d_1, respectively, changes its sign.

In ns^2 subshells the Cooper minimum appears usually at an energy above the threshold of the next subshell, and therefore the dipole matrix elements with correlations taken into account are complex. Two kinds of their behavior on the complex plane, with the photon energy increasing, are shown in the insert of Fig. 11 by arrows. Both d_1 and d_3 never turn out to be zero and both go either above or below the origin. As a consequence, the phase difference $(\Delta_3 - \Delta_1)$ changes smoothly across the Cooper minimum region and is negative or positive, respectively. Its magnitude is of order 1, whereas $(\delta_3 - \delta_1)$ is of order $(\alpha Z)^2$; therefore, near the Cooper minimum Δ is essentially defined by $(\Delta_3 - \Delta_1)$. As a matter of fact, Δ is negative in the $5s^2$ subshell of Xe (Cheng et al., 1980; solid curve in Fig. 11) and positive in the $6s^2$ subshell of Hg (Schönhense et al., 1982a; chain curve in Fig. 11).

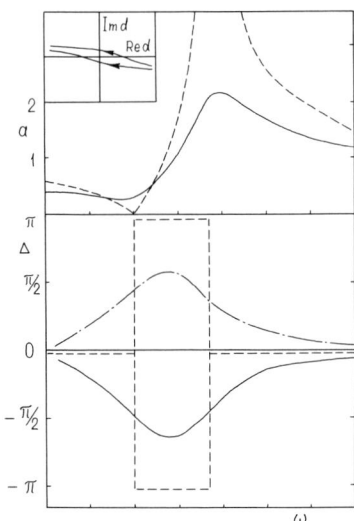

FIG. 11. An illustration of typical behavior of the parameters a and Δ as a function of the photon nergy ω for real (----) and complex (——,—·—·) dipole matrix elements. Inset: behavior of the dipole matrix element on the complex plane.

From this behavior of the parameters a and Δ one can conclude that in alkali atoms (real dipole matrix elements) in the Cooper minimum β reaches -1, A reaches $+1$, and η is everywhere small, of order $(\alpha Z)^2$. In ns^2 subshells, when more than one channel is opened (complex dipole matrix elements), β and A do not reach -1 and $+1$, respectively, but instead η can be of order 1. These general statements are illustrated in Figs. 9 and 10 for Cs and Xe, respectively, and they are supported also by new measurements of Schönhense et al. (1982a) for the $6s^2$ subshell of Hg, presented in Fig. 12. In this experiment the polarization of photoelectrons ejected from the ns^2 subshell by unpolarized light has been measured for the first time. From the values of β and η measured by Schönhense et al. (1982a) the parameters a and Δ have been deduced supposing that the Cooper minimum appears between the first two experimental points in Fig. 12 (Shannon and Codling, 1978) above the ionization threshold of the $5d_{5/2}$ and $5d_{3/2}$ subshells. In accord with the general statements, η is here of order 1 and positive. The significant deviation of the RRPA calculations of Johnson et al. (1982) from the measurements (see Fig. 12a) is explained by the fact that the RRPA fails to reproduce the correct position of the Cooper minimum, which in the RRPA appears below the $5D_{3/2}$ threshold.

When more than three measurements are performed for the subshell with $l = 0$, the additional information can be used to check the self-consistency of experimental results. As an example, Fig. 13 shows the photoionization

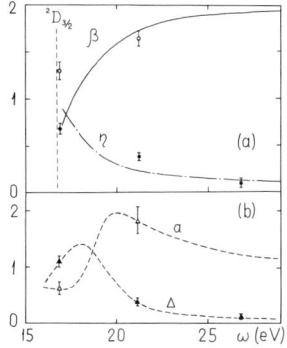

FIG. 12. (a) Experimental results for β (○) and η (●) in the 6s² subshell of Hg (Schönhense et al., 1982a); (———,— · — ·) the RRPA results of Johnson et al. (1982) for β and η, respectively. (b) Parameters a and Δ deduced from the measured points. (----) Tentative behavior of these parameters if the Cooper minimum lies near $\omega = 18$ eV.

cross section σ and the parameters β, A, and η measured for the 6s² subshell of Hg in the region of the 5d⁹ 6s²(²D$_{3/2}$) 6p(³D$_1$) autoionization resonance (Brehm, 1966; Schäfers et al., 1982). The magnitude of the η parameter can be calculated from Eq. (76) using the experimental results for β and A. Calculated in this way, values of η (also shown in Fig. 13) appear to be several times higher than the measured ones, which means that, as was mentioned by the authors, the measured values of η are probably too low due to the high electron scattering cross section in the low-energy range. The measurement of η was intended only to find out the sign of the phase-shift difference.

From the experimental results for σ, β, A, and η Schäfers et al. (1982) have deduced the dipole matrix elements d_1 and d_3 and the phase shift difference Δ in the region of autoionization resonances above the 6s² subshell threshold of Hg. At present there is no calculation in this energy region, although Stewart (1970) have considered a general theory of polarization phenomena in the region of autoionization resonances in ns² subshells.

C. Atoms with One Outer p Electron

Al, Ga, In, and Tl atoms with one np electron in the outer subshell are convenient objects for both theoretical and experimental investigations. The spin–orbit splitting of their ground states increases from 0.014 eV in Al up to 0.966 eV in Tl (Kozlov, 1981). Therefore, at temperatures around 1000°C in Al vapors the populations of the ²P$_{3/2}$ and ²P$_{1/2}$ states are close to the statistical values, in Ga they are approximately equal, and in In and Tl,

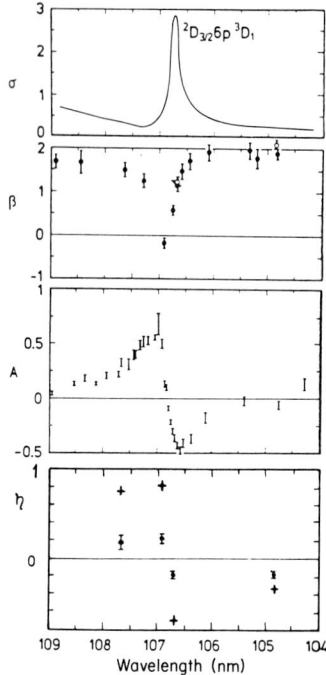

FIG. 13. Photoionization of the $6s^2$ subshell of Hg in the autoionization region (from Schäfers *et al.*, 1982). Experimental points for β: from Brehm and Höfler (1979) (●) and from Niehaus and Ruf (1972) (○); for A and η: from Schäfers *et al.* (1982); cross section σ data are from Brehm (1966). (+) The values of η calculated from Eq. (76) using β and A from this figure.

respectively, 90% and practically 100% of the atoms are in the lower $^2P_{1/2}$ state. Thus, only in Tl is it possible to investigate polarization phenomena without energy analysis of photoelectrons. Figure 14 shows the experimental photoionization cross section (Krylov and Kozlov, 1979; Kozlov, 1981) and the $A^{1/2}$ parameter (Heinzmann *et al.*, 1975, 1976) for Tl atoms together with theoretical results of Cherepkov (1977, 1980b) obtained using the open-shell RPAE method (Cherepkov and Chernysheva, 1977). The photoionization cross section is greatly influenced by autoionization resonances. According to Beutler and Demeter (1934), the 200.7, 161, 149, and 130.4 nm lines are identified as transitions into the $^4P_{3/2}$, $^2D_{3/2}$, $^2P_{1/2}$, and $^2S_{1/2}$ final states, respectively, of the $6s\,6p^2$ configuration. Since in RPAE calculations the LS-coupling has been used, it was possible to consider only ^2D and ^2S resonances. For the former there is satisfactory agreement with the experiment, whereas for the latter the calculated width and the peak value of the cross section were found to be equal to the respective experimental

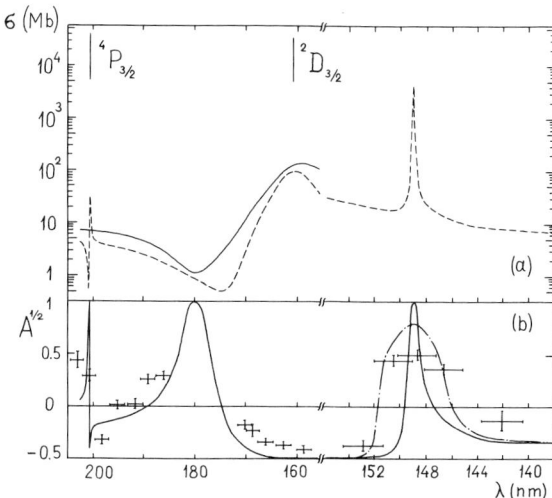

FIG. 14. Photoionization of the Tl atom. (a) Cross section σ calculated in the RPAE by Cherepkov (1980b) (——) and measured by Krylov and Kozlov (1979) (----). (b) The $A^{1/2}$ parameter measured by Heinzmann et al. (1975, 1976), and calculated in the RPAE by Cherepkov (1980b) (——). In the $^4P_{3/2}$ autoionization resonance the experimental cross section has been used to calculate $A^{1/2}$. The dot–dash curve was obtained from the full curve by folding it with a 3.1-nm instrumental resolution of Heinzmann et al. (1975).

values for the 149 nm line. Therefore, it was supposed that the identification of Beutler and Demeter should be corrected (Cherepkov, 1980b). If the 149 nm line corresponds to the $^2P_{1/2} \rightarrow {}^2S_{1/2}$ transition, then from the analysis of the experimental profile the 130.4 nm line can be attributed to the $^2P_{1/2} \rightarrow {}^2P_{3/2}$ transition. This assignment has been supported also by Connerade and Baig (1981) using another argument, whereas the 149 nm line has been attributed by them to the mixture of 60% $^2P_{1/2}$ and 35% $^2S_{1/2}$ states. In view of the poor LS characterization of this line, we note that in any case this is a transition into the $j = \frac{1}{2}$ state.

The width of the 149 nm line on the calculated curve for $A^{1/2}$ is greater than on the cross section curve, but the experimental resolution was still greater. Therefore, the calculated curve was convoluted with the experimental resolution. As seen from Fig. 14, there is satisfactory agreement between theory and experiment. It should be noted that the measurement of a photoionization cross section with the 0.03-nm resolution does not enable the asymmetry of this resonance to be determined (Krylov and Kozlov, 1979), whereas the measurement of the degree of electron polarization with the 3.1-nm resolution (Heinzmann et al., 1975) enables this to be done although the width of this resonance is equal to 0.08 nm only. Therefore, the measurement of the degree of spin polarization serve to determine the Fano profiles (Fano, 1961) with high precision and far from a resonance.

Figure 15 shows the $\beta^{1/2}$ and $\eta^{1/2}$ parameters for Tl atoms calculated in the RPAE. Both change sharply on the width of each resonance, and in every resonance the $\eta^{1/2}$ parameter has two zeroes. This behavior follows from Eq. (15) and from general properties of the dipole matrix elements and phase shift in a resonance. According to the Fano theory (Fano, 1961), in every resonance the dipole matrix element d_l changes its sign in the cross section minimum, and the phase shift δ_l is smoothly increasing in π, leading to one zero of the sine of the phase shift difference.

In RPAE calculations the phase shift δ_l is always equal to the Hartree–Fock one and has no resonance increasing in π. Instead, the dipole matrix element in the RPAE is complex, and if we present it in the form of Eq. (75), the additional phase shift Δ_l will have the increase in π over a resonance. Figure 16 shows the real and imaginary parts of the dipole matrix element $\langle \epsilon d \| D(\omega) \| 6p \rangle$ in the RPAE in the region of the 6s 6p²(²D) resonance. Its trajectory on the complex plane (see insert in Fig. 16) is quite close to a circle going through the origin. At the origin Δ_l has a jump in π which corresponds to the change of sign of a real matrix element of the Fano theory, and furthermore there is the smooth increase of Δ_l in π along the trajectory.

The analogous resonance behavior of the $\beta^{3/2}$, $A^{3/2}$, and $\eta^{3/2}$ parameters has been established in the region of autoionization resonances between the ²P$_{3/2}$ and ²P$_{1/2}$ thresholds of rare gases (Lee, 1974; Johnson et al., 1980) and ²D$_{5/2}$ and ²D$_{3/2}$ thresholds of Pd atoms (Radajević and Johnson, 1983). It is

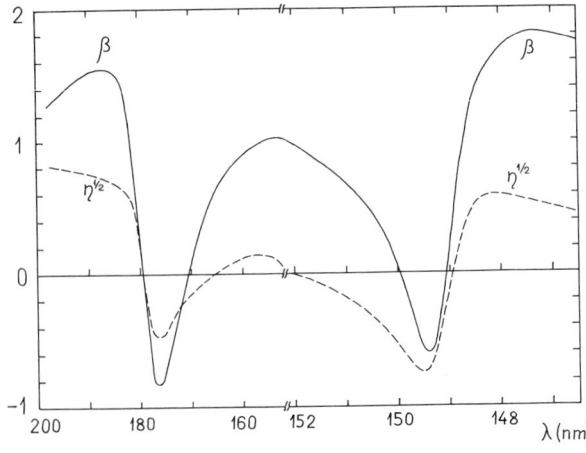

FIG. 15. Parameters β and $\eta^{1/2}$ in Tl calculated in the RPAE (Cherepkov, 1980b).

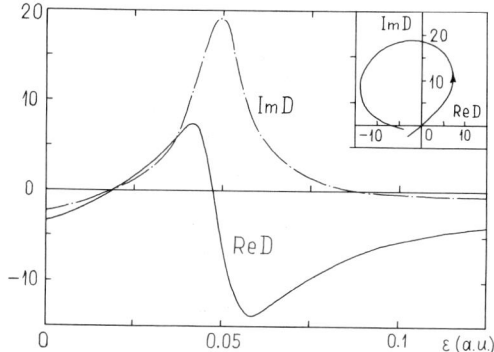

FIG. 16. The real and imaginary parts of the dipole matrix element $\langle \epsilon d \| D(\omega) \| 6p \rangle$ versus photoelectron energy ϵ in the RPAE in the region of the 6s $6p^2(^2D)$ autoionization resonance. Insert: the same on the complex plane.

worthwhile to mention also that polarization of photoelectrons ejected from excited Cs atoms has been calculated by Jaouen *et al.* (1980) and by Cherepkov (1981c).

D. METHOD FOR IDENTIFICATION OF AUTOIONIZATION RESONANCES

The Zeeman and Stark effects, normally used for the identification of discrete levels, could not be used for the identification of autoionization resonances since the natural width of the latter (> 10 cm^{-1}) is much greater than the Zeeman and Stark splitting in the laboratory-attainable fields (~ 1 cm^{-1}). Therefore, the autoionization resonances have until now been identified only theoretically using the photoabsorption cross section and an analogy with the discrete spectrum. As was shown by Cherepkov (1980b), the measurement of the degree of spin polarization of photoelectrons ejected by circularly polarized light, or in some cases even of the angular asymmetry parameter β^j, can be used for experimental identification of autoionization resonances.

Consider the $ns^2 \, np \to ns \, np^2$ autoionization resonances in atoms of the Al group. Experiment has shown (Hudson and Kieffer, 1971; Kozlov, 1981) that the photoionization cross section in these resonances is usually two or even three orders of magnitude greater than the background. In the atoms under consideration only one dissociation channel given in Table III is opened for every resonance. Therefore, it is possible to neglect in Eqs. (53)–(56) the contribution of all matrix elements except the resonance one.

TABLE III

LIMIT VALUES OF PARAMETERS A^j AND β^j IN
RESONANCES FOR DIFFERENT VALUES OF
TOTAL ANGULAR MOMENTUM OF THE
INITIAL AND FINAL STATES

J_{initial} $ns^2\, np$	J_{final} $ns\, np^2$	Dissociation Channel	A^j	β^j
$\frac{1}{2}$	$\frac{1}{2}$	$np_{1/2} \rightarrow \epsilon s_{1/2}$	1.0	0
$\frac{1}{2}$	$\frac{3}{2}$	$np_{1/2} \rightarrow \epsilon d_{3/2}$	−0.5	1.0
$\frac{3}{2}$	$\frac{1}{2}$	$np_{3/2} \rightarrow \epsilon s_{1/2}$	−0.5	0
$\frac{3}{2}$	$\frac{3}{2}$	$np_{3/2} \rightarrow \epsilon d_{3/2}$	−0.2	−0.8
$\frac{3}{2}$	$\frac{5}{2}$	$np_{3/2} \rightarrow \epsilon d_{5/2}$	0.7	0.8

The values of the parameters β^j and A^j in resonances in this limit are presented in Table III. They depend on both the total angular momentum of the initial state and the total angular momentum of a resonance. This enables the measurement of the parameters A^j and β^j to be used for the identification of autoionization resonances. As to the orbital angular momentum L and the total spin S, they could not be determined by this method.

If the initial atomic states with $j = l \pm \frac{1}{2}$ are populated in proportion to their statistical weights, as in Al vapors, the degree of spin polarization of photoelectrons far from resonances will be small. In a resonance either $|d_{l+1}|^2$ or $|d_{l-1}|^2$ is two or three orders of magnitude greater than another one. Therefore, the nonresonant contribution can be neglected, and the degree of polarization and β^j will be close to the values presented in Table III even without energy analysis of photoelectrons.

The existing identification of $ns^2\, np \rightarrow ns\, np^2$ autoionization resonances in Al and Ga, where the LS-coupling works well, is beyond doubt. One of the transitions $^2P_{1/2,3/2} \rightarrow {}^2S_{1/2}$ in In atoms was not observed until now, and the observed 175.7 nm line has been attributed to the $^2P_{3/2} \rightarrow {}^2S_{1/2}$ transition by Garton (1954) and to the $^2P_{1/2} \rightarrow {}^2S_{1/2}$ transition by Marr and Heppinstall (1966). Figure 17 shows the photoionization cross section of In vapors (Kozlov, 1981) and the values of $\overline{A}$ estimated from Eq. (77) using the experimental cross section. It is seen from the figure that measuring only the sign of the degree of spin polarization of photoelectrons is sufficient for the unambiguous identification of this line.

Exactly the same situation applies in the case of the 130.4 nm line of Tl atoms discussed above. The identification of Beutler and Demeter (1934) corresponds to the positive sign of the degree of polarization, whereas the

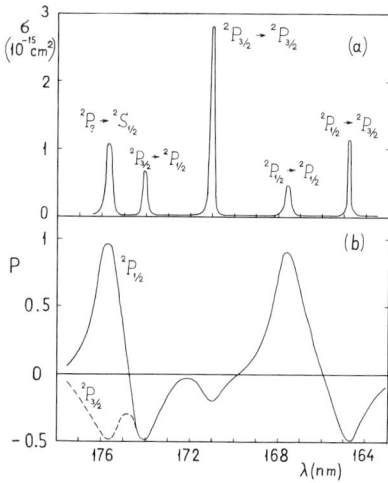

FIG. 17. (a) The photoionization cross section of the In atom, obtained by experiment (Kozlov, 1981; Hudson and Kieffer, 1971). (b) The degree of polarization of the total electron current ejected by circularly polarized light; estimation from the experimental cross section. The curves in the region of the $5s^2\,5p \to 5s\,5p^2(^2S_{1/2})$ autoionization resonance correspond to the $^2P_{1/2}$ (———) and $^2P_{3/2}$ (----) initial states.

identification proposed by Cherepkov (1980b) gives the negative sign. As for the 149 nm line, it is in both cases attributed to the transition into the final state with the same $J = \frac{1}{2}$ and with different L, but the value of L could not be established experimentally.

The measurement of β^j can also be used for the identification of autoionization resonances. However, β^j varies sharply just near the resonance maximum of the cross section (see Fig. 15), and it is necessary to find its value at a point with a large first derivative, whereas the A^j parameter has a broad extremum in a resonance. Therefore, much better resolution is needed for the β^j measurement being used for identification purposes.

In the resonances corresponding to the $ns^2\,np \to ns\,np\,n'p$ transitions the photoionization cross section is only several times higher than the background. Therefore, the limit values of the parameters (given in Table III) could not be used in these resonances. The values of $\overline{A}$ still can be estimated from the measured cross section using, in particular, the fact that the resonance appears only in one channel.

The same consideration is valid for atoms with one d electron in the outer shell, and also for some resonances in atoms with more than one electron in the outer subshell. In any case a resonance should have only one dissociation channel for the method described above to be applicable.

TABLE IV
CHARACTERISTICS OF POLARIZED ELECTRON SOURCES

Method	P (%)	Light source			Quantum yield Y	P^2Y
		Wavelength (nm)	Width (nm)	σ (Mb)		
Photoionization of a polarized Li beam (Alguard et al., 1979)	85	~200		1.8	~10^{-4}	~10^{-4}
Fano effect in Rb (Drachenfels et al., 1977)	65	~270		0.02	~10^{-6}	~10^{-6}
Fano effect in Cs (Wainwright et al., 1978)	63	~300		0.04	~10^{-6}	~10^{-6}
Fano effect in Cs (Möllenkamp and Heinzmann, 1982)	84	~300		0.04	~10^{-6}	~10^{-6}
Photoemission from GaAs (Pierce et al., 1980)	⩽43	790		—	0.03	0.006
Resonance photoionization of Al atoms	⩽90	193.2	⩽0.3	430	⩽0.1	⩽0.1
Resonance photoionization of Al atoms	⩽95	176.54	⩽0.04	930	⩽0.1	⩽0.1
Resonance photoionization of Ga atoms	⩽98	151.98	⩽0.07	8000	⩽0.1	⩽0.1
Resonance photoionization of Ga atoms	⩽95	161.0	⩽0.6	700	⩽0.1	⩽0.1
Resonance photoionization of In atoms	⩽95	175.74	⩽0.7	900	⩽0.1	⩽0.1
Resonance photoionization of Tl atoms	⩽98	149.0	⩽0.2	4000	⩽0.1	⩽0.1

E. Sources of Polarized Electrons

Polarization phenomena considered here can be used for construction of polarized electron sources required for a wide variety of experiments in different branches of physics. The idea of obtaining polarized electrons from photoionization of a polarized alkali-atom beam by unpolarized light proposed at first by Fues and Hellman (1930) was not fully realized until 40 years later (Baum and Koch, 1969; Hughes et al., 1972). The great progress achieved during the past decade in the development of polarized electron sources has been reviewed recently by Celotta and Pierce (1980). The main purpose of this section is to give a short account of possible ways for producing polarized electrons from atomic photoionization processes.

As was shown previously, photoelectrons ejected at a definite angle from any atomic subshell are usually polarized, but it is difficult to obtain a high electron current in this way. Of practical interest is the case when highly polarized total electron current is obtained at low photon energies, at which intense sources of light are available. At present only the Fano effect in alkali atoms fulfill these conditions. Some characteristics of sources discussed are summarized in Table IV. The main disadvantage of sources based on the Fano effect is their low quantum yield (the yield of electrons per incident photon) connected with the low photoionization cross section of alkali atoms in the Cooper minimum.

For comparison, Table IV contains also the parameters of the source based on photoemission of electrons from GaAs crystal by circularly polarized light, which gives at present the highest intensity of a polarized electron beam. However, this source gives a relatively low degree of spin polarization (the theoretical upper limit is 50%, since the $p_{3/2} \to s_{1/2}$ transition is excited). Therefore, the Fano sources still can compete with them (Möllenkamp and Heinzmann, 1982). At present these two types of sources seem to be the most practical for use in experiments.

It is possible also to obtain the electron beam of high polarization by irradiating atoms of the Al group by circularly polarized light with the wavelength corresponding to the autoionization resonances $ns^2 np(^2P_{1/2}) \to ns\, np^2(^2P_{1/2}, {}^2S_{1/2})$ (Cherepkov, 1977). From Eqs. (13) and (22) one finds for the degree of polarization of the total electron current

$$P = A^{1/2} = \frac{x - 0.5}{1 + x} \tag{80}$$

where $x = \sigma_s/\sigma_d$ is the ratio of the partial cross sections for the transitions $np_{1/2} \to \epsilon s_{1/2}$ and $np_{1/2} \to \epsilon d_{3/2}$. In the resonances $\sigma_s \gg \sigma_d$, $x \gg 1$ and $P \to 1$. The complete polarization of photoelectrons corresponding to the

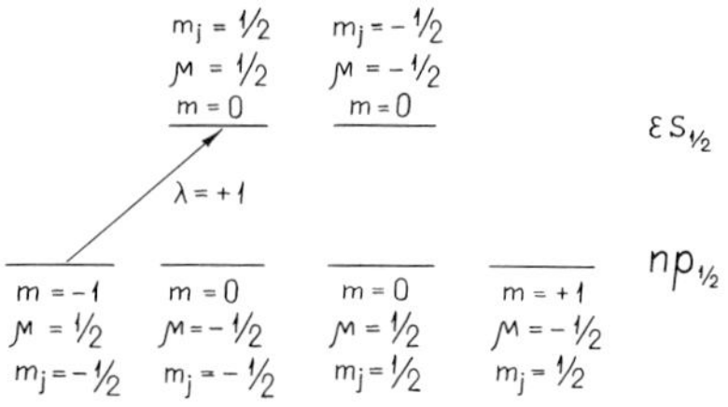

FIG. 18. The level diagram illustrating the total polarization of photoelectrons from the $n\mathrm{p}_{1/2} \to \epsilon\mathrm{s}_{1/2}$ transition, ejected by right circularly polarized light ($\lambda = +1$).

$n\mathrm{p}_{1/2} \to \epsilon\mathrm{s}_{1/2}$ transition follows from the dipole selection rules and is illustrated in Fig. 18. The main advantage of this kind of source is the very high photoionization cross section in the resonances (Kozlov, 1981), which leads to a high quantum yield. In many aspects these sources resemble the sources based on the Fano effect, but they require a very narrow light source in the vacuum ultraviolet (VUV) region, the width of which should be comparable with the width of autoionization resonances. Up to now even the degree of polarization of photoelectrons in these resonances has not been measured with appropriate resolution. Still, rapid progress in developing new sources of light in the VUV region can make this method practical in the future.

IV. Spin Polarization of Molecular Photoelectrons

Whereas the spin polarization of atomic photoelectrons has already been studied for a long time, the first investigations of spin polarization of molecular photoelectrons appeared in 1980 and 1981 (Heinzmann et al., 1980a, 1981; Cherepkov, 1981a,b). This section outlines the basic theoretical results concerning the structure of the angular distribution of molecular photoelectrons with defined spin polarization. At present there are no calculations of polarization phenomena in molecules and the experimental results are scarce.

A. General Derivation

Consider the most general case of photoionization of molecules oriented in some direction **n** (**n** is the unit vector) by linearly or circularly polarized light, and define a molecular coordinate system (primed coordinates and projections of orbital momenta and spins) with the Z' axis directed along **n**, and a laboratory system (unprimed coordinates and projections of orbital momenta and spins) with the Z axis directed along $\mathbf{e}$ for linearly polarized light or along the direction of the photon beam for circularly polarized and unpolarized light. As in atoms, the spin polarization of molecular photoelectrons appears through the spin–orbit (or spin–axis) interaction, which manifests itself in the multiplet splitting of molecular levels and in the difference between continuous-spectrum wave functions corresponding to different projections of the electron spin on the molecular axis.

The initial-state wave function of an N-electron molecule Ψ_0^N corresponds usually to the zero values of all quantum numbers, whereas the final-state (ion) wave function Ψ_k^{N-1} is characterized by the set of appropriate quantum numbers k. The photoelectron wave function $\varphi_{\mathbf{p}\mu_1'}^-(\mathbf{r}')$ in the molecular coordinate system can be written as

$$\varphi_{\mathbf{p}\mu_1'}^-(\mathbf{r}') = 4\pi \sum_{l_1,m_1'} (i)^{l_1} \exp\{-i\delta_{l_1 m_1'}\} F_{\epsilon l_1 m_1' \mu_1'}(\mathbf{r}')$$
$$\times Y^*_{l_1 m_1'}(\theta_{\mathbf{p}}', \varphi_{\mathbf{p}}') \chi_{\mu_1'} \qquad (81)$$

Since all wave functions are defined in the molecular system, it is necessary to transform also the dipole operator of Eq. (1) into the molecular frame with the help of the rotation matrix $D^j_{MM'}(\Omega)$, where $\Omega \equiv \{\alpha, \beta, \gamma\}$ are the Euler angles (Rose, 1957). From the other side, the photoelectron is observed in the laboratory system; therefore, the spherical function and the spinor in Eq. (81) should be transformed analogously from the molecular into the laboratory system. Finally, one obtains for the dipole matrix element

$$\langle \varphi_{\mathbf{p}\mu_1}^- \Psi_k^{N-1} | \hat{d}_\lambda | \Psi_0^N \rangle = 4\pi \sum_{l_1, m_1, m_1', \mu_1', \lambda'}$$
$$\times (-1)^{l_1} \exp\{i\delta_{l_1 m_1'}\} Y_{l_1 m_1}(\hat{\boldsymbol{\kappa}})$$
$$\times D^{l_1}_{m_1 m_1'}(\Omega) D^{1*}_{\lambda \lambda'}(\Omega) D^{1/2*}_{\mu_1 \mu_1'}(\Omega)$$
$$\times \langle k l_1 m_1' \mu_1' | \hat{d}_{\lambda'} | 0 \rangle \qquad (82)$$

where matrix element $\langle k l_1 m_1' \mu_1' | \hat{d}_{\lambda'} | 0 \rangle$ is defined completely in the molecular system (Cherepkov, 1981a). Inserting Eq. (82) into Eq. (7) and performing standard transformations, we arrive at the expression for the angular distribution of photoelectrons with defined spin orientation:

$$I_k^\lambda(\boldsymbol{\kappa}, \mathbf{s}, \mathbf{n}) = \frac{\sigma_k(\omega)}{2} \sum_{L,M_L}^{2l_{max}} \sum_{S,M_S}^{1} \sum_{T,M_T,M_T'}^{2l_{max}+3} A^{\lambda T M_T M_T'}_{L M_L S M_S}$$
$$\times Y_{LM_L}(\hat{\boldsymbol{\kappa}}) Y_{SM_S}(\hat{\mathbf{s}}) D^T_{M_T M_T'}(\Omega) \qquad (83)$$

where l_{max} is the highest orbital momentum retained in the expansion (81),

σ_k is the photoionization cross section:

$$\sigma_k(\omega) = \frac{4\pi^2\alpha\omega}{3n_g} \sum_{\text{d.s.}} \sum_{l_1,m_1'} \sum_{\lambda',\mu_1'} |\langle 0|\hat{d}_{\lambda'}^*|kl_1 m_1' \mu_1'\rangle|^2 \quad (84)$$

n_g is the total statistical weight of the initial state, and

$$A_{LM_LSM_S}^{\lambda T M_T M_T'} = \frac{3}{B} \sum_{\text{d.s.}} \sum [J, K, T](2[l_1, l_2, L])^{1/2}(i)^{l_1-l_2}$$

$$\times \begin{pmatrix} l_1 & l_2 & L \\ 0 & 0 & 0 \end{pmatrix} \exp\{-i(\delta_{l_1 m_1'} - \delta_{l_2 m_2'})\}$$

$$\times (-1)^{\lambda+\lambda'-m_2'+M_L+M_T+\mu_1'+1/2-S} \begin{pmatrix} l_1 & l_2 & L \\ -m_1' & m_2' & -M_L' \end{pmatrix}$$

$$\times \begin{pmatrix} 1 & 1 & J \\ \lambda & -\lambda & 0 \end{pmatrix} \begin{pmatrix} 1 & 1 & J \\ \lambda' & -\lambda'' & -M_J' \end{pmatrix} \begin{pmatrix} \tfrac{1}{2} & \tfrac{1}{2} & S \\ -\mu_1' & \mu_2' & -M_S' \end{pmatrix}$$

$$\times \begin{pmatrix} L & J & K \\ M_L & 0 & -M_L \end{pmatrix} \begin{pmatrix} L & J & K \\ M_L' & M_J' & -M_K' \end{pmatrix} \begin{pmatrix} K & S & T \\ M_K' & M_S' & -M_T' \end{pmatrix}$$

$$\times \begin{pmatrix} K & S & T \\ M_L & M_S & -M_T \end{pmatrix} \langle 0|\hat{d}_{\lambda'}^*|kl_1 m_1' \mu_1'\rangle\langle kl_2 m_2' \mu_2'|\hat{d}_{\lambda''}|0\rangle \quad (85)$$

Here the outer summation extends over all degenerate states (d.s.) of both molecule and ion, the inner summation extends over l_1, l_2, m_1', m_2', λ', λ'', μ_1', μ_2', J, M_J', K, M_K', M_L', M_S', and

$$B = \sum_{\text{d.s.}} \sum_{l_1,m_1'} \sum_{\lambda',\mu_1'} |\langle 0|\hat{d}_{\lambda'}^*|kl_1 m_1' \mu_1'\rangle|^2 \quad (86)$$

The parameters A are normalized by the condition $A_{0000}^{0000} = 1$.

Expressions (83) and (85) constitute a convenient basis for subsequent analysis since they take the form of a product of the dynamical (dipole matrix elements and phase shift differences) and the geometrical (spherical functions, the rotation matrix, and the 3j-symbols) factors.

As in atoms, it is convenient to write the angular distribution (83) in vector form independent of a particular choice of the laboratory system using definitions (16). The vector structure of all allowed terms (but not coefficients) can be written also without knowing Eq. (83) from the conditions to be invariant under space (except for optically active molecules) and time inversion (Brehm, 1971). Of all vectors introduced above, **κ**, **e**, and **n** are polar vectors whereas **s** and $\mathbf{s}_y$ are axial vectors, and all vectors except **n** change sign under time inversion. To write all allowed terms correctly, the following conditions should be fulfilled:

(1) **s** can enter only linearly;
(2) κ and **n** can stay in any degree;
(3) **e** can enter only in even degree whereas s_y can enter either in even or in odd degree, and higher than the second degree of each vector can appear only for ionization of oriented molecules;
(4) the vector combination can be noninvariant under the time inversion if it contains κ;
(5) in optically active (chiral) molecules terms noninvariant under the space inversion are allowed.

The first two conditions immediately follow from Eq. (83). The condition that **e** can enter only in even degree follows from the fact that the angular distribution is quadratic in the product of the dipole moment **d** and the polarization vector **e**, whereas $s_y = \pm [\mathbf{e}_x \mathbf{e}_y]$ is itself quadratic in the polarization vector **e** and therefore can stay in odd degree. Higher than the second degrees of s_y and **e** appear as reference directions relative to which the orientation of a molecule is defined.

The fourth condition reflects the fact that only the product of a coefficient and a vector combination should be invariant under time inversion but not each of them separately.[1] The operation of time inversion includes the replacement of wave functions by complex conjugate ones (Landau and Lifshitz, 1976), which leads to the change of sign of phase shifts. If a coefficient is proportional to the sine of the phase shift difference as η is [see Eq. (15)], it changes sign under time inversion. Therefore, the vector combination after this coefficient should also change sign under time inversion in order for their product to be invariant. Indeed, the vector combination $(\mathbf{s}[\kappa \mathbf{s}_y])(\kappa \mathbf{s}_y)$ after η in Eq. (17) changes sign under time inversion. The total cross section is always proportional to the sum of squared moduli of dipole matrix elements, whereas the sine of the phase shift difference can appear only in the angular distribution of photoelectrons, which contains the interference of different partial waves. Therefore, the vector combination, which is not invariant under time inversion, should contain κ.

Terms with odd degree of s_y do not appear for linearly polarized and unpolarized light, whereas terms with even degree of s_y appear for both linearly polarized and unpolarized light and can be obtained by replacement

---

[1] This fact has not been taken into account by Pratt et al. (1973). Still, due to mistakes in defining the transformation properties of vectors ξ and ζ under the time inversion mentioned by Gomes and Byrne (1980), they have correctly found the set of polarization parameters $C_{\mu\nu}$ different from zero.

of s_y by **e** or **k**, respectively, in the corresponding vector product [compare Eqs. (17)–(19)].

Finally, the fifth condition is obvious since chiral molecules have neither a center of inversion nor a plane of symmetry. Of course, it is supposed that only dextrorotatory or levorotatory optical isomers are investigated.

All five conditions follow from Eq. (83), which also gives the coefficients before each vector combination.

B. Molecules Having a Plane of Symmetry

Averaging Eq. (83) over orientations of molecules, one obtains the angular distribution of photoelectrons with defined spin orientation, ejected from unoriented molecules. For molecules having a plane of symmetry, its structure is the same as in atoms and is given by Eqs. (17)–(19). The parameters β^k, A^k, γ^k, η^k are connected with $A_{LM_LSM_S}^{\lambda TM_TM_T'}$ by the following relations (Cherepkov, 1981a,b):

$$\beta^k = \frac{2\sqrt{5}}{2 - 3\lambda^2} A_{2000}^{2000}, \qquad A^k = \sqrt{3} A_{0010}^{1000}$$

$$\gamma^k = -\sqrt{15} A_{2010}^{1000} = \sqrt{5}(A_{211-1}^{1000} + A_{2-111}^{1000}) \qquad (87)$$

$$\eta^k = \frac{3i\sqrt{5}}{2(2 - 3\lambda^2)} (A_{2-111}^{2000} - A_{211-1}^{2000})$$

The expression for β^k coincides with the expressions derived previously (Tully et al., 1968; Buckingham et al., 1970).

To observe the polarization of molecular photoelectrons, it is necessary to separate photoelectrons corresponding to different fine-structure components of the molecular ion state. As in atoms [see Eq. (32)], the polarization parameters A^k, γ^k, and η^k have the opposite sign for different fine-structure components and are inversely proportional to the statistical weights of these states. This fact has been proved experimentally for the CH_3Br molecule (Heinzmann et al., 1981).

From the expression for A^k it follows (Cherepkov, 1981a) that the greatest attainable degree of polarization of the total photoelectron current from molecules is equal to 50% as compared with 100% in atoms. In general, the polarization of molecular photoelectrons is expected to be lower due to the fact that, even for absorption of polarized light, there is always a sum over all polarization states of a photon in the molecular system [sum over λ' and λ'' in Eq. (85)] which leads to a greater number of allowed transitions as compared with the atomic case.

In contrast to the atomic case, in the molecular case the sum over l_1 and l_2 in Eq. (85) is not restricted. Therefore, the number of dipole matrix elements and phase shift differences is, in principle, infinite, whereas the number of experimentally measurable parameters for unoriented molecules is still equal to five, and the problem of extracting theoretical quantities from the measured parameters in molecules becomes much more complicated. To have more experimentally measured parameters, it is necessary to consider the photoionization of oriented molecules. The angular distribution of photoelectrons from oriented molecules has already been investigated theoretically by Kaplan and Markin (1969), Dill (1976), Dill *et al.* (1976), and Davenport (1976), and experimentally by Liebsch (1976), Smith *et al.* (1976), and Baird (1981), whereas the spin polarization of photoelectrons has been investigated only theoretically (Cherepkov, 1981b).

Integration of Eq. (83) over electron ejection angles gives the total cross section for photoionization of molecules oriented in some direction **n**, but randomly oriented around their axes, with defined spin polarization of photoelectrons, by circularly polarized light:

$$\sigma_k^{\pm 1}(\mathbf{s}, \mathbf{n}, \omega) = \frac{\sigma_k(\omega)}{2} \{1 + A_{0000}^{\pm 1200} P_2(\mathbf{n}\hat{\mathbf{s}}_\gamma) + A^k(\mathbf{s}\mathbf{s}_\gamma)$$
$$+ \sqrt{3} A_{0010}^{\pm 1200}[3(\mathbf{s}\mathbf{s}_\gamma)(\mathbf{n}\mathbf{s}_\gamma)^2 - \tfrac{1}{2}(\mathbf{s}\mathbf{s}_\gamma) - \tfrac{3}{2}(\mathbf{n}\mathbf{s})(\mathbf{n}\mathbf{s}_\gamma)]\} \quad (88)$$

The polarization is defined by two parameters only, since photoelectrons appear to be polarized in the plane containing the vectors **n** and $\mathbf{s}_\gamma$. For molecules oriented along the photon beam the degree of polarization in this direction can reach 100%, whereas for the photon beam perpendicular to the molecular axis this degree of polarization for linear molecules is equal to zero irrespective of dynamical parameters (Cherepkov, 1981b). For absorption of linearly polarized or unpolarized light the photoelectrons are unpolarized for all molecules.

The number of terms in the angular distribution of photoelectrons ejected from oriented molecules is, in principle, infinite, since the sum over L and T in Eq. (83) is not restricted. Its investigation gives the most complete information on molecules. From the invariance of the molecular wave function under reflection in the plane of symmetry it follows that $L + T$ should be even. In molecules having a center of symmetry, both L and T should be even separately. Table V gives the vector structure of the first terms of Eq. (83) for oriented molecules. For example, terms proportional to $(\kappa \mathbf{n})$ and $(\mathbf{n}\mathbf{e})(\kappa \mathbf{e})$ lead to a difference between electron currents in the κ and $-\kappa$ directions, which enables one to define the orientation of molecules adsorbed at a surface from the angular distribution measurements. As follows from calculations of Davenport (1976) and Dill et al. (1976), for CO this difference reaches several times, that is, the terms discussed above give the main contribution to the angular distribution of photoelectrons.

The term proportional to $(\mathbf{n}[\kappa\mathbf{s}_\gamma])$ leads to the circular dichroism in the angular distribution (that is, to a difference between the intensity of electron fluxes ejected at a definite angle by right and left circularly polarized light) for molecules which do not have a center of inversion, whereas the term proportional to $(\mathbf{n}[\kappa\mathbf{s}_\gamma])(\mathbf{n}\mathbf{s}_\gamma)(\kappa\mathbf{s}_\gamma)$ leads to the same effect in molecules having a center of inversion (Cherepkov, 1982).

C. Optically Active Molecules

The angular distribution of photoelectrons with defined spin polarization ejected by circularly polarized light from unoriented chiral molecules has the form (Cherepkov, 1983)

$$I_k^{\pm 1}(\kappa, \mathbf{s}) = \frac{\sigma_k(\omega)}{8\pi} \left\{ 1 - \frac{\beta^k}{2} P_2(\kappa\mathbf{s}_\gamma) + A^k(\mathbf{s}\mathbf{s}_\gamma) \right.$$
$$- \eta^k(\mathbf{s}[\kappa\mathbf{s}_\gamma])(\kappa\mathbf{s}_\gamma) - \gamma^k[\tfrac{3}{2}(\kappa\mathbf{s})(\kappa\mathbf{s}_\gamma) - \tfrac{1}{2}(\mathbf{s}\mathbf{s}_\gamma)]$$
$$+ D^k(\kappa\mathbf{s}_\gamma) + C^k(\mathbf{s}[\kappa\mathbf{s}_\gamma]) + B_1^{k,\pm 1}(\kappa\mathbf{s})$$
$$\left. + B_2^{k,\pm 1}(\kappa\mathbf{s}_\gamma)(\mathbf{s}\mathbf{s}_\gamma) + B_3^{k,\pm 1}(\kappa\mathbf{s}_\gamma)^2(\kappa\mathbf{s}) \right\} \quad (89)$$

It contains five new terms and, correspondingly, five new independent parameters:

$$D^k = A_{1000}^{1000}, \qquad C^k = \frac{3i}{2}(A_{1-111}^{1000} - A_{111-1}^{1000})$$

$$B_3^{k,\lambda} = \tfrac{5}{2}\sqrt{21}A_{3010}^{\lambda 000}, \quad B_1^{k,\lambda} = -\tfrac{3}{2}[A_{111-1}^{\lambda 000} + A_{1-111}^{\lambda 000} - \sqrt{\tfrac{7}{3}}A_{3010}^{\lambda 000}] \quad (90)$$

$$B_2^{k,\lambda} = A_{1010}^{\lambda 000} + \tfrac{3}{2}(A_{111-1}^{\lambda 000} + A_{1-111}^{\lambda 000}) - \sqrt{21}A_{3010}^{\lambda 000}$$

For absorption of linearly polarized and unpolarized light, terms of Eq. (89) linear in $\mathbf{s}_\gamma$ disappear, whereas in other terms $\mathbf{s}_\gamma$ should be replaced by $\mathbf{e}$ or $\mathbf{k}$, respectively. The D^k parameter characterizes the circular dichroism in the angular distribution of photoelectrons, which has been considered previously by Ritchie (1976a,b). The parameters $B_1^{k,\lambda}$, $B_2^{k,\lambda}$, and $B_3^{k,\lambda}$ characterize the longitudinal polarization of photoelectrons, which appears also for absorption of linearly polarized and unpolarized light. Parameter C^k characterizes the component of the transverse polarization of photoelectrons, which has an opposite sign for right and left circularly polarized light. Thus, the angular distribution of photoelectrons with given spin polarization ejected from chiral molecules is defined by 10 independent parameters,

TABLE V

Vector Structure of First Terms in Angular Distribution of Photoelectrons with Defined Spin Orientation Ejected from Oriented Molecules by Circularly Polarized Light

Term in Eq. (83)	Vector structure of first terms
	Arbitrary molecules
$A^{\lambda 100}_{1000}$	$(ns_\gamma)(\kappa s_\gamma)$
$A^{\lambda 1\pm 10}_{1\pm 100}$	$(n[\kappa s_\gamma])$, (κn), $(\kappa s_\gamma)(ns_\gamma)$
$A^{\lambda 100}_{1010}$	$(ns_\gamma)(\kappa s_\gamma)(ss_\gamma)$
$A^{\lambda 1\pm 10}_{10 1\pm 1}$	$(\kappa s_\gamma)(n[ss_\gamma])$, $(\kappa s_\gamma)(ns)$, $(ns_\gamma)(\kappa s_\gamma)(ss_\gamma)$
$A^{\lambda 1\pm 10}_{1\pm 110}$	$(ss_\gamma)(n[\kappa s_\gamma])$, $(ss_\gamma)(\kappa n)$, $(ns_\gamma)(\kappa s_\gamma)(ss_\gamma)$
$A^{\lambda 200}_{2000}$	1, $(ns_\gamma)^2$, $(\kappa s_\gamma)^2$, $(ns_\gamma)^2(\kappa s_\gamma)^2$
$A^{\lambda 2\pm 10}_{2\pm 100}$	$(ns_\gamma)(\kappa s_\gamma)(n\kappa)$, $(ns_\gamma)^2(\kappa s_\gamma)^2$, $(n[\kappa s_\gamma])(ns_\gamma)(\kappa s_\gamma)$
	Chiral molecules only
$A^{\lambda 200}_{1000}$	$(\kappa s_\gamma)(ns_\gamma)^2$, (κs_γ)
$A^{\lambda 100}_{2000}$	$(ns_\gamma)P_2(\kappa \hat{s}_\gamma)$
$A^{\lambda 100}_{2010}$	$(ns_\gamma)(ss_\gamma)P_2(\kappa \hat{s}_\gamma)$
$A^{\lambda 200}_{1010}$	$(\kappa s_\gamma)(ss_\gamma)P_2(ns_\gamma)$
$A^{\lambda 2\pm 10}_{1\pm 100}$	$(ns_\gamma)(n[\kappa s_\gamma])$, $(n\kappa)(ns_\gamma)$
$A^{\lambda 1\pm 10}_{20 1\pm 1}$	$(ns)P_2(\kappa \hat{s}_\gamma)$, $(ns_\gamma)(ss_\gamma)P_2(\kappa s_\gamma)$, $(s[ns_\gamma])P_2(\kappa s_\gamma)$

whereas in atoms and nonchiral molecules it is defined by only 5 independent parameters.

In the total cross section of Eq. (88) for oriented chiral molecules one additional term appears:

$$(A^{\pm 11 10}_{0011} - A^{\pm 11 -10}_{001-1})(n[ss_\gamma]) \qquad (91)$$

whereas some additional terms in the angular distribution of photoelectrons ejected from oriented chiral molecules are listed in Table V.

It is worthwhile to note that all effects considered here appear already in electric-dipole approximation, whereas the well-known phenomena of optical rotation and circular dichroism in photoabsorption (Hansen and Bouman, 1980) appear in the electric dipole–magnetic dipole interference terms and correspond to the next order in α. As a consequence, a relative contribution of each term considered here can be of the order of 1. The magnitude of a given coefficient depends on the structure of a molecule, on quantum numbers of the ionized subshell, on the magnitude of the spin–orbit interaction, and on the photon energy. To resolve experimentally the photoelectrons corresponding to different fine-structure components of the molecular ion state, it is better to investigate molecules containing at least one heavy atom. The effects in angular distribution, like the circular dichroism, are not connected with the spin–orbit interaction and can be observed without energy resolution between photoelectrons from different fine-structure levels.

Coefficients of all terms specific for chiral molecules are proportional to the differences of pairs of dipole matrix elements, which differ from each other by signs of all projections of orbital momenta and spins. For nonchiral molecules having a plane of symmetry these differences are identically equal to zero, whereas for chiral molecules they are different from zero. A realistic estimation of these differences is quite complicated (Cherepkov, 1982), whereas from the crude estimation of Ritchie (1976a) it follows that they can be of the order of 1.

There should be a similarity between polarization phenomena in elastic electron scattering from chiral molecules, discussed by Farago (1981), and polarization of photoelectrons ejected from them by unpolarized light, like the similarity between the Mott scattering of electrons from atoms and the polarization of atomic photoelectrons. Estimation of an asymmetry factor η in elastic electron scattering from chiral molecules (which depends on the degree of dissymmetry in the structure of the molecule), made by Rich et al. (1982) for the particular case of twisted ethylene, gives $\eta \sim 10^{-2}$, and higher values of this factor cannot be ruled out. Effects discussed in this section should be proportional to the same parameter η and, thus, are quite detectable. If the fine-structure components of the ionic state are not resolved, the polarization effects will be of order $\eta(\alpha Z)^2$, whereas the circular dichroism in the angular distribution still remains of order η. The helicity induced in an initially unpolarized electron beam after elastic scattering from chiral molecules is of order $\eta(\alpha Z)^2$ (Rich et al., 1982), that is, lower than effects in the angular distribution of photoelectrons with defined spin orientation provided the multiplet structure is resolved. Therefore, angle- and spin-resolved photoelectron spectroscopy is the most promising way of searching for new effects connected with the absence of mirror symmetry in the structure of chiral molecules.

V. Conclusion

It was the aim of this review to show that spin polarization of atomic and molecular photoelectrons ejected at a definite angle from unpolarized targets, is not an exception but rather a rule irrespective of photon energy. Therefore, angle- and spin-resolved photoelectron spectroscopy opens new possibilities in investigations of atomic and molecular structure. If the photoelectron spectroscopy gives only one parameter, the partial photoionization cross section, and the angle-resolved photoelectron spectroscopy adds to it the angular asymmetry parameter β, then the angle- and spin-resolved photoelectron spectroscopy makes it possible to define at once three additional parameters, which can be connected with spin projections onto three mutually perpendicular directions. For unoriented chiral molecules the number of measurable parameters increases up to 10. Only angle- and spin-resolved photoelectron spectroscopy enables one to perform the complete quantum-mechanical experiment for atoms. Spin polarization measurements offer the first possibility for experimental identification of autoionization resonances. Photoionization of atoms allows one to produce a beam of highly spin-polarized electrons. And, obviously, angle- and spin-resolved photoelectron spectroscopy gives the data for the most exhaustive comparison of first-principle theories with experiment.

Investigations of spin polarization phenomena require quite sophisticated apparatus for their measurement and elaborate theories for their explanation. The progress in construction of intense sources of light during the past decade, and especially the appearance of the synchrotron radiation sources, has made angle- and spin-resolved photoelectron spectroscopy a practical tool for studying atoms and molecules. On the other hand, new theoretical methods like the random phase approximation with exchange and its relativistic counterpart, developed for atomic calculations during the past decade, have made it possible to predict and to calculate the polarization of photoelectrons with high reliability. Therefore, polarization phenomena in atoms have been carefully investigated and are now relatively well understood. Analogous investigations of molecules are just starting, and there is no doubt that they will give new insight into the molecular structure (Schäfers *et al.*, 1983). In particular, angle- and spin-resolved photoelectron spectroscopy may prove to be a very efficient tool for studying chiral molecules.

ACKNOWLEDGMENTS

The author gratefully acknowledges Profs. M. Ya. Amusia, J. Kessler, U. Heinzmann, and Dr. M. Yu. Kuchiev for helpful discussions of different parts of this review, and Prof. W. R. Johnson for sending unpublished results.

REFERENCES

Alguard, M. J., Clendelin, J. E., Ehrlich, R. D., Hughes, V. W., Ladish, J. S., Lubell, M. S., Schüler, K. P., Baum, G., Raith, W., Miller, R. H., and Lysenko, W. (1979). *Nucl. Instrum. Methods* **163**, 29.

Amusia, M. Ya. (1981). *Adv. At. Mol. Phys.* **17**, 1.
Amusia, M. Ya., and Cherepkov, N. A. (1975). *Case Stud. At. Phys.* **5**, 47.
Amusia, M. Ya., and Ivanov, V. K. (1977). *Izv. Akad. Nauk SSSR, Ser. Fiz.* **41**, 2509 (in Russian).
Amusia, M. Ya., Cherepkov, N. A., and Chernysheva, L. V. (1971). *Zh. Eksp. Teor. Fiz.* **60**, 160.
Baird, R. J. (1981). *J. Electron. Spectrosc.* **24**, 55.
Baum, G., and Koch, U. (1969). *Nucl. Instrum. Methods* **71**, 189.
Baum, G., Lubell, M. S., and Raith, W. (1970). *Phys. Rev. Lett.* **25**, 267.
Baum, G., Lubell, M. S., and Raith, W. (1972). *Phys. Rev. A* **5**, 1073.
Berkowitz, J. (1979). "Photoabsorption, Photoionization and Photoelectron Spectroscopy." Academic Press, New York.
Bethe, H. (1933). *Hand. Phys.* **24**, 273.
Beutler, H., and Demeter, W. (1934). *Z. Phys.* **91**, 202.
Blum, K. (1981). "Density Matrix Theory and Applications." Plenum, New York.
Brehm, B. (1966). *Z. Naturforsch.* **21a**, 196.
Brehm, B. (1971). *Z. Phys.* **242**, 195.
Brehm, B., and Höfler, K. (1978). *Phys. Lett. A* **68A**, 437.
Buckingham, A. D., Orr, B. J., and Sichel, J. M. (1970). *Philos. Trans. R. Soc. London, Ser. A* **268**, 147.
Celotta, R. J., and Pierce, D. T. (1980). *Adv. At. Mol. Phys.* **16**, 101.
Chang, J.-J., and Kelly, H. P. (1972). *Phys. Rev. A* **5**, 1713.
Cheng, K. T., Huang, K.-N., and Johnson, W. R. (1980). *J. Phys. B* **13**, L45.
Cherepkov, N. A. (1972). *Phys. Lett. A* **40A**, 119.
Cherepkov, N. A. (1973). *Zh. Eksp. Teor. Fiz.* **65**, 933; *Sov. Phys.—JETP (Engl. Transl.)* **38**, 463, 1974.
Cherepkov, N. A. (1977). *J. Phys. B* **10**, L653.
Cherepkov, N. A. (1978). *Phys. Lett. A* **66A**, 204.
Cherepkov, N. A. (1979). *J. Phys. B* **12**, 1279.
Cherepkov, N. A. (1980a). *J. Phys. B* **13**, L689.
Cherepkov, N. A. (1980b). *Opt. Spectrosc.* **49**, 1067.
Cherepkov, N. A. (1981a). *J. Phys. B* **14**, 2165.
Cherepkov, N. A. (1981b). *J. Phys. B* **14**, L623.
Cherepkov, N. A. (1981c). *Proc. Int. Conf. Phys. Electron. At. Collisions, 12th, 1981* p. 26.
Cherepkov, N. A. (1982). *Chem. Phys. Lett.* **87**, 344.
Cherepkov, N. A. (1983). *J. Phys. B* **16**, 1543.
Cherepkov, N. A., and Chernysheva, L. V. (1977). *Izv. Akad. Nauk. SSSR, Ser. Fiz.* **41**, 2158. *Bull. Acad. Sci. USSR, Phys. Ser. (Engl. Transl.)* **41**, No. 12, 46, 1977.
Connerade, J. P., and Baig, M. A. (1981). *J. Phys. B* **14**, 29.
Cook, T. B., Dunning, F. B., Foltz, G. W., and Stebbings, R. F. (1977). *Phys. Rev. A* **15**, 1526.
Cooper, J. W. (1962). *Phys. Rev.* **128**, 681.
Cooper, J., and Zare, R. N. (1968). *J. Chem. Phys.* **48**, 942.
Davenport, J. W. (1976). *Phys. Rev. Lett.* **36**, 945.
Delone, N. B., and Fedorov, M. V. (1979). *Usp. Fiz. Nauk* **127**, 651.
Dehmer, J. L., and Dill, D. (1976). *Phys. Rev. Lett.* **37**, 1049.
Dehmer, J. L., Chupka, W. A., Berkowitz, J., and Jivery, W. T. (1975). *Phys. Rev. A* **12**, 1966.
Dill, D. (1973). *Phys. Rev. A* **7**, 1976.
Dill, D. (1976). *J. Chem. Phys.* **65**, 1130.
Dill, D., Siegel, J., and Dehmer, J. L. (1976). *J. Chem. Phys.* **65**, 3158.

Drachenfels, v. W., Koch, U. T., Müller, T. M., Paul, W., and Schaefer, H. R. (1977). *Nucl. Instrum. Methods* **140**, 47.
Fano, U. (1961). *Phys. Rev.* **124**, 1866.
Fano, U. (1969a). *Phys. Rev.* **178**, 131.
Fano, U. (1969b). *Phys. Rev.* **184**, 250.
Fano, U., and Cooper, J. W. (1968). *Rev. Mod. Phys.* **40**, 441.
Farago, P. S. (1981). *J. Phys. B* **14**, L743.
Fues, E., and Hellman, H. (1930). *Phys. Z.* **31**, 465.
Garton, W. R. S. (1954). *Proc. Phys. Soc., London, Sect. A* **67**, 864.
Geiger, J. (1977). *Z. Phys. A* **282**, 129.
Gomes, P. R. S., and Byrne, J. (1980). *J. Phys. B* **13**, 3975.
Gustafsson, T. (1977). *Chem. Phys. Lett.* **51**, 383.
Hall, H. (1936). *Rev. Mod. Phys.* **8**, 358.
Hansen, A. E., and Bouman, T. D. (1980). *Adv. Chem. Phys.* **44**, 545.
Heinzmann, U. (1978). *J. Phys. B* **11**, 399.
Heinzmann, U. (1980a). *J. Phys. B* **13**, 4353.
Heinzmann, U. (1980b). *J. Phys. B* **13**, 4367.
Heinzmann, U. (1980c). *Appl. Opt.* **19**, 4087.
Heinzmann, U., and Schäfers, F. (1980). *J. Phys. B* **13**, L415.
Heinzmann, U., Kessler, J., and Lorenz, J. (1970). *Z. Phys.* **240**, 42.
Heinzmann, U., Heuer, H., and Kessler, J. (1975). *Phys. Rev. Lett.* **34**, 441.
Heinzmann, U., Heuer, H., and Kessler, J. (1976). *Phys. Rev. Lett.* **36**, 1444.
Heinzmann, U., Schäfers, F., Thimm, K., Wolcke, A., and Kessler, J. (1979a). *J. Phys. B* **12**, L679.
Heinzmann, U., Schönhense, G., and Kessler, J. (1979b). *Phys. Rev. Lett.* **42**, 1603.
Heinzmann, U., Schäfers, F., and Hess, B. A. (1980a). *Chem. Phys. Lett.* **69**, 284.
Heinzmann, U., Schönhense, G., and Kessler, J. (1980b). *J. Phys. B* **13**, L153.
Heinzmann, U., Wolcke, A., and Kessler, J. (1980c). *J. Phys. B* **13**, 3149.
Heinzmann, U., Osterheld, B., Schäfers, F., and Schönhense, G. (1981). *J. Phys. B* **14**, L79.
Huang, K.-N. (1980). *Phys. Rev. A* **22**, 223.
Huang, K.-N. (1982). *Phys. Rev. Lett.* **48**, 1811.
Huang, K.-N., and Starace, A. F. (1979). *Phys. Rev. A* **19**, 2335.
Huang, K.-N., Johnson, W. R., and Cheng, K. T. (1981). *At. Data & Nucl. Data Tables* **26**, 33.
Hudson, R. D., and Kieffer, L. J. (1971). *At. Data* **2**, 205.
Hughes, V. W., Long, R. L., Lubell, M. S., Posner, M., and Raith, W. (1972). *Phys. Rev. A* **5**, 195.
Hultberg, S., Nagel, B., and Olsson, P. (1968). *Ark. Fys.* **38**, 1.
Jacobs, V. L. (1972). *J. Phys. B* **5**, 2257.
Jaouen, M., Declemy, A., Rachman, A., and Laplanche, G. (1980). *J. Phys. B* **13**, L699.
Johnson, W. R., and Cheng, K. T. (1978). *Phys. Rev. Lett.* **40**, 1167.
Johnson, W. R., and Cheng, K. T. (1979). *Phys. Rev. A* **20**, 978.
Johnson, W. R., and Lin, C. D. (1979). *Phys. Rev. A* **20**, 964.
Johnson, W. R., Cheng, K. T., Huang, K.-N., and Le Dourneuf, M. (1980). *Phys. Rev. A* **22**, 989.
Johnson, W. R., Radojević, V., Deshmukh, V. P., and Cheng, K. T. (1982). *Phys. Rev. A* **25**, 337.
Kabachnik, N. M., and Sazhina, I. P. (1976). *J. Phys. B* **9**, 1681.
Kaminski, H., Kessler, J., and Kollath, K. J. (1979). *J. Phys. B* **12**, L383.
Kaminski, H., Kessler, J., and Kollath, K. J. (1980). *Phys. Rev. Lett.* **45**, 1161.

Kaplan, I. G., and Markin, A. P. (1969). *Dokl. Akad. Nauk SSSR* **184**, 66; *Sov. Phys.—Dokl. (Engl. Transl.)* **14**, 36 1969.
Kelly, H. P., and Carter, S. L. (1980). *Phys. Scr.* **21**, 448.
Kessler, J. (1969). *Rev. Mod. Phys.* **41**, 3.
Kessler, J. (1976). "Polarized Electrons." Springer-Verlag, Berlin, Heidelberg and New York.
Kessler, J. (1981). *Comments At. Mol. Phys.* **10**, 47.
Kessler, J. (1982). *In* "Physics of Electronic and Atomic Collisions" (S. Datz, ed.), p. 467. North-Holland Publ., Amsterdam.
Klar, H. (1980). *J. Phys. B* **13**, 3117.
Klar, H., and Kleinpoppen, H. (1982). *J. Phys. B* **15**, 933.
Kollath, K. J. (1980). *J. Phys. B* **13**, 2901.
Kozlov, M. G. (1981). "Absorption Spectra of Metal Vapours in Vacuum Ultraviolet." Nauka, Moscow (in Russian).
Krylov, B. E., and Kozlov, M. G. (1979). *Opt. Spectrosc.* **47**, 838.
Landau, L. D., and Lifshitz, E. M. (1976). "Quantum Mechanics (Non-Relativistic Theory)," 3rd ed. Pergamon, Oxford.
Lee, C. M. (1974). *Phys. Rev. A* **10**, 1598.
Liebsch, A. (1976). *Phys. Rev. B* **13**, 544.
Marr, G. V. (1974). *In* "Vacuum Ultraviolet Radiation Physics" (E. E. Koch, R. Haensel, and C. Kunz, eds.), p. 168. Pergamon-Vieweg, Braunschweig.
Marr, G. V., and Heppinstall, R. (1966). *Proc. Phys. Soc., London* **87**, 547.
Möllenkamp, R., and Heinzmann, U. (1982). *J. Phys. E* **15**, 692.
Mott, N. F., and Massey, H. S. W. (1965). "The Theory of Atomic Collisions," Chapter IX. Oxford Univ. Press (Clarendon), London and New York.
Nagel, B. C. H. (1960). *Ark. Fys.* **18**, 1.
Nagel, B. C. H., and Olsson, P. O. M. (1960). *Ark. Fys.* **18**, 29.
Niehaus, A., and Ruf, M. W. (1972). *Z. Phys.* **252**, 84.
Norcross, D. W. (1973). *Phys. Rev. A* **7**, 606.
Ong, W., and Manson, S. T. (1979). *Phys. Rev. A* **20**, 2364.
Parzyński, R. (1980). *Acta Phys. Pol. A* **57**, 49, 59.
Pierce, D. T., Celotta, R. J., Wang, G.-C., Unertl, W. N., Galejs, A., Kuyatt, C. E., and Mielczarek, S. R. (1980). *Rev. Sci. Instrum.* **51**, 478.
Pratt, R. H., Levee, R. D., Pexton, R. L., and Aron, W. (1964). *Phys. Rev. A* **134**, 916.
Pratt, R. H., Ron, A., and Tseng, H. K. (1973). *Rev. Mod. Phys.* **45**, 273.
Radojević, V., and Johnson, W. R. (1983). *J. Phys. B* **16**, 177.
Rau, A. R. P., and Fano, U. (1971). *Phys. Rev. A* **4**, 1751.
Rich, A., Van House, J., and Hegstrom, R. A. (1982). *Phys. Rev. Lett.* **48**, 1341.
Ritchie, B. (1976a). *Phys. Rev. A* **13**, 1411.
Ritchie, B. (1976b). *Phys. Rev. A* **14**, 359.
Rose, M. E. (1957). "Elementary Theory of Angular Momentum." Wiley, New York.
Samson, J. A. R. (1982). *In* "Handbuch der Physik" (W. Mehlhon, ed.), Vol. 31, p. 123. Springer-Verlag, Berlin.
Samson, J. A. R., and Gardner, J. L. (1974). *Phys. Rev. Lett.* **33**, 671.
Sauter, F. (1931). *Ann. Phys. (Leipzig)* [5] **11**, 454.
Schäfers, F., Schönhense, G., and Heinzmann, U. (1982). *Z. Phys. A* **304**, 41.
Schäfers, F., Baig, M. A., and Heinzmann, U. (1983). *J. Phys. B* **16**, L1.
Schönhense, G. (1980). *Phys. Rev. Lett.* **44**, 640.
Schönhense, G., Heinzmann, U., Kessler, J., and Cherepkov, N. A. (1982a). *Phys. Rev. Lett.* **48**, 603.
Schönhense, G., Schäfers, F., Heinzmann, U., and Kessler, J. (1982b). *Z. Phys. A* **304**, 31.

Seaton, M. J. (1951). *Proc. R. Soc. London, Ser. A* **208**, 408, 418.
Shannon, S. P., and Codling, K. (1978). *J. Phys. B* **11**, 1193.
Smith, R. J., Anderson, J., and Lapeyre, G. J. (1976). *Phys. Rev. Lett.* **37**, 1081.
Sobel'man, I. I. (1972). "An Introduction to the Theory of Atomic Spectra." Pergamon, Oxford.
Starace, A. F. (1982). *In* "Handbuch der Physik" (W. Mehlhorn, ed.), Vol. 31, p. 1. Springer-Verlag, Berlin.
Stewart, H. A. (1970). *Phys. Rev. A* **6**, 2260.
Torop, L., Morton, J., and West, J. B. (1976). *J. Phys. B* **9**, 2035.
Tully, J. C., Berry, R. S., and Dalton, B. J. (1968). *Phys. Rev.* **176**, 95.
Wainwright, P. F., Alguard, M. J., Baum, G., and Lubell, M. S. (1978). *Rev. Sci. Instrum.* **49**, 571.
Walker, T. E. H., and Waber, J. T. (1973). *J. Phys. B* **6**, 1165.
Walker, T. E. H., and Waber, J. T. (1974). *J. Phys. B* **7**, 674.
Weisheit, J. C. (1972). *Phys. Rev. A* **5**, 1621.
West, J. B., and Morton, J. (1978). *At. Data Nucl. Data Tables* **22**, 103.
White, M. G., Southworth, S. H., Kobrin, P., Poliakoff, E. D., Rosenberg, R. A., and Shirley, D. A. (1979). *Phys. Rev. Lett.* **43**, 1661.

INDEX

A

Adiabatic energy splitting, expressions for, 7–10
Adiabatic fixed-nuclei approximation, 318–319
Adiabaticity factors in molecular collisions, 353
Analyzing power vs. polarizing power, 198–201
AO–CC equations, approximate treatments for, 39–48, *see also* Two-center atomic orbital basis
Argon
  complete density matrix calculations for, 214–218
  differential cross section for excitation of $^1P_1$ state of, 216
  ϵ and Δ parameters for excitation of $^1P_1$ state of, 217–218
  scattering angle for $^1P_1$ excitation of, 218
Atomic and molecular ion spectroscopy, 166–176
Atomic and pseudostate expansions, 35–48
  approximate treatments for AO–CC equations in, 39–48
  multichannel VPS method in, 40–43
  numerical calculations in, 37–39
  two-center atomic orbital basis in, 35–37
Atomic collisions, photoionization in, 395
Atomic hydrogen, *see also* Hydrogen
  electron capture by protons from, 100–113
  ionization asymmetry of, 243–245
  spin-dependent ionization asymmetries in, 243–245
Atomic hydrogen excitation
  angular differential cross sections for, 85–86
  cross sections for, 83–84, 91–92
  energy-loss spectroscopy in measurement of, 84
  by helium ions, 96–99
  by protons, 82–92
  total cross sections for, 117–121
Atomic hydrogen protons, differential elastic scattering cross sections for, 127
Atomic levels, fine-structure splitting of, 397–401
Atomic photoelectrons, spin polarization of, 395–443
Autoionization resonances, identification of, 424–432
Axial resonance spectrometer in electron geonium experiment, 152

B

Barium, superelastic scattering intensity in, 213
Barium ion at center of rf trap, 169
Basis orbitals, choice of in molecular orbital basis, 18–19
BEA theory, *see* Binary encounter approach
Beryllium ions, hyperfine resonance of, 174
Binary encounter approach to electron capture in atom-atom collisions, 78
Bohr-Lindhard classical model, 60–61
Born approximation
  close-coupling approximation and, 81
  continuum distorted wave, 107
  continuum intermediate step, 107
  distorted wave, *see* Distorted wave Born approximation
  first, *see* First Born approximation
  for $H^+ + H \rightarrow H^+ + H(n=2)$, 86
  for $H^+ + H \rightarrow H^+(\theta) + He^*(n=2)$, 93

450 INDEX

internuclear potential and, 104
in ion-atom collision technique, 79
ionization asymmetry factor and, 245
proton impact energies and, 87
for proton impact excitation of atomic hydrogen, 86–87
second, see Second Born approximation
for total excitation of atomic hydrogen to $n = 2$ level, 96
vibrational excitation and, 311–312, 382
Born distorted wave approximation, in atomic hydrogen ionization, 119–120
Born-Oppenheimer approximation, 284
 failure of in $A^1\Sigma^+$ state of LiH, 300
 molecular collision dynamics and, 347
 validity of, 301
Bound electronic state, decay of, 32
Brinkman-Kramers approximation, 42–43, 115
Brinkman-Kramers cross section, 49
Buckingham potential, 277, 297
Buckingham potential curves, ground-state reduced, 298–299
Bulk-phase collision data, thermal averaging in, 355

C

Captured electrons, final-state distributions of, 34
Carbon dioxide molecule
 <2-eV region in, 334
 intermediate energies in, 334–335
 vibrational excitation in, 332–335
Carbon monoxide molecule
 rotational rainbows in, 367–368
 vibrational excitation in, 330–332
Carbon tetrafluoride, vibrational excitation of, 382–385
CDW approximation, see Continuum distorted wave Born approximation
Charge exchange, at high velocities, 48–50
Charge-exchange collisions, reduced velocity as natural expansion parameter in, 3
Charge-exchange process, classical descriptions of, 50–61
Clamped nuclei, zero-order adiabatic approximation for, 267

Classical trajectory–Monte Carlo calculations, 51–56, 59–60
 for differential electron capture, 113
 for elastic scattering differential cross section, 124–127
 and electron capture in proton–hydrogen atom collisions, 109
 for $H + Z$ system, 54–56
 in ion-atom collisions, 78–79
 in proton–atomic hydrogen collisions, 120
Clebsch-Gordan coefficient, 190
Close-coupling approximation, 81–82
 for pseudostates, 81
Coherence parameters
 influence on polarization of emitted light, 210–214
 spin-dependent interactions and, 204–205
 and symmetry property of scattering amplitude, 222–223
Coherent superposition state
 of substates, 206
 without well-defined quantum number, 209
Coincidence experiment
 coordinate system used in, 210
 electron-photon, see Electron-photon coincidence experiment
Collisions, see Atomic collisions; Electron-molecule collisions; Ion-atom collisions; Ion-ion collisions; Molecular collisions; Proton–atomic hydrogen collisions
Collision system, as coordinate system, 237
Comparison equation method, 8
Compensated experiment trap, for positron geonium state, 157
Configuration interaction method, 285
Continuum distorted-wave Born approximation, 49, 107
Continuum intermediate state Born approximation, 107
Continuum states, electron capture into, 121
Cooper minimum region, spin polarization degree and, 404, 423–424
Coordinate system
 collision system as, 237
 excitation process in, 205
Cos ϵ and cos Δ parameters
 measurement of, 213
 physical importance of, 207–210

Coulomb interaction, and H + Z collision process, 50
Coulomb-projected Born approximation, 80, see also Born approximation
 for atomic hydrogen proton impact excitation, 87
 for helium atom, 94
Coupled-state approximations, 353
Coupled-state impact parameter calculations, 81
CPB approximation, see Coulomb-projected Born approximation
Cross sections, calculation of by expansion methods using molecular orbital basis, 10–35
Crossed-beam experiments
 equipment for, 357–360
 schematic drawing of, 70
CTMC, see Classical trajectory–Monte Carlo calculations

D

Decay models, diabatic PSS equations and, 32–35
Diabatic basis states, molecular orbital basis and, 19–20
Diabatic perturbed stationary state equations
 approximate treatments of, 24
 decay models in, 32–35
 multichannel Landau-Zener model and, 29–32
 separable-interaction model and, 25–27
Diatom-diatom scattering, 378–380
Diatomic combinations, excited-state reduced RKRV potential curves for, 293
Diatomic hybrids, reduced ground-state RKRV potential curves for, 280
Diatomic molecule excitation, model for, 310
Diatomic molecules
 adiabatic internuclear potential of, 305
 construction of internuclear potentials of, 294–296
 physical system of, 266
 reduced ground-state RKRV potentials for, 275–281
 reduced potential curve method for, 265–305

vibrational excitation in, 380–381
Differential excitation cross section measurements, state-selected, 99
Dipole matrix elements, quantum defect difference and, 418
Distorted-wave Born approximation
 approximation symmetry property of, 220–222
 distorting potentials in, 105
 Hg analysis of, 219–225
Drude electron gas model, 347
DWBA, see Distorted wave Born approximation

E

Effective core potentials method, 304
Eikonal approximation, 49
 distortion in, 80
Eikonal distorted wave calculation
 for $H^+ + H \rightarrow H^+ + H(n = 2)$, 87
 protein-helium excitation and, 95
Elastic scattering, 123–127, see also Scattering amplitudes; Scattering experiments
 as simplest collisional process, 123
Elastic scattering differential cross section, for helium protons, 124–125
Electron(s)
 in 8–12-eV region, 328
 inelastic collisions between light atoms and, 189–192
 polarized, see Polarized electrons
 scattered, see Scattered electrons; Scattering
 superelastic scattering of, on laser-excited atoms, 212–214
 transversally polarized, 227–228
Electron-atom collisions, spin-dependent phenomena in, 187–261
Electron capture, 1–62, 100–117
 atomic and pseudostate expansions in, 35–48
 charge exchange at high velocities in, 48–50
 classical scattering models in, 108–109
 into continuum states, 121
 cross section for, 49
 as multiple-step process, 107

multistate two-centered calculations in, 106
primary particle detection in, 74–77
in proton-atomic hydrogen collisions, 100–113
by protons from helium, 113–117
Electron g factor, measurement of, 150
Electron geonium experiment, 151–156
resonance data in, 154–155
results and conclusions in, 155–156
sideband cooling in, 154
Electronic motion, Schrödinger equation for, 4
Electronic spin reorientation transitions, in $^{24}Mg^+$, 173
Electron impact, vibrational excitation of molecules by, 309–340, *see also* Electron-molecule collisions
Electron-light atom inelastic collisions, exchange effects in, 189–192
Electron-molecule collisions
applications in, 323–340
hybrid approximation in, 319
for hydrogen halides, 335–338
local complex potential in, 321–323
resonance models in, 321–323
R-matrix method in, 319–320
Electron-nonrotating vibrator system, wave function for, 317
Electron-photon coincidence experiments, 191–192
without spin selection, 204–225
with polarized electrons, 236–241
Empirical potential functions, 272
classification of, 297–298
Energy loss, spectra differential in, 75
Energy-loss data, energy range in, 84
Energy-loss differential cross sections, 121–123
Energy-loss spectrometer, schematic drawing of, 75
Energy-loss spectrometry, in atomic hydrogen excitation measurement, 84
Energy-loss spectrum, for He^+ ions, 75–76
Exchange effects, in inelastic collisions between electrons and light atoms, 189–192
Excitation, vibrational, *see* Vibrational excitation

Excitation process
in coordinate system, 205
in LS-coupling scheme, 200
spin-orbit interaction during, 233
Excited states, reduced potential curve method and, 291–293
Expansion methods, cross-section calculation with, using molecular orbital basis, 10–35
Explicit spin-dependent interactions, defined, 189

F

Fano effect, 404–407
in Cs atom, 422
in $5s^2$ subshell of xenon, 422
in ns and ns^2 subshells, 421–424
in polarized electron production, 243
FBA, *see* First Born approximation
Fine-structure effects
LS-coupling limit and, 197–198
inside target atoms, 232
First Born approximation
amplitude in, 221
as continuum distorted wave approximation, 107
First Born approximation differential ionization cross sections, 121
for $He^+ + H \rightarrow He^+(\theta) + H(n = 2)$, 98–99
First Mainz experiment, 160–162
Fizeau-type velocity selections, 358
Fluorides, reduced ground-state Hulburt-Hirshfelder potential curves of, 282
Fraunhoffer integral technique, 81
in atomic hydrogen excitation experiments, 90
Free particle Green's function, 48
Frequency standards, microwave or optical transitions of stored atomic ions in, 175–176

G

Gating pulses and electronics, for sodium ionization experiment, 255

INDEX

Geonium atom, 151, 156–159
Geonium apparatus, schematic drawing of, 152
g factor, measurement of, 150
Glauber approximation, 49, 80, 121
  for atomic hydrogen proton impact excitation, 88
  ionization asymmetry factor and, 245
  for proton-helium excitation, 94–95
Green's function, free-particle, 48
Ground-state hyperfine splitting, 247
Ground-state reduced Buckingham potential curves, for rare gases, 298–299

H

Halogen compounds, reduced ground-state Hulburt-Hirschfelder potential curves of, 283
Hamiltonian
  for complete electron and molecule system, 315
  rational, 313
Hamilton-Jacobi equation, 57
Harmonic vibration frequency, 273
Harmonic spectroscope vibrational constant, 273
H_2–CO potential, 350
Heavy atom excitation, 192–204
  by polarized electrons, 225–236
$H^+ + H$ electron capture, multistate calculations for, 106–107, see also Electron capture
$H^+ + H \rightarrow H(\theta) + H^+$, angular differential cross section for electron capture in, 110–112
$H^+ + H(1s) \rightarrow H(\theta,\Sigma nl) + H^+$, differential cross section for, 109
$H^+ + He \rightarrow H(\theta) + He^+$, differential cross section for electron capture in, 114
$He^+ + H$, angular differential cross sections for, 97
$He^+ + H \rightarrow He^+(\theta) + H(n = 2)$, angular differential cross sections for, 97–98
He – He system, *ab initio* calculation for, 297
Helium
  ionization of, 122
  rotation-resolved time-of-flight spectra for, 371–372
Helium atom excitation, by protons, 92–96
Helium-hydrogen cross sections, exact vs. approximate, 374–378
Helium ions, atomic hydrogen excitation by, 96–99
Helium protons
  elastic scattering differential cross sections for, 124–126
  electron capture by, 113–116
Helium-sodium molecule
  rotational excitation in, 367–368
  Tang-Toennies model potential for, 374–375
High-overtone mode-selective vibrational excitation, of CF_4, 382–385
Hulburt-Hirschfelder potential function, 278, 281
Hydrogen, atomic, *see* Atomic hydrogen
Hydrogen atom-charged ion collisions, classical descriptions of, 50–61
Hydrogen atoms
  collision with fully stripped atoms, 1–62
  electron capture in collision of, 1–62
Hydrogen chloride, vibrational excitation in, 336
Hydrogen fluoride, vibrational excitation in, 336
Hydrogen halides, vibrational excitation in, 336–338
Hydrogen ion, vibrational excitation in, 340
Hydrogen molecule, vibrational excitation in, 324–325
Hydrogen sulfide molecule, vibrational excitation in, 339–340
Hyperfine ground states, electron-spin polarization as function of magnetic field in, 251
Hyperfine resonances, of trapped $^{25}Mg^+$ and $^9Be^+$ ions, 174
H + Z charge-exchange problem
  analytical models for, 56–61
  over-barrier capture model for, 56–60
H + Z collision system
  classical trajectory–Monte Carlo calculations for, 54–56
  Coulomb interaction and, 50
  in undistorted wave approximation, 44

I

ICR spectrometer, see Ion cyclotron resonance spectrometer
Incoherent superposition, of substates, 206
Independent parameters, number of in spin-dependent phenomena, 198–201
Inelastic electron-atom collisions, spin-dependent phenomena in, 187–261
Inelastic molecular collisions, theoretical approach to, 346–347
Infinite order sudden approach, 352–353
Integrated Stokes parameters, see also Stokes parameter(s)
  probing of spin-orbit interaction with, 229–234
  spin-orbit coupling effect of, 233
Interhalogens, reduced ground-state RRKV potential curves for, 281
Intermediate ions, excitation of, 192–204
Internuclear potential, Born approximation and, 104
Ion-atom collisions
  binary encounter approach in, 78
  classical techniques in, 78–79
  excitation problem in, 99
  extra electron in, 92
  theoretical techniques in, 78–82
Ion-atom system interactions, 67
  experimental methods in, 68–77
Ion creation, 147–148
Ion cyclotron resonance spectrometer, 149
Ion detection, 148
Ion detector, time of flight spectrum of ions arriving at, 254
Ion-ion collisions, transport from, 145
Ionization, 117–123
  of atomic hydrogen, see Atomic hydrogen ionization
  as spin-dependent phenomenon, 241–259
Ionization asymmetry data, 242–359
Ionization asymmetry factor, 245, 256
Ionization cross section measurement, crossed-beam techniques for, 72–73
Ion motion, kinematics of, 142
Ion samples, polarization of, 148–149
Ion storage exchange collision method, 136–137, 171
Ion storage techniques, 137–149
Ion trapping potential vs. proton cyclotron frequency, 161
Ion traps
  in radiative lifetime measurements, 180
  trapping efficiency of, 149
  types of, 137–149
IOS approach, see Infinite order sudden approach
ISEC method, see Ion storage exchange collision method
I_z-conserving energy sudden approximation, 353

K

Kingdon ion trap, 137, 145–147

L

Landau-Zener model, 27–32
Landau-Zener transitions, 42
Laser cooling, of stored ions, 167–170
Laser-excited atoms, scattering from, 212–214
Lennard-Jones potential, of rare gases, 297
Lennard-Jones potential curves, ground-state reduced, 298
Lennard-Jones potential function, 277
Lepton spectroscopy, 149–159
  historical perspective on, 149–150
Light atom-electron inelastic collisions, exchange effects in, 189–192
Lithium, ionization asymmetry of, 246–247
Lithium hydroxide, $A^1\Sigma^+$ state of, 300–301
Local complex potential, in electron-molecule collisions, 321–323
LS-coupling
  in aluminum and gallium, 430
  cos ϵ and cos Δ measurements relating to, 213
  defined, 189
  excitation process in, 200
  limit in, 197–191
  polarization transfer and, 230
Lyman-α radiation in crossed-beam experiments, 71

M

Magnesium ion
  electronic spin reorientations in, 173
  hyperfine resonance of, 174
Magnetic bottle as ion trap, 137
Magnetic coupling, in electron geonium experiment, 153
Magnetostatic ion trap, 137
Mainz experiments in mass spectroscopy of stable ions, 160–163
Mass spectroscopy, of stable ions, 159–166
Mathieu equation, 139
Mercury, 6^3P_1 state of, 202–203
Mercury analysis of DWBA, 219–225
Mercury excitation, spin-orbit phenomena in elastic collisions during, 201–204
Mercury ions, 40.5 GHz hyperfine transition of, 175
Mercury resonances, classification of close to 6^3P_1 threshold, 234–236
Microwave and rf atomic ion spectroscopy, 171–175
Microwave Zeeman resonances, 178
Molecular beam detectors
  properties of, 361
  for scattering experiments, 360–362
  in vibrational and rotational excitation, 357–362
Molecular collisions
  ab initio and model potentials in, 347–351
  cross-section computations in, 351–353
  dynamics of, 347
  experimental techniques in, 354–362
  theoretical approach to, 346
  vibrational and rotational excitation in, 345–389
Molecular constants
  estimation of, 296–297
  experimental value errors in, 298
Molecular detectors, 357–362
Molecular eigenfunctions and eigenenergies, 311–314
Molecular ion spectroscopy, 176
Molecular orbital basis
  alternate translational factors in, 20–22
  choice of basis orbitals in, 18–19
  in cross-section calculation of expansion methods, 14–22
  diabatic basic states in, 19–20
  multichannel Landau-Zener model and, 27–32
  numerical calculations with, 22–42
  perturbed stationary state approximation in, 16–17
  pseudocrossing in, 19
Molecular orbital coupled-channel calculations, 23
Molecular photoelectrons, spin polarization of, 395–443
Molecule excitation, model of, 309–310
Monoenergetic mass-selected ions, acceleration of, 74
Monte Carlo method, in charge transfer descriptions, 51–56, *see also* Classical trajectory—Monte Carlo calculations
Mott analyzer 243–244
Multichannel analyzers, gating pulses in, 253
Multichannel charge-exchange problem, 28
Multichannel Landau-Zener model, 27–32
Multichannel Vainshtein-Presnyakov-Sobel'man method, 40–43
Multiply charged ions, charge-exchange collisions with atoms, 3
Multistate two-centered calculations, in electron capture, 106
Multistate two-center coupled-state approximation, 109–110

N

Negative atomic ions, negative ion spectroscopy and, 176–180
Neon, complete density matrix calculations for, 214–219
Nitric oxide molecule, vibrational excitation of, 338
Nitrogen molecule, resonance analysis of, 326–327
Nitrous oxide molecule, vibrational excitation of, 339
Nozzle beams, properties of, 359
ns and ns^2 subshell, Fano effect in, 421–424
Nuclear spin reorientation transitions, in $^{25}Mg^+$, 173

O

OE approximation, *see* Optical eikonal approximation
OHCE (one and a half centered) model, 36–37
One-electron two-Coulomb centers system, *see also* (Z_1eZ_2) system
  basic properties of, 3–10
  eigenvalues problem for, 3
Optical atomic ion spectroscopy, 166–167
Optical crossed-beam experimental apparatus, 70
Optical eikonal approximation calculation, 109
Optical frequency standards, advantage of, 175
Optically active molecules, spin polarization of, 440–442
Optical-pumping double-resonance methods, 172
Optical sideband cooling, of stored ions, 167
Over-barrier capture model, for H + Z charge-exchange problem, 56–60
Oxides, reduced ground-state Hulburt-Hirschfelder potential curves for, 282
Oxygen molecule
  rotational rainbows in, 367–368
  vibrational excitation in, 338

P

Parameters $\cos \epsilon$ and $\cos \Delta$, physical importance of, 207–210
Parameters λ and χ, in helium atom excitation, 209
Paul ion trap, 137–143
  theoretical stability diagram for, 139
Penning ion trap, 137, 143–145
  in ion mass measurements, 159
  laser cooling in, 168
  leptons in, 149
  magnetic resonance and, 155
  in Mainz experiments, 160–161
  in University of Washington experiment, 164
Perturbed stationary state approximation, molecular orbital basis and, 16–17

Photodetachment data, for SeH⁻ near threshold, 179
Photoelectrons, angular distribution of, 441
Photoionization
  in atomic collisions, 395
  of $6s^2$ subshell of Hg in autoionization region, 426
  of thallium atom, 427
Photoionization cross section of indium atom, 431
Pierce electron gun, 253
Polarization of emitted light, coherence parameters and, 210–214
Polarization production/monitoring, ion samples in, 148–149
Polarized atoms, measurement of ionization processes with, 191, 242
Polarized electron-polarized sodium scattering apparatus, schematic of, 250
Polarized electrons
  electron-photon coincidence experiments with, 236–241
  excitation of heavy atoms by, 225–236
  Fano effect in production of, 243
  ferromagnetic EuS in production of, 246
  measurement of ionization process with, 242
  production of, 191
  sources of, 432–434
Polarized electron scattering, real independent parameters in, 199
Polarizing power vs. analyzing power, 198–201
Positron geonium experiment, 156–159
Positron geonium state, preparation of, 156–157
Positron g factor, measurement of, 150
Potassium, ionization asymmetry of, 247–249
Potential curve(s)
  ab initio calculation of, 268–269
  construction of, 267–271
  reduced, *see* Reduced potential curve
  Rydberg-Klein-Rees-Vanderslice method for, 269–271
  theoretical, 267–269
Potential functions, empirical, 272
Precision axial resonance spectrometer, 152
Primary particle detection, experimental methods involving, 74–77

Prolate spheroidal coordinates, 4-5
Proton-atomic hydrogen collisions
  CTMC theoretical results in, 120
  electron capture and, 100-103
  targets in, 100-103
  scattering models in, 108-109
  theoretical applications in, 67
Proton-atomic hydrogen crossed-beam experiment, 70
Proton-atomic hydrogen ionization cross sections, in UDWA, 120
Proton-atomic hydrogen resonant electron capture problem, CDW approximation in, 108
Proton cyclotron frequency
  vs. number of trapped protons, 163
  vs. trapping potential, 161
Proton cyclotron resonances, graphs of, 160, 163, 165
Proton excitation, of helium atoms 92-96
Protons, *see also* Proton-atomic hydrogen collisions
  atomic hydrogen excitation by, 82-92
  atomic hydrogen ionization by, 117
Pseudocrossing, radial coupling matrix element charges near, 19
Pseudostates, close-coupling procedure in, 81-82
PSS approximation, *see* Perturbed stationary state approximation
Pump, laser in rotational scattering cross section, 363

Q

Quadring trap, 164
Quantum-mechanical experiment, complete, 410-413
Quantum-mechanical techniques, in ion-atom collisions, 79-82

R

Rabi analyzer 249
  atomic beam signal of, 252
Race-track ion trap, 149
Radiative lifetime measurements, ion traps in, 180

Radiation-pressure cooling, of stored ions, 167
Random phase approximation with exchange, 416-426
  phase shift in, 428
Rare gas atoms, spin polarization in, 416-421
Rare gases, ground-state reduced Lennard-Jones and Buckingham type potential curves for, 298-299
Reduced density matrix
  defined, 188, 195
  of scattered electrons, 192-197
Reduced ground-state Hulburt-Hirschfelder potential curves, 281-283
Reduced ground-state RKRV potential curves, 275-283
Reduced internuclear distance, defined, 273
Reduced Lennard-Jones potential curves, for rare gases 298-299
Reduced potential curve(s)
  ab initio-calculated, 284-286
  behavior of, 274-275
  changes in with growing atomic numbers 275
  of diatomic molecules, 271-293
  for different molecules, 275
  ground-state, 275-283
  inaccurate *ab initio* calculations of, 286
Reduced potential curve geometry, sensitivity to changes in value of molecular constants, 284
Reduced potential curve method, 265-305
  applications of, 266, 294-302
  defined, 266
  excited states and, 291-293
  limitations of, 302-303
  mathematical foundations of, 290-291
  misunderstandings about, 302-303
  verification of, 307
Reduced potential energy, defined, 273
Reduced quantities
  defined, 271-272
  properties of, 272-275
Reduced RKRV potential curves, 276-282, 287
Reduced theoretical potential curves, 284-290
  RKRV curve and, 287

Reflection invariance in scattering plane, 260–261
Relativistic random phase approximation, 416–424
Release distance, in Bohr-Lindhard classical model, 60
Resonance data, in electron geonium experiment, 154–155
Resonance models, in electron-molecule collisions, 321–323
rf heating, kinetic energy and background neutrals in, 141
rf trap, Ba^+ ion at center of, 169, *see also* Paul ion trap
RKRV method, *see* Rydberg-Klein-Rees-Vanderslice method
Rotational energy sudden approximation, 353
Rotational excitation
 experimental techniques in, 353–362
 in molecular collisions, 345–389
Rotational Hamiltonian, 313
Rotationally inelastic collisions, summary of, 386–388
Rotational rainbow oscillations, 367
Rotational rainbows
 classical, 365
 in Na_2–Ar and Na_2–Ne, 364–374
Rotational scattering cross sections, 362–380
Rotational wavefunctions, moment of inertia and, 314
RPAE, *see* Random phase approximation with exchange
RPC method, *see* Reduced potential curve method
RRPA, *see* Relativistic random phase approximation
Runge-Lentz vector, 50
Rydberg-Klein-Rees-Vanderslice potential curve method, 269–271, 294
 ab initio-calculated, 286
 deviation from, 301
 reduced theoretical potential curve and, 287–288
 rotational levels in, 298–299

S

Scattered electrons, reduced density matrix of, 192–197

Scattering
 electron-electron, 378–380
 superelastic, 212–214
Scattering amplitudes
 coherence parameters and, 222
 for $E = 180$ eV, 222
 expressions for, 259–260
Scattering asymmetry, spin polarization and, 223
Scattering experiments
 molecular beam detectors for, 360–362
 summary of, 385–389
Scattering formalism, 315–323
Scattering models, in electron capture, 108–109
Scattering plane, reflection invariance in, 260–261
Schrödinger equation
 for center-of-mass system, 10
 for electronic motion at fixed internuclear distance, 4
 molecular eigenfunctions and, 312
 scattering solution for, 79
 true solution for, 79–80
Secondary particle detection, experimental methods for, 69–74
Second Born approximation as continuum intermediate state approximation, 107
Selective vibrational excitation, of CF_4 by ions, 382–385, *see also* Vibrational excitation
Self-consistent field, vatiational, 285
Separable-interaction model, 25–26
Seven-state close coupling calculations, 88
Sideband cooling, in electron geonium experiment, 154
Sideband excitation probe, in positron geonium experiment, 157
Simple ion-atom systems, interactions in, 67–129, *see also* Ion-atom system interactions
Slater-type atomic orbitals, 285
Sodium, ionization asymmetry of, 247–259
Sodium-argon molecule
 rotational rainbows in, 364–374
 rotational scattering cross sections for, 362–364
Sodium fluoride molecule, vibrational excitation of, 340

Sodium ionization experiment, block diagram of gating pulses and electronics for, 255
Sodium-23 ground state, hyperfine splitting and magnetic sublevels of, 251
Sodium molecule, rotational cross sections in, 362–364
Sodium-neon molecule, rotational rainbows in, 364–374
Space charge ion traps, 149
Spectroscopy, *see also* Atomic and molecular ion spectroscopy; Mass spectroscopy; Optical atomic ion spectroscopy
  microwave and rf, *see* Microwave and rf spectroscopy
  molecular ion, 176
  negative ion, 176–180
Spin-dependent electron atom scattering processes, 188
Spin-dependent interactions, coherence parameters and, 204–225
Spin-dependent ionization, asymmetric, 243–259
Spin-dependent phenomena, 187–261
  electron-photon coincidence experiments and, 236–041
  excitation of heavy atoms by polarized electrons, 224–236
  independent parameters in, 198–201
  ionization, 241–259
Spin effects, in inelastic collisions between electrons and two-electron atoms, 195
Spin-orbit coupling effect, integrated Stokes parameter and, 233
Spin-orbit interaction
  during excitation, 233–234
  inside target atoms, 237
  probing of by integrated Stokes parameters, 229–234
  simultaneous inclusion of in discrete and continuous spectra, 407–410
Spin-orbit phenomena, in mercury excitation, 201–204
Spin polarization
  angular dependence of, 420
  asymmetry studies and, 192–204
  of atomic and molecular photoelectrons, 394–443
  of atoms with one outer p electron, 425–428
  equations for degree of, 401–404
  from fine-structure splitting of atomic levels, 397–401
  of molecular photoelectrons, 434–442
  of molecules with plane of symmetry, 438–440
  of optically active molecules, 440–442
  theory of, 397–415
Spin polarization effects, causes of, 189
Spin polarization parameter
  for $3p^6$ subshell of argon, 419
  for $5p^6$ subshell of xenon, 420
Spin polarization phenomena, comparison with experiments and applications, 415–434
Spin-sensitive parameters, 188, 204
Stark effect, 429
Stokes parameter(s), 225–236
  defined, 211
  integrated, 229–234
  measurements of, 228
Stored ions
  atomic and molecular ion spectroscopy of, 166–176
  frequency standard applications of, 175–176
  high-resolution spectroscopy of, 135–180
  mass spectroscopy of, 159–166
  microwave and rf atomic ion spectroscopy of, 171–175
  optical atomic ion spectroscopy of, 166–167
  optical sideband cooling of, 167
  radiation-pressure cooling of, 167
Sturmian functions, 89
Sudden approximations
  I_z-conserving energy, 353
  infinite order 352–353
Sudden collision factorization relations, 353–354
Superelastic cross section, nonvanishing of, 213
Superelastic scattering intensity, in barium, 213

T

TAC, *see* Time-to-amplitude converter
Tang-Toennies model potential, 374
Target atoms
  fine-structure interaction inside, 232

spin-orbit interaction inside, 237
Technetium molecule, ground-state potential curve of, 295
Threshold laws, validity of, 257–259
Time-to-amplitude converter, 73
Transversally polarized electrons, 227–228, *see also* Polarized electrons
TSAE method, *see* Two-state two-center atomic expansion method
Two-center atomic orbital basis, 35–37
Two-center Coulomb system, reduction to hydrogen-like system, 5
Two-state two-center atomic expansion method, 109
    for $H^+ + H \rightarrow H(1s) + He^+(1s)$, 116
    static potentials with, 99
    variable nuclear charge in, 110–111

U

Unitarized distorted wave approximation, 40–48
    ion-ion cross sections in, 105
    n and l distributions of captured electrons in, 47

V

Vainshtein-Presynakov-Sobel'man approximation, 40–43, 90
    for atomic hydrogen excitation by helium ions, 97
    in ion-atom collisions, 80
    for proton-helium excitation, 95

Vibrational excitation, 309–340, 380–385
    in diatomics, 380–381
    experimental work in, 323–340, 354–362
    intermediate energies in, 329–330
    in molecular collisions, 345–389
Vibrational Hamiltonian, 313
Vibrationally inelastic collisions, summary of, 386–388
Vibrational-state-resolved low-energy ion-molecule scattering experiments, first, 346
Vibration close coupling, 317–318
Vibration-rotation close coupling, 315

W

Wannier law, 259
Washington University experiment, 164–166
Water molecule, vibrational excitation of, 339

Z

Zeeman effect, 429
Zero-order adiabatic approximation, of clamped nuclei, 267
$(Z_1 e Z_2)$ system, *see also* One-electron two-Coulomb centers system
    adiabatic energy splitting expressions for, 7–10
    eigenvalue problem for, 3–7
    electronic energies of, 6
    Runge-Lentz vector in, 50

Contents of Previous Volumes

Volume 1

Molecular Orbital Theory of the Spin Properties of Conjugated Molecules, *G. G. Hall and A. T. Amos*

Electron Affinities of Atoms and Molecules, *B. L. Moiseiwitsch*

Atomic Rearrrangement Collisions, *B. H. Bransden*

The Production of Rotational and Vibrational Transitions in Encounters between Molecules, *K. Takayanagi*

The Study of Intermolecular Potentials with Molecular Beams at Thermal Energies. *H. Pauly and J. P. Toennies*

High-Intensity and High-Energy Molecular Beams, *J. B. Anderson, R. P. Andres, and J. B. Fenn*

AUTHOR INDEX—SUBJECT INDEX

Volume 2

The Calculation of van der Waals Interactions, *A. Dalgarno and W. D. Davison*

Thermal Diffusion in Gases, *E. A. Mason, R. J. Mum, and Francis J. Smith*

Spectroscopy in the Vacuum Ultraviolet, *W. R. S. Garton*

The Measurement of the Photoionization Cross Sections of the Atomic Gases, *James A. R. Samson*

The Theory of Electron–Atom Collisions, *R. Peterkop and V. Veldre*

Experimental Studies of Excitation in Collisions between Atomic and Ionic Systems, *F. J. deHeer*

Mass Spectrometry of Free Radicals, *S. N. Foner*

AUTHOR INDEX—SUBJECT INDEX

Volume 3

The Quantal Calculation of Photoionization Cross Sections, *A. L. Stewart*

Radiofrequency Spectroscopy of Stored Ions I: Storage, *H. G. Dehmelt*

Optical Pumping Methods in Atomic Spectroscopy, *B. Budick*

Energy Transfer in Organic Molecular Crystals: A Survey of Experiments, *H. C. Wolf*

Atomic and Molecular Scattering from Solid Surfaces, *Robert E. Stickney*

Quantum Mechanics in Gas Crystal-Surface van der Waals Scattering, *F. Chanoch Beder*

Reactive Collisions between Gas and Surface Atoms, *Henry Wise and Bernard J. Wood*

AUTHOR INDEX—SUBJECT INDEX

Volume 4

H. S. W. Massey—A Sixtieth Birthday Tribute, *E. H. S. Burhop*

Electronic Eigenenergies of the Hydrogen Molecular Ion, *D. R. Bates and R. H. G. Reid*

Applications of Quantum Theory to the Viscosity of Dilute Gases, *R. A. Buckingham and E. Gal*

Positrons and Positronium in Gases, *P. A. Fraser*

Classical Theory of Atomic Scattering, *A. Burgess and I. C. Percival*

Born Expansions, *A. R. Holt and B. L. Moiseiwitsch*

Resonances in Electron Scattering by Atoms and Molecules, *P. G. Burke*

Relativistic Inner Shell Ionization, *C. B. O. Mohr*

Recent Measurements on Charge Transfer, *J. B. Hasted*

Measurements of Electron Excitation Functions, *D. W. O. Heddle and R. G. W. Keesing*

Some New Experimental Methods in Collision Physics, *R. F. Stebbings*

Atomic Collision Processes in Gaseous Nebulae, *M. J. Seaton*

Collisions in the Ionosphere, *A. Dalgarno*

The Direct Study of Ionization in Space, *R. L. F. Boyd*

AUTHOR INDEX—SUBJECT INDEX

Volume 5

Flowing Afterglow Measurements of Ion-Neutral Reactions, *E. E. Ferguson, F. C. Fehsenfeld, and A. L. Schmeltekopf*

Experiments with Merging Beams, *Roy H. Neynaber*

Radiofrequency Spectroscopy of Stored Ions II: Spectroscopy, *H. G. Dehmelt*

The Spectra of Molecular Solids, *O. Schnepp*

The Meaning of Collision Broadening of Spectral Lines: The Classical Oscillator Analog, *A. Ben-Reuven*

The Calculation of Atomic Transition Probabilities, *R. J. S. Crossley*

Tables of One- and Two-Particle Coefficients of Fractional Parentage for Configurations $s^l s'^m p^q$, *C. D. H. Chisholm, A. Dalgarno, and F. R. Innes*

Relativistic Z-Dependent Corrections to Atomic Energy Levels, *Holly Thomis Doyle*

AUTHOR INDEX—SUBJECT INDEX

Volume 6

Dissociative Recombination, *J. N. Bardsley and M. A. Biondi*

Analysis of the Velocity Field in Plasma from the Doppler Broadening of Spectral Emission Lines, *A. S. Kaufman*

The Rotational Excitation of Molecules by Slow Electrons, *Kazuo Takayanagi and Yukikazyu Itikawa*

The Diffusion of Atoms and Molecules, *E. A. Mason and R. T. Marrero*

Theory and Application of Sturmain Functions, *Manuel Rotenberg*

Use of Classical Mechanics in the Treatment of Collisions between Massive Systems, *D. R. Bates and A. E. Kingston*

AUTHOR INDEX—SUBJECT INDEX

Volume 7

Physics of the Hydrogen Master, *C. Audion, J. P. Schermann, and P. Grivet*

Molecular Wave Functions: Calculation and Use in Atomic and Molecular Processes, *J. C. Browne*

Localized Molecular Orbitals, *Harel Weinstein, Ruben Paunez, and Maurice Cohen*

General Theory of Spin-Coupled Wave Functions for Atoms and Molecules, *J. Gerratt*

Diabatic States of Molecules—Quasi-Stationary Electronic States, *Thomas F. O'Malley*

Selection Rules within Atomic Shells, *B. R. Judd*

Green's Function Technique in Atomic and Molecular Physics, *Gy. Csanak, H. S. Taylor, and Robert Yaris*

A Review of Pseudo-Potentials with Emphasis on Their Application to Liquid Metals, *A. J. Greenfield*

AUTHOR INDEX—SUBJECT INDEX

Volume 8

Interstellar Molecules: Their Formation and Destruction, *D. McNally*

Monte Carlo Trajectory Calculations of Atomic and Molecular Excitation in Thermal Systems, *James C. Keck*

Nonrelativistic Off-Shell Two-Body Coulomb Amplitudes, *Joseph C. Y. Chen and Augustine C. Chen*

Photoionization with Molecular Beams, *R. B. Cairns, Halstead Harrison, and R. I. Schoen*

The Auger Effect, *E. H. S. Burhop and W. N. Asaad*

AUTHOR INDEX—SUBJECT INDEX

Volume 9

Correlation in Excited States of Atoms, *A. W. Weiss*

The Calculation of Electron–Atom Excitation Cross Sections, *M. R. H. Rudge*

Collision-Induced Transitions between Rotational Levels, *Takeshi Oka*

The Differential Cross Section of Low-Energy Electron–Atom Collisions, *D. Andrick*

Molecular Beam Electronic Resonance Spectroscopy, *Jens C. Zorn and Thomas C. English*

Atomic and Molecular Processes in the Martian Atmosphere, *Michael B. McElroy*

AUTHOR INDEX—SUBJECT INDEX

Volume 10

Relativistic Effects in the Many-Electron Atom, *Lloyd Armstrong, Jr and Serge Feneuille*

The First Born Approximation, *K. L. Bell and A. E. Kingston*

Photoelectron Spectroscopy, *W. C. Price*

Dye Lasers in Atomic Spectroscopy, *W. Lange, J. Luther, and A. Steudel*

Recent Progress in the Classification of the Spectra of Highly Ionized Atoms, *B. C. Fawcett*

A Review of Jovian Ionospheric Chemistry, *Wesley T. Huntress, Jr.*

SUBJECT INDEX

Volume 11

The Theory of Collisions between Charged Particles and Highly Excited Atoms, *I. C. Percival and D. Richards*

Electron Impact Excitation of Positive Ions, *M. J. Seaton*

The R-Matrix Theory of Atomic Process, *P. G. Burke and W. D. Robb*

Role of Energy in Reactive Molecular Scattering: An Information–Theoretic Approach, *R. B. Bernstein and R. D. Levine*

Inner Shell Ionization by Incident Nuclei, *Johannes M. Hansteen*

Stark Broadening, *Hans R. Greim*

Chemiluminescence in Gases, *M. F. Folde and B. A. Thrush*

AUTHOR INDEX—SUBJECT INDEX

Volume 12

Nonadiabatic Transitions between Ionic and Covalent States, *R. K. Janev*

Recent Progress in the Theory of Atomic Isotope Shift, *J. Bauche and R. -J. Champeau*

Topics on Multiphoton Processes in Atoms, *P. Lambropoulos*

Optical Pumping of Molecules, *M. Broyer, G. Gouedard, J. C. Lehman, and J. Vigue*

Highly Ionized Ions, *Ivan A. Sellin*

Time-of-Flight Scattering Spectroscopy, *Wilhelm Raith*

Ion Chemistry in the D Region, *George C. Reid*

AUTHOR INDEX—SUBJECT INDEX

Volume 13

Atomic and Molecular Polarizabilities—A Review of Recent Advances, *Thomas M. Miller and Benjamin Bederson*

Study of Collisions by Laser Spectroscopy, *Paul R. Berman*

Collision Experiments with Laser-Excited Atoms in Crossed Beams, *I. V. Hertel and W. Stoll*

Scattering Studies of Rotational and Vibrational Excitation of Molecules, *Manfred Faubel and J. Peter Toennies*

Low-Energy Electron Scattering by Complex Atoms: Theory and Calculations, *R. K. Nesbet*

Microwave Transitions of Interstellar Atoms and Molecules, *W. B. Sommerville*

AUTHOR INDEX—SUBJECT INDEX

Volume 14

Resonances in Electron, Atom, and Molecule Scattering, *D. E. Golden*

The Accurate Calculation of Atomic Properties by Numerical Methods, *Brian C. Webster, Michael J. Jamieson, and Ronald F. Stewart*

(e, 2e) Collisions, *Erich Weigold and Ian E. McCarthy*

Forbidden Transition in One- and Two-Electron Atoms, *Richard Marrus and Peter J. Mohr*

Semiclassical Effects in Heavy-Particle Collisions, *M. S. Child*

Atomic Physics Tests of the Basic Concepts in Quantum Mechanics, *Francis M. Pipkin*

Quasi-Molecular Interference Effects in Ion–Atom Collisions, *S. V. Bobashev*

Rydberg Atoms, *S. A. Edelstein and T. F. Gallagher*

UV and X-Ray Spectroscopy in Astrophysics, *A. K. Dupree*

AUTHOR INDEX—SUBJECT INDEX

Volume 15

Negative Ions, *H. S. W. Massey*

Atomic Physics from Atmospheric and Astrophysical Studies, *A. Dalgarno*

Collisions of Highly Excited Atoms, *R. F. Stebbings*

Theoretical Aspects of Positron Collisions in Gases, *J. W. Humberston*

Experimental Aspects of Positron Collisions in Gases, *T. C. Griffith*

Reactive Scattering: Recent Advances in Theory and Experiment, *Richard B. Bernstein*

Ion–Atom Charge Transfer Collisions at Low Energies, *J. B. Hasted*

Aspects of Recombination, *D. R. Bates*

The Theory of Fast Heavy-Particle Collisions, *B. H. Bransden*

Atomic Collision Processes in Controlled Thermonuclear Fusion Research, *H. B. Gilbody*

Inner-Shell Ionization, *E. H. S. Burhop*

Excitation of Atoms by Electron Impact, *D. W. O. Heddle*

Coherence and Correlation in Atomic Collisions, *H. Kleinpoppen*

Theory of Low-Energy Electron–Molecule Collisions, *P. G. Burke*

AUTHOR INDEX—SUBJECT INDEX

Volume 16

Atomic Hartree–Fock Theory, *M. Cohen and R. P. McEachran*

Experiments and Model Calculations to Determine Interatomic Potentials, *R. Düren*

Sources of Polarized Electrons, *R. J. Celotta and D. T. Pierce*

Theory of Atomic Processes in Strong Resonant Electromagnetic Fields, *S. Swain*

Spectroscopy of Laser-Produced Plasmas, *M. H. Key and R. J. Hutcheon*

Relativistic Effects in Atomic Collisions Theory, *B. L. Moiseiwitsch*

Parity Nonconservation in Atoms: Status of Theory and Experiment, *E. N. Fortson and L. Wilets*

INDEX

Volume 17

Collective Effects in Photoionization of Atoms, *M. Ya. Amusia*

Nonadiabatic Charge Transfer, *D. S. F. Crothers*

Atomic Rydberg States, *Serge Feneuille and Pierre Jacquinot*

Superfluorescence, *M. F. H. Schuurmans, Q. H. F. Vrehen, D. Polder, and H. M. Gibbs*

Applications of Resonance Ionization Spectroscopy in Atomic and Molecular Physics, *M. G. Payne, C. H. Chen, G. S. Hurst, and G. W. Foltz*

Inner-Shell Vacancy Production in Ion–Atom Collisions, *C. D. Lin and Patrick Richard*

Atomic Processes in the Sun, *P. L. Dufton and A. E. Kingston*

INDEX

Volume 18

Theory of Electron–Atom Scattering in a Radiation Field, *Leonard Rosenberg*

Positron–Gas Scattering Experiments, *Talbert S. Stein and Walter E. Kauppila*

Nonresonant Multiphoton Ionization of Atoms, *J. Morellec, D. Normand, and G. Petite*

Classical and Semiclassical Methods in Inelastic Heavy-Particle Collisions, *A. S. Dickinson and D. Richards*

Recent Computational Developments in the Use of Complex Scaling in Resonance Phenomena, *B. R. Junker*

Direct Excitation in Atomic Collisions: Studies of Quasi-One-Electron Systems, *N. Andersen and S. E. Nielsen*

Model Potentials in Atomic Structure, *A. Hibbert*

Recent Developments in the Theory of Electron Scattering by Highly Polar Molecules, *D. W. Norcross and L. A. Collins*

Quantum Electrodynamic Effects in Few-Electron Atomic Systems, *G. W. F. Drake*

INDEX